T0271385

PRACTICAL SIGNAL PROCESSING AND ITS APPLICATIONS

With Solved Homework Problems

ADVANCED SERIES IN ELECTRICAL AND COMPUTER ENGINEERING

Editors: W.-K. Chen *(University of Illinois, Chicago, USA)*
          Y.-F. Huang *(University of Notre Dame, USA)*

The purpose of this series is to publish work of high quality by authors who are experts in their respective areas of electrical and computer engineering. Each volume contains the state-of-the-art coverage of a particular area, with emphasis throughout on practical applications. Sufficient introductory materials will ensure that a graduate and a professional engineer with some basic knowledge can benefit from it.

Published:

For the complete list of titles in this series, please visit
http://www.worldscientific.com/series/asece

Advanced Series in Electrical and Computer Engineering – Vol. 17

PRACTICAL SIGNAL PROCESSING AND ITS APPLICATIONS

With Solved Homework Problems

Sharad R Laxpati

Vladimir Goncharoff

University of Illinois at Chicago, USA

W⃝ World Scientific

NEW JERSEY · LONDON · SINGAPORE · BEIJING · SHANGHAI · HONG KONG · TAIPEI · CHENNAI

Published by

World Scientific Publishing Co. Pte. Ltd.

5 Toh Tuck Link, Singapore 596224

USA office: 27 Warren Street, Suite 401-402, Hackensack, NJ 07601

UK office: 57 Shelton Street, Covent Garden, London WC2H 9HE

Library of Congress Cataloging-in-Publication Data
Names: Laxpati, S. R., author. | Goncharoff, Vladimir, author.
Title: Practical signal processing and its applications : with solved homework problems /
    by Sharad R. Laxpati (University of Illinois at Chicago, USA),
    Vladimir Goncharoff (University of Illinois at Chicago, USA).
Description: [Hackensack] New Jersey : World Scientific, [2017] |
    Series: Advanced series in electrical and computer engineering ; volume 17
Identifiers: LCCN 2017036466 | ISBN 9789813224025 (hc : alk. paper)
Subjects: LCSH: Signal processing--Textbooks.
Classification: LCC TK5102.9 .L39 2017 | DDC 621.382/2--dc23
LC record available at https://lccn.loc.gov/2017036466

British Library Cataloguing-in-Publication Data
A catalogue record for this book is available from the British Library.

For any available supplementary material, please visit
http://www.worldscientific.com/worldscibooks/10.1142/10551#t=suppl

Desk Editor: Suraj Kumar

Typeset by Stallion Press
Email: enquiries@stallionpress.com

Printed in Singapore

Dedication

We dedicate this work to our spouses, Maureen Laxpati and
Marta Goncharoff, in sincere appreciation of their love and support.

Preface

The purpose of this book is two-fold: to emphasize the similarities in the mathematics of continuous and discrete signal processing, and as the title suggests to include practical applications of theory presented in each chapter. It is an enlargement of the notes we have developed over four decades while teaching the course 'Discrete and Continuous Signals & Systems' at the University of Illinois at Chicago (UIC). The textbook is intended primarily for sophomore and junior-level students in electrical and computer engineering, but will also be useful to engineering professionals for its background theory and practical applications. Students in related majors at UIC may take this course, generally during their junior year, as a technical elective. Prerequisites are courses on differential equations and electrical circuits, but most students in other majors acquire sufficient background in introductory mathematics and physics courses.

There is a plethora of texts on signal processing; some of them cover mostly analog signals, some mostly digital signals, and others include both digital and analog signals within each chapter or in separate chapters. We have found that we can give students a better understanding in less time by presenting analog and digital signal processing concepts in parallel (students like this approach). The mathematics of digital signal processing is not much different from the mathematics of analog signal processing: both require an understanding of signal transforms, the frequency domain, complex number algebra, and other useful operations. Thus, we wrote most chapters in this textbook to emphasize parallelism between analog and digital signal processing theories: there is a topic-by-topic, equation-by-equation match between digital/analog chapter pairs {2, 3}, {4, 5} and {9, 10}, and a somewhat looser correspondence between chapter pairs {7, 8} and {11, 12}. We hope that because of this textbook organization, even when reading only the analog or only the digital chapters of the textbook, readers will be able to quickly locate and understand the corresponding parallel-running descriptions in the other chapters. However, this textbook is designed to teach students *all* the material in Chapters 1-10 during a one-semester course.

Sampling theory (Ch. 6) is presented at an early stage to explain the close relationship between continuous- and discrete-time domains. The Fourier series is introduced as the special case of Fourier transform operating on a periodic waveform, and the DFT is introduced as the special case of discrete-time Fourier transform operating on a periodic sequence; this is a more satisfactory approach in our opinion. Chapters {11, 12} provide useful applications of Z and Laplace transform analysis; as time permits, the instructor may include these when covering Chapters {9, 10}.

To maintain an uninterrupted flow of concepts, we avoid laborious derivations without sacrificing mathematical rigor. Readers who desire mathematical details will find them in the footnotes and in cited reference texts. For those who wish to immediately apply what they have learned, plenty of MATLAB® examples are given throughout. And, of course, students will appreciate the Appendix with its 100 pages of fully-worked-out homework problems.

This textbook provides a fresh and different approach to a first course in signal processing at the undergraduate level. We believe its parallel continuous-time/discrete-time approach will help students understand and apply signal processing concepts in their further studies.

Overview of material covered:

- Chapter 1: Overview of the goals, topics and tools of signal processing.

- Chapters 2, 3: Time domain signals and their building blocks, manipulation of signals with various time-domain operations, using these tools to create new signals.

- Chapters 4, 5: Fourier transform to the frequency domain and back to time domain, operations in one domain and their effect in the other, justification for using the frequency domain.

- Chapter 6: Relationship between discrete-time and continuous-time signals in both time and frequency domains; sampling and reconstruction of signals.

- Chapters 7, 8: Time and frequency analysis of linear systems, ideal and practical filtering.

- Chapters 9, 10: Generalization of the Fourier transform to the Z/Laplace transform, and justification for doing that.

- Chapters 11, 12: Useful applications of Z/S domain signal and system analysis.

- Appendix: Solved sample problems for material in each chapter.

The flowchart in Fig. 1.4, p. 13, shows this textbook's organization of material that makes it possible to follow either discrete- or continuous-time signal processing, or follow each chapter in numerical sequence. We recommend the following schedule for teaching a 15-week semester-long university ECE course on introductory signal processing:

	Chapter	# lectures	
	1	1	Introductory lecture
Continuous-time	3	4	
Continuous-time	5	6	
Discrete-time	2	2	
Discrete-time	4	4	
both	6	4	Expand if necessary
Continuous-time	8	5	
Discrete-time	7	5	
Continuous-time	10 (& 12)	5	Examples from Ch.12 as needed
Discrete-time	9 (& 11)	5	Examples from Ch.11 as needed

41 lectures total

We are indebted to our UIC faculty colleagues for their comments about and use of the manuscript in the classroom, and to the publisher's textbook reviewers. We also thank our many students who, over the years, have made teaching such a rewarding profession for us, with special thanks to those students who have offered their honest comments for improving this textbook's previous editions.

Sharad R. Laxpati and Vladimir Goncharoff

Contents

List of Tables

List of Figures

Chapter 1

Introduction to Signal Processing

1.1 Analog and Digital Signal Processing

Welcome to your first course on signal processing! Electromagnetic signal waveforms are the communication medium of today's fast-paced interconnected world, and the information they convey may be represented in either analog or digital form. In analog form, a signal is a continuously varying waveform, usually a function of time. The name analog is used because the waveform is analogous to some physical parameter: e.g., instantaneous pressure, velocity, light intensity or temperature. Analog signals are usually the most direct measure of an actual event in nature. Signals in analog form may be processed using analog electronic circuitry, and for this reason analog signal processing is a topic firmly rooted in the electrical engineering field. The term digital refers to another form of signals – those that may be represented using lists of numbers. Advances in computer engineering have revolutionized signal processing by making it possible to replace analog circuitry with hardware that performs calculations in real-time. Digital signal processing is accomplished by a computer program whose algorithm operates on one list of numbers to produce another. However, the mathematics of digital signal processing is not radically different from the mathematics of analog signal processing: they both require an understanding of signal transforms, the frequency domain, complex number algebra and other useful operations. For that reason, most chapters in this textbook emphasize the parallelism between analog and digital signal processing theories: there is a topic-by-topic, equation-by-equation correspondence between digital-analog chapter pairs {2, 3}, {4, 5} and {9, 10}, and somewhat looser correspondence between chapter pairs {7, 8} and {11, 12}. Thus, when reading an analog chapter, you will be able to quickly locate and understand a parallel-running description in the corresponding digital chapter (and vice versa).

1.2 Signals and their Usefulness

Signals convey information. Just as sailors once signaled between ships using flags, a time-varying electrical signal may be used to represent information and transfer it between electronic devices (and people). Consider the following applications of using signals:

1.2.1 *Radio communications*

Radio waves are perturbations in the electromagnetic field that can travel through some materials at nearly the speed of light. These signals make possible today's vital wireless communications technologies: emergency services, personal mobile telephone and data use, broadcast radio and television, RFID[a] and Bluetooth®, to name a few.

1.2.2 *Data storage*

At times, we wish to communicate via signals that do not require real-time transmission. Such signals are designed to be efficiently and reliably stored until they are needed. On-demand communications signals include audio and video recordings, electronic textbooks, photographs, web pages, banking records, etc.

1.2.3 *Naturally-occurring signals*

Not all signals are man-made. For example, electrical signals generated by the human heart are used by doctors to diagnose cardiac diseases. Light from distant stars, or the vibration of the ground during an earthquake, are

---

[a] RFID: Radio Frequency Identification, such as that embedded in some credit cards for data and power transfer over short distances.

examples of naturally-generated signals. Weather reports often include the analysis and prediction of temperature, humidity and wind speed signals.

1.2.4 *Other signals*

As electrical and computer engineering students, you are aware of circuit voltage and current waveforms as descriptions of a circuit's operation. The prices of stocks at a stock exchange vs. time are another example of signals that are useful in our daily lives. Security video cameras generate surveillance images that are displayed, stored, and perhaps later analyzed.

1.3 Applications of Signal Processing

Now that you know what *signals* are, and how these signals may be useful for us, we will mention a few operations that are commonly done on signals; that is, what the applications are of signal processing:

- Modulation/demodulation – placing/removing information onto/from a carrier signal when communicating over longer distances, typically using wireless signals;

- Telemetry and navigation – sensing specialized reference signals to determine one's location, and determining what path to take to a desired destination;

- Compression – reducing the amount of redundant information, with or without degradation, so that a signal may be more efficiently stored or transmitted;

- Enhancement – operating on a signal to improve its perceived quality (as in audiovisual signals) or another characteristic that is deemed desirable;

- Filtering – blocking/attenuating some signal components while passing/boosting other signal components, to achieve some useful purpose;

- Coding – converting a signal to a different format so that it may be more immune to interference, or better suited for storage or transmission;

- Encryption – converting a signal to a different format so that the information it conveys may be hidden from those not authorized to receive it;

- Feature Extraction – identifying or estimating a desired representative signal component;

- Control – generating and injecting a signal to properly guide a system's operation.

1.4　Signal Processing: Practical Implementation

Analog signals are processed using analog circuits, which include these basic building blocks: input transducer to detect a physical parameter and convert it to a representative electrical signal, amplifier to boost the signal level, adder/subtractor, multiplier, filter to alter the signal's frequency content, output transducer to convert the electrical signal to a physical parameter, and a power supply to provide the electrical energy required by these blocks to properly function. Within each analog circuit building block one may find operational amplifiers, transistors, diodes, resistors, capacitors and inductors. Usually the simplest signal processing tasks are done using analog electronic circuits, because that approach is least expensive to implement.

Digital signals are either in the form of binary control signals, such as the output of an on/off switch, or in the form of binary symbol groups representing numerical codes. In the case of numerical codes, these are processed using programs running on digital computers designed specifically for the task at hand. For example: a mobile phone digitizes the speaker's voice to produce a stream of binary numbers; specialized hardware then extracts perceptually-important parameters from this stream that are compressed and coded for transmission. Finally, the coded signal modulates a radio frequency carrier wave so that it is efficiently and reliably transmitted to the nearest cell tower. The entire process requires both digital signal processing algorithms executed on a computer, as well as radio-frequency analog circuitry. Thus, analog and digital processing both play a role in mobile telephone communications.

1.5 Basic Signal Characteristics

When we analyze signals that are functions of time, then time is the independent variable. The analog waveform in Fig. 1.1, for example, represents pressure variations in air when a person is speaking. This type of signal is defined at all instants of time, and will from now on be referred to as a *continuous-time signal*. In the previous paragraph, we had referred to digital signals representing numerical codes. These may be thought of as sequences of numbers[b] that are operated on by digital computers. To make it possible for these lists of numbers to be stored in digital computer

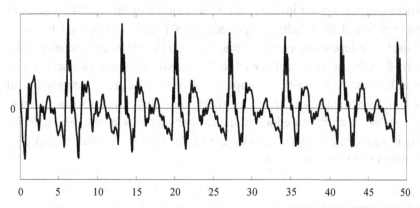

Figure 1.1. Shown is a 50-millisecond span of a continuous-time speech signal. Its nearly-periodic nature is the result of vocal cord vibrations during vowel sounds.

---

[b] These are usually represented in base 2 (binary), composed of symbols 1 and 0.

memory, the lists must be of finite length and each number in the list must have a finite number of bits[c] (having finite precision). For academic purposes, when analyzing digital signals, it is convenient to remove the finite-list-length and finite-precision restrictions. When such a signal is a function of time, then it is called a *discrete-time signal*.[d] For example, Fig. 1.2 shows the discrete-time signal points obtained by sampling a sine wave at uniform increments of time.

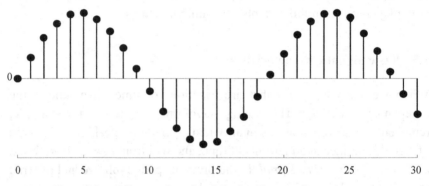

Figure 1.2.   A discrete-time signal obtained by sampling a sine wave.

Of course, time is not always the independent variable. In image signal processing, each color image pixel is represented by an intensity values of red, green and blue light components as a function of spatial location. Thus, the independent variables may be $\{x, y\}$ in rectangular coordinates, or $\{r, \theta\}$ using polar coordinates. Finally, the signal values themselves may be either real or complex. Complex-valued signals are extremely useful as a mathematical tool for compactly representing a transformed version of a signal in an intermediate domain. For this reason, the remainder of this introductory chapter provides the reader with a refresher of complex numbers and their operations.

---

[c] bit = **b**inary dig**it**.

[d] Discrete-time signals are discrete in time, but continuous in amplitude (having infinite precision). When the amplitudes of discrete-time signals are finite-length binary codewords, then these are called *digital* signals.

1.6 Complex Numbers

1.6.1 *Complex number math refresher*

Define imaginary number $j = \sqrt{-1}$ (electrical engineers use j instead of i, since the symbol i is used to represent current). Let $c = a + jb$, where $\{a, b\}$ are real numbers. This is called the *rectangular* or *Cartesian* form of c. Then:

(a) the *real part* of c is: $Re\{c\} = a$
(b) the *imaginary part* of c is: $Im\{c\} = b$
(c) the *complex conjugate* of c is: $c^* \equiv Re\{c\} - jIm\{c\} = a - jb$
(d) the *magnitude squared* of c is: $|c|^2 \equiv c\,c^* = a^2 + b^2$
(e) the *magnitude* of c is: $|c| \equiv \sqrt{c\,c^*} = \sqrt{a^2 + b^2}$
(f) the *phase angle* of c is: $\angle c$ or $arg(c) \equiv Tan^{-1}(b/a)$ (see[e] below)
(g) the *polar form* of c is: $c = |c|e^{j\angle c}$
　　(Engineers often write $c = |c|\angle c$ as shorthand for $c = |c|e^{j\angle c}$)

Polar form of $e^{j\theta}$:　　　　　　　　　$|e^{j\theta}| = 1, \; \angle e^{j\theta} = \theta: \; e^{j\theta} = 1\angle\theta$

Rectangular form of $e^{j\theta}$:　　　　$e^{\pm j\theta} = \cos(\theta) \pm j\sin(\theta)$
　　　　　　　　　　　　　　　　　(Euler's formula)

Cosine and sine expressions:　　$\cos(\theta) = \frac{1}{2}\left(e^{j\theta} + e^{-j\theta}\right)$

　　　　　　　　　　　　　　　$\sin(\theta) = \frac{1}{2j}\left(e^{j\theta} - e^{-j\theta}\right)$

Define complex numbers $c_1 = a_1 + jb_1 = r_1e^{j\theta_1} = r_1\angle\theta_1, c_2 = a_2 + jb_2 = r_2e^{j\theta_2} = r_2\angle\theta_2$. Offsets of π are added to phase angles θ_1 and θ_2 to ensure that r_1 and r_2 are positive real values. Then:

(a) $c_1c_2 = r_1r_2e^{j(\theta_1+\theta_2)} = r_1r_2\angle(\theta_1 + \theta_2)$
(b) $c_1 \pm c_2 = (a_1 \pm a_2) + j(b_1 \pm b_2)$

---

[e] In order for $|c|e^{j\angle c} = c$, $\angle c = Tan^{-1}(b/a)$ must be calculated to give results over range $(-\pi, \pi]$ (or some other range spanning 2π radians). To do this the sign of each a and b must be considered.

(c) $|c_1 c_2| = |c_1||c_2| = r_1 r_2$ 　　　(in general, $|c_1 + c_2| \neq |c_1| + |c_2|$)

(d) $\angle(c_1 c_2) = \angle c_1 + \angle c_2 = \theta_1 + \theta_2$ 　(in general, $\angle(c_1+c_2) \neq \angle c_1 + \angle c_2$)

(e) $|c_1/c_2| = |c_1|/|c_2| = r_1/r_2$

(f) $\angle(c_1/c_2) = \angle c_1 - \angle c_2 = \theta_1 - \theta_2$

(g) $(c_1 c_2)^* = c_1^* c_2^*$

(h) $(c_1 + c_2)^* = c_1^* + c_2^*$

1.6.2 *Complex number operations in MATLAB®*

In most chapters, we provide examples of MATLAB® code that are useful for demonstrating the theory being presented. MATLAB® (MATrix LABoratory)[f] has become the world's de facto standard programming environment for engineering education, and it is especially well-suited for signal processing computations. Numerous online tutorials and textbooks are available for students who are new to MATLAB.® Here are MATLAB® operations that implement calculations on complex numbers:

a) finding the real part of c: 　　　　　　　`real(c)`

b) finding the imaginary part of c: 　　　　`imag(c)`

c) finding the complex conjugate of c: 　　`conj(c)`

d) finding complex conjugate transpose of c: 　`c'` 　　(see [g])

e) finding the magnitude of c: 　　　　　　`abs(c)`

f) finding the phase angle of c: 　　　　　`angle(c)` 　(see [h])

g) finding $e^{j\theta}$: 　　　　　　　　　　`exp(j*theta)`

h) finding $A\angle\theta$: 　　　　　　　　　`A*exp(j*theta)`

1.6.3 *Practical applications of complex numbers*

Complex numbers are often used as a means to an end; for example, when factoring a real-valued cubic polynomial one may need to perform some

---

[f] MATLAB® is a proprietary product from MathWorks, Inc. Many universities purchase site licenses for their students, and a relatively inexpensive student version is also available. Or, one may use a free clone such as FreeMat for most operations.

[g] This is useful for vectors and matrices. When c is a scalar, `c'` = `conj(c)`.

[h] Function **angle** returns the *principal value* of phase angle (within the interval $[-\pi, \pi]$).

intermediate calculations in the complex domain before arriving at the roots (even when they are purely real!). However, by introducing some extra special operations, any numerical calculations done with complex numbers may also be done with two sets of real numbers. For example, the product of two complex numbers $(a + jb)$ and $(c + jd)$ is the complex result $(e + jf)$, where the two real numbers $\{e, f\}$ at the output are related to the four real numbers $\{a, b, c, d\}$ at the input using the operations: $e = ac - bd$, $f = ad + bc$ (this is how digital computers perform complex multiplication). Using complex numbers makes the notation more compact than with real numbers, while keeping basically the same algebraic rules and operations.

Because a complex number has real and imaginary parts, complex number notation is also useful for compactly representing signals having two independent components. For example, representing the location of point A on a plane at coordinates (x_A, y_A) as the complex number $c_A = x_A + jy_A$ makes the representation of location both more compact and simpler to manipulate; this is evident when calculating the distance between points A and B as $|c_A - c_B|$. However, the main application of complex numbers in this textbook is dealing with sinusoidal signals. To show this we begin with the Taylor series for e^x, which is valid for any x (real or complex):

$$e^x = 1 + x + \frac{x^2}{2!} + \frac{x^3}{3!} + \frac{x^4}{4!} + \frac{x^5}{5!} + \frac{x^6}{6!} + \frac{x^7}{7!} + \frac{x^8}{8!} + \cdots \qquad (1.1)$$

Next, for some real constant θ, we let $x = j\theta$ so that it is imaginary:

$$e^{j\theta} = 1 + j\theta - \frac{\theta^2}{2!} - j\frac{\theta^3}{3!} + \frac{\theta^4}{4!} + j\frac{\theta^5}{5!} - \frac{\theta^6}{6!} - j\frac{\theta^7}{7!} + \frac{\theta^8}{8!} + \cdots$$

$$= \left\{ 1 - \frac{\theta^2}{2!} + \frac{\theta^4}{4!} - \frac{\theta^6}{6!} + \frac{\theta^8}{8!} - \cdots \right\} + j\left\{ \theta - \frac{\theta^3}{3!} + \frac{\theta^5}{5!} - \frac{\theta^7}{7!} + \cdots \right\}, \text{ or}$$

$$e^{j\theta} = \cos(\theta) + j\sin(\theta), \qquad (1.2)$$

as expressed by the Maclaurin series for $\cos(\theta)$ and $\sin(\theta)$. As θ increases, $e^{j\theta}$ describes a circular path about the origin in a counter-clockwise direction when plotted on a complex plane: $Im\{e^{j\theta}\}$ vs.

$Re\{e^{j\theta}\}$. The real component of $e^{j\theta}$ is $\cos(\theta)$ and the imaginary component of $e^{j\theta}$ is $\sin(\theta)$; this relationship between circular motion and sinusoidal waveforms is at the heart of trigonometry.

The expression $e^{j\theta} = \cos(\theta) + j\sin(\theta)$ is a special case of Euler's Formula,[i] which relates trigonometric functions $\cos(\theta)$ and $\sin(\theta)$ to complex numbers. For example, with this formula it is easy to show that:

$$\cos(\theta) = \frac{1}{2}e^{j\theta} + \frac{1}{2}e^{-j\theta} \qquad (1.3)$$

$$\sin(\theta) = \frac{1}{2j}e^{j\theta} - \frac{1}{2j}e^{-j\theta}. \qquad (1.4)$$

To represent a real sinusoidal signal in time having constant amplitude A, constant frequency ω, and constant phase ϕ, one may multiply by A and replace θ with $\omega t + \phi$:

$$A\cos(\omega t + \phi) = \frac{A}{2}e^{j(\omega t+\phi)} + \frac{A}{2}e^{-j(\omega t+\phi)}$$

$$= \frac{A}{2}e^{j\omega t}e^{j\phi} + \frac{A}{2}e^{-j\omega t}e^{-j\phi}$$

$$= Ce^{j\omega t} + \left(Ce^{j\omega t}\right)^* = 2Re\{Ce^{j\omega t}\}, \quad (1.5)$$

where complex constant $C = Ae^{j\phi}/2$. We see that once frequency ω is known, only the complex constant C need be specified to completely describe any real sinusoidal function of time. More commonly we use $2C = Ae^{j\phi}$, which is called a "phasor" (phase-vector), to compactly describe $A\cos(\omega t + \phi)$.[j] Given phasor $Ae^{j\phi}$, the corresponding sinusoidal signal in time may be found as:

$$Re\{Ae^{j\phi} \cdot e^{j\omega t}\} = A\cos(\omega t + \phi). \qquad (1.6)$$

With this background, we are ready to see the benefits of using complex phasor notation to represent real sinusoidal time signals. First, consider the sum of two such sinusoids having the same frequency ω:

---

[i] Recommended book on the topic: *An Imaginary Tale: The Story of* $\sqrt{-1}$ by Paul J. Nahin, Princeton University Press, 2010.

[j] Together with knowledge of constant frequency value ω.

Introduction to Signal Processing

$$x(t) = A_1 \cos(\omega t + \phi_1) + A_2 \cos(\omega t + \phi_2). \tag{1.7}$$

It is known that the sum of two sinusoids at the same frequency yields a pure sinusoid at that frequency, albeit having different amplitude and phase: $x(t) = A_3 \cos(\omega t + \phi_3)$. The values of parameters A_3 and ϕ_3 may be found using various trigonometric identities and tedious manipulations. However, the calculations are much simpler when using complex numbers and phasors:

$$x(t) = Re\{A_1 e^{j\phi_1} e^{j\omega t}\} + Re\{A_2 e^{j\phi_2} e^{j\omega t}\}$$

$$= Re\{A_1 e^{j\phi_1} e^{j\omega t} + A_2 e^{j\phi_2} e^{j\omega t}\}$$

$$= Re\{(A_1 e^{j\phi_1} + A_2 e^{j\phi_2}) e^{j\omega t}\}$$

$$= Re\{(A_3 e^{j\phi_3}) e^{j\omega t}\}, \text{ or}$$

$$A_3 e^{j\phi_3} = A_1 e^{j\phi_1} + A_2 e^{j\phi_2}. \tag{1.8}$$

Therefore: $A_3 = |A_3 e^{j\phi_3}| = |A_1 e^{j\phi_1} + A_2 e^{j\phi_2}|$,

$$\phi_3 = \angle\{A_1 e^{j\phi_1} + A_2 e^{j\phi_2}\}, \quad \text{and}$$

$$x(t) = |A_1 e^{j\phi_1} + A_2 e^{j\phi_2}| \cos(\omega t + \angle\{A_1 e^{j\phi_1} + A_2 e^{j\phi_2}\}). \tag{1.9}$$

These solutions for A_3 and ϕ_3 are easily done graphically on a phasor diagram, as shown in Fig. 1.3. Another application of complex numbers to sinusoidal signal analysis is this: as we will see in later chapters, when a time-invariant linear system has an input signal $x(t)$ that is a sinusoid, the output signal $y(t)$ will be a sinusoid at the same frequency (but with possibly different amplitude and phase). How may we model this process? As it turns out, the transformation from $A_x \cos(\omega t + \phi_x)$ to $A_y \cos(\omega t + \phi_y)$ is easily done using multiplication in the phasor domain:

$$A_y e^{j\phi_y} = A_x e^{j\phi_x} \left\{ \frac{A_y e^{j\phi_y}}{A_x e^{j\phi_x}} \right\} = A_x e^{j\phi_x} \left\{ \frac{A_y}{A_x} e^{j(\phi_y - \phi_x)} \right\}$$

$$= A_x e^{j\phi_x}\{H\}, \tag{1.10}$$

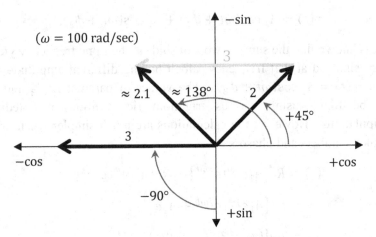

Figure 1.3. Phasor diagram graphical solution for $2\cos(100t + 45°) + 3\sin(100t - 90°) \approx 2.1\cos(100t + 138°)$.

where complex constant $H = (A_y/A_x)\,e^{j(\phi_y - \phi_x)}$. The output signal $y(t)$ may then be found as $Re\{A_x e^{j\phi_x} \cdot H e^{j\omega t}\}$.

1.7 Textbook Organization

This textbook presents its information by chapter *pairs*: one is digital and the other is analog. Once a concept is understood in one domain then it is much easier to comprehend in the other domain. We recommend that the reader begin study of signals in the continuous-time domain with Chapters 3 & 5, followed by Chapters 2 & 4 for a review of the same operations and transforms on discrete-time signals. Chapter 6 relates the continuous- and discrete-time domains using the theory of sampling, which explains the parallelism between analog and digital signal processing. The analog-digital learning approach continues by studying continuous-time systems and the Laplace transform in Chapters 8 & 10, followed by discrete-time systems and the Z transform in Chapters 7 & 9. Finally, to whet the reader's appetite for further study in the field of signal processing, Chapters 11 & 12 describe some practical applications of the Z and Laplace transforms.

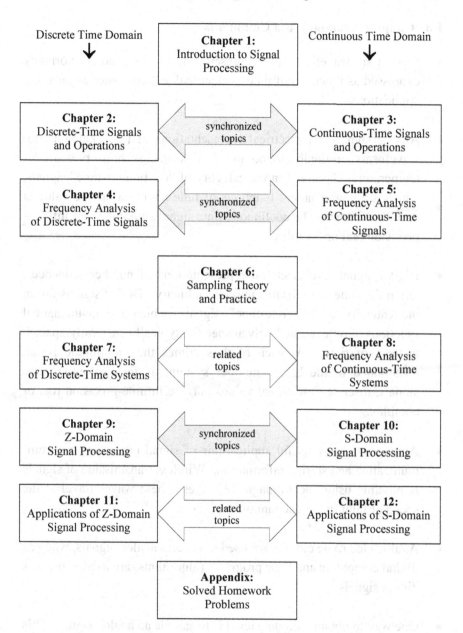

Figure 1.4. Textbook chapter organization, showing the parallelism between discrete-time and continuous-time domains.

1.8   Chapter Summary and Comments

- Signals are waveforms that represent information, and are normally expressed as functions of time in electrical and computer engineering applications.

- Analog signals are electrical representations of (are analogous to) waveforms originally found in other forms, such as pressure or temperature. They fall in the category of "continuous-time" signals, which are mathematical functions of time having a unique value at each time instant. The continuous-time signal amplitudes are infinite-precision real or complex.

- Digital signals are electrical representations of number sequences. Often the code used for this purpose is binary. Digital signals fall in the category of "discrete-time" signals, which are mathematical functions of time defined only at specific, typically uniformly-spaced, instants of time. At each of those times their amplitude is not constrained to the binary number system; as for continuous-time signals, discrete-time signal values may be infinite-precision real or complex.

- Among the many useful applications of signal processing are communicating and storing information. Wireless transmission of signals is possible using electromagnetic waves. These waves travel at the speed of light in the vacuum of free space.

- Analog electronic circuits are used to process analog signals, whereas digital computers and their programs (algorithms) are used to process digital signals.

- One way to obtain a digital signal is to sample an analog signal. This is called analog-to-digital conversion. Digital-to-analog conversion does the opposite.

- Analog circuits are often the simplest and least expensive signal processing tools. This is because most signals that we need to process naturally occur in analog form, and the result of signal processing is typically also analog. Circuits process electrical waveforms directly, without the need for the intermediate steps of analog-to-digital and/or digital-to-analog conversion.

- The digital revolution has replaced many analog circuits with digital computers. Digital computers are advantageous because of their potentially smaller size, lower sensitivity to noise, greater function-ality, and capability to easily change function (programmability).

- Analog and digital signal processing are based on similar theories and mathematics; these include calculus, differential and difference equa-tions, and complex numbers.

- This textbook teaches introductory analog and digital signal processing concepts in parallel; but for Chapters 1 and 6, all chapters come in pairs to present similar concepts for discrete-time and continuous-time signal processing.

1.9 Homework Problems

P1.1 Label each of the following signals as being "continuous-time" or "discrete-time":
  a) the current temperature in downtown Chicago
  b) the world's population at any given time
  c) the average number of automobiles using the interstate highway system on any day of the year
  d) the number of automobiles using the interstate highway system in the time period: (one hour ago) $\leq t \leq$ (now)
  d) a list describing the results of flipping a coin 100 times
  e) a thermostat output signal to turn an air conditioner on/off
  f) voltage across an inductor being measured 200 times per sec

P1.2   List four applications of signal processing that affect you daily.

P1.3   What is an advantage of processing signals in analog format
       (instead of digital)?

P1.4   What is an advantage of processing signals in digital format
       (instead of analog)?

P1.5   Solve the following without using a calculator:
  a)   Find the real part of $e^{-j\pi/2}$
  b)   Find the imaginary part of $-e^{j\pi/2}$
  c)   Find the magnitude of $1 + 3e^{j\pi/2}$
  d)   Find the phase of -5 in radians
  e)   Find the phase of $1 - j$ in degrees
  f)   Express $-1 + 2j$ in polar form
  g)   Express $-\sqrt{2}e^{j\pi/4}$ in rectangular form
  h)   Express $2e^{j\pi/3} - 1$ in rectangular form
  i)   Find $\left(e^{j2.34}/\sqrt{2}\right)\left(e^{j2.34}/\sqrt{2}\right)^{*}$
  j)   Express e^c in polar form, in terms of $a = Re\{c\}$ and
       $b = Im\{c\}$

P1.6   Solve the following using MATLAB® (show your code and the
       result):
  a)   Find the real part of $\pi e^{j\pi/6}$
  b)   Find the imaginary part of $7.1\angle 22°$
  c)   Find the magnitude of $-4.5 + j1.52$
  d)   Find the magnitude of $e^{e^{2-j}}$
  e)   Find the phase of $1 + 3j$ in radians
  f)   Find the phase of $1 + 3j$ in degrees
  g)   Express $2 - \sqrt{-2}$ in polar form
  h)   Express $-2e^{-j90°}$ in rectangular form
  i)   Express $e^{-j\pi/7} - \pi/7$ in rectangular form
  j)   Express $(2 + 3j)^{*}(1 - 5j)^{*}$ in rectangular form

Chapter 2

Discrete-Time Signals and Operations

2.1 Theory

2.1.1 *Introduction*

Consider $x(n)$, defined to be an ordered sequence of numbers, where index n is an integer in the range $-\infty$ to $+\infty$. The purpose of n is to keep track of the relative ordering of values in sequence x. If we associate a specific time value with n, such as nT seconds, then sequence $x(n)$ becomes a discrete-time signal: $x(n_0)$ represents a number that is associated with the time instant $n_0 T$ sec. The word discrete is used because x is defined at only discrete values of time.

n:	...	-2	-1	0	1	2	3	4	...
$x(n)$:	...	6.34	-3.21	-23.3	35.8	0.95	-1.83	19.7	...

Figure 2.1. Example of a sequence $x(n)$ as a function of its index variable n.

In some cases, a sequence such as $x(n)$ is obtained by sampling a continuous-time signal $x(t)$ at uniformly-spaced increments of time: $x(n) = x(t)|_{t=nT\ sec}$. For this reason, the properties of continuous-time signal $x(t)$ are closely reflected by the properties of sequence $x(n)$. In fact, it may even be possible to recover $x(t)$, at all values of t, from its discrete-time samples; this is the topic of Ch. 6. But even when $x(n)$ is not derived from sampling a continuous-time signal, similarities in

17

continuous-time and discrete-time signal processing exist and should not be ignored. Chapter 3 may be referred to for a topic-by-topic, equation-by-equation presentation of material in the continuous time domain that is synchronized to this chapter.

2.1.2 *Basic discrete-time signals*

2.1.2.1 *Impulse function*

The most basic discrete-time signal is the impulse, or Kronecker Delta function:

$$\delta(n) \equiv \begin{cases} 1, & n = 0; \\ 0, & n \neq 0. \end{cases} \qquad (2.1)^{a}$$

  Whenever the argument of function $\delta(\cdot)$ is equal to zero, the result is 1; otherwise, the function produces 0. As an example, $\delta(n-2)$ is equal to 1 at $n = 2$; it is equal to 0 at $n \neq 2$. We say that $\delta(n-2)$ is an impulse function shifted right by two samples. Figure 2.2 shows the impulse sequence $\delta(n)$ without any shift, and Fig. 2.3 shows $\delta(n-4)$ that is $\delta(n)$ shifted to the right by 4 samples:

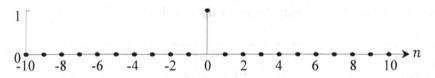

Figure 2.2. Impulse sequence $\delta(n)$.

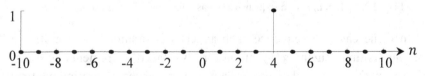

Figure 2.3. Delayed impulse sequence $\delta(n-4)$.

---

[a] A more general definition of Kronecker Delta is as a function of two integers: $\delta_{i,j} \equiv 0 \ (i \neq j) \ or \ 1 \ (i = j)$. Our $\delta(n)$ is therefore equal to $\delta_{n,0}$.

Now we introduce the *sifting*, or *sampling* property of a Kronecker impulse function:

$$\sum_{n=-\infty}^{\infty} x(n)\delta(n - n_0) = x(n_0). \qquad (2.2)$$

The sifting property concept is this: because $\delta(n - n_0) = 0$ everywhere except at $n = n_0$, only the value of $x(n)$ at $n = n_0$ is what matters in the product term $x(n)\delta(n - n_0)$. Therefore, $x(n)$ is *sampled* at $n = n_0$ to give $x(n)\delta(n - n_0) = x(n_0)\delta(n - n_0)$, which is an impulse function scaled by the constant $x(n_0)$. The resulting scaling factor of the impulse function has been obtained by *sifting* the value $x(n)|_{n=n_0}$ out from the entire sequence of numbers that is called "$x(n)$".[b]

2.1.2.2 *Periodic impulse train*

It is useful to form a periodic signal by repeating an impulse every M samples:

$$\delta_M(n) \equiv \sum_{k=-\infty}^{\infty} \delta(n - kM). \qquad (2.3)$$

Because the impulses in the plot look like the teeth of a comb, an impulse train is also referred to as a *comb function*. The shortest period that $\delta_M(n)$ can practically have is one sample, but $\delta_1(n) = 1 \; \forall n$ is the same as a constant sequence.

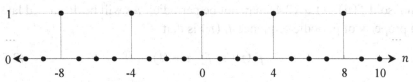

Figure 2.4. Impulse train $\delta_4(n)$.

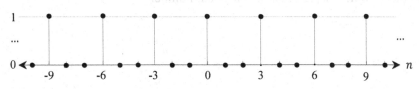

Figure 2.5. Impulse train $\delta_3(n)$.

---

[b] We will use the terms *discrete-time signal, function* and *sequence* interchangeably in this text.

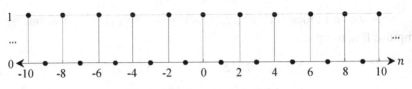

Figure 2.6. Impulse train $\delta_2(n)$.

2.1.2.3 *Sinusoid*

The discrete-time version of a sinusoid is defined as:

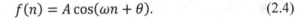

$$f(n) = A\cos(\omega n + \theta). \qquad (2.4)$$

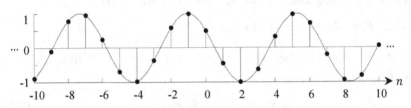

Figure 2.7. A sinusoidal sequence ($A = 1$, $\omega = 1$, $\theta = \pi/3$).

This cosine sequence has amplitude A, radian frequency ω and phase θ. Note that although the underlying continuous sinusoid (dashed line in Fig. 2.7) is periodic in time with period $2\pi/\omega$ sec, the discrete-time sinusoid $f(n)$ in Eq. (2.4) may not be periodic! As will be discussed later a property of periodic sequence $f_p(n)$ is that

$$f_p(n) = f_p(n \pm N). \qquad (2.5)$$

(where N is some integer). For the discrete-time sinusoid $A\cos(\omega n + \theta)$ to be periodic, $2\pi/\omega$ must be a rational number.[c]

---

[c] The period is then the smallest integer $N > 0$ that is a multiple of $2\pi/\omega$: $A\cos(\omega n + \theta) = A\cos(\omega(n + N) + \theta) = A\cos(\omega n + k2\pi + \theta)$, or $N = k(2\pi/\omega)$.

2.1.2.4 *Complex exponential*

The complex exponential sequence is useful as an eigenfunction[d] for linear system analysis:

$$g(n) = e^{j\omega n}. \tag{2.6}$$

By invoking Euler's formula, $e^{j\phi} = \cos(\phi) + j\sin(\phi)$, we see that the real and imaginary components of the complex exponential sequence are discrete-time sinusoids: $Re\{e^{j\omega n}\} = \cos(\omega n), Im\{e^{j\omega n}\} = \sin(\omega n)$.

2.1.2.5 *Unit step function*

The discrete-time unit step function $u(n)$ is defined as:

$$u(n) \equiv \begin{cases} 0, & n < 0; \\ 1, & n \geq 0. \end{cases} \tag{2.7}$$

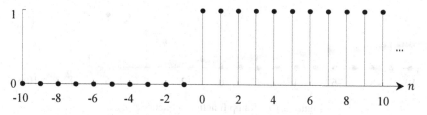

Figure 2.8.   Unit step function sequence $u(n)$.

2.1.2.6 *Signum function*

The discrete-time signum[e] function sgn(n) is defined as:

$$\text{sgn}(n) \equiv \begin{cases} -1, & n < 0; \\ 0, & n = 0; \\ 1, & n > 0. \end{cases} \tag{2.8}$$

---

[d] When the input to a linear system is an eigenfunction, the system output is the same signal only multiplied by a constant.

[e] *signum* is the word for "sign" in Latin.

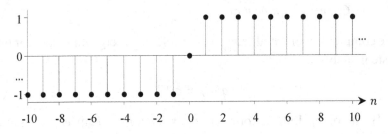

Figure 2.9.   Signum function sequence sgn(n).

2.1.2.7 *Ramp function*

The discrete-time ramp function $r(n)$ is defined as:

$$r(n) \equiv \begin{cases} 0, & n < 0; \\ n, & n \geq 0. \end{cases} \tag{2.9}$$

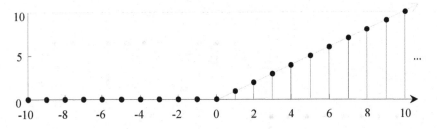

Figure 2.10.   Ramp function sequence $r(n)$.

2.1.2.8 *Rectangular pulse*

The discrete-time rectangular pulse function $\text{rect}_K(n)$ is defined as:

$$\text{rect}_K(n) \equiv \begin{cases} 1, & |n| \leq K; \\ 0, & \text{elsewhere.} \end{cases} \tag{2.10}$$

Using this definition $\text{rect}_K(n)$ always has an odd[f] number ($= 2K + 1$) of nonzero samples, and it is centered at index $n = 0$.

---

[f] To obtain a rectangular pulse that is an even number of samples wide, one may add an impulse function on one side or the other of $\text{rect}_K(n)$.

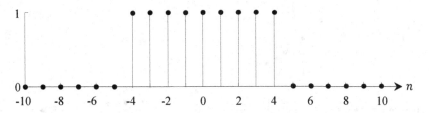

Figure 2.11. Rectangular pulse sequence $rect_4(n)$.

2.1.2.9 *Triangular pulse*

Define the discrete-time triangular pulse function $\Delta_K(n)$ to be:

$$\Delta_K(n) \equiv \begin{cases} 1 - |n/K|, & |n| \leq K; \\ 0, & \text{elsewhere.} \end{cases} \qquad (2.11)$$

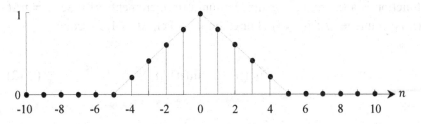

Figure 2.12. Triangular pulse sequence $\Delta_5(n)$.

$\Delta_K(n)$ always has an odd number ($=2K - 1$) of nonzero samples, and it is centered at index $n = 0$.

2.1.2.10 *Exponential decay*

The sequence a^n, where a is a real constant and $0 < |a| < 1$, decays to zero as $n \rightarrow \infty$ and grows without bound as $n \rightarrow -\infty$. Define[g] the exponentially decaying sequence $\equiv u(n)a^n$, which is shown in Fig. 2.13.

---

[g] An equivalent expression for the decaying exponential signal $u(n)a^n$ is $u(n)e^{-bn}$, where $b = -\ln(a) > 0$.

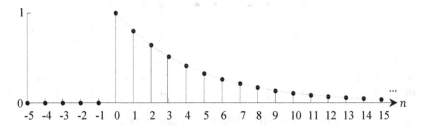

Figure 2.13. Exponentially decaying sequence $u(n)(0.8)^n$.

2.1.2.11 *Sinc function*

The sinc function appears often in signal processing, and we will define[h] it as $\text{sinc}(\phi) = \sin(\phi)/\phi$. Note[i] that $\text{sinc}(0) = 1$. The discrete-time sinc function is a sampling of $\text{sinc}(t)$ at uniform increments of time: $t = n\Delta t$ (n is an integer, Δt is real). For example, when $\Delta t = 1.0$ second:

$$\text{sinc}(n) \equiv \sin(n)/n \qquad (2.12)$$

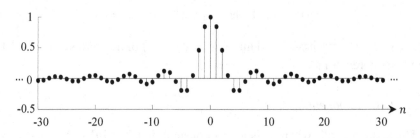

Figure 2.14. Discrete-time sequence $\text{sinc}(n)$.

---

[h] Some authors define $\text{sinc}(x) = \sin(\pi x)/\pi x$, which equals zero at integer values of x. MATLAB® has a built-in sinc function defined that way.

[i] Because one cannot directly evaluate $\text{sinc}(0) = \sin(0)/0 = 0/0$, apply l'Hospital's rule:
$$\lim_{\phi \to 0} \sin(\phi)/\phi = \lim_{\phi \to 0} (d\sin(\phi)/d\phi)/(d\phi/d\phi) = \lim_{\phi \to 0} \cos(\phi)/1 = 1.$$

2.1.3 *Signal properties*

2.1.3.1 *Energy and power sequences*

The *energy* of a discrete-time signal is defined as the sum of its magnitude-squared values:

$$E_x \equiv \sum_{n=-\infty}^{\infty}|x(n)|^2 = \sum_{n=-\infty}^{\infty} x(n)x^*(n). \qquad (2.13)$$

Clearly, E_x is always real and non-negative. The energy value is a measure of how much work this signal[j] would do if applied as a voltage across a 1Ω resistor. Signals having finite energy are called energy signals. Examples of discrete-time energy signals are $\delta(n)$, $\text{rect}_K(n)$, $\Delta_K(n)$, $u(n)a^n$ (with $|a| < 1$)[k], and $\text{sinc}(n)$.

For periodic and some other infinite-duration sequences the summation result of Eq. (2.13) will be infinite: $E_x = \infty$. In that case a more meaningful measure may be *power*, which is the average energy per sample of the sequence $x(n)$:

$$P_x \equiv \lim_{M\to\infty}\left(\frac{1}{2M+1}\sum_{n=-M}^{M}|x(n)|^2\right)$$

$$= \lim_{M\to\infty}\left(\frac{1}{2M+1}\sum_{n=-M}^{M} x(n)x^*(n)\right). \qquad (2.14)$$

Signals having finite, nonzero power are called *power signals*. Examples of discrete-time power signals are $u(n)$, $\text{sgn}(n)$ and $e^{j\omega n}$, whose power values are $\{1/2, 1, 1\}$ respectively. Power signals have infinite energy, and energy signals have zero power. Some discrete signals fall in neither category, such as the ramp sequence $r(n)$ that has both infinite energy and infinite power.[l] Periodic sequences (Sec. 2.1.3.3) have power = (Energy in one period) / (number of samples in one period). This may be expressed as $P_{x_p} \equiv \frac{1}{N}\sum_{n=0}^{N-1}|x_p(n)|^2$, where periodic $x_p(n)$ has N

---

[j] In this case, the signal $x(t)$ that is represented by its samples $x(n) = x(t)|_{t=nT_s}$.

[k] Proof that $u(n)a^n$ is an energy signal: $\text{Energy}\{u(n)a^n\} = \sum_{n=0}^{\infty}|a^n|^2 = \sum_{n=0}^{\infty}(a^2)^n = \frac{1}{1-a^2} < \infty$, when $|a| < 1$.

[l] Such sequences may still play a useful role in signal processing, as we shall see in Ch. 9.

samples per period. Based on this relation one may show that the power of periodic sinusoidal sequence $A\cos(2\pi n/N + \theta)$ is $|A|^2/2$ when $N > 2$.[m]

2.1.3.2 *Summable sequences*

A sequence $x(n)$ is said to be *absolutely summable* when:

$$\sum_{n=-\infty}^{\infty}|x(n)| = \sum_{n=-\infty}^{\infty}\sqrt{x(n)x^*(n)} < \infty. \qquad (2.15)$$

A sequence $x(n)$ is said to be *square-summable* when:

$$\sum_{n=-\infty}^{\infty}|x(n)|^2 = \sum_{n=-\infty}^{\infty}x(n)x^*(n) = E_x < \infty. \qquad (2.16)$$

We see, therefore, that square-summable sequences are energy signals.

2.1.3.3 *Periodic sequences*

A sequence $x(n)$ is *periodic* when there are nonzero integers m for which the following is true:

$$x_p(n) = x_p(n + m). \qquad (2.17)$$

The *fundamental period* of $x_p(n)$ is the smallest positive integer $m = m_0$ that makes Eq. (2.17) true. A periodic sequence has the property that $x_p(n) = x_p(n + km_0)$, for any integer k.

The fundamental frequency, in repetitions or cycles per second (Hertz), of a periodic discrete-time signal is the reciprocal of its fundamental period: $f_0 = 1/m_0$ Hz (assuming samples in sequence $x_p(n)$ are spaced at 1 sample/sec). Frequency may also be measured in radians per second, as we do in this textbook, which is defined as $\omega = 2\pi f$.[n] Thus, the fundamental frequency $\omega_0 = 2\pi/m_0$ rad/sec.

---

[m] This also holds for sinusoidal sequences that are not periodic: $A\cos(\omega n + \theta)$ for any $\omega \neq 2\pi/k$ $(k \in \mathbb{Z})$.

[n] The term *radians* is derived from *radius* of a circle. When a wheel having radius $r = 1$ rolls along the ground for one revolution, the horizontal distance travelled is equal to the wheel's circumference $c = 2\pi r$ (or 2π radii). Thus $f = 1$ cycle per second corresponds

2.1.3.4 *Sum of periodic sequences*

The sum of two or more periodic sequences is also periodic due to their discrete-time nature.[o] These periodic additive components are *harmonically related*: that is, the frequency of each component is an integer multiple of fundamental frequency ω_0. The fundamental period of the sum of harmonically-related signals having fundamental frequency ω_0 is $m_0 = 2\pi/\omega_0$ samples.

For example: $x(n) = \cos(5\pi n) + \sin(1.5\pi n)$ is periodic, having fundamental frequency $\omega_0 = 0.5\pi$ rad/sec[p] and period $m_0 = 4$ samples, whereas $y(n) = \cos(5n) + \sin(\pi n)$ is not periodic because one of its components ($\cos(5n)$) is not periodic.

Each additive term in a harmonically-related sum of sinusoids having frequency $\omega = k\omega_0$ is identified by its *harmonic index*: $k = \omega/\omega_0$. Thus, in the example above, the term $\cos(5\pi n)$ in $x(n)$ is at the 10th harmonic of $\omega_0 = 0.5\pi$ rad/sec.[q]

2.1.3.5 *Even and odd sequences*

A sequence $x(n)$ is *even* if it is unchanged after time reversal (replacing n with $-n$):

---

to $\omega = 2\pi(1) = 2\pi$ radians/sec. We use rad/sec measure to describe frequency since it simplifies notation somewhat (e.g., $\cos(\omega n)$ as compared to $\cos(2\pi f n)$), but in practice engineers almost always specify frequency as f in cycles per second, or Hertz (Hz).

[o] For example, the sum of sequence $x_{p1}(n)$ having period m_1 samples, and sequence $x_{p2}(n)$ having period m_2 samples, will be a periodic sequence having period $m_1 m_2$ samples or less.

[p] Notice that even though the fundamental frequency of $x(n)$ is 0.5π rad/sec there is no individual component having that frequency.

[q] The fundamental frequency is the 1st harmonic of a periodic signal, although it is rarely referred to as such.

$$x_e(n) = x_e(-n) \qquad (2.18)$$

A sequence $x(n)$ is *odd* if it is negated by time reversal:

$$x_o(n) = -x_o(-n). \qquad (2.19)^r$$

Every sequence may be expressed as a sum of its even and odd components:

$$y(n) = y_e(n) + y_o(n), \qquad (2.20)$$

where $\qquad\qquad y_e(n) = \tfrac{1}{2}(y(n) + y(-n)), \qquad (2.21)$

and $\qquad\qquad y_o(n) = \tfrac{1}{2}(y(n) - y(-n)). \qquad (2.22)$

Signals that are even described in this chapter include $\delta(n)$, $\delta_K(n)$, $\cos(\omega n)$, $\mathrm{rect}_K(n)$, $\Delta_K(n)$ and $\mathrm{sinc}(n)$. Some odd signals are $\sin(\omega n)$ and $\mathrm{sgn}(n)$. Signals $u(n)$, $r(n)$, $e^{j\omega n}$, and $u(n)a^n$ are neither even nor odd.

2.1.3.6 *Right-sided and left-sided sequences*

If $x(n) = 0$ for $n < m$, where m is some finite index value, then $x(n)$ is a *right-sided sequence*. For example, $u(n + 3)$ is right-sided. If $x(n) = 0$ for $n > m$, where m is some finite integer, then $x(n)$ is a *left-sided sequence*. Sequence $\cos(n)u(-n)$ is left-sided.

2.1.3.7 *Causal, anticausal sequences*

A sequence $x(n)$ is *causal* if $x(n) = 0$ for all $n < 0$.[s] A sequence $x(n)$ is *anticausal* if $x(n) = 0$ for all $n > 0$. Both causal and anticausal sequences may have a nonzero value at $n = 0$, hence only one sequence is both

---

[r] A consequence of Eq. (2.19) is that every odd sequence has $x_o(0) = 0$, because only zero has the property that $0 = -0$.

[s] The term *causal* comes from the behavior of real-world systems. If in response to an impulse at $n = 0$ a system outputted a signal prior to $n = 0$, it would be basing its output on the knowledge of a future input. Nature's cause-and-effect relationship is that a

causal and anticausal: $\delta(n) \times$ constant. Every sequence may be expressed as the sum of causal and anticausal components, as for example: $x(n) = x(n)u(n) + x(n)(1 - u(n))$. Causal sequences are a subclass of right-sided sequences, and anticausal sequences are a subclass of left-sided sequences. Causal sequences described in this chapter are $\delta(n)$, $u(n)$, $r(n)$, and $u(n)a^n$. Time-reversing a causal sequence makes it anticausal, and vice versa, so that $\delta(-n) = \delta(n)$, $u(-n)$, $r(-n)$, and $u(-n)a^{-n}$ are all anticausal sequences.

2.1.3.8 *Finite-length and infinite-length sequences*

If $x(n) = 0$ for $|n| > M$, where M is some finite positive integer, then $x(n)$ is a *finite-length sequence*. When no such value of M exists then $x(n)$ is an *infinite-length sequence*. Finite-length sequences are both right-sided and left-sided. Finite-length signals described in this chapter are $\delta(n)$, $\text{rect}_K(n)$ and $\Delta_K(n)$. The signals $\delta_K(n)$, $A\cos(\omega n + \theta)$, $\text{sinc}(n)$, $\text{sgn}(n)$, $u(n)$, $r(n)$, $e^{j\omega n}$, and $u(n)a^n$ all have infinite length.

2.1.4 *Signal operations*

2.1.4.1 *Time shift*

When sequence $x(n)$ is shifted to the right by m samples to give $x(n - m)$, we call this an m-sample *time delay* (since the index values n and m are most often associated with time). On the other hand, a shift to the left is called a *time advance*.

Figures 2.15 and 2.16 demonstrate these concepts. Note that the endpoints of $\text{rect}_4(n)$ are at sample indexes $n = \{-4, 4\}$, while the endpoints of $\text{rect}_4(n - 2)$ are at $n - 2 = \{-4, 4\}$ and the endpoints of $\text{rect}_4(n + 1)$ are at $n + 1 = \{-4, 4\}$.

---

response at present time can only be caused by past and present stimuli. When driven by a causal input sequence (such as an impulse located at $n = 0$), real-world systems will produce a causal output sequence.

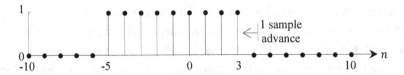

Figure 2.15. Pulse $\text{rect}_4(n + 1)$, a time-shifted version of sequence $\text{rect}_4(n)$.

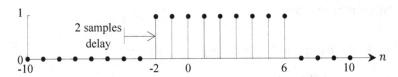

Figure 2.16. Pulse $\text{rect}_4(n - 2)$, a time-shifted version of sequence $\text{rect}_4(n)$.

In Fig. 2.17, note that time-delayed impulse $\delta(n - 4)$ is located at the index $n = 4$ ($\delta(n)$ is located at $n = 0$, $\delta(n - 4)$ is located at $n - 4 = 0$):

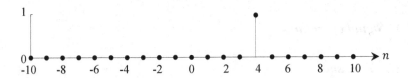

Figure 2.17. Delayed impulse sequence $\delta(n - 4)$.

When, in a functional expression for sequence $x(n)$, every occurrence of index n is replaced by $n - m$, the result is that $x(n)$ gets delayed by m samples. This is demonstrated in Fig. 2.18:

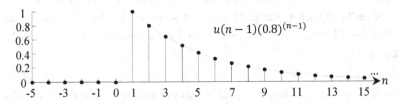

Figure 2.18. Delayed exponentially decaying sequence.

2.1.4.2 *Time reversal*

Replacing every occurrence of n with $-n$ in an expression for $x(n)$ causes the sequence to flip about $n = 0$ (only $x(0)$ stays where it originally was). Time-reversing a causal sequence makes it anticausal, and vice-versa. Time-reversing an even sequence has no effect, while time-reversing an odd sequence negates the original. Here is what happens when the sequence $u(n-2)$ is reversed in time:

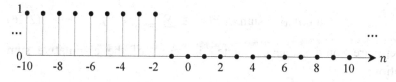

Figure 2.19. Sequence $u(-n-2)$.

Figure 2.19 demonstrates an interesting (and often confusing) concept: the sequence shown could either be a delayed unit step function that was then time-reversed, or a time-reversed unit step function that was then time-advanced:

$$u((-n) - 2) = u(-(n+2)). \tag{2.23}$$

2.1.4.3 *Time scaling*

Time scaling of discrete-time signals is done either by keeping every k^{th} sample in a sequence and discarding the rest (called "down-sampling by factor k"), or by inserting $k - 1$ zeros between samples in a sequence (called "up-sampling by factor k"). The two operations have the effect of compressing the signal toward index 0, or expanding the signal away from index 0, respectively ($k > 0$). There is usually loss of information when down-sampling a sequence, so that $x(n)$ cannot be reconstructed from its down-sampled version $x(kn)$. On the other hand, a sequence that had been upsampled by factor k can be exactly recovered using the operation of down-sampling by factor k. To summarize, down-sampling $x(n)$ by factor k to obtain sequence $y(n)$ is:

$$y(n) = x(nk). \tag{2.24}$$

Up-sampling $x(n)$ by factor k to obtain sequence $y(n)$ is:

$$y(n) = \begin{cases} x(n/k), & n/k \in \mathbb{Z}; \\ 0, & n/k \notin \mathbb{Z}. \end{cases} \tag{2.25}$$

2.1.4.4 *Cumulative sum and backward difference*

The cumulative sum of discrete-time signal $x(n)$ is itself a function of independent time index n:

$$\text{Cumulative sum of } x(n) \equiv \sum_{k=-\infty}^{n} x(k). \tag{2.26}$$

Here are some sequences stated in terms of the cumulative sum operation:

$$u(n) = \sum_{k=-\infty}^{n} \delta(k), \tag{2.27}$$

$$r(n+1) = \sum_{k=-\infty}^{n} u(k), \tag{2.28}$$

$$\text{rect}_K(n) = \sum_{k=-\infty}^{n} \{\delta(k+K) - \delta(k-(K+1))\}. \tag{2.29}$$

Related to the cumulative sum is the backward difference operator:

$$\Delta\{x(n)\} \equiv x(n) - x(n-1). \tag{2.30}$$

Here are sequences stated in terms of the backward difference operation:

$$\delta(n) = \Delta\{u(n)\}, \tag{2.31}$$

$$u(n-1) = \Delta\{r(n)\}, \tag{2.32}$$

$$\delta(n+K) - \delta(n-(K+1)) = \Delta\{\text{rect}_K(n)\}. \tag{2.33}$$

2.1.4.5 *Conjugate, magnitude and phase*

The conjugate, magnitude and phase of signal $g(n)$ are found according to the normal rules of complex number algebra:[t]

---

[t] Note that the operation is performed on the entire sequence $g(n)$.

$$g^*(n) = Re\{g(n)\} - j\, Im\{g(n)\}, \tag{2.34}$$

$$|g(n)| = \sqrt{Re\{g(n)\}^2 + Im\{g(n)\}^2} = \sqrt{g(n)g^*(n)}, \tag{2.35}$$

$$\angle g(n) = \text{Tan}^{-1}\left\{\frac{Im\{g(n)\}}{Re\{g(n)\}}\right\}. \tag{2.36)^u}$$

2.1.4.6 *Equivalent signal expressions*

Basic signals presented in the previous section may related to one another via algebraic operations, or combined to give other signals of interest. For example, here are different ways of describing the rectangular pulse and signum functions[v]:

$$\text{rect}_k(n) = u(n+k) - u(n - (k+1)), \tag{2.37}$$

$$\text{rect}_k(n) = u(n+k)u(-n+k), \tag{2.38}$$

$$\text{sgn}(n) = 2u(n) - 1 - \delta(n), \tag{2.39}$$

$$\text{sgn}(n) = u(n) - u(-n). \tag{2.40}$$

In fact, scaled and delayed versions of sequence $\delta(n)$ will add to form any sequence $x(n)$:

$$\begin{aligned}
x(n) = \cdots &+ x(-2)\delta(n - (-2)) \\
&+ x(-1)\delta(n - (-1)) \\
&+ x(0)\delta(n - (0)) \\
&+ x(1)\delta(n - (1)) \\
&+ x(2)\delta(n - (2)) \cdots, \tag{2.41}
\end{aligned}$$

or $$x(n) = \sum_{k=-\infty}^{\infty} x(k)\delta(n-k).$$

---

[u] Evaluate the Tan^{-1} function to produce a phase angle in all four quadrants; this is done by noting the signs of imaginary and real components of $g(n)$ prior to division.

[v] Plot these to convince yourself that the expressions are equivalent!

2.1.5 *Discrete convolution*

One of the most useful operations for signal processing and linear system analysis is the *convolution* operation,[w] defined here for discrete-time signals:

Discrete Convolution

$$x(n) * y(n) \equiv \sum_{k=-\infty}^{\infty} x(k)y(n-k) \qquad (2.42)$$

By replacing $n - k$ with m, we see that an equivalent form of this summation is

$$\sum_{m=-\infty}^{\infty} y(m)x(n-m) = y(n) * x(n) \qquad (2.43)$$

Thus, the convolution operation is commutative:

$$x(n) * y(n) = y(n) * x(n) \qquad (2.44)$$

2.1.5.1 *Convolution with an impulse*

Let's see what happens when we convolve signal $x(n)$ with the impulse function $\delta(n)$:

$$x(n) * \delta(n) = \sum_{k=-\infty}^{\infty} x(k)\delta(n-k)$$

$$= \sum_{k=-\infty}^{\infty} x(n)\delta(n-k)$$

$$= x(n) \sum_{k=-\infty}^{\infty} \delta(n-k) = x(n). \qquad (2.45)$$

This result, $x(n) * \delta(n) = x(n)$, shows us that convolving with an impulse function is an identity operation in the realm of convolution. More interesting and useful for signal processing, however, are the following relations that are also easily derived:

$$x(n) * \delta(n - n_1) = x(n - n_1). \qquad (2.46)$$

---

[w] The convolution operation is also called a *convolution product* because of its similarity to multiplicative product notation.

$$x(n) * \delta(n + n_2) = x(n + n_2). \tag{2.47}$$

Thus, a signal may be delayed or advanced in time by convolving it with a time-shifted impulse function. As will be discussed in Ch. 7, convolution plays a critical role in describing the behavior of linear shift-invariant systems.

2.1.5.2 *Convolution of two pulses*

Define "pulse" $p(n)$ to be a finite-length sequence whose time span is $n_1 \le n \le n_2$, having pulse width (maximum number of nonzero samples) $w = n_2 - n_1 + 1$. Given two pulses $p_a(n)$ and $p_b(n)$, we can make some statements about their convolution product $x(n) = p_a(n) * p_b(n)$:

- $x(n)$ will also be a pulse; the pulse width of x equals sum of the widths of p_a and p_b minus 1: $w_x = w_{p_a} + w_{p_b} - 1$;

- The left edge of $x(n)$ will fall at time index that is the sum of {time index at left edge of $p_a(n)$} and {time index at left edge of $p_b(n)$}; [x]

- The right edge of $x(n)$ will fall at time index that is the sum of {time index at right edge of $p_a(n)$} and {time index at right edge of $p_b(n)$}. [y]

---

[x] Assume $p_a(n) = 0$ outside of $n_1 \le n \le n_2$, and that $p_b(n) = 0$ outside of $n_3 \le n \le n_4$. Then $x(n) = p_a(n) * p_b(n) = \sum_{k=-\infty}^{\infty} p_a(k)p_b(n-k) = \sum_{k=n_1}^{n_2} p_a(k)p_b(n-k)$ due to the limited time span of $p_a(n)$. By considering the limited time span of $p_b(n)$ we note that $x(n) = 0$ when $n - n_1 < n_3$ ($n < n_1 + n_3$) or when $n - n_2 > n_4$ ($n > n_2 + n_4$). Thus $x(n) = 0$ outside of $n_1 + n_3 \le n \le n_2 + n_4$.

[y] See footnote above.

2.1.6 *Discrete-time cross-correlation*

Cross-correlation is used to measure the similarity between two signals, and for energy signals this operation is defined by an expression similar to convolution:

Discrete-Time Cross-Correlation

$$\phi_{xy}(n) \equiv \textstyle\sum_{k=-\infty}^{\infty} x^*(k)y(n+k). \qquad (2.48)^z$$

Unlike convolution, cross-correlation is not commutative: $\phi_{xy}(n) \neq \phi_{yx}(n)$.[aa] The cross-correlation between two energy sequences may be calculated using convolution: [bb]

$$\phi_{xy}(n) = x^*(-n) * y(n). \qquad (2.49)$$

For power sequences the summation in Eq. (2.48) will not converge at all values of n. We therefore modify the definition of cross-correlation for power signals: [cc]

$$R_{xy}(n) \equiv \lim_{M\to\infty} \frac{1}{2M+1} \textstyle\sum_{k=-M}^{M} x^*(k)y(n+k). \qquad (2.50)$$

For periodic power signals $x_p(n)$ and $y_p(n)$, both having period N, this definition of cross-correlation simplifies to:

$$R_{xy}(n) \equiv \frac{1}{N} \textstyle\sum_{k=0}^{N-1} x_p^*(k)y_p(n+k). \qquad (2.51)$$

The cross-correlation between energy sequence $x(n)$ and itself is called the *autocorrelation* of $x(n)$:

Discrete-Time Autocorrelation

$$\phi_{xx}(n) \equiv \textstyle\sum_{k=-\infty}^{\infty} x^*(k)x(n+k). \qquad (2.52)$$

---

[z] Another common notation for $\phi_{xy}(n)$ is $x(n) \star y(n)$.

[aa] When both $x(n)$ and $y(n)$ are real sequences then $\phi_{xy}(n) = \phi_{yx}(-n)$.

[bb] $\phi_{xy}(n) = \sum_{k=-\infty}^{\infty} x^*(k)y(n+k) = \sum_{m=-\infty}^{\infty} x^*(-m)y(n-m) = \sum_{m=-\infty}^{\infty} g(m)y(n-m) = g(n) * y(n)$, where $g(m) = x^*(-m)$.

[cc] Notation: $R_{xy}(n)$ represents the cross-correlation between power signals, and $\phi_{xy}(n)$ the cross-correlation between energy signals.

The autocorrelation of signal $x(n)$ is a measure of similarity between $x(n)$ and delayed versions of itself. From these definitions, Energy $\{x(n)\} = \phi_{xx}(0)$ and Power$\{x(n)\} = R_{xx}(0)$.

2.2 Practical Applications

2.2.1 Discrete convolution to calculate the coefficient values of a polynomial product

When multiplying together polynomials $P_1(x) = a_0 + a_1 x + a_2 x^2 + \cdots + a_N x^N$ and $P_2(x) = b_0 + b_1 x + b_2 x^2 + \cdots + b_M x^M$ we obtain another polynomial $P_3(x) = c_0 + c_1 x + c_2 x^2 + \cdots + c_{M+N} x^{M+N}$. How are coefficients of $P_3(x)$ related to the coefficients of $P_1(x)$ and $P_2(x)$? We have been taught in school to calculate this product by hand using simple rules, but a concisely-stated solution is that the coefficients of $P_3(x)$ are related to those of $P_1(x)$ and $P_2(x)$ through discrete convolution!

Let's write the coefficients of these polynomials using our familiar discrete signal notation:

$$P_1(x) = a(0) + a(1)x + a(2)x^2 + \cdots + a(N)x^N$$

$$= \sum_{n=-\infty}^{\infty} a(n)x^n, \qquad (2.53)$$

where coefficient sequence $a(n)$ is assumed $= 0$ outside of index range $0 \le n \le N$. In the same way, we define sequence $b(n)$ to describe polynomial $P_2(x)$ and sequence $c(n)$ to describe polynomial $P_3(x)$. The coefficients of $P_3(x)$ are then found as:

$$c(n) = a(n) * b(n). \qquad (2.54)$$

Example 2.1
Find the product of polynomials $f(x) = -x^4 + 3x^3 + 3$ and $g(x) = 6x^5 - x^3 + 2x^2 + x + 5$:

Using MATLAB® (see Section 2.3), we define arrays that store the coefficients of these two polynomials:

```
% MATLAB code example Ch2-1
% Multiplying together polynomials:
    f = [-1 3 0 0 3];
    g = [6 0 -1 2 1 5];
    h = conv(f,g);
    disp(h)
```

ans =   -6   18   1   -5   23   -2   12   6   3   15

From this we conclude that $h(x) = f(x)g(x) =$
$-6x^9 + 18x^8 + x^7 - 5x^6 + 23x^5 - 2x^4 + 12x^3 + 6x^2 + 3x + 15.$

2.2.2 *Synthesizing a periodic signal using convolution*

Given pulse $p(n)$ of length N samples. A periodic signal having sequence $p(n)$ as one of its periods may be synthesized by convolving $p(n)$ with an impulse train whose period is N samples:

$$p(n) * \delta_N(n) = p(n) * \sum_{k=-\infty}^{\infty} \delta(n - kN)$$

$$= \sum_{k=-\infty}^{\infty} p(n) * \delta(n - kN)$$

$$= \sum_{k=-\infty}^{\infty} p(n - kN) = x_p(n), \qquad (2.55)$$

which is a summation of time-shifted copies of pulse $p(n)$. The spacing of these pulses is N samples so that none of them overlap with neighboring pulses but instead sum up to give periodic sequence $x_p(n)$.[dd] Therefore we have taken advantage of the convolution operator, together with an impulse train, to compactly describe periodic sequence $x_p(n)$ in terms of one of its periods. Note that pulse $p(n)$ may be *any N consecutive samples* of $x_p(n)$, not necessarily those starting at time index $n = 0$. (This model for creating a periodic sequence will be used in Chapter 4 to derive an expression for the Fourier Series.)

[dd] Subscript p in $x_p(n)$ specifies that the sequence is periodic (not related to the label of pulse $p(n)$).

Example 2.2

Using the discrete-time convolution operation, find a compact expression for periodic $x(n) = \{..., 0, 0, 0, 1, 1, 1, 0, 0, 0, 1, 1, 1, ...\}$ when it is given that $x(-1) = 0$ and $x(0) = 1$:

Answer: $x(n) = \text{rect}_1(n - 1) * \delta_6(n)$.

2.2.3 *Normalized cross-correlation*

A normalized form of the cross-correlation measure between real-valued sequences $x(n)$ and $y(n)$ is shown in Eq. (2.56). The result of this formula is that $-1 \leq C_{xy}(n) \leq 1$, for all values of n.[ee]

$$C_{xy}(n) = \frac{\phi_{xy}(n)}{\sqrt{\phi_{xx}(0)\phi_{yy}(0)}} = \frac{\phi_{xy}(n)}{\sqrt{E_x E_y}} = \frac{\sum_{k=-\infty}^{\infty} x(k)y(n+k)}{\sqrt{\sum_{k=-\infty}^{\infty} x^2(k) \sum_{k=-\infty}^{\infty} y^2(k)}}. \quad (2.56)$$

The beauty of this normalized measure is that a threshold value may be set (e.g., 0.75) above which two discrete-time signals are judged to be similar in "shape," and this threshold is independent of the energy levels of either $x(n)$ or $y(n)$.

When one of the two sequences has finite length (assume it is $x(n)$, such that $x(n) = 0$ outside of the interval $-N \leq n < N$), then the expression for $C_{xy}(n)$ given above may be written as:

$$C_{xy}(n) = \frac{\sum_{k=-N}^{N} x(k)y(n+k)}{\sqrt{(\sum_{k=-N}^{N} x^2(k))(\sum_{k=-\infty}^{\infty} y^2(k))}}. \quad (2.57)$$

Since only $2N + 1$ samples of sequence $y(n)$ are used to calculate cross correlation in the numerator, we can modify the calculation of E_y in the denominator to include only those samples. This is called the

---

[ee] In probability theory $\rho_{xy} = C_{xy}(0)$ is called the correlation coefficient between samples of random variable X and random variable Y, when these samples are stored as sequences $x(n)$ and $y(n)$.

short-time normalized cross-correlation between finite-length sequence $x(n)$ and energy sequence $y(n)$:

$$STC_{xy}(n) = \frac{\sum_{k=-N}^{N} x(k)y(n+k)}{\sqrt{(\sum_{k=-N}^{N} x^2(k))(\sum_{k=-N}^{N} y^2(n+k))}}. \tag{2.58}$$

The following graphs demonstrate some of the pattern-matching features of both $C_{xy}(n)$ and $STC_{xy}(n)$:

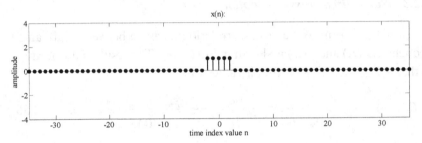

Figure 2.20. Rectangular pulse sequence $x(n)$.

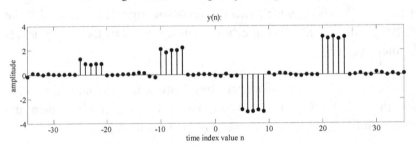

Figure 2.21. Signal $y(n)$, composed of noise plus rectangular pulses at various delays and amplitudes.

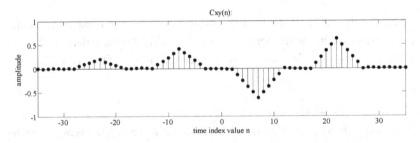

Figure 2.22. Normalized cross-correlation $C_{xy}(n)$ between $x(n)$ and $y(n)$. Notice that rectangular pulses in $y(n)$ (Fig. 2.20) were detected as peaks of the triangular pulses.

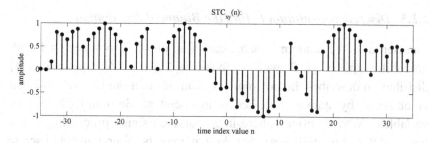

Figure 2.23. Short-time normalized cross-correlation $STC_{xy}(n)$ between $x(n)$ and $y(n)$. Rectangular pulses in $y(n)$ were detected as locations where $|STC_{xy}(n)| \cong 1$.

2.2.4 *Waveform smoothing by convolving with a pulse*

When signal $x(n)$ is convolved with $\text{rect}_K(n)/(2K+1)$, a rectangular pulse having width of $2K+1$ samples, each nonzero sample having amplitude $1/(2K+1)$ (resulting in the sum of all samples = 1, just like for an impulse function),

$$\left(\frac{1}{2K+1}\right) \text{rect}_K(n) * x(n) = \left(\frac{1}{2K+1}\right) \sum_{k=-\infty}^{\infty} \text{rect}_K(k)x(n-k)$$

$$= \left(\frac{1}{2K+1}\right) \sum_{k=-K}^{K}(1)x(n-k)$$

$$= \left(\frac{1}{2K+1}\right) \sum_{m=n-K}^{n+K} x(m) = y(n), \qquad (2.59)$$

the result is the average value of discrete-time sequence x near sample index n. This operation is called *moving window averaging* and has the effect of smoothing a signal, or reducing rapidity of amplitude change vs. time. The wider the rectangular pulse (higher integer value of K), the more smoothing results.[ff] Pulses other than rectangular may also be used. Waveform smoothing is the time-domain result of low-pass filtering, a frequency-domain operation that will be introduced in Ch. 7.

[ff] One may say that zero smoothing results when $x(n)$ is convolved with impulse function $\delta(n)$.

2.2.5 *Discrete convolution to find the Binomial distribution*

The Bernoulli(p) random variable can be in one of two states: $= 1$ with probability p, or $= 0$ with probability $1 - p$. A Binomial(n, p) distribution describes the outcome probabilities of a random variable that is obtained by adding together n independent Bernoulli(p) random variables. Why mention it here in a course on signal processing? It is because the Binomial(n, p) distribution may be found using discrete convolution. Here is the method: define sequence[gg] $P_X(n)$ to describe the probability that a Bernoulli(p) random variable will have outcome $= n$:

$$P_X(n) = \begin{cases} p, & n = 1; \\ 1 - p, & n = 0; \\ 0, & otherwise. \end{cases} \quad (2.60)$$

If we add together n such independent Bernoulli(p) random variables $(X_1, X_2, \ldots, X_n)$ to give us random variable Y, then the probability that Y will have outcome $= n$ is found this way:

$$P_Y(n) = P_{X_1}(n) * P_{X_2}(n) * \cdots * P_{X_n}(n). \quad (2.61)$$

In this case $P_{X_1}(n) = P_{X_2}(n) = \cdots = P_{X_n}$, but Eq. (2.61) holds true for random variables if they are independent of one another.

Example 2.3
Calculate and plot a Binomial($50, 0.5$) distribution $P_Y(n)$.

```
% MATLAB code example Ch2-2
% Calculating and plotting a Binomial(50,0.5)
distribution:
    n = 50;
    p = 0.5;
    X = [p, 1-p];
    Y = 1;
    for k = 1:n
```

[gg] In probability theory, $P_X(x)$ – the probability that random variable X has value x – is called the probability mass function (a discrete version of the probability density function). Argument x is usually not restricted to an integer value as we had done by writing it in sequence notation $P_X(n)$.

```
Y = conv(Y,X);   % discrete convolution
                 % of two sequences
end
stem(0:n,Y,'filled')
xlabel('outcome value')
ylabel('probability value')
title(['Binomial(n,p) distribution for n = 50',...
       'and p = 0.5'])
```

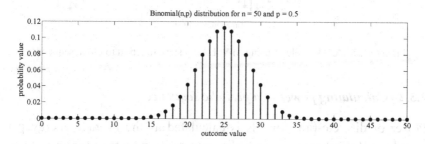

Figure 2.24. MATLAB® plot of a Binomial(50, 0.5) distribution.

2.3 Useful MATLAB® Code

2.3.1 *Plotting a sequence*

MATLAB's **stem** instruction is made for plotting a sequence of values without connecting the points:

```
% MATLAB code example Ch2-3
n = -30:30;
x = sin(n*pi/15);
stem(n,x,'filled')
xlabel('time index value n')
ylabel('amplitude')
title('Plot of sin(n\pi/15) vs. n')
```

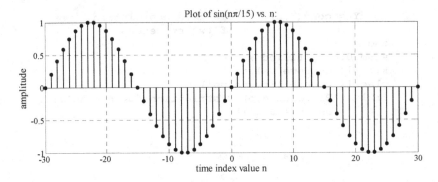

Figure 2.25. An example of using MATLAB's stem function to plot a sequence.

2.3.2 *Calculating power of a periodic sequence*

Power of discrete-time signals is calculated as the average energy per sample, and the power of a periodic sequence can be found from just one of its periods. Assume sequence $p(n)$ stores one period (M samples) of periodic sequence $f_p(n) = p(n) * \delta_M(n)$. Then the formula and corresponding MATLAB® code to calculate P_{f_p} is:

$$P_{f_p} = \tfrac{1}{M}\sum_{n=1}^{M}\left|f_p(n)\right|^2 = \tfrac{1}{M}\sum_n |p(n)|^2. \qquad (2.62)$$

```
Pfp = sum(abs(p).^2)/M   or   Pfp = mean(abs(p).^2)
```

2.3.3 *Discrete convolution*

One of the most important and speed-optimized functions built into MATLAB® is discrete convolution (which, as discussed in Ch. 4, is based on the FFT algorithm). In MATLAB® code the convolution between two finite-length sequences $x(n) * y(n)$ is implemented as: `conv(x,y)`.

2.3.4 *Moving-average smoothing of a finite-length sequence*

As previously mentioned in Section 2.2.1.4, one may *smooth* a sequence of numbers (reduce their short-term variability) by convolving them with a rectangular pulse function. This may be done to better visualize the slowly-varying components of a discrete signal, as in financial market price trend analysis. Of particular interest is the case where we convolve with a discrete rectangular pulse whose samples sum to 1; this is called a *moving average filter*:

$$MA_y(n) = y(n) * \frac{1}{2K+1} \text{rect}_K(n) \qquad (2.63)$$

In MATLAB® this operation is easily implemented as follows:

```
M = 2*K+1;   % M is an odd integer
MAy = conv(y,ones(1,M)/M);
```

Example 2.4

```
% MATLAB code example Ch2-4
% Moving window average smoothing a noisy sinusoidal
% sequence:
    n = -30:30;
xnoisy = sin(n*pi/15) + randn(size(n))/5;
figure
stem(n,xnoisy,'filled')
xlabel('time index value n')
ylabel('amplitude')
title('Before Smoothing:')
```

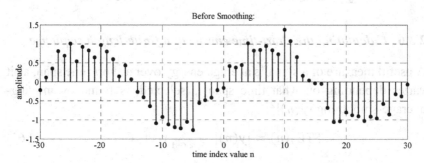

Figure 2.26. A noisy sinusoidal sequence before smoothing.

```
% MATLAB code example Ch2-4, cont.
K = 5;
M = 2*K + 1;
xnoisy_smoothed = conv(xnoisy,ones(1,M)/M);
figure
stem(n,xnoisy_smoothed(K+(1:length(n))),'filled')
```

```
xlabel('time index value n')
ylabel('amplitude')
title('After Smoothing:')
```

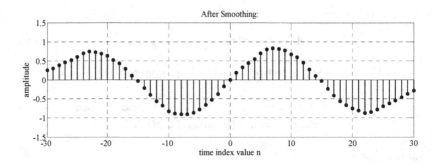

Figure 2.27. A noisy sinusoidal sequence after smoothing.

2.3.5 *Calculating energy of a finite-length sequence*

MATLAB® is a matrix- (or array) oriented processing language, so it may very efficiently perform operations such as calculating energy of a sequence:

$$E_y = \sum_n |y(n)|^2 \quad \rightarrow \quad \texttt{Ey = sum(abs(y).\^2)}. \qquad (2.64)$$

2.3.6 *Calculating the short-time energy of a finite-length sequence*

A useful measure of a signal's average energy over a short time span, as it varies depending on what time span is selected, is found by moving-average-smoothing $|y(n)|^2$:

$$\text{STE}_y(n) = |y(n)|^2 * \frac{1}{2K+1} \text{rect}_K(n). \qquad (2.65)$$

In MATLAB® this operation is implemented as follows:

```
M = 2*K+1;  % M is an odd integer
STEy = conv(abs(y).^2,ones(1,M)/M);
```

Example 2.5

```
% MATLAB code example Ch2-5
% Estimating short-time energy of a sequence:
```

```
n = -30:30;
x = cos(n*pi/15) + rand(size(n));
figure
stem(n,x,'filled')
xlabel('time index value n')
ylabel('amplitude')
title('Original sequence x(n):')
```

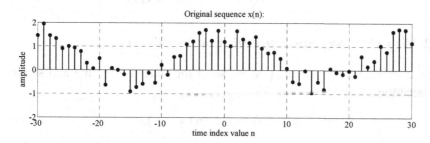

Figure 2.28. Sequence $x(n)$.

```
% MATLAB code example Ch2-5, cont.
K = 5; M = 2*K + 1;
STEx = conv(abs(x).^2,ones(1,M)/M);
Figure; stem(n,STEx(K+(1:length(n))),'filled')
xlabel('time index value n')
ylabel('energy')
title('Short-time energy of x(n):')
```

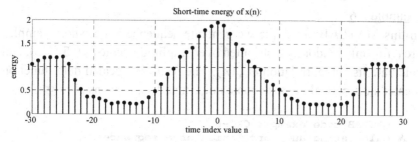

Figure 2.29. Calculated short-time energy of the sequence $x(n)$ in Fig. 2.27.

2.3.7 *Cumulative sum and backward difference operations*

MATLAB® has built-in functions to calculate both the cumulative sum and backward difference operations:

Cumulative sum of $x(n)$:

$$\sum_{k=-\infty}^{n} x(k) \rightarrow \texttt{cumsum(x)}. \tag{2.66}$$

Backward difference of $y(n)$:

$$y(n) - y(n-1) \rightarrow \texttt{diff(y)}. \tag{2.67}$$

2.3.8 *Calculating cross-correlation via convolution*

Discrete-time cross-correlation is useful for pattern matching and signal detection:

$$\phi_{xy}(n) = x^*(-n) * y(n). \tag{2.68}$$

In MATLAB® the cross-correlation between two finite-length signals is efficiently calculated via convolution:

```
phi_xy = conv(conj(flipud(x(:))),y);
```

Similarly, the autocorrelation of finite-length sequence $x(n)$ is found as:

```
phi_xx = conv(conj(flipud(x(:))),x);
```

Example 2.6
In this MATLAB® example we generate sequence $x(n)$ whose samples are uniformly randomly distributed over amplitude range $[-0.5, \ 0.5]$, and calculate its autocorrelation $\phi_{xx}(n)$. The result is a scaled delta function located at $n = 0$, along with some noise that disappears as $N \rightarrow \infty$:

```
% MATLAB code example Ch2-6
% Calculating autocorrelation of a sequence:
N = 1e4;
x = rand(1,N+1) - 0.5;
phi_xx = conv(conj(flipud(x(:))),x);
plot(-N:N,phi_xx)
```

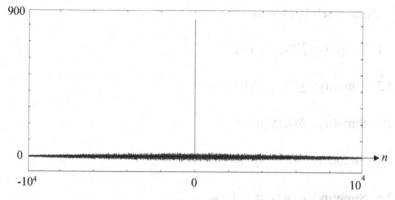

Figure 2.30. Autocorrelation of random noise.

2.4 Chapter Summary and Comments

- Discrete-time signals, also known as *sequences*, are defined as lists of numbers that are ordered by time index values. In this sense a discrete-time signal is only defined at specific time instants. One way to think of it is to imagine that a discrete-time signal is a list of amplitude values that are obtained by sampling a continuous-time signal at uniformly-spaced times (e.g., at integer values of t).

- In this chapter, we define some basic sequences that may be used as building blocks to synthesize more complicated sequences.

- Operations such as addition, multiplication, convolution, time shift, time reversal, backward time difference, etc., are useful for manipulating and combining discrete-time signals. In later chapters, we will make use of these operations for "signal processing."

- Almost every discrete-time signal and operation has a continuous-time domain equivalent. These relationships are evident in the parallel presentations in Chapters 2 and 3, and may be explained using the theory of sampling (Ch. 6).

2.5 Homework Problems

P2.1 Simplify: $\sum_{n=-\infty}^{\infty} \delta(n) =$

P2.2 Simplify: $\sum_{n=-\infty}^{\infty} \delta(n)x(n) =$

P2.3 Simplify: $\delta(n)x(n) =$

P2.4 Simplify: $\delta(n-1)x(n) =$

P2.5 Simplify: $\delta(n)x(n-1) =$

P2.6 Simplify: $\sum_{n=-\infty}^{\infty} \delta(n+1)x(n) =$

P2.7 Simplify: $\sum_{n=-\infty}^{\infty} \delta(n)x(n+1) =$

P2.8 Simplify: $\sum_{n=-\infty}^{\infty} \delta(n-1)x(n+1) =$

P2.9 Simplify: $\sum_{k=-\infty}^{\infty} \delta(k-n)x(n) =$

P2.10 Given sequence $x(n) = n \, \text{rect}_3(n-3)$:
   a) Find $x(4)$
   b) Find $x(-4)$
   c) Find $w(10)$ when $w(n) = x(n) * \delta_7(n)$
   d) Find $y(-10)$ when $y(n) = \sum_{k=0}^{7} x(k)\delta_7(n-k)$

P2.11 Is $x(n) = \sin(5n)$ periodic? If so, what is its period?

P2.12 Is $y(n) = \cos(\pi n/8)$ periodic? If so, what is its period?

P2.13 Express $y(n) = 4\cos(5n)$ as a sum of complex exponential sequences.

P2.14 State in terms of an impulse function:
   $u(n)u(-n) =$

P2.15 State as a sum of one or more impulse functions:
$$u(n+1)u(-n-1) =$$

P2.16 State in terms the sum of one or more impulse functions:
$$u(n-1) - u(n-2) =$$

P2.17 State in terms of the sum of two unit step functions:
$$\text{sgn}(n) =$$

P2.17 State in terms of the sum of two unit step functions:
$$\text{rect}_4(n) =$$

P2.18 State in terms of a rectangular pulse function:
$$u(n-5) - u(n) =$$

P2.19 State in terms of a triangular pulse function:
$$r(n-3) - 2r(n) + r(n+3) =$$

P2.20 State in terms of a triangular pulse function:
$$r(n) - 2r(n+2) + r(n+4) =$$

P2.21 Calculate the energy of $f(n) = 4(2^{-n})u(n)$.

P2.22 Calculate the energy of $f(n) = 4(2^{-n})(u(n) - u(n-100))$.

P2.23 Simplify: $\text{sinc}(\pi n) =$

P2.24 Given: $g(n) = \text{sinc}(\pi n/2)$. Calculate $g(n)$ for $-5 \le n \le 5$.

P2.25 Simplify: $\text{sinc}(\pi n/2) \cdot \text{sinc}(\pi(n-1)/2) =$

P2.26 What is the energy of $u(n)$?

P2.27 What is the power of $\text{sgn}(n)$?

P2.28 What is the energy of $\text{rect}_5(n)$?

P2.29 What is the power of $\Delta_5(n)$?

P2.30 Label each of the following sequences as {odd, even, or neither}:
    a) $\text{rect}_3(n)$
    b) $\text{rect}_3(n) - \delta(n)$
    c) $\delta(n-1) - \delta(n+1)$
    d) $\delta(n-1)$
    e) $\text{sgn}(n)$
    f) $\text{sgn}(n) - \delta(n)$

P2.31 Find the even part of each sequence listed in Problem P2.30.

P2.32 Find the odd part of each sequence listed in Problem P2.30.

P2.33 The result of multiplying $x(n)$ by $u(n-2)$ will be:
    (check all that apply)
    a) causal
    b) right-sided
    c) anticausal
    d) left-sided

P2.34 The result of multiplying $x(n)$ by $\text{rect}_3(n+2)$ will be:
    (check all that apply)
    a) causal
    b) right-sided
    c) anticausal
    d) left-sided

P2.35 List three sequences that are both causal and anticausal.

P2.36 The product of $x(n)$ and $u(n)$ is causal: (True or False)

P2.37 The product of $x(-n)$ and $u(-n)$ is anticausal: (True or False)

P2.38 The product of $x(n)$ and $1 - u(n)$ is anticausal: (True or False)

P2.39 Time-shifting $x(n)u(n)$ to the left by 5 samples guarantees that it is not causal: (True or False)

P2.40 Every causal sequence is right-sided: (True or False)

P2.41 Every left-sided sequence is anticausal: (True or False)

P2.42 The product of a right-sided sequence and a left-sided sequence is a finite-width sequence: (True or False)

P2.43 Find the cumulative sum of the sequence $\delta(n + 4)$.

P2.44 Find the cumulative sum of the sequence $\delta(n + 1) - \delta(n - 2)$.

P2.45 Find the backward difference of $\text{sgn}(n)$.

P2.46 Find the backward difference of $\text{rect}_3(n)$.

P2.47 Simplify: $u(2n) =$

P2.48 Simplify: $u\big(2(n - 4)\big) =$

P2.49 Express the following as a rectangular pulse sequence:
$u(n + 3) - u(n) =$

P2.50 Express the following as a sum of two unit step functions:
$\text{rect}_5(n - 5) =$

P2.51 Express the following as a product of two unit step functions:
$\text{rect}_5(n - 5) =$

P2.52 Simplify: $\delta(n) * 2\text{sinc}(n + 3) =$

P2.53 Simplify: $2\delta(n) * 2\text{sinc}(n + 3) =$

P2.54 Simplify: $2\delta(n - 2) * 2\text{sinc}(n + 3) =$

P2.55   Simplify: $\delta(n-4) * x(n) =$

P2.56   Simplify: $\delta_4(n) + \delta_4(n-2) =$

P2.57   Periodic sequence $x_p(n)$ has period 9 samples, and its sample values over time span $0 \le n \le 8$ are $\{4, 3, 2, 1, 0, 0, 1, 2, 3\}$. Express $x_p(n)$ as a convolution product of impulse train with triangular pulse function.

P2.58   Periodic sequence $y_p(n)$ has period 9 samples, and its sample values over time span $0 \le n \le 8$ are $\{1, 2, 3, 4, 3, 2, 1, 0, 0\}$. Express $y_p(n)$ as a convolution product of impulse train with time-shifted triangular pulse function.

P2.59   Finite-width sequence $x_1(n) = 0$ except over $-2 \le n \le 5$, and finite-width sequence $x_2(n) = 0$ except over $6 \le n \le 10$. Find the time index of the left-most nonzero sample, time index of the right-most nonzero sample, and width (max. number of nonzero samples) of the pulse that is defined by convolution product $x_1(n) * x_2(n)$.

Chapter 3

Continuous-Time Signals and Operations

3.1 Theory

3.1.1 *Introduction*

Consider $x(t)$, a single-valued function of independent variable t, where t is a real number in the range $-\infty$ to $+\infty$. In signal processing applications t would normally represent time, but this is not necessarily the case. For example, signal $x(a)$ could be a measure of temperature at different altitudes of the earth's atmosphere; in that case independent variable a has distance units. Whatever the interpretation of independent variable t, in this chapter we assume $x(t)$ to be defined over the continuum of values $-\infty \leq t \leq \infty$. That is why $x(t)$ is called a continuous (or continuous-time) signal.

In some cases, a sequence of values is obtained by sampling a continuous-time signal $x(t)$ at uniformly-spaced increments of time: $x(n) = x(t)|_{t=nT\ sec}$. Chapter 2 deals with such sequences of numbers, which are called discrete-time signals. One would expect the properties of continuous-time signal $x(t)$ to be closely reflected by the properties of sequence of samples $x(n)$. In fact, it may even be possible to recover $x(t)$, at all values of t, from its discrete-time samples; this is the topic of Ch. 6: Sampling Theory and Practice. But even when $x(n)$ is not derived from sampling a continuous-time signal $x(t)$, similarities in continuous-time and discrete-time signal processing exist and should not be ignored. Chapter 2 may be referred to for a topic-by-topic, equation-by-equation presentation of material in the discrete time domain that is synchronized to this chapter.

3.1.2 *Basic continuous-time signals*

3.1.2.1 *Impulse function*

The impulse function $\delta(t)$, also called the Dirac Delta, has the property:

$$\int_{-\infty}^{\infty} \delta(t) \, dt = \int_{0-}^{0+} \delta(t) \, dt = 1. \tag{3.1}$$

We may think of $\delta(t)$ as a unit of area that is concentrated at a single point: $t = 0$. Some describe $\delta(t)$ as a pulse that is infinitely tall and infinitesimally narrow, having unity area.[a] The impulse function[b] $\delta(t)$ is therefore drawn as an upward-pointing arrow, with a number next to the arrow specifying the area of the impulse, as shown in Fig. 3.1:

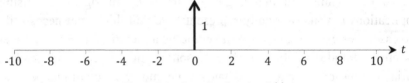

Figure 3.1.   Impulse function $\delta(t)$.

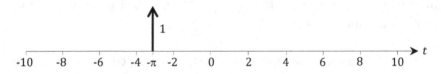

Figure 3.2.   Shifted impulse function $\delta(t + \pi)$.

---

[a] In terms of the rectangular pulse rect(t) that is defined later in this chapter, $\delta(t) = \lim_{T \to 0} (1/T) \, \text{rect}(t/T)$. Note that $(1/T) \, \text{rect}(t/T)$ has area equal to one, regardless of the value of positive real constant T.

[b] Strictly speaking, the Dirac impulse is not a function (for the reason that, in spite of being equal to zero $\forall t$ except at $t = 0$, it has nonzero area). Instead $\delta(t)$ is a mathematical object called a "generalized function," since it has function-like behavior when interacting with other signals.

The location of an impulse function in time is the value of t that makes the argument of $\delta(\cdot)$ equal zero. Thus, $\delta(t - t_0)$ is located at $t = t_0$. Figure 3.2 shows $\delta(t + \pi)$, which is $\delta(t)$ shifted to the left by π seconds.

In Figs. 3.1 and 3.2, both impulses have area equal to 1. Time shift does not change an impulse function's area $\left(\int_{-\infty}^{\infty} \delta(t - t_0)\, dt = 1 \right)$, it only moves it to a different point in time. As with any function, to change the area of an impulse we multiply it by a constant. Thus the area of $c\delta(t)$ is equal to $\int_{-\infty}^{\infty} c\delta(t)\, dt = c \int_{-\infty}^{\infty} \delta(t)\, dt = c$.

Now we introduce the most important property of a Dirac impulse function, called the "sifting" or "sampling" property:

$$\int_{-\infty}^{\infty} x(t)\delta(t - t_0)\, dt = x(t_0). \tag{3.2}$$

The sifting property concept is this: because $\delta(t - t_0) = 0$ everywhere except at $t = t_0$, only the value of $x(t)$ at $t = t_0$ is what matters in the product term $x(t)\delta(t - t_0)$. Therefore $x(t)$ is sampled at $t = t_0$ to give $x(t)\delta(t - t_0) = x(t_0)\delta(t - t_0)$, which is an impulse function with area $= x(t_0)$. The constant $x(t_0)$ has been obtained by sampling, or sifting it out from the waveform that is called "$x(t)$". Figure 3.3 illustrates why $x(t)\delta(t - t_0) = x(t_0)\delta(t - t_0)$:

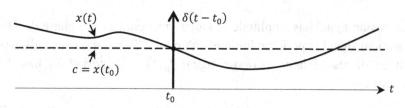

Figure 3.3. Multiplying $\delta(t - t_0)$ by signal $x(t)$ gives the same product as does multiplying $\delta(t - t_0)$ by the constant $c = x(t_0)$.

3.1.2.2 *Periodic impulse train*

It is useful to form a periodic signal by repeating an impulse every T seconds:

$$\delta_T(t) \equiv \textstyle\sum_{k=-\infty}^{\infty} \delta(t - kT). \tag{3.3}$$

Because the impulses in the plot look like the teeth of a comb, an impulse train is also referred to as a *comb* function. Time span T, the period of $\delta_T(t)$, may be any positive real number.

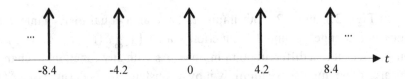

Figure 3.4. Impulse train $\delta_{4.2}(t)$. (When not specified, assume each impulse area = 1.)

3.1.2.3 *Sinusoid*

The sinusoidal signal is defined as:

$$f(t) = A\cos(\omega t + \theta) \tag{3.4}$$

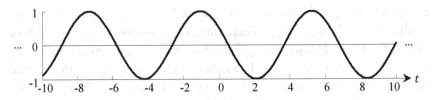

Figure 3.5. A sinusoidal signal ($A = 1, \omega = 1, \theta = \pi/3$).

This cosine signal has amplitude A, radian frequency ω and phase θ. The sinusoidal signal is periodic in time with period $T_0 = 2\pi/\omega$ seconds. As will be discussed later, periodic signal $f_p(t)$ with period T_0 has the property:

$$f_p(t) = f_p(t \pm T_0). \tag{3.5}$$

Thus, a sinusoidal signal may be shifted in time by T_0 seconds, or integer multiples thereof, without any change.

3.1.2.4 *Complex exponential*

The complex exponential signal is useful as an eigenfunction[c] for linear system analysis:

$$g(t) = e^{j\omega t}. \tag{3.6}$$

By invoking Euler's formula, $e^{j\phi} = \cos(\phi) + j\sin(\phi)$, we see that the real and imaginary components of the complex exponential signal are sinusoids: $Re\{e^{j\omega t}\} = \cos(\omega t)$, $Im\{e^{j\omega t}\} = \sin(\omega t)$.

3.1.2.5 *Unit step function*

The unit step function $u(t)$ is defined as:

$$u(t) \equiv \begin{cases} 0, & t < 0; \\ 1/2, & t = 0; \\ 1, & t > 0. \end{cases} \tag{3.7}$$

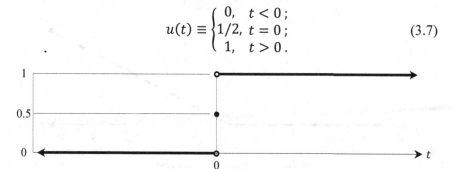

Figure 3.6.   Unit step function $u(t)$.

3.1.2.6 *Signum function*

The signum[d] function $\text{sgn}(t)$ is defined as:

$$\text{sgn}(t) \equiv \begin{cases} -1, & t < 0; \\ 0, & t = 0; \\ 1, & t > 0. \end{cases} \tag{3.8}$$

---

[c] When the input to a linear system is an eigenfunction, the system output is the same signal only multiplied by a constant.

[d] "Signum" is the word for "sign" in Latin.

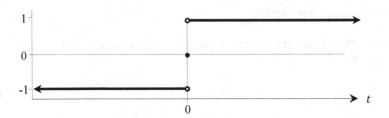

Figure 3.7. Signum function sgn(t).

3.1.2.7 *Ramp function*

The ramp function $r(t)$ is defined as:

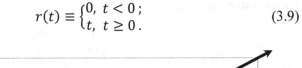

$$r(t) \equiv \begin{cases} 0, & t < 0; \\ t, & t \geq 0. \end{cases} \qquad (3.9)$$

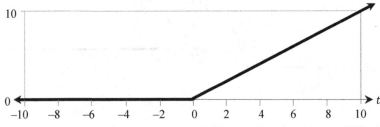

Figure 3.8. Ramp function $r(t)$.

3.1.2.8 *Rectangular pulse*

The rectangular pulse function rect(t) is defined as:

$$\text{rect}(t) \equiv \begin{cases} 1, & |t| < 1/2 \; ; \\ 1/2, & |t| = 1/2 \; ; \\ 0, & |t| > 1/2 \; . \end{cases} \qquad (3.10)$$

The rectangular pulse function has a width of 1 and a height of 1; some refer to it as a "box" function.

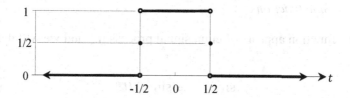

Figure 3.9. Rectangular pulse function rect(t).

3.1.2.9 *Triangular pulse*

Define the triangular pulse function $\Delta(t)$ to be:

$$\Delta(t) \equiv \begin{cases} 1 - 2|t|, & |t| \leq 1/2; \\ 0, & |t| > 1/2. \end{cases} \tag{3.11}$$

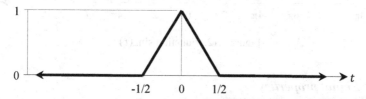

Figure 3.10. Triangular pulse function $\Delta(t)$.

3.1.2.10 *Exponential decay*

The waveform e^{-bt}, where b is a real constant and $b > 0$, decays to zero as $t \to \infty$ and grows without bound as $t \to -\infty$. Define the exponentially decaying signal[e] $\equiv u(t)e^{-bt}$, which is shown in Fig. 3.11.

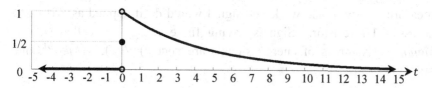

Figure 3.11. Exponentially decaying signal $u(t)e^{-0.223t}$.

---

[e] An equivalent expression for the decaying exponential signal $u(t)e^{-bt}$ is $u(t)a^t$, where $a = e^{-b}$ ($0 < a < 1$).

3.1.2.11 *Sinc function*

The sinc function appears often in signal processing, and we will define it as:

$$\text{sinc}(t) \equiv \sin(t)/t. \tag{3.12}$$

As seen in Fig. 3.12, the $\text{sinc}(t)$ function has zero crossings at every π seconds, except at $t = 0$ where the ratio $\sin(t)/t$ evaluates to 1.[f]

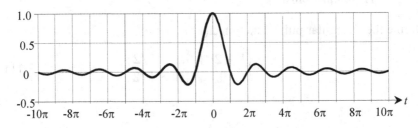

Figure 3.12. Function $\text{sinc}(t)$.

3.1.3 *Signal properties*

3.1.3.1 *Energy and power signals*

The *energy* of a continuous-time signal is defined as the area of its magnitude-squared value:

$$E_x \equiv \int_{-\infty}^{\infty} |x(t)|^2 \, dt = \int_{-\infty}^{\infty} x(t)x^*(t) \, dt. \tag{3.13}$$

Clearly, E_x is always real and non-negative. The energy value is a measure of how much work this signal would do if applied as a voltage across a 1Ω resistor. Signals having finite energy are called *energy signals*. Examples of energy signals are $\text{rect}(t)$, $\Delta(t)$, $u(t)e^{-bt}$ and $\text{sinc}(t)$.

---

[f] Because one cannot directly evaluate $\text{sinc}(0) = \sin(0)/0 = 0/0$, apply l'Hospital's rule: $\lim_{\phi \to 0} \sin(\phi)/\phi = \lim_{\phi \to 0} (d \sin(\phi)/d\phi)/(d\phi/d\phi) = \lim_{\phi \to 0} \cos(\phi)/1 = 1.$

For periodic and some other signals the integration result of Eq. (3.13) will be infinite: $E_x = \infty$. In that case a more meaningful measure may be the *power* of $x(t)$, its average energy per unit time:

$$P_x \equiv \lim_{T \to \infty} \left(\frac{1}{T} \int_{-T/2}^{T/2} |x(t)|^2 \, dt \right)$$

$$= \lim_{T \to \infty} \left(\frac{1}{T} \int_{-T/2}^{T/2} x(t)x^*(t) \, dt \right). \qquad (3.14)$$

Signals having finite, nonzero power are called *power signals*. Examples of continuous-time power signals are $u(t)$, $\text{sgn}(t)$ and $e^{j\omega t}$, whose power values are $\{1/2, 1, 1\}$ respectively. Power signals have infinite energy, and energy signals have zero power. Some signals fall in neither category: for example, $r(t)$, $\delta(t)$ and $\delta_T(t)$ all have both infinite energy and infinite power.[g]

Periodic signals have power = (Energy in one period) / (time span of one period). This may be expressed as $P_{x_p} \equiv \frac{1}{T_0} \int_{T_0} |x_p(t)|^2 \, dt$, where periodic signal $x_p(t)$ has period equal to T_0 sec. Based on this relation it is easy to show that the power of sinusoid $A \cos(\omega t + \theta)$ is $|A|^2/2$.[h]

3.1.3.2 *Integrable signals*

A signal $x(t)$ is said to be absolutely integrable when:

$$\int_{-\infty}^{\infty} |x(t)| \, dt = \int_{-\infty}^{\infty} \sqrt{x(t)x^*(t)} \, dt < \infty. \qquad (3.15)$$

A signal $x(t)$ is said to be square-integrable when:

$$\int_{-\infty}^{\infty} |x(t)|^2 \, dt = \int_{-\infty}^{\infty} x(t)x^*(t) \, dt = E_x < \infty. \qquad (3.16)$$

We see, therefore, that square-integrable waveforms are energy signals.

---

[g] These signals are still very useful as mathematical tools, however, as will be shown in later chapters.

[h] The energy in one period of a sinusoid having amplitude A and period T_0 is $|A|^2 T_0/2$.

3.1.3.3 *Periodic signals*

A signal $x(t)$ is *periodic* when there are some nonzero time shifts T for which the following is true:

$$x_p(t) = x_p(t + T). \tag{3.17}$$

The *fundamental period* of $x_p(t)$ is the smallest positive value $T = T_0$ that makes Eq. (3.17) true. A periodic signal has the property $x_p(t) = x_p(t + kT_0)$, for any integer k.

The *fundamental frequency*, in repetitions or cycles per second (Hertz), of a periodic signal is the reciprocal of its fundamental period: $f_0 = 1/T_0$ Hz. Frequency may also be measured in *radians per second*, as we do in this textbook, which is defined as $\omega = 2\pi f$.[i] Thus, fundamental frequency $\omega_0 = 2\pi/T_0$ rad/sec.

3.1.3.4 *Sum of periodic signals*

The sum of two or more periodic signals is only periodic if the frequencies of these additive components are *harmonically related*: that is, the frequency of each component is an integer multiple of a fundamental frequency ω_0. As a result, for any two harmonically-related components in the sum having frequencies $\omega_1 = k_1\omega_0$ and $\omega_2 = k_2\omega_0$, the ratio $\omega_1/\omega_2 = k_1/k_2$ is a rational number. The fundamental period of the sum of harmonically-related signals having fundamental frequency ω_0 is $T_0 = 2\pi/\omega_0$ sec.

---

[i] The term *radians* is derived from *radius* of a circle. When a wheel having radius $r = 1$ rolls along the ground for one revolution, the horizontal distance travelled is equal to the wheel's circumference $c = 2\pi r$ (or 2π radii). Thus $f = 1$ cycle per second corresponds to $\omega = 2\pi(1) = 2\pi$ radians/sec. We use rad/sec measure to describe frequency since it simplifies notation somewhat (e.g. $\cos(\omega t)$ as compared to $\cos(2\pi f t)$), but in practice engineers almost always specify frequency as f in Hertz (Hz).

For example: $x(t) = \cos(5t) + \sin(1.5t)$ is periodic, having fundamental frequency $\omega_0 = 0.5$ rad/sec[j] and period $T_0 = 4\pi$ sec, whereas $y(t) = \cos(5t) + \sin(\pi t)$ is not periodic because $\omega_1/\omega_2 = 5/\pi$ is irrational.

Each additive term in a harmonically-related sum of sinusoids having frequency $\omega = k\omega_0$ is identified by its *harmonic index*: $k = \omega/\omega_0$. Thus, in the example above, the term $\sin(1.5t)$ in $x(t)$ has harmonic index 3 (is "the 3rd harmonic" of $\omega_0 = 0.5$ rad/sec).[k]

3.1.3.5 *Even and odd signals*

A signal $x(t)$ is even if it is unchanged after time reversal (replacing t with $-t$):

$$x_e(t) = x_e(-t). \tag{3.18}$$

A signal $x(t)$ is odd if it is negated by time reversal:

$$x_o(t) = -x_o(-t). \tag{3.19}[l]$$

Every signal may be expressed as a sum of its even and odd components:

$$y(t) = y_e(t) + y_o(t), \tag{3.20}$$

where

$$y_e(t) = \tfrac{1}{2}(y(t) + y(-t)), \tag{3.21}$$

and

$$y_o(t) = \tfrac{1}{2}(y(t) - y(-t)). \tag{3.22}$$

Signals that are even and are described in this chapter are $\delta(t)$, $\delta_T(t)$, $\cos(\omega t)$, $\text{rect}(t)$, $\Delta(t)$ and $\text{sinc}(t)$. Some odd signals are $\sin(\omega t)$ and $\text{sgn}(t)$. Signals $u(t)$, $r(t)$, $e^{j\omega t}$, and $u(t)e^{-bt}$ are neither even nor odd.

---

[j] Notice that even though the fundamental frequency of $x(t)$ is 0.5 rad/sec there is no individual component having that frequency.

[k] The fundamental frequency is the 1st harmonic of a periodic signal, although it is rarely referred to as such.

[l] A consequence of Eq 3.19 is that every odd signal has $x_o(0) = 0$, because only value zero has the property that $0 = -0$.

3.1.3.6 *Right-sided and left-sided signals*

If $x(t) = 0$ for $t < T$, where T is some finite time value, then $x(t)$ is a *right-sided signal*. For example, $u(t + 3)$ is right-sided because it equals zero for $t < -3$. If $x(t) = 0$ for $t > T$, where T is some finite time value, then $x(t)$ is a *left-sided signal*. The signal $\cos(t)u(-t)$ is left-sided because it equals zero for $t > 0$.

3.1.3.7 *Causal, anticausal signals*

A signal $x(t)$ is *causal* if $x(t) = 0$ for all $t < 0$.[m] A signal $x(t)$ is *anticausal* if $x(t) = 0$ for all $t > 0$. Both causal and anticausal signals may have a nonzero value at $t = 0$, hence only one signal is both causal and anticausal: $a\delta(t)$, where a is a constant. Every signal may be expressed as the sum of causal and anticausal components, as for example: $x(t) = x(t)u(t) + x(t)(1 - u(t))$. Causal signals are a subclass of right-sided signals, and anticausal signals are a subclass of left-sided signals. Causal signals described in this chapter are $\delta(t)$, $u(t)$, $r(t)$, and $u(t)e^{-bt}$. Reversing a causal signal in time makes it anticausal, and vice versa. Therefore $\delta(-t) = \delta(t)$, $u(-t)$, $r(-t)$, and $u(-t)e^{bt}$ are all anticausal signals.

3.1.3.8 *Finite-length and infinite-length signals*

If $x(t) = 0$ for $|t| > T$, where T is some finite time value, then $x(t)$ is a *finite-length signal*. When no such value of T exists then $x(t)$ is an *infinite-length signal*. Finite-length signals are both right-sided and left-sided. Finite-length signals described in this chapter are $\delta(t)$, $rect(t)$ and $\Delta(t)$. The signals $\delta_T(t)$, $A\cos(\omega t + \theta)$, $sinc(t)$, $sgn(t)$, $u(t)$, $r(t)$, $e^{j\omega t}$, and $u(t)e^{-bt}$ all have infinite length.

---

[m] The term "causal" comes from the behavior of real-world systems. If, in response to an impulse at $t = 0$, a system outputted a signal prior to $t = 0$, it would be basing its output on the knowledge of a future input. Nature's cause-and-effect relationship is that a response at present time can only be caused by past and present stimuli. When driven by a causal input signal (such as an impulse at $t = 0$), real-world systems will produce a causal output signal.

3.1.4 *Continuous-time signal operations*

3.1.4.1 *Time delay*

When signal $x(t)$ is shifted to the right by t_0 seconds to give $x(t - t_0)$, we call this a t_0-second *time delay*. On the other hand, a shift to the left is called a *time advance*. Figures 3.13 and 3.14 demonstrate these concepts:[n]

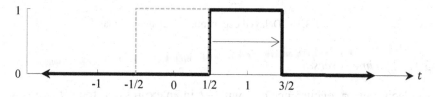

Figure 3.13. Signal $\text{rect}(t - 1)$, which is $\text{rect}(t)$ after 1-sec delay.

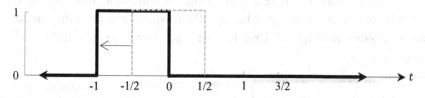

Figure 3.14. Signal $\text{rect}(t + 1/2)$, which is $\text{rect}(t)$ after 1/2-sec advance.

Note that the endpoints of $\text{rect}(t)$ are at time values $t = \{-1/2, 1/2\}$, while the endpoints of $\text{rect}(t - 1)$ are at $t - 1 = \{-1/2, 1/2\}$ and the endpoints of $\text{rect}(t + 1/2)$ are at $t + 1/2 = \{-1/2, 1/2\}$. In Fig. 3.15, the time-delayed impulse $\delta(t - 4)$ is located at $t = 4$ (since $\delta(t)$ is located at $t = 0$, $\delta(t - 4)$ is located at $t - 4 = 0$):

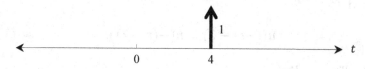

Figure 3.15. Delayed impulse function $\delta(t - 4)$.

---

[n] For ease of graphing, when the exact value of $\text{rect}(t)$ at a discontinuity is not important we will draw it as shown.

When, in a functional expression for signal $x(t)$, every occurrence of t is replaced by $t - t_0$, the result is that $x(t)$ gets delayed by t_0 seconds. This is demonstrated in Fig. 3.16:

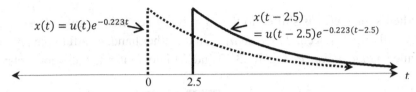

Figure 3.16. Delayed exponentially decaying signal.

3.1.4.2 *Time reversal*

Replacing every occurrence of t with $-t$ in an expression for $x(t)$ causes the signal to flip about $t = 0$ (only point $x(0)$ stays where it originally was). Time-reversing a causal signal makes it anticausal, and vice-versa. Time-reversing an even signal has no effect, while time-reversing an odd signal negates the original. Here is what happens when signal $u(t - 2)$ is reversed in time:

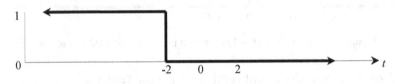

Figure 3.17. Signal $u((-t) - 2)$.

Fig. 3.17 demonstrates an interesting (and often confusing) concept: the signal shown could either be a delayed unit step function that was then time-reversed, or a time-reversed unit step function that was then time-advanced:

$$u((-t) - 2) = u(-(t + 2)). \qquad (3.23)$$

3.1.4.3 *Time scaling*

Time scaling a continuous-time signal is done either by replacing time variable t with at (called "time-compression by factor a"), or by replacing time variable t with t/a (called "time-expansion by factor a"). The two operations have the effect of compressing the signal toward $t = 0$, or expanding the signal away from $t = 0$, respectively ($a > 0$).

Regarding compressing a signal in time: one may be tempted to think that more and more information may be compressed in time, ultimately reaching the situation that all of the world's knowledge is represented by a near-zero-length signal. Theoretically this is possible, but it comes at a price: when a signal is compressed in time the rapidity of changes in its wave shape is increased by the same factor. Practical communications systems are limited in how rapidly signals they pass may vary vs. time, so that the dream of transmitting an infinite amount of information in zero time cannot be achieved in practice.[o]

3.1.4.4 *Cumulative integral and time differential*

The cumulative integral of continuous-time signal $x(t)$ is itself a function of independent time index t:

$$\text{Cumulative Integral } \{x(t)\} \equiv \int_{-\infty}^{t} x(\tau)\, d\tau. \tag{3.24}$$

Here are some signals stated in terms of the cumulative integral operation:

$$u(t) = \int_{-\infty}^{t} \delta(\tau)\, d\tau, \tag{3.25}$$

$$r(t) = \int_{-\infty}^{t} u(\tau)\, d\tau, \tag{3.26}$$

$$\text{rect}(t) = \int_{-\infty}^{t} \left\{ \delta\left(\tau + \tfrac{1}{2}\right) - \delta\left(\tau - \tfrac{1}{2}\right) \right\} d\tau. \tag{3.27}$$

Related to the cumulative integral is the differential operator:

$$\text{Differential of } x(t) \equiv \frac{dx(t)}{dt}. \tag{3.28}$$

Here are some signals stated in terms of the differential operation:

---

[o] As we will see in Chapters 6–8, the rapidity of signal change vs. time is determined by a system's frequency bandwidth.

$$\delta(t) = \frac{d}{dt}\{u(t)\}, \tag{3.29}$$

$$u(t) = \frac{d}{dt}\{r(t)\}, \tag{3.30}$$

$$\delta\left(t + \frac{1}{2}\right) - \delta\left(t - \frac{1}{2}\right) = \frac{d}{dt}\{rect(t)\}. \tag{3.31}$$

3.1.4.5 *Conjugate, magnitude and phase*

The conjugate, magnitude and phase of signal $g(t)$ are found according to the normal rules of complex algebra:[p]

$$g^*(t) = Re\{g(t)\} - j\,Im\{g(t)\}, \tag{3.32}$$

$$|g(t)|^2 = Re\{g(t)\}^2 + Im\{g(t)\}^2 = g(t)g^*(t), \tag{3.33}$$

$$\angle g(t) = \text{Tan}^{-1}\left\{\frac{Im\{g(t)\}}{Re\{g(t)\}}\right\}. \tag{3.34}[q]$$

3.1.4.6 *Equivalent signal expressions*

Basic signals presented in the previous section may be related to one another via algebraic operations, or combined to give other signals of interest. For example, here are different ways of describing the rectangular, signum and unit step functions in terms of one another:[r]

$$rect(t) = u(t + 1/2) - u(t - 1/2), \tag{3.35}$$

$$rect(t) = u(t + 1/2)\,u(-t + 1/2), \tag{3.36}$$

$$sgn(t) = 2u(t) - 1, \tag{3.37}$$

$$sgn(t) = u(t) - u(-t), \tag{3.38}$$

$$u(t) = 1 - u(-t). \tag{3.39}$$

---

[p] Note that the operation is performed on the entire signal $g(t)$.

[q] Evaluate the Tan^{-1} function to produce a phase angle in all four quadrants; this is done by noting the signs of imaginary and real components of $g(t)$ prior to division.

[r] Plot these to convince yourself that the expressions are equivalent!

3.1.5 *Convolution*

One of the most useful operations for signal processing and linear system analysis is the convolution operation,[s] defined here for continuous-time signals:

$$\boxed{\text{Convolution}} \quad x(t) * y(t) \equiv \int_{\tau=-\infty}^{\infty} x(\tau)y(t-\tau)\,d\tau. \tag{3.40}$$

By replacing $t - \tau$ with ϕ, we obtain an equivalent form of this integral:

$$\int_{\phi=-\infty}^{\infty} y(\phi)x(t-\phi)\,d\phi = y(t) * x(t). \tag{3.41}$$

Thus the convolution operation is commutative:

$$x(t) * y(t) = y(t) * x(t). \tag{3.42}$$

3.1.5.1 *Convolution with an impulse*

Let's see what happens when we convolve signal $x(t)$ with the impulse function $\delta(t)$:

$$x(t) * \delta(t) = \int_{\tau=-\infty}^{\infty} x(\tau)\delta(t-\tau)\,d\tau,$$

$$= \int_{\tau=-\infty}^{\infty} x(t)\delta(t-\tau)\,d\tau,$$

$$= x(t) \int_{\tau=-\infty}^{\infty} \delta(t-\tau)\,d\tau = x(t). \tag{3.43}$$

This result, $x(t) * \delta(t) = x(t)$, shows us that convolving with an impulse function is an identity operation in the realm of convolution. More interesting and useful for signal processing, however, are the following relations that are also easily derived:

$$x(t) * \delta(t - t_1) = x(t - t_1), \tag{3.44}$$

$$x(t) * \delta(t + t_2) = x(t + t_2). \tag{3.45}$$

---

[s] The convolution operation is also called a "convolution product" because of its similarity to multiplicative product notation.

Thus, we see that a signal may be delayed or advanced in time by convolving it with a time-shifted impulse function. As will be discussed in Ch. 8, convolution plays a critical role in describing the behavior of linear time-invariant systems.

3.1.5.2 *Convolution of two pulses*

Define "pulse" $p(t)$ to be a signal that is zero outside of a finite time span $t_1 \leq t \leq t_2$, having pulse width $w = t_2 - t_1$. Given two pulses $p_a(t)$ and $p_b(t)$, we can state the following about their convolution product $x(t) = p_a(t) * p_b(t)$:

- $x(t)$ will also be a pulse; the pulse width of x equals the sum of pulse widths of p_a and p_b: $w_x = w_{p_a} + w_{p_b}$;

- The left edge of $x(t)$ will lie at a time that is the sum of {time at left edge of $p_a(t)$} and {time at left edge of $p_b(t)$};[t]

- The right edge of $x(t)$ will lie at a time that is the sum of {time at right edge of $p_a(t)$} and {time at right edge of $p_b(t)$}.[u]

3.1.6 *Cross-correlation*

Cross-correlation is used to measure the similarity between two signals, and for energy signals this operation is defined by an expression similar to convolution:

| Cross-Correlation | $\phi_{xy}(t) \equiv \int_{-\infty}^{\infty} x^*(\tau)y(t+\tau)\, d\tau.$ | (3.46)[v] |

---

[t] Assume $p_a(t) = 0$ outside of $t_1 \leq t \leq t_2$ and that $p_b(t) = 0$ outside of $t_3 \leq t \leq t_4$. Then $x(t) = p_a(t) * p_b(t) = \int_{-\infty}^{\infty} p_a(\tau)p_b(t-\tau)d\tau = \int_{t_1}^{t_2} p_a(\tau)p_b(t-\tau)d\tau$ due to the limited time span of $p_a(t)$. By considering the limited time span of $p_b(t)$ we note that $x(t) = 0$ when $t - t_1 < t_3$ ($t < t_1 + t_3$) or $t - t_2 > t_4$ ($t > t_2 + t_4$). Thus $x(t) = 0$ outside of time range $t_1 + t_3 \leq t \leq t_2 + t_4$.

[u] See footnote above.

[v] Another common notation for $\phi_{xy}(t)$ is $x(t) \star y(t)$.

Unlike convolution, cross-correlation is not commutative: $\phi_{xy}(t) \neq \phi_{yx}(t)$.[w] The cross-correlation between two energy signals may be calculated using convolution:[x]

$$\phi_{xy}(t) = x^*(-t) * y(t). \qquad (3.47)$$

For power signals the integral in Eq. (3.46) will not converge at all values of t. We therefore modify the definition of cross-correlation for power signals:[y]

$$R_{xy}(t) \equiv \lim_{T \to \infty} \frac{1}{2T} \int_{-T}^{T} x^*(\tau) y(t + \tau) \, d\tau. \qquad (3.48)$$

For periodic power signals $x_p(t)$ and $y_p(t)$, both having period T_0, this definition of cross-correlation simplifies to:

$$R_{xy}(t) \equiv \frac{1}{T_0} \int_0^{T_0} x^*(\tau) y(t + \tau) \, d\tau \qquad (3.49)$$

The cross-correlation between energy signal $x(t)$ and itself is called the autocorrelation of $x(t)$:

$$\boxed{\text{Autocorrelation}} \qquad \phi_{xx}(t) \equiv \int_{\tau=-\infty}^{\infty} x^*(\tau) x(t + \tau) \, d\tau \qquad (3.50)$$

The autocorrelation of signal $x(t)$ is a measure of similarity between $x(t)$ and delayed versions of itself. From these definitions, energy $E_x = \phi_{xx}(0)$, and power $P_x = R_{xx}(0)$.

---

[w] When both $x(t)$ and $y(t)$ are real signals then $\phi_{xy}(t) = \phi_{yx}(-t)$.

[x] $\phi_{xy}(t) = \int_{\tau=-\infty}^{\infty} x^*(\tau) y(t + \tau) \, d\tau = \int_{\beta=-\infty}^{\infty} x^*(-\beta) y(t - \beta) \, d\beta = \int_{\beta=-\infty}^{\infty} g(\beta) y(t - \beta) \, d\beta = g(t) * y(t)$, where $g(\beta) = x^*(-\beta)$.

[y] Notation: $R_{xy}(t)$ represents the cross-correlation between power signals, and $\phi_{xy}(t)$ the cross-correlation between energy signals.

3.2 Practical Applications

3.2.1 *Synthesizing a periodic signal using convolution*

Given pulse $p(t)$ having length T sec. A periodic signal having pulse $p(t)$ as one of its periods may be synthesized by convolving $p(t)$ with an impulse train whose period is T sec:

$$x_p(t) = p(t) * \delta_T(t) = p(t) * \sum_{k=-\infty}^{\infty} \delta(t - kT)$$

$$= \sum_{k=-\infty}^{\infty} p(t) * \delta(t - kT) = \sum_{k=-\infty}^{\infty} p(t - kT), \qquad (3.51)$$

which is a summation of time-shifted copies of pulse $p(t)$. The spacing of these pulses is T seconds so that none of them overlap with neighboring pulses but instead sum up to give periodic waveform $x_p(t)$.[z] This model for creating a periodic signal will be used in Chapter 5 to derive an expression for the Fourier transform of periodic signals.

3.2.2 *Waveform smoothing by convolving with a pulse*

When signal $x(t)$ is convolved with $(1/T)\,\text{rect}(t/T)$, a rectangular pulse having area $= 1$,

$$x(t) * \frac{1}{T}\,\text{rect}\left(\frac{t}{T}\right) = \frac{1}{T}\left\{\text{rect}\left(\frac{t}{T}\right) * x(t)\right\}$$

$$= \frac{1}{T}\int_{-\infty}^{\infty} \text{rect}\left(\frac{\tau}{T}\right) x(t - \tau)\, d\tau = \frac{1}{T}\int_{-T/2}^{T/2}(1)x(t - \tau)\, d\tau$$

$$= \frac{1}{T}\int_{t+T/2}^{t-T/2} x(\beta)\,(-d\beta) = \frac{1}{T}\int_{t-T/2}^{t+T/2} x(\beta)\, d\beta, \qquad (3.52)$$

we see that the result is the average value of $x(t)$ in the vicinity of t. The effect is to smooth, or reduce rapidity of amplitude changes vs. time, of the signal $x(t)$. The wider the rectangular pulse (higher value of T), the

---

[z] Subscript p in $x_p(t)$ specifies that the sequence is periodic (not related to the label of pulse $p(t)$).

more smoothing results.[aa] "Smoothing pulses" other than the rectangular pulse may also be used. Waveform smoothing is the time-domain result of lowpass filtering, a frequency-domain operation that will be introduced in Chapter 8.

Example 3.1
Pulse $\Delta(t)$ is shown below before and after smoothing. The smoothing pulse convolved with $\Delta(t)$ was $5\text{rect}(5t)$ (having area = 1)[bb]:

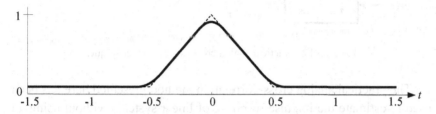

Figure 3.18. Triangular pulse function $\Delta(t)$, before (dotted line) and after (solid line) smoothing via convolution with pulse $5\text{rect}(5t)$.

3.2.3 *Practical analog cross-correlation*

After a change of variables, $\phi_{xy}(t) \equiv \int_{\tau=-\infty}^{\infty} x^*(\tau)y(t+\tau)\,d\tau$ becomes $\phi_{xy}(\tau) \equiv \int_{t=-\infty}^{\infty} x^*(t)y(t-\tau)\,dt$. When both $x(t)$ and $y(t)$ are real their cross-correlation $\phi_{xy}(\tau) = \int_{t=-\infty}^{\infty} x(t)y(t-\tau)\,dt$, which is the area of the product of $x(t)$ and a delayed version of $y(t)$. Similarly, for power signals we may write:

$$R_{xy}(\tau) \equiv \lim_{T\to\infty} \frac{1}{T}\int_{-T/2}^{T/2} x(t)y(t-\tau)\,dt, \qquad (3.53)$$

which is the time average of $x(t)y(t-\tau)$. In practice, time averaging may be approximated using a low-pass filter (Ch. 8). The diagram below,

---

[aa] It is intuitively pleasing, therefore, that zero smoothing results when convolving $x(t)$ with zero-width impulse $\delta(t)$.

[bb] Why unity area? Because then the average value of the waveform remains unchanged, and the two waveforms may be plotted and compared to one another on the same graph.

therefore, describes a practical method to calculate the cross-correlation between two power signals. By slowly linearly increasing delay τ as function of time, in response the output of this circuit will sweep out an approximation to $R_{xy}(\tau)$:

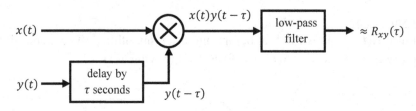

Figure 3.19. Practical analog cross-correlation technique.

The practical analog cross-correlation method presented here has been used to estimate the impulse response of linear systems without using an impulse function.[cc]

3.2.4 *Normalized cross-correlation as a measure of similarity*

A useful normalized form of the cross-correlation between real[dd] waveforms $x(t)$ and $y(t)$ is:

$$C_{xy}(t) = \frac{\phi_{xy}(t)}{\sqrt{\phi_{xx}(0)\phi_{yy}(0)}} = \frac{\phi_{xy}(t)}{\sqrt{E_x E_y}}. \tag{3.54}$$

It may be shown that $-1 \le C_{xy}(t) \le 1$, for all values of t. When $y(t) = ax(t - t_0)$ then $C_{xy}(t_0) = 1$ (constant $a > 0$) or $C_{xy}(t_0) = -1$ (constant $a < 0$). $C_{xy}(t_0) = 0$ occurs for the case that $x(t - t_0)$ and $y(t)$

[cc] Goncharoff, V., J. E. Jacobs, and D. W. Cugell. "Wideband acoustic transmission of human lungs." Medical and Biological Engineering and Computing 27.5 (1989): 513-519.
[dd] Because both $x(t)$ and $y(t)$ are real, $\phi_{xy}(t) = \phi_{yx}(-t)$, $C_{xy}(t) = C_{yx}(-t)$, and thus only time reversal distinguishes the similarity between $x(t)$ and $y(t)$ from the similarity between $y(t)$ and $x(t)$.

are *orthogonal* [ee] to one another – that is, they cannot be made any more similar-looking by multiplying one of them by a nonzero constant. The beauty of this normalization is that a threshold value may be set (e.g., 0.75) above which two continuous-time signals are judged to be similar in appearance, and this threshold is independent of the amplitude scaling factors of either $x(t)$ or $y(t)$.

3.2.5 *Application of convolution to probability theory*

In probability theory, probability density function $f_X(x)$ is used to describe the probabilistic nature of continuous random variable X. (The probability that a sample of X will fall in amplitude range $x_1 < x < x_2$ is the area of $f_X(x)$ over the range $x_1 < x < x_2$.) Given two independent continuous random variables X and Y having arbitrary probability density functions $f_X(x)$ and $f_Y(y)$, respectively, what is the probability density function of their sum $X + Y$? The answer is a convolution product:

$$f_{X+Y}(\tau) = f_X(\tau) * f_Y(\tau). \qquad (3.55)$$

3.3 Useful MATLAB® Code

A continuous-time signal is defined at every instant of time, so that even over finite time spans there are infinitely many points of data. To represent an analog signal using a digital computer having a finite amount of storage memory, one must either reduce these signal data to a few parameters (e.g., for a sinusoid: amplitude A, frequency ω, and phase θ), which is a *parametric* representation, or sample the signal at certain times over a limited time span to give a list of numbers, which is a *sampled* representation. Parametric representations of signals are the most accurate and compact, but are only possible for certain idealized waveforms. Typically, real-world signals cannot be represented parametrically without some approximation error. Therefore, signals discussed in this chapter

---

[ee] The definition of *orthogonal* signals is this: $x(t)$ and $y(t)$ are orthogonal over all time when $\int_{-\infty}^{\infty} x^*(t)y(t)dt = \int_{-\infty}^{\infty} x(t)y^*(t)dt = 0$.

will be manipulated and plotted in a sampled representation using MATLAB®. In the following demonstration code, we chose the number of sample points, uniformly spaced across the time range being analyzed, large enough so that sampling effects are not noticeable.[ff]

3.3.1 *Plotting basic signals*

The basic plotting command in MATLAB® is **plot(x,y)**, which plots a waveform by connecting points $(x_1, y_1), (x_2, y_2), \ldots, (x_N, y_N)$ using straight line segments. Arrays **x** and **y** store the ordered data rectangular components $\{x_1, x_2, \ldots, x_n\}$ and $\{y_1, y_2, \ldots, y_n\}$ respectively:

```
% Example_Ch3_1
% Plotting ten cycles of a sine wave
Npts = 1000;
x = linspace(0,10,Npts);
omega = 2*pi;
y = sin(omega*x);
plot(x,y)
xlabel('time in seconds')
ylabel('amplitude')
title('sin(\omegat), \omega = 2\pi rad/sec')
```

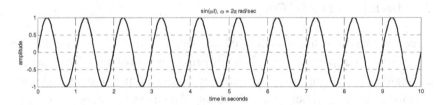

Figure 3.20. MATLAB® plot of $y(t) = \sin(2\pi t)$.

MATLAB® permits the user to define functions, which can simplify the plotting of combinations of building-block signals. The following is an example of using this method to plot the signal $y(t) = u(t + 1.5) + \text{rect}(t/2) + \Delta(t) - u(t - 1.5)$:

```
function Example_Ch3_2
% Plotting y(t) = u(t+1.5)+rect(t/2)+Delta(t)-u(t-1.5)
```

[ff] The effects of sampling an analog signal are discussed in detail in Chapter 6.

```
Npts = 10000;
t = linspace(-2,2,Npts);
y = u(t+1.5) + rect(t/2) + Delta(t) - u(t-1.5);
plot(t,y)
axis([-2 2 -1 3])
xlabel('time in seconds')
ylabel('amplitude')
title(['A plot of y(t) = u(t+1.5) + '], ...
      ['rect(t/2) + \Delta(t) - u(t-1.5)'])
grid on
end

function f = u(t)
% define the unit step function
    f = zeros(size(t));
    f(t>0) = 1;
    f(t==0) = 1/2;
end

function f = rect(t)
% define the rectangular pulse function
    f = zeros(size(t));
    f(abs(t)<1/2) = 1;
    f(abs(t)==1/2) = 1/2;
end

function f = Delta(t)
% define the triangular pulse function
    f = zeros(size(t));
    indexes = find(abs(t)<1/2);
    f(indexes) = 1 - 2*abs(t(indexes));
end
```

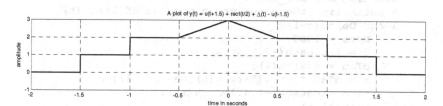

Figure 3.21. MATLAB® plot of $y(t) = u(t + 1.5) + \text{rect}(t/2) + \Delta(t) - u(t - 1.5)$.

3.3.2 *Estimating continuous-time convolution*

Continuous-time convolution product $x(t) * y(t) = \int_{-\infty}^{\infty} x(\tau)y(t - \tau)d\tau$ may be expressed as:

$$\lim_{\Delta\tau\to 0} \sum_{k=-\infty}^{\infty} x(k\Delta\tau)y(t-k\Delta\tau)\Delta\tau. \qquad (3.56)$$

Even if $\Delta\tau > 0$, this summation will give reasonably good results when the two signals are relatively slowly-varying in time ($x(t) \approx x(t+\Delta)$ and $y(t) \approx y(t+\Delta)$). Let $z(t) = x(t) * y(t)$. Then, for a suitably small time increment $\Delta\tau$,

$$z(n\Delta\tau) \approx \sum_{k=-\infty}^{\infty} x(k\Delta\tau)y(n\Delta\tau - k\Delta\tau)\Delta\tau$$

$$= \Delta\tau \sum_{k=-\infty}^{\infty} x(k\Delta\tau)y\big((n-k)\Delta\tau\big). \qquad (3.57)$$

In MATLAB® the **conv** function calculates discrete-time convolution[gg], as defined by the following summation:

$$\sum_{k=-\infty}^{\infty} x(k)y(n-k) \qquad (3.58)$$

Therefore, an approximation to $z(t) = x(t) * y(t)$ at times $t = n\Delta\tau$ may be obtained by sampling signals $x(t)$ and $y(t)$ at every $t = n\Delta\tau$, then using MATLAB's® built-in discrete convolution function to process the two sequences of numbers, and finally scaling the result by $\Delta\tau$. Implied is the requirement that the signals $x(t)$ and $y(t)$ be of finite length (zero outside of some range) so that their sampled representations may be stored in finite-length arrays without any omissions. In the following example we demonstrate this method to convolve $2\text{rect}(t)$ with $\Delta(t-1)$:

```
function Example_Ch3_3
% Estimating the convolution product of 2rect(t)
% and Delta(t-1)
    Npts = 1000;
    t = linspace(-2,2,Npts);
    dTau = t(2)-t(1);
    z = dTau * conv(2*rect(t),Delta(t-1));
    t1 = linspace(-4,4,length(z));
    plot(t1,z)
    xlabel('time in seconds')
    ylabel('amplitude')
    title('2rect(t) convolved with \Delta(t-1)')
    axis([-3 3 0 1])
    grid on
    end
```

[gg] Discrete convolution is presented in Chapter 2.

```
function f = rect(t)
% define the rectangular pulse function
    f = zeros(size(t));
    f(abs(t)<1/2) = 1;
    f(abs(t)==1/2) = 1/2;
end

function f = Delta(t)
% define the triangular pulse function
    f = zeros(size(t));
    indexes = find(abs(t)<1/2);
    f(indexes) = 1 - 2*abs(t(indexes));
end
```

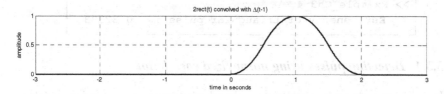

Figure 3.22. MATLAB® plot of $2\text{rect}(t)$ convolved with $\Delta(t-1)$.

3.3.3 *Estimating energy and power of a signal*

The energy of finite-length signal $x(t)$ may be estimated from its samples using the relation:

$$E_x = \int_{-\infty}^{\infty}|x(t)|^2 dt = \lim_{\Delta\tau\to0} \sum_{k=-\infty}^{\infty}|x(k\Delta\tau)|^2\,\Delta\tau$$

$$\approx \Delta\tau \sum_{k=-\infty}^{\infty}|x(k\Delta\tau)|^2. \tag{3.59}$$

In the following MATLAB® example we verify that the energy of the triangular pulse $\Delta(t)$ is indeed $= 1/3$ as the theory[hh] predicts:

```
function Example_Ch3_4
% Estimating energy of the triangular pulse
    Npts = 1e6;
```

[hh] $\text{Energy}\{\Delta(t)\} = \int_{-\infty}^{\infty}|\Delta(t)|^2 dt = \int_{-1/2}^{1/2}(1-2t)^2 dt$
$\qquad = 2\int_{0}^{1/2}(1-2t)^2 dt = 2\int_{0}^{1/2}(1-4t+4t^2)dt$
$\qquad = 2(t-2t^2+(4/3)t^3)|_0^{1/2} = 2(1/2-1/2+1/6) = 1/3.$

```
    t = linspace(-1/2,1/2,Npts);
    dTau = t(2)-t(1);
    E = dTau * sum(Delta(t).^2);
    disp(['Est. energy of triangular pulse is: ', ...
        num2str(E)])
end

function f = Delta(t)
% define the triangular pulse function
    f = zeros(size(t));
    indexes = find(abs(t)<1/2);
    f(indexes) = 1 - 2*abs(t(indexes));
end
```

```
>> Example_Ch3_4
    Est. energy of triangular pulse is: 0.33333
```

3.3.4 *Detecting pulses using normalized correlation*

In this example, we synthesize a signal containing a sum of triangular pulses at various delays and scaling factors, plus noise. Then, using the approximation to $C_{xz}(t)$ as shown below, we calculate the similarity between this synthesized waveform $z(t)$ and a noise-free triangular pulse $x(t)$. The resulting normalized cross-correlation similarity measure $C_{xz}(t)$ is plotted vs. t (which tells us at what times the synthesized waveform shape is most similar to a triangular pulse):

$$C_{xz}(t) = \frac{\phi_{xz}(t)}{\sqrt{E_x E_z}} = \frac{x^*(-t) * z(t)}{\sqrt{E_x E_z}}, \quad \text{so} \quad C_{xz}(n\Delta\tau)$$

$$\approx \frac{\sum_{k=-\infty}^{\infty} x^*(-k\Delta\tau) z\big((n-k)\Delta\tau\big)}{\sqrt{\sum_{k=-\infty}^{\infty} |x(k\Delta\tau)|^2 \sum_{k=-\infty}^{\infty} |z(k\Delta\tau)|^2}} \tag{3.60}$$

```
function Example_Ch3_5
% Detecting the presence of triangular pulses in a
%   noisy waveform
% Note: the convolution implementation of cross-
%   correlation is used to take advantage of MATLAB's
%   fast discrete convolution algorithm.

    Npts = 1e5;
    t = linspace(-3,3,Npts);
    x = Delta(t);
```

```
figure(1); plot(t,x); grid on
xlabel('time in seconds');
title('Triangular pulse function x(t):')

y = -10*Delta(t+2) - 5*Delta(t) + 10*Delta(t-2);
figure(2); plot(t,y); grid on
xlabel('time in seconds'); title('Signal y(t):')
z = y + 10*randn(size(t));
figure(3); plot(t,z); grid on
xlabel('time in seconds');
title('Signal z(t) = y(t) + noise:')

Cxz = conv(fliplr(conj(x)),z)/ ...
        sqrt(sum(abs(x).^2)*sum(abs(z).^2));
t1 = linspace(-6,6,length(Cxz));
figure(4); plot(t1,Cxz); grid on
v = axis; axis([-3 3 v(3:4)])
xlabel('time in seconds');
title('Normalized cross correlation C_x_z(t):')
end

function f = Delta(t)
% define the triangular pulse function
   f = zeros(size(t));
   indexes = find(abs(t)<1/2);
f(indexes) = 1 - 2*abs(t(indexes));
end
```

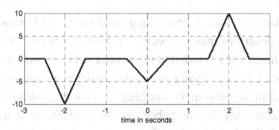

Figure 3.23. Original signal $y(t)$ that is composed of three triangular pulses.

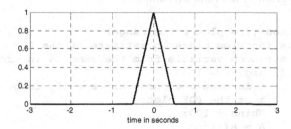

Figure 3.24. Triangular pulse $x(t)$ used for waveform matching.

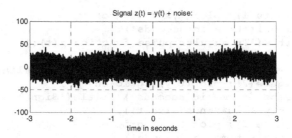

Figure 3.25. Signal $z(t) = y(t) +$ noise added.

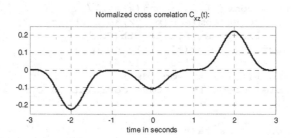

Figure 3.26. Normalized cross-correlation result $C_{xz}(t)$, showing locations and polarities of triangular pulses that were detected in the noise waveform $z(t)$.

3.3.5 *Plotting estimated probability density functions*

One application of signal processing is dealing with random signals. Randomness describes many aspects of real-world behavior, especially in signal communications: unwanted noise interference, communication path delay, number of users trying to send messages at the same time, etc. When a random waveform is sampled to give a random variable, this random variable may be described by its *probability density function* (PDF). The following MATLAB® function estimates and plots the PDF from a histogram of the random samples:

```
function [] = plot_pdf(A,range)
% This function plots an estimate of the PDF
% of a random variable from the samples in array A
    if nargin == 1
        % in case user does not specify range
        % of the PDF plot:
        Nbins = 100;
        A = A(:);
        Hmin = min(A);
```

```
            Hmax = max(A);
            Span = Hmax - Hmin;
            range = linspace(Hmin- ...
                    Span/10,Hmax+Span/10,Nbins);
        end
        dr = range(2) - range(1);
        PDF = hist(A(:),range)/dr/length(A);
        plot(range,PDF)
    end
```

In Section 3.2.5 we learned that when two independent continuous random variables are added together, the PDF of their sum is the convolution product of their individual PDF's. To demonstrate this concept: the following MATLAB® function generates 10^6 samples of random variable X whose PDF $f_X(x) = \text{rect}(x)$, independently generates 10^6 samples of random variable Y whose PDF $f_Y(y) = \text{rect}(y)$, and then adds together these two lists of sample values to give random variable Z whose PDF $f_Z(z)$ should equal to the convolution product of the other two PDF's: $\Delta(a/2) = \text{rect}(a) * \text{rect}(a)$.[ii]

```
% MATLAB Example_Ch3_6
% Demonstrate that the PDF of the sum of two
%   independent random variables is the convolution
%   product of their individual PDF's
    Npts = 1e6;
    X = rand(1,Npts)-1/2;  % PDF of X is rect(x)
    Y = rand(1,Npts)-1/2;  % PDF of Y is rect(y)
    % (Note that Y is independent of X)

% create new random variable as the sum of 2 others
Z = X + Y;
% range of amplitude values & plot resolution
range = linspace(-2,2,200);

figure(1); plot_pdf(X,range)
axis([-1.5,1.5,0,1.5]); grid on;
xlabel('amplitude value')
title('Estimated PDF of random variable X')

figure(2); plot_pdf(Y,range)
axis([-1.5,1.5,0,1.5]); grid on;
xlabel('amplitude value')
title('Estimated PDF of random variable Y')
```

[ii] Independent variable a used to avoid confusing notation $\Delta(z/2) = \text{rect}(x) * \text{rect}(y)$.

```
figure(3); plot_pdf(Z,range)
axis([-1.5,1.5,0,1.5]); grid on;\
xlabel('amplitude value')
title('Estimated PDF of random variable Z = X+Y')
```

The resulting plots confirm the premise $f_Z(a) = f_X(a) * f_Y(a)$. It is as if nature performed the convolution!

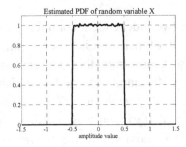

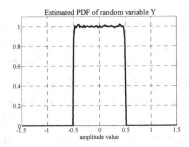

Figure 3.27. Estimated PDF of r.v. X.   Figure 3.28. Estimated PDF of r.v. Y.

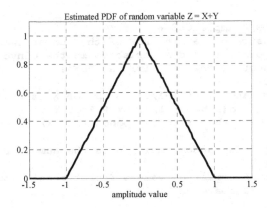

Figure 3.29. Estimated PDF of random variable $Z = X + Y$, demonstrating the fact that $f_Z(a) = f_X(a) * f_Y(a)$.

3.4 Chapter Summary and Comments

- This chapter introduces basic continuous-time signal types that are often used to present the concepts of signal processing. Even though most of them cannot be generated (for example, even a simple sine wave has infinite time duration that makes it impossible to observe in our lifetime), such signals are useful models for waveforms that are used in practice.

- The Dirac delta function is not well-defined by what it is (zero width, infinite height, unity area located at the single point $t = 0$), but is well-defined by what it does when convolved with other signals. In signal processing the main purposes of a delta function are to represent non-zero area at a single point, to achieve waveform time-shift via convolution, to determine a signal's value at a specific time (Ch. 6), and to serve as input signal for linear system analysis (Ch. 8).

- The basic signals that we define in this chapter may be used as building blocks to synthesize more complicated signals.

- Operations such as addition, multiplication, convolution, time shift, time reversal, differentiation, etc., are useful for manipulating and combining continuous-time signals. In later chapters, we will make use of these operations for "signal processing."

- Almost every continuous-time signal and operation has a discrete-time domain equivalent. These relationships are evident in the parallel presentations in Chapters 2 and 3, and may be explained using the theory of sampling (the topic of Ch. 6).

3.5 Homework Problems

P3.1 Simplify: $\int_{-\infty}^{\infty} \delta(t)dt$

P3.2 Simplify: $\int_{-\infty}^{\infty} \delta(t)x(t)dt$

P3.3 Simplify: $\delta(t)x(t) =$

P3.4 Simplify: $\delta(t-1)x(t) =$

P3.5 Simplify: $\delta(t)x(t-1) =$

P3.6 Simplify: $\int_{-\infty}^{\infty} \delta(t+1)x(t)dt$

P3.7 Simplify: $\int_{-\infty}^{\infty} \delta(t)x(t+1)dt$

P3.8 Simplify: $\int_{-\infty}^{\infty} \delta(t-1)x(t+1)dt$

P3.9 Simplify: $\int_{-\infty}^{\infty} \delta(t-\beta)x(\beta)dt =$

P3.10 Given waveform $x(t) = t \, \text{rect}(t/3 - 1/2)$.
   a) Find the value of $x(4)$.
   b) Find the value of $x(1)$.
   c) When $w(t) = x(t) * \delta_6(t)$, find the value of $w(10)$.
   d) When $y(t) = x(t+3/2) * \delta_6(t)$, find the value of $y(-10)$.

P3.11 Is $x(t) = \sin(5t) + \cos(4.9t)$ periodic? If so, what is its period?

P3.12 Is $y(t) = \cos(\pi t/8) + \cos(3t)$ periodic? If so, what is its period?

P3.13 Express $y(t) = 4\cos(5t)$ as a sum of complex exponential signals.

P3.14 Simplify: $u(t) + u(-t) =$

P3.15 State in terms of a rectangular pulse function:
   $u(t+1)u(-t-1) =$

P3.16 State in terms of a rectangular pulse function:
   $u(t-1) - u(t-2) =$

P3.17 State in terms of two unit step functions: $-\text{sgn}(t) =$

P3.18 State in terms of two unit step functions: $\text{rect}(t/2) =$

P3.19 State in terms of a rectangular pulse function:
$$u(t-5) - u(t) =$$

P3.20 State in terms of a triangular pulse function:
$$r(t+3) - 2r(t) + r(t-3) =$$

P3.21 State in terms of a triangular pulse function:
$$r(t/2) - 2r(t/2 + 2) + r(t/2 + 4) =$$

P3.22 Calculate the area of $f(t) = 2e^{-t}u(t)$.

P3.23 Calculate the energy of $f(t) = 2e^{-t}(u(t) - u(t-3))$.

P3.24 Given: $g(t) = \text{sinc}(\pi t/2)$. Calculate $g(t)$ at times $t = \{0, 1, 2, 3, 4\}$ seconds.

P3.25 Find an expression for: $\frac{d}{dt}\text{sinc}(t)$

P3.26 What is the energy of $\Delta(t)$?

P3.27 What is the power of $\text{sgn}(t)$?

P3.28 What is the energy of $\text{rect}(t/4)$?

P3.29 What is the power of $\Delta(5t)$?

P3.30 Label each of the following signals as {odd, even, or neither}:
a) $\text{rect}(t)$
b) $\text{rect}(t + 1/2)$
c) $\delta(t-1) - \delta(t+1)$
d) $\delta(t-1)$
e) $\text{sgn}(t)$
f) $\text{sgn}(t)\,\text{rect}(t)$

P3.31 Find the even part of each of the following signals:
a) $\text{rect}(t)$

b)   rect$(t + 1/2)$
c) $\delta(t - 1) - \delta(t + 1)$
d) $\delta(t - 1)$
e)   sgn(t)
f)   sgn(t) rect(t)

P3.32 Find the odd part of each of the following signals:
a)   rect(t)
b)   rect$(t + 1/2)$
c) $\delta(t - 1) - \delta(t + 1)$
d) $\delta(t - 1)$
e)   sgn(t)
f)   sgn(t) rect(t)

P3.33 The result of multiplying $x(t)$ by $u(t + 2)$ will be:
(check all that apply)
a)   causal
b)   right-sided
c)   anticausal
d)   left-sided

P3.34 The result of multiplying $x(t)$ by rect$(t + 3)$ will be:
(check all that apply)
a)   causal
b)   right-sided
c)   anticausal
d)   left-sided

P3.35 List three signals that are both causal and anticausal.

P3.36 The product of $x(t)$ and $u(t)$ is guaranteed to be causal:
(True or False?)

P3.37 The product of $x(-t)$ and $u(t)$ is guaranteed to be anticausal:
(True or False?)

P3.38 The product of $x(t)$ and $1 - u(t)$ is guaranteed to be anticausal: (True or False?)

P3.39 Time-shifting $x(t)u(t)$ to the left by 5 seconds guarantees that it is not causal: (True or False?)

P3.40 Every causal signal is right-sided: (True or False?)

P3.41 Every left-sided signal is anticausal: (True or False?)

P3.42 The product of a right-sided signal and a left-sided signal is a finite-width signal: (True or False?)

P3.43 Find the cumulative integral of the signal $\delta(t + 4)$.

P3.44 Find the cumulative integral of the signal $\delta(t + 1) - \delta(t - 2)$.

P3.45 Find the time derivative of $\text{sgn}(t)$.

P3.46 Find the time derivative of $\text{rect}(t)$.

P3.47 Simplify: $u(2t) =$

P3.48 Simplify: $u\big(2(t - 4)\big) =$

P3.49 Express the following as a rectangular pulse signal:
$u(t + 3) - u(t) =$

P3.50 Express the following as a sum of unit step functions:
$\text{rect}(t - 5) =$

P3.51 Express the following as a product of unit step functions:
$\text{rect}(t - 5) =$

P3.52 Simplify: $\delta(t) * \text{sinc}(t) =$

P3.53 Simplify: $2\delta(t-2) * 2\text{sinc}(t+3) =$

P3.54 Simplify: $\delta(t-4) * x(t) =$

P3.55 Simplify: $\delta_4(t) + \delta_4(t-2) =$

P3.56 Given: periodic signal $x_p(t)$ has period 2 seconds, and every 2 seconds it toggles between the values 0 and 1. At time $t = 0$ sec there is a 0-to-1 transition. Express $x_p(t)$ as a convolution product of impulse train with rectangular pulse function.

P3.57 Given: periodic signal $x_p(t)$ has period 2 seconds, and every 2 seconds it toggles between the values 0 and 1. At time $t = 1/2$ sec there is a 0-to-1 transition. Express $x_p(t)$ as a convolution product of impulse train with rectangular pulse function.

P3.58 Given: finite-width signal $x_1(t) = 0$ except over $-2 \le t \le 5$ sec, and finite-width signal $x_2(t) = 0$ except over $6 \le t \le 10$. Find the time at the left edge, time at the right edge, and time width of the pulse that is given by convolution product $x_1(t) * x_2(t)$.

Chapter 4

Frequency Analysis of
Discrete-Time Signals

4.1 Theory

4.1.1 *Discrete-Time Fourier Transform (DTFT)*

Frequency analysis may be used to better understand, compare, and manipulate signals in time. It is done by decomposing a time-varying signal into a summation of sinusoids. With that representation, systems that process time-varying signals may be developed based on the understanding of how they affect a signal's individual sinusoidal components.

Frequency analysis is based on the theory developed by Joseph Fourier (1768–1830) in connection with his research on heat. We will view the discrete-time Fourier transform as a mapping of a sequence of numbers in the time domain to its uniquely associated components in the frequency domain. The mathematical relationship between these two functions $f(n)$ and $F(e^{j\omega})$ is as follows:

$$
\boxed{\text{Discrete-Time Fourier Transform}} \quad F(e^{j\omega}) = \sum_{n=-\infty}^{\infty} f(n)e^{-j\omega n} \qquad (4.1)^{\text{a}}
$$

---

[a] Why use the notation $F(e^{j\omega})$ instead of $F(\omega)$? The reasons are: (a) to remind us that we are dealing with the result of a discrete-time (not continuous-time) Fourier transform, and (b) to underline the fact that $F(e^{j\omega}) = F(e^{j(\omega\pm 2\pi)})$ is a periodic function of ω having period 2π radians/sec.

| Inverse Discrete-Time Fourier Transform | $f(n) = \frac{1}{2\pi} \int_{2\pi} F(e^{j\omega}) e^{j\omega n} d\omega.$ (4.2)[b] |

Or, as written in shorthand notation:

$$F(e^{j\omega}) = \mathcal{F}\{f(n)\}, \quad (4.3)$$

$$f(n) = \mathcal{F}^{-1}\{F(e^{j\omega})\}. \quad (4.4)$$

$F(e^{j\omega})$ is called a frequency spectrum because it describes $f(n)$ as a summation of sinusoids having frequency values over the entire spectrum of possibilities. Likewise, $|F(e^{j\omega})|$ is called the *magnitude spectrum* and $\angle F(e^{j\omega})$ is called the *phase spectrum*.

We are interested in signals whose Fourier transforms are *invertible*, meaning that both the Fourier transform summation and its inverse integral exist to give a unique mapping between $f(n)$ and $F(e^{j\omega})$. When that is true, the following notation is used:

$$f(n) \Leftrightarrow F(e^{j\omega}) \quad (4.5)$$

Practically speaking, discrete-time signals that may be measured and recorded are those having invertible Fourier transforms.[c] Engineers often call these *real-world* signals in the sense that they have finite amplitudes, and have finite energy. But many signals of interest for theoretical analysis do not fall in this category. For example: an infinitely-long sequence of numbers that are all equal, such as that corresponding to samples of a d.c. voltage, has a Fourier summation (Eq. (4.1)) that does not converge at $\omega = 0$ (and at all other integer multiples of 2π). Likewise, the Fourier summation of a periodic discrete-time signal does not converge for some values of ω. Fortunately, in most cases of interest the problem of Fourier

---

[b] The integration may be done over any frequency interval of width 2π rad/sec, which is the period of periodic spectrum $F(e^{j\omega})$.

[c] Identifying the class of signals whose Fourier transforms exist and are invertible is an advanced mathematical topic.

summation non-convergence may be overcome, and the Fourier transform made invertible, by permitting the use of generalized functions such as the Dirac[d] impulse $\delta(x)$ in the frequency domain.

Note that, although we discuss Fourier analysis in terms of time variation and frequency content, n and ω are two independent variables that may have different interpretations in other fields of science, engineering and mathematics where Fourier analysis is used. As a matter of fact, in some applications such as digital image processing, Fourier analysis is even applied to multidimensional functions; e.g., a function of $\{n, m\}$ is mapped to a function of $\{e^{j\omega n}, e^{j\omega m}\}$.

4.1.2 *Fourier transforms of basic signals*

Let us now derive, step-by-step, the Fourier transforms of a few of the basic sequences introduced previously in Ch. 2. For many other functions, it is straightforward to find the Fourier transform by evaluating Eq. (4.1), which may be simplified with the help of a table of summations.[e] In other instances, application of Fourier transform properties is an excellent tool available to us. Examples worked out in the following section will provide experience in such techniques.

4.1.2.1 *Exponentially decaying signal*

Let $x(n) = u(n)a^n$ (assuming $|a| < 1$). Its Discrete-Time Fourier transform $X(e^{j\omega})$ is then found from the definition in Eq. (4.1):

$$X(e^{j\omega}) = \mathcal{F}\{u(n)a^n\} = \sum_{n=-\infty}^{\infty} u(n)a^n e^{-j\omega n}$$

$$= \sum_{n=0}^{\infty} a^n e^{-j\omega n} = \sum_{n=0}^{\infty} (ae^{-j\omega})^n = \frac{1}{1-ae^{-j\omega}}, \qquad (4.6)$$

or $u(n)a^n$ and $1/(1 - ae^{-j\omega})$ comprise a "Fourier transform pair":

---

[d] The Dirac impulse function and its properties are covered in Chapter 3.

[e] Dwight; Abramowitz *et al.*

$$a^n u(n) \Leftrightarrow \frac{1}{1-ae^{-j\omega}} \quad (|a| < 1) \qquad (4.7)^{\text{f}}$$

Note that $X(e^{j\omega}) = 1/(1 - ae^{-j\omega})$ is periodic in ω, with period 2π rad/sec, as are all spectra produced by the discrete-time Fourier transform (Eq. (4.1)). This periodicity is the direct result of the discrete nature of $x(n)$.[g] Only one period of a periodic signal is needed to describe the entire signal, and thus the inverse discrete-time Fourier transform's (Eq. (4.2)) range of integration is one period of $X(e^{j\omega})$. This causal signal, $x(n) = u(n)a^n$, when time reversed becomes the anticausal $x(-n) = u(-n)a^{-n}$, whose discrete-time Fourier transform is: [h]

$$a^{-n}u(-n) \Leftrightarrow \frac{1}{1-ae^{j\omega}} \quad (|a| < 1) \qquad (4.8)$$

In preparation for the next derivation, we will calculate the Fourier transform of the following sequence that is symmetric about $n = 0$:

$$\mathcal{F}\{u(n)a^n + u(-n)a^{-n} - \delta(n)\} = \mathcal{F}\{u(n)a^n\} + \mathcal{F}\{u(-n)a^{-n}\} - \mathcal{F}\{\delta(n)\}$$

$$= \frac{1}{1-ae^{-j\omega}} + \frac{1}{1-ae^{j\omega}} - 1 \,(|a| < 1), \qquad (4.9)$$

which simplifies to give another useful Fourier transform pair:

$$a^{|n|} \Leftrightarrow \frac{1-a^2}{1-2a\cos(\omega)+a^2} \quad (|a| < 1) \qquad (4.10)$$

---

[f] The Geometric Series $\sum_{n=0}^{\infty} r^n = \frac{1}{1-r}$ when $|r| < 1$. When $r = ae^{-j\omega}$, $|r| = |ae^{-j\omega}| = |a||e^{-j\omega}| = |a|$. $\therefore |r| < 1 \rightarrow |a| < 1$.

[g] We shall see that, with any Fourier transform pair, when a signal is discrete (as if multipled by an impulse train) in one domain then its transformed version is periodic in the other domain.

[h] $\mathcal{F}\{u(-n)a^{-n}\} = \sum_{n=-\infty}^{\infty} u(-n)a^{-n}e^{-j\omega n} = \sum_{n=-\infty}^{0} a^{-n}e^{-j\omega n} = \sum_{n=-\infty}^{0}(ae^{j\omega})^{-n} = 1/(1 - ae^{j\omega})$ when $|a| < 1$.

4.1.2.2 *Constant value*

As mentioned previously, a constant waveform such as $x(n) = 1$ presents us with the problem that its Fourier transform summation does not converge at values of ω that are integer multiples of 2π rad/sec. This summation may be solved with the help of generalized functions, however. First, consider the spectrum $F(e^{j\omega}) = 2\pi\delta_{2\pi}(\omega)$, which has a Dirac impulse located every integer multiple of 2π.[i] What is its inverse Fourier transform?

$$\mathcal{F}^{-1}\{2\pi\delta_{2\pi}(\omega)\} = \frac{1}{2\pi}\int_{-\pi}^{\pi} 2\pi\delta_{2\pi}(\omega)e^{j\omega n}d\omega = \frac{1}{2\pi}\int_{-\pi}^{\pi} 2\pi\delta(\omega)e^{j\omega n}d\omega$$

$$= \frac{1}{2\pi}\int_{-\pi}^{\pi} 2\pi\delta(\omega)e^{j0n}d\omega = \int_{-\pi}^{\pi}\delta(\omega)d\omega = 1. \quad (4.11)$$

Therefore, if the Fourier transform is invertible then $\mathcal{F}\{1\} = 2\pi\delta_{2\pi}(\omega)$. Let us verify this with the help of generalized function $x(n) = \lim_{a\to 1} a^{|n|} = 1$:

$$X(e^{j\omega}) = \mathcal{F}\left\{\lim_{a\to 1} a^{|n|}\right\} = \lim_{a\to 1}\left(\frac{1-a^2}{1-2a\cos(\omega)+a^2}\right)$$

$$= \begin{cases} \infty & \omega = 2\pi k, k \in \mathbb{Z}; \\ 0 & \text{otherwise.} \end{cases} \quad (4.12)^{[j]}$$

What kind of function $X(e^{j\omega})$ is zero everywhere except at points $\omega = 2\pi k, k \in \mathbb{Z}$? One candidate is the Dirac delta function impulse train $\delta_{2\pi}(\omega)$ times a constant (having unspecified impulse areas). To find the area of each impulse, we note that the area of periodic spectrum $(1 - a^2)/(1 - 2a\cos(\omega) + a^2)$ over one period is independent of a:

$$\int_{-\pi}^{\pi}\frac{1-a^2}{1-2a\cos(\omega)+a^2}\,d\omega = 2\pi. \quad (4.13)$$

We confirm, therefore, that $x(n) = 1$ and $X(e^{j\omega}) = 2\pi\delta_{2\pi}(\omega)$ are a Fourier transform pair:

---

[i] The Dirac impulse train is presented in Chapter 3.

[j] $\mathbb{Z}$ is the symbol that represents the set of integers (Zahlen is a German word for numbers).

$$1 \Leftrightarrow 2\pi\delta_{2\pi}(\omega) \qquad (4.14)$$

4.1.2.3 *Impulse function*

By substituting $\delta(n)$ for $f(n)$ in the discrete-time Fourier transform definition (Eq. 4.1) we obtain:

$$F\left(e^{j\omega}\right) = \sum_{n=-\infty}^{\infty} \delta(n)e^{-j\omega n} = \sum_{n=-\infty}^{\infty} \delta(n)e^{-j\omega 0}$$

$$= \sum_{n=-\infty}^{\infty} \delta(n) = 1. \qquad (4.15)$$

Using similar arguments as before, we may also show that the inverse Fourier transform of 1 is $\delta(n)$. We now have another Fourier transform pair:

$$\delta(n) \Leftrightarrow 1 \qquad (4.16)$$

4.1.2.4 *Delayed impulse function*

Similarly, and just as easily, we may find the Fourier transform of $\delta(n - n_0)$:

$$\mathcal{F}\{\delta(n - n_0)\} = \sum_{n=-\infty}^{\infty} \delta(n - n_0)e^{-j\omega n} = \sum_{n=-\infty}^{\infty} \delta(n - n_0)e^{-j\omega n_0}$$

$$= e^{-j\omega n_0} \sum_{n=-\infty}^{\infty} \delta(n - n_0) = e^{-j\omega n_0}. \qquad (4.17)$$

This gives us the Fourier transform pair:

$$\delta(n - n_0) \Leftrightarrow e^{-j\omega n_0} \qquad (4.18)$$

4.1.2.5 *Signum function*

The signum function is defined in Chapter 2 as:

$$\text{sgn}(n) \equiv \begin{cases} -1, & n < 0; \\ 0, & n = 0; \\ 1, & n > 0. \end{cases} \tag{4.19}$$

Because the discrete Fourier transform of $\text{sgn}(n)$ does not converge in the ordinary sense, we must consider using functions that approach $\text{sgn}(n)$ in the limit. One such generalized function, among many that may be considered, is:

$$\text{sgn}(n) = \lim_{a \to 1} \{u(n)a^n - u(-n)a^{-n}\}. \tag{4.20}$$

Solving for $\mathcal{F}\{\text{sgn}(n)\}$ in this formulation:

$$\mathcal{F}\{\text{sgn}(n)\} = \lim_{a \to 1}\left(\sum_{n=-\infty}^{\infty}\{u(n)a^n - u(-n)a^{-n}\}e^{-j\omega n}\right)$$

$$= \lim_{a \to 1}\left(\sum_{n=-\infty}^{\infty} u(n)a^n e^{-j\omega n} - \sum_{n=-\infty}^{\infty} u(-n)a^{-n}e^{-j\omega n}\right)$$

$$= \lim_{a \to 1}\left(\frac{1}{1-ae^{-j\omega}} - \frac{1}{1-ae^{j\omega}}\right)$$

$$= \lim_{a \to 1}\left(\frac{ae^{-j\omega} - ae^{j\omega}}{(1 - ae^{-j\omega})(1 - ae^{j\omega})}\right)$$

$$= \lim_{a \to 1}\left(\frac{-2ja\sin(\omega)}{1 - 2a\cos(\omega) + a^2}\right)$$

$$= \frac{j\sin(\omega)}{\cos(\omega) - 1}$$

$$= \left(1 + e^{-j\omega}\right)/\left(1 - e^{-j\omega}\right). \tag{4.21}$$

One can also show[k] that $\mathcal{F}^{-1}\{(1 + e^{-j\omega})/(1 - e^{-j\omega})\} = \text{sgn}(n)$. Therefore, we obtain the Fourier transform pair:

$$\boxed{\text{sgn}(n) \Leftrightarrow \frac{1+e^{-j\omega}}{1-e^{-j\omega}}} \tag{4.22}$$

---

[k] $\mathcal{F}^{-1}\left\{\frac{1+e^{-j\omega}}{1-e^{-j\omega}}\right\} = \frac{1}{2\pi}\int_{-\pi}^{\pi}\left(\frac{1+e^{-j\omega}}{1-e^{-j\omega}}\right)e^{j\omega n}d\omega$, which with numerical methods may easily be confirmed to give: $\{1, n > 0;\ 0, n = 0;\ -1, n < 0\} = \text{sgn}(n)$.

4.1.2.6 *Unit step function*

As with sgn(n), the Fourier transform of $u(n)$ can be found by expressing it in the form of a generalized function. Or, now that $\mathcal{F}\{\text{sgn}(n)\}$, $\mathcal{F}\{\delta(n)\}$ and $\mathcal{F}\{1\}$ are known, we may express $u(n)$ in terms of these functions and quickly find its Fourier transform using the linear superposition property:[1]

$$u(n) = \tfrac{1}{2}\left(\text{sgn}(n) + \delta(n) + 1\right)$$

$$\mathcal{F}\{u(n)\} = \frac{1}{2}\,\mathcal{F}\{\text{sgn}(n) + \delta(n) + 1\}$$

$$= \frac{1}{2}\mathcal{F}\{\text{sgn}(n)\} + \frac{1}{2}\mathcal{F}\{\delta(n)\} + \frac{1}{2}\mathcal{F}\{1\}$$

$$= \tfrac{1}{2}\left(\frac{1+e^{-j\omega}}{1-e^{-j\omega}}\right) + \tfrac{1}{2}(1) + \tfrac{1}{2}\left(2\pi\,\delta_{2\pi}(\omega)\right)$$

$$= \tfrac{1}{2}\left(\frac{1+e^{-j\omega}}{1-e^{-j\omega}} + 1\right) + \pi\,\delta_{2\pi}(\omega)$$

$$= \tfrac{1}{2}\left(\frac{1+e^{-j\omega}+1-e^{-j\omega}}{1-e^{-j\omega}}\right) + \pi\,\delta_{2\pi}(\omega)$$

$$= \tfrac{1}{2}\left(\frac{2}{1-e^{-j\omega}}\right) + \pi\,\delta_{2\pi}(\omega) = \frac{1}{1-e^{-j\omega}} + \pi\,\delta_{2\pi}(\omega). \qquad (4.23)$$

Therefore, because this Fourier transform may be shown to be invertible, we have the Fourier transform pair:

$$\boxed{\; u(n) \Leftrightarrow \pi\,\delta_{2\pi}(\omega) + \frac{1}{1-e^{-j\omega}} \;} \qquad (4.24)$$

4.1.2.7 *Complex exponential function*

To find the Fourier transform of complex exponential signal $e^{j\omega_0 n}$, which is essentially a phasor at frequency ω_0, we begin by solving for the inverse Fourier transform of frequency-shifted comb function $\delta_{2\pi}(\omega - \omega_0)$ using Eq. (4.2):

---

[1] Linear superposition and other properties of the Fourier transform are discussed in Section 4.1.3.1.

$$\mathcal{F}^{-1}\{\delta_{2\pi}(\omega - \omega_0)\} = \frac{1}{2\pi}\int_{-\pi}^{\pi}\delta_{2\pi}(\omega - \omega_0)\,e^{j\omega n}d\omega$$

$$= \frac{1}{2\pi}\int_{-\pi}^{\pi}\delta(\omega - \omega_0)\,e^{j\omega_0 n}d\omega$$

$$= \frac{e^{j\omega_0 n}}{2\pi}\int_{-\pi}^{\pi}\delta(\omega - \omega_0)\,d\omega = \frac{e^{j\omega_0 n}}{2\pi}. \qquad (4.25)$$

After multiplying both sides by 2π, we get the desired result:
$2\pi\,\mathcal{F}^{-1}\{\delta_{2\pi}(\omega - \omega_0)\} = \mathcal{F}^{-1}\{2\pi\,\delta_{2\pi}(\omega - \omega_0)\} = e^{j\omega_0 n}$, or

$$\boxed{e^{j\omega_0 n} \Leftrightarrow 2\pi\,\delta_{2\pi}(\omega - \omega_0)} \qquad (4.26)$$

4.1.2.8 *Sinusoid*

The Fourier transform of $A\cos(\omega_0 n + \theta)$ is derived by expressing it in terms of complex exponential functions using Euler's identity:

$$A\cos(\omega_0 n + \theta) = A\left(\tfrac{1}{2}e^{j(\omega_0 n + \theta)} + \tfrac{1}{2}e^{-j(\omega_0 n + \theta)}\right). \qquad (4.27)$$

Then, taking advantage of $e^{j\omega_0 t} \Leftrightarrow 2\pi\,\delta_{2\pi}(\omega - \omega_0)$:

$$\mathcal{F}\{A\cos(\omega_0 n + \theta)\} = \frac{A}{2}\mathcal{F}\{e^{j(\omega_0 n + \theta)} + e^{-j(\omega_0 n + \theta)}\}$$

$$= \frac{A}{2}\mathcal{F}\{e^{j\omega_0 n}e^{j\theta} + e^{-j\omega_0 n}e^{-j\theta}\}$$

$$= \frac{A}{2}e^{j\theta}\,\mathcal{F}\{e^{j\omega_0 n}\} + \frac{A}{2}e^{-j\theta}\mathcal{F}\{e^{-j\omega_0 n}\}$$

$$= Ae^{j\theta}\pi\,\delta_{2\pi}(\omega - \omega_0) + Ae^{-j\theta}\pi\,\delta_{2\pi}(\omega + \omega_0),$$

$$(4.28)$$

or
$$\boxed{\begin{array}{c} \cos(\omega_0 n + \theta) \Leftrightarrow \\[6pt] e^{-j\theta}\pi\delta_{2\pi}(\omega + \omega_0) + e^{j\theta}\pi\delta_{2\pi}(\omega - \omega_0) \end{array}} \qquad (4.29)$$

Setting $\theta = \{0, -\pi/2\}$ gives us these two Fourier transform pairs:

$$\cos(\omega_0 n) \Leftrightarrow \pi\delta_{2\pi}(\omega + \omega_0) + \pi\delta_{2\pi}(\omega - \omega_0) \qquad (4.30)$$

$$\sin(\omega_0 n) \Leftrightarrow j\pi\delta_{2\pi}(\omega + \omega_0) - j\pi\delta_{2\pi}(\omega - \omega_0) \qquad (4.31)$$

Note that, as demonstrated above, the Fourier transform of a real even signal is real, and the Fourier transform of a real odd signal is imaginary.[m]

4.1.2.9 *Rectangular pulse function*

The Fourier transform of a rectangular function is derived by direct summation, applying the definition given by Eq. (4.1):

$$F\left(e^{j\omega}\right) = \sum_{n=-\infty}^{\infty} \text{rect}_K(n)e^{-j\omega n}$$

$$= \sum_{n=-K}^{K}(1)e^{-j\omega n} = 1 + 2\sum_{n=1}^{K}\cos(n\omega) \qquad (4.32)$$

We now have the Fourier transform pair:

$$\text{rect}_K(n) \Leftrightarrow 1 + 2\sum_{k=1}^{K}\cos(k\omega) \qquad (4.33)$$

These and some other discrete-time Fourier transform pairs are summarized in Table 4.1.

---

[m] With sequences: even $\times$ even = even, even $\times$ odd = odd, odd $\times$ odd = even. Also, $\sum_{n=-\infty}^{\infty} x_e(n) = x_e(0) + 2\sum_{n=1}^{\infty} x_e(n)$, and $\sum_{n=-\infty}^{\infty} x_o(n) = 0$. Therefore $\mathcal{F}\{x(n)\} = \sum_{n=-\infty}^{\infty} x(n)e^{-j\omega n} = \sum_{n=-\infty}^{\infty}\left(x_e(n) + x_o(n)\right)(\cos(\omega n) - j\sin(\omega n)) = \sum_{n=-\infty}^{\infty} x_e(n)\cos(\omega n) + \sum_{n=-\infty}^{\infty} x_o(n)\cos(\omega n) - j\sum_{n=-\infty}^{\infty} x_e(n)\sin(\omega n) - j\sum_{n=-\infty}^{\infty} x_o(n)\sin(\omega n) = x_e(0) + 2\sum_{n=1}^{\infty} x_e(n)\cos(\omega n) + 0 - j0 - 2j\sum_{n=1}^{\infty} x_o(n)\sin(\omega n)$. $\therefore \mathcal{F}\{x(n)\} = x(0) + 2\sum_{n=1}^{\infty} x_e(n)\cos(\omega n) - 2j\sum_{n=1}^{\infty} x_o(n)\sin(\omega n)$, which clearly shows why real and even $x(n)$ has real $\mathcal{F}\{x(n)\}$, and real and odd $x(n)$ has imaginary $\mathcal{F}\{x(n)\}$.

Table 4.1. Table of discrete-time Fourier transform pairs.

Entry #	$f(n)$	$F(e^{j\omega})$	Refer to Eq. #
1.	$\frac{1}{2\pi}\int_{2\pi} F(e^{j\omega})e^{j\omega n}d\omega$	$\sum_{n=-\infty}^{\infty} f(n)e^{-j\omega n}$	(4.1), (4.2)
2.	$a^n u(n)$	$\dfrac{1}{1 - ae^{-j\omega}};\quad |a| < 1$	(4.7)
3.	$a^{-n} u(-n)$	$\dfrac{1}{1 - ae^{j\omega}};\quad |a| < 1$	(4.8)
4.	$a^{|n|}$	$\dfrac{1 - a^2}{1 - 2a\cos(\omega) + a^2};$ $|a| < 1$	(4.10)
5.	1	$2\pi\delta_{2\pi}(\omega)$	(4.14)
6.	$\delta(n)$	1	(4.16)
7.	$\delta(n - n_0)$	$e^{-j\omega n_0}$	(4.18)
8.	$\text{sgn}(n)$	$\dfrac{1 + e^{-j\omega}}{1 - e^{-j\omega}} = \dfrac{2}{j\omega} * \delta_{2\pi}(\omega)$	(4.22)[n]
9.	$u(n)$	$\pi\,\delta_{2\pi}(\omega) + 1/(1 - e^{-j\omega})$ $= \left(\pi\,\delta(\omega) + \dfrac{1}{j\omega}\right) * \delta_{2\pi}(\omega) + \dfrac{1}{2}$	(4.24)[o]

(*Continued*)

---

[n] Numerical methods confirm that $(1 + e^{-j\omega})/(1 - e^{-j\omega}) = \frac{2}{j\omega} * \delta_{2\pi}(\omega)$.

[o] Since $u(n) = \frac{1}{2}(1 + \text{sgn}(n) + \delta(n))$, $\mathcal{F}\{u(n)\} = \frac{1}{2}\left(2\pi\delta_{2\pi}(\omega) + \frac{2}{j\omega} * \delta_{2\pi}(\omega) + 1\right) = \pi\delta_{2\pi}(\omega) + \frac{1}{j\omega} * \delta_{2\pi}(\omega) + \frac{1}{2} = \left(\pi\delta(\omega) + \frac{1}{j\omega}\right) * \delta_{2\pi}(\omega) + \frac{1}{2}$.

Table 4.1. (*Continued*)

Entry #	$f(n)$	$F(e^{j\omega})$	Refer to Eq. #
10.	$e^{j\omega_0 n}$	$2\pi\,\delta_{2\pi}(\omega - \omega_0)$	(4.26)
11.	$\cos(\omega_0 n + \theta)$	$e^{-j\theta}\pi\delta_{2\pi}(\omega + \omega_0)$ $+e^{j\theta}\pi\delta_{2\pi}(\omega - \omega_0)$	(4.29)
12.	$\cos(\omega_0 n)$	$\pi\delta_{2\pi}(\omega + \omega_0)$ $+\pi\delta_{2\pi}(\omega - \omega_0)$	(4.30)
13.	$\sin(\omega_0 n)$	$j\pi\delta_{2\pi}(\omega + \omega_0)$ $-j\pi\delta_{2\pi}(\omega - \omega_0)$	(4.31)
14.	$\text{rect}_K(n)$ pulse width $= 2K + 1$ samples	$1 + 2\sum_{k=1}^{K}\cos(k\omega)$ $= T\,\text{sinc}(\omega T/2) * \delta_{2\pi}(\omega)$ $(T = 2K + 1)$	(4.33)[p]
15.	$\dfrac{W}{\pi}\text{sinc}(Wn)$	$\text{rect}\left(\dfrac{\omega}{2W}\right) * \delta_{2\pi}(\omega)$ "bandwidth" = W rad/sec [q]	Lathi, p.852
16.	$\dfrac{W}{2\pi}\text{sinc}^2(Wn/2)$	$\Delta\left(\dfrac{\omega}{2W}\right) * \delta_{2\pi}(\omega)$ "bandwidth" = W rad/sec [r]	Lathi, p.852
17.	$e^{-n^2/(2\sigma^2)}$	$\sigma\sqrt{2\pi}e^{-\sigma^2\omega^2/2} * \delta_{2\pi}(\omega)$	See [s] below

(*Continued*)

---

[p] The equality is confirmed using numerical methods, and using sampling theory (Ch.6).

[q] The word "bandwidth" is in quotes here because it is found by considering only one period of the periodic spectrum. Also, note that rect$(\omega/2W)$ is a continuous, not discrete, function; the continuous rect(x) is introduced in Chapter 3.

[r] The continuous triangular pulse function $\Delta(x)$ was introduced in Chapter 3.

[s] This is confirmed using numerical methods, and using sampling theory (Ch.6).

Table 4.1. (*Continued*)

Entry #	$f(n)$	$F(e^{j\omega})$	Refer to Eq. #
18.	$\Delta_M(n)$ "pulse width" = 2M samples [t]	$1 + 2\sum_{k=1}^{M-1}\left(1 - \frac{k}{M}\right)\cos(k\omega)$ $= (T/2)\,\text{sinc}^2(\omega T/4) *$ $\delta_{2\pi}(\omega)$, where $T = 2M$.	See [u] below
19.	$\delta_M(n) =$ $\sum_{k=-\infty}^{\infty}\delta(n - kM)$ Impulse train in time, period = M samples	$\omega_0\,\delta_{\omega_0}(\omega) =$ $\omega_0\sum_{k=-\infty}^{\infty}\delta(\omega - k\omega_0)$ Impulse train in frequency, period $\omega_0 = 2\pi/M$ rad/sec	(4.49) Stuller, p.239

4.1.3 *Fourier transform properties*

In most applications, we will need to derive Fourier transforms of many different functions. Frequently it is convenient to use some of the mathematical properties of the Fourier summation given in Eq. (4.1). Therefore, in this section we will discuss some of the important properties and then show in a table all other properties that may be useful. Once again you will find that it is straightforward to verify these entries in Table 4.2.

4.1.3.1 *Linearity*

Let us consider two different functions of time, $f_1(n)$ and $f_2(n)$, whose Fourier transforms are known to be $F_1(e^{j\omega})$ and $F_2(e^{j\omega})$, respectively: $f_1(n) \Leftrightarrow F_1(e^{j\omega})$ and $f_2(n) \Leftrightarrow F_2(e^{j\omega})$. Then from the DTFT definition

---

[t] The words "pulse width" are in quotes here because they refer to the width, in seconds, of the continuous-time pulse that has been sampled to give the discrete-time pulse $\Delta_M(n)$ when assuming a sampling rate of 1 sample/sec. In terms of the total number of sample points that are nonzero, $\Delta_M(n)$ has a pulse width of $2M - 1$ samples.

[u] The equality is confirmed using numerical methods, and using sampling theory (Ch. 6).

in Eq. (4.1), for two arbitrary constants a and b, the following relationship[v] holds true:

$$af_1(n) + bf_2(n) \Leftrightarrow aF_1(e^{j\omega}) + bF_2(e^{j\omega}) \qquad (4.34)$$

With this property, we may find Fourier transforms of signals that are linear combinations of signals having known transforms. The following example illustrates this property and shows the plot of the function and its transform.

Example 4.1
Find the Fourier transform of $x(n)$ that is shown below:

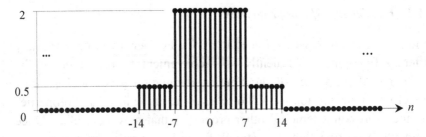

Figure 4.1. Sequence $x(n)$ to be transformed to the frequency domain in Example 4.1.

To solve this problem, express $x(n)$ as the sum of two rectangular pulses:

$$x(n) = 0.5 \, \text{rect}_{14}(n) + 1.5 \, \text{rect}_7(n). \qquad (4.35)$$

We see that $x(n) = af_1(n) + bf_2(n)$: $a = 0.5$, $f_1(n) = \text{rect}_{14}(n)$, $b = 1.5$ and $f_2(n) = \text{rect}_7(n)$. Therefore, applying the linearity property, we can say that $\mathcal{F}\{x(n)\} = a\mathcal{F}\{f_1(n)\} + b\mathcal{F}\{f_2(n)\}$:

---

[v] This is called the "Linear Superposition" property. It combines the scaling property ($\mathcal{F}\{af(n)\} = a\mathcal{F}\{f(n)\}$) and the additive property ($\mathcal{F}\{f_1(n) + f_2(n)\} = \mathcal{F}\{f_1(n)\} + \mathcal{F}\{f_2(n)\}$) into a single expression.

$$\mathcal{F}\{x(n)\} = 0.5\,\mathcal{F}\{\text{rect}_{14}(n)\} + 1.5\,\mathcal{F}\{\text{rect}_7(n)\}$$

$$= 0.5\left(1 + 2\sum_{k=1}^{14} \cos(k\omega)\right) + 1.5\left(1 + 2\sum_{k=1}^{7} \cos(k\omega)\right)$$

$$= 2 + 4\sum_{k=1}^{7} \cos(k\omega) + \sum_{k=8}^{14} \cos(k\omega). \tag{4.36}$$

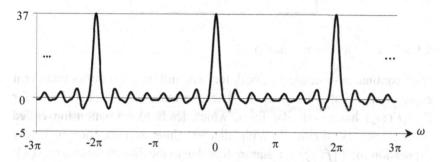

Figure 4.2. From Example 4.1: $\mathcal{F}\{x(n)\} = X(e^{j\omega}) = 2 + 4\sum_{k=1}^{7} \cos(k\omega) + \sum_{k=8}^{14} \cos(k\omega)$.

4.1.3.2 *Time shifting*

In practice, quite often we encounter a signal whose Fourier transform is known but for the fact that it is delayed (or advanced) in time. Here we ask a question – is there a simple relationship between the two Fourier transforms?

Consider our experience with sinusoidal signal analysis in a basic circuit theory course. We know that a time shifted sinusoid at a frequency ω_0 rad/sec is a sinusoid at the same frequency and amplitude, but now has a phase offset from its original value. Since the Fourier transform of an arbitrary time function is a linear combination of various frequencies,[w]

---

[w] The inverse discrete-time Fourier transform integral may be written as:

$f(n) = (1/2\pi)\int_{2\pi} F(e^{j\omega})(\cos(\omega n) + j\sin(\omega n))d\omega =$

$(1/2\pi)\int_{-\pi}^{\pi} F(e^{j\omega})\cos(\omega n)\,d\omega + (j/2\pi)\int_{-\pi}^{\pi} F(e^{j\omega})\sin(\omega n)\,d\omega =$

$(1/2\pi)\int_{-\pi}^{\pi}\left(F_e(e^{j\omega}) + F_o(e^{j\omega})\right)\cos(\omega n)\,d\omega + (j/2\pi)\int_{-\pi}^{\pi}\left(F_e(e^{j\omega}) +\right.$

$\left. F_o(e^{j\omega})\right)\sin(\omega n)\,d\omega = (1/2\pi)\int_{-\pi}^{\pi} F_e(e^{j\omega})\cos(\omega n)\,d\omega +$

$(j/2\pi)\int_{-\pi}^{\pi} F_o(e^{j\omega})\sin(\omega n)\,d\omega = (1/\pi)\int_{0}^{\pi} F_e(e^{j\omega})\cos(\omega n)\,d\omega +$

$(j/\pi)\int_{0}^{\pi} F_o(e^{j\omega})\sin(\omega n)\,d\omega.$ In other words, $f(n)$ is a weighted sum of sines and each of these frequency components will now have a corresponding phase offset associated with it:

$$f(n \pm n_0) \Leftrightarrow F(e^{j\omega})e^{\pm j\omega n_0} \qquad (4.37)^x$$

4.1.3.3 *Time/frequency duality*

With continuous-time signals, both forward and inverse Fourier transform expressions involve integration. The expressions for $\mathcal{F}\{f(t)\}$ and for $\mathcal{F}^{-1}\{F(\omega)\}$ have a similar form, which leads to a relationship called "time/frequency duality." With discrete-time signals, however, the expression for $\mathcal{F}\{f(n)\}$ is a summation due to the discrete nature of $f(n)$. This also results in the spectrum $F(e^{j\omega})$ being periodic. For these reasons, although the relationship of time/frequency duality exists it is not obvious. We will revisit this concept later in the chapter when discussing the discrete-time Fourier Series and Discrete Fourier Transform (DFT).

4.1.3.4 *Convolution*

Many signal processing applications require either multiplying or convolving together two signals. As we shall now see, multiplication and convolution are closely related through the Fourier transform. What may be a difficult-to-simplify convolution of two signals in one domain is simply expressed as a multiplicative product in the other domain. Here we show the correspondence between multiplication and convolution in time and frequency domains.

---

cosines at all positive frequencies in the range of integration (where the weights for sines are $(j/\pi)F_o(e^{j\omega})$ and the weights for cosines are $(1/\pi)F_e(e^{j\omega})$).

^{x} Proof: $\mathcal{F}\{f(n \pm n_0)\} = \sum_{n=-\infty}^{\infty} f(n \pm n_0)e^{-j\omega n} = \sum_{m=-\infty}^{\infty} f(m)e^{-j\omega(m \mp n_0)} = \sum_{m=-\infty}^{\infty} f(m)e^{-j\omega m}e^{\pm j\omega n_0} = e^{\pm j\omega n_0} \sum_{m=-\infty}^{\infty} f(m)e^{-j\omega m} = e^{\pm j\omega n_0}\mathcal{F}\{f(n)\} = e^{\pm j\omega n_0}F(e^{j\omega})$.

If $f_1(n) \Leftrightarrow F_1(e^{j\omega})$ and $f_2(n) \Leftrightarrow F_2(e^{j\omega})$, then:

$$f_1(n) * f_2(n) \Leftrightarrow F_1(e^{j\omega})F_2(e^{j\omega}) \qquad (4.38)^y$$

$$f_1(n)f_2(n) \Leftrightarrow \frac{1}{2\pi} F_1(e^{j\omega}) \circledast F_2(e^{j\omega}) \qquad (4.39)^z$$

The "$\circledast$" operator in Eq. (4.39) above is called a *circular*, or *periodic convolution* that needs to be done because both $F_1(e^{j\omega})$ and $F_2(e^{j\omega})$ are periodic (a regular convolution integral will not converge):

$$F_1(e^{j\omega}) \circledast F_2(e^{j\omega}) \equiv \frac{1}{2\pi} \int_{2\pi} F_1(e^{j\phi})F_2(e^{j(\omega-\phi)})d\phi. \qquad (4.40)$$

The circular convolution integral has limits over one period only (any band of frequencies spanning 2π rad/sec), since both $F_1(e^{j\omega})$ and $F_2(e^{j\omega})$ are completely described by one of their periods.

Example 4.2
Find the Fourier transform of $x(n) = \text{rect}_5(n) * \text{sinc}(2\pi n/11)$:

Let $f_1(n) = \text{rect}_5(n) \Leftrightarrow 1 + 2\sum_{k=1}^{5} \cos(k\omega)$ and $f_2(n) = \text{sinc}\left(\frac{2\pi n}{11}\right) \Leftrightarrow \frac{11}{2} \text{rect}\left(\frac{\omega}{4\pi/11}\right) * \delta_{2\pi}(\omega)$; then $\mathcal{F}\{f_1(n) * f_2(n)\} =$

---

y Proof: $\mathcal{F}\{f_1(n) * f_2(n)\} = \sum_{n=-\infty}^{\infty}\{\sum_{k=-\infty}^{\infty} f_1(k)f_2(n-k)\}e^{-j\omega n} =$
$\sum_{k=-\infty}^{\infty} f_1(k)\{\sum_{n=-\infty}^{\infty} f_2(n-k)e^{-j\omega n}\} = \sum_{k=-\infty}^{\infty} f_1(k)\mathcal{F}\{f_2(n-k)\} =$
$\sum_{k=-\infty}^{\infty} f_1(k)\left(e^{-j\omega k}F_2(e^{j\omega})\right) = \left(\sum_{k=-\infty}^{\infty} f_1(k)e^{-j\omega k}\right)F_2(e^{j\omega}) = F_1(e^{j\omega})F_2(e^{j\omega}).$
z Proof: $\{f_1(n)f_2(n)\} = \sum_{n=-\infty}^{\infty} f_1(n)f_2(n)e^{-j\omega n} = \sum_{n=-\infty}^{\infty} F^{-1}\{F_1(e^{j\omega})\}f_2(n)e^{-j\omega n} =$
$\sum_{n=-\infty}^{\infty}\left((1/2\pi)\int_{-\pi}^{\pi} F_1(e^{j\beta})e^{j\beta n}d\beta\right)f_2(n)e^{-j\omega n} =$
$(1/2\pi)\int_{-\pi}^{\pi} F_1(e^{j\beta})\left(\sum_{n=-\infty}^{\infty} e^{j\beta n}f_2(n)e^{-j\omega n}\right)d\beta =$
$(1/2\pi)\int_{-\pi}^{\pi} F_1(e^{j\beta})\left(\sum_{n=-\infty}^{\infty} f_2(n)e^{-j(\omega-\beta)n}\right)d\beta =$
$(1/2\pi)\int_{-\pi}^{\pi} F_1(e^{j\beta})F_2(e^{j(\omega-\beta)})d\beta \equiv (1/2\pi) F_1(e^{j\omega}) \circledast F_2(e^{j\omega}).$

$$F_1(e^{j\omega})F_2(e^{j\omega}) = \{1 + 2\sum_{k=1}^{5}\cos(k\omega)\} \times \{\frac{11}{2}\,\text{rect}\left(\frac{\omega}{4\pi/11}\right) *$$
$$\delta_{2\pi}(\omega)\}.$$

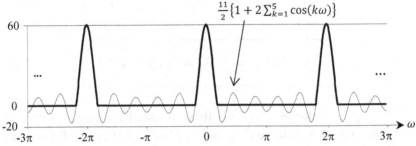

Figure 4.3. The spectrum $X(e^{j\omega}) = \mathcal{F}\{x(n)\}$ in Example 4.2.

4.1.3.5 *Modulation*

In communication applications, we find that an information signal $f(n)$, such as speech, may be wirelessly transmitted over radio waves by first multiplying it with a sinusoid at a relatively high frequency ω_0 (called the carrier frequency). Therefore, it is important to understand the frequency-domain effects of multiplication with a sinusoid in time. We find, as shown below, that the modulation process shifts $F(e^{j\omega})$ by $\pm\omega_0$ in frequency. Consider a signal $m(n) = f(n)\cos(\omega_0 n)$. From Eq. 4.1, we have:

$$\mathcal{F}\{m(n)\} = M(e^{j\omega}) = \sum_{n=-\infty}^{\infty} f(n)\cos(\omega_0 n)e^{-j\omega n}$$

$$= \sum_{n=-\infty}^{\infty} f(n)\left(\tfrac{1}{2}e^{j\omega_0 n} + \tfrac{1}{2}e^{-j\omega_0 n}\right)e^{-j\omega n}$$

$$= \sum_{n=-\infty}^{\infty} f(n)\left(\tfrac{1}{2}e^{-j(\omega-\omega_0)n} + \tfrac{1}{2}e^{-j(\omega+\omega_0)n}\right)$$

$$= \tfrac{1}{2}\sum_{n=-\infty}^{\infty} f(n)\,e^{-j(\omega-\omega_0)n} + \tfrac{1}{2}\sum_{n=-\infty}^{\infty} f(n)\,e^{-j(\omega+\omega_0)n}$$

$$= \tfrac{1}{2}F(e^{j(\omega-\omega_0)}) + \tfrac{1}{2}F(e^{j(\omega+\omega_0)}). \tag{4.41}$$

This is called the modulation property of the Fourier transform:

$$f(n)\cos(\omega_0 n) \Leftrightarrow \tfrac{1}{2}F(e^{j(\omega-\omega_0)}) + \tfrac{1}{2}F(e^{j(\omega+\omega_0)}) \tag{4.42}$$

When allowing for an arbitrary cosine phase shift ϕ, the modulation property is:

$$f(n)\cos(\omega_0 n + \phi) \Leftrightarrow \frac{1}{2}e^{j\phi}F\left(e^{j(\omega-\omega_0)}\right) + \frac{1}{2}e^{-j\phi}F\left(e^{j(\omega+\omega_0)}\right)$$

(4.43)

Example 4.3
Plot $\mathcal{F}\{\text{rect}_{10}(n)\cos((\pi/6)n)\}$.

To begin, $\text{rect}_{10}(n) \Leftrightarrow 1 + 2\sum_{k=1}^{10}\cos(k\omega)$ from Entry #14 in Table 4.1. This periodic spectrum is plotted in Fig. 4.4:

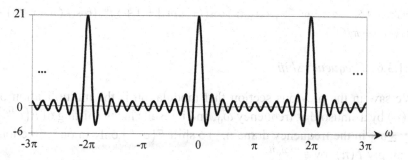

Figure 4.4. Plot of $\mathcal{F}\{\text{rect}_{10}(n)\} = 1 + 2\sum_{k=1}^{10}\cos(k\omega)$.

The modulation property (Eq. (4.42)) tells us that $\mathcal{F}\{f(n)\cos((\pi/6)n)\} = (1/2)F\left(e^{j(\omega-\pi/6)}\right) + (1/2)F\left(e^{j(\omega+\pi/6)}\right)$.
Therefore, $\mathcal{F}\{\text{rect}_{10}(n)\cos((\pi/6)n)\} = (1/2)\left(1 + 2\sum_{k=1}^{10}\cos(k(\omega - \pi/6))\right) + (1/2)(1 + 2\sum_{k=1}^{10}\cos(k(\omega + \pi/6))) = 1 + \sum_{k=1}^{10}(\cos(k(\omega, -\pi/6)) + \cos(k(\omega + \pi/6)))$ as in Fig. 4.5.

Modulation, therefore, shifts[aa] the energy/power of $f(n)$ from the original ("baseband") frequency range that is occupied by $F(e^{j\omega})$ to the frequency range that is in vicinity of ω_0. If ω_0 is such that the shifted copies of $F(e^{j\omega})$ are non-overlapping in frequency, then $f(n)$ may easily

---

[aa] Modulation, or the frequency shifting of $F(e^{j\omega})$, is also commonly referred to as frequency heterodyning.

be recovered from the modulated signal $f(n)\cos(\omega_0 n)$ through the process of *filtering*, which will be shown in a later chapter.

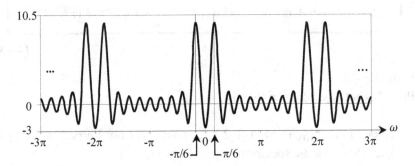

Figure 4.5. Plot of $\mathcal{F}\{\text{rect}_{10}(n)\cos((\pi/6)n)\} = 1 + \sum_{k=1}^{10}\left(\cos\left(k(\omega - \pi/6)\right) + \cos\left(k(\omega + \pi/6)\right)\right)$.

4.1.3.6 *Frequency shift*

We saw in the previous section that modulation is the multiplication of $f(n)$ by a sinusoid at frequency ω_0, and it results in the shifting of $F(e^{j\omega})$ by $\pm\omega_0$ in the frequency domain. To shift $F(e^{j\omega})$ only in one direction, multiply $f(n)$ by $e^{\pm j\omega_0 n}$:

$$\boxed{f(n)e^{\pm j\omega_0 n} \Leftrightarrow F\left(e^{j(\omega\mp\omega_0)}\right)} \tag{4.44}$$

4.1.3.7 *Time scaling*

Unlike continuous-time signals, discrete-time (or *sampled*) signals cannot be easily stretched or compressed in time by an arbitrary scale factor. There are two time scaling operations that are simple to implement on sampled signals, however, and we present them next.

Down-sampling by integer factor a:
Define $g(n) = f(an)$. The sequence $g(n)$ exists only when scaling factor a is an integer (otherwise $a \cdot n$ is not an integer, making $f(an)$ undefined).

Considering only positive, nonzero values of a,[bb] we see that $g(n)$ is a *down-sampled* version of $f(n)$: beginning at time index $n = 0$, and going from there in both positive and negative directions, only every a^{th} sample of $f(n)$ is retained. Obviously, some information is lost by down-sampling $f(n)$.[cc] What effect does down-sampling have in the frequency domain? We answer this question by taking the discrete-time Fourier transform (Eq. (4.1)) of $f(an)$ to obtain:

$$f(an) \Leftrightarrow \frac{1}{a} X\left(e^{j\omega/a}\right), \quad a \in \mathbb{Z}^+,$$

$$\text{where } X\left(e^{j\omega}\right) = F\left(e^{j\omega}\right) * \sum_{m=0}^{a-1} \delta\left(\omega - \frac{m}{a}2\pi\right). \quad (4.45)$$

To understand the spectrum of $f(an)$, first consider $X(e^{j\omega})$ that is defined above in Eq. (4.45). Because of being formed by adding together a frequency-shifted copies of $F(e^{j\omega})$, with frequency shifts that are equally spaced over 2π, $X(e^{j\omega})$ is periodic with period equal to $2\pi/a$. Subsequent frequency stretching $X(e^{j\omega})$ by factor a gives $X(e^{j\omega/a})$ a period of 2π (rad/sec). Eq. (4.45) leads to the observation that when a time signal is down-sampled by some factor (compressed in time by removing samples) its spectrum is expanded by the same factor. In the frequency domain, down-sampling results in possibly destructive[dd] overlap-addition of the shifted and stretched copies of $F(e^{j\omega})$.

[bb] Negative values of integer a result in time reversal, which is not our goal here, and $a = 0$ creates a constant-level signal $g(n) = f(0)$ that is not very interesting.

[cc] Except when the samples being eliminated all happen to be zero (it is possible – see the following section on *up-sampling*).

[dd] The loss of frequency-domain information, which is a consequence of overlap-adding spectra, is equivalent to the amount of information lost in the time-domain by eliminating all but every a^{th} sample in $f(n)$. The *information content* of a signal is a fascinating topic that is typically presented in engineering courses on probability theory.

Up-sampling by integer factor a:

$$\text{Next define } g(n) = \begin{cases} f(n/a), & n/a = \text{integer;} \\ 0, & \text{otherwise.} \end{cases} \quad (a \in \mathbb{Z})$$

By this definition, sequence $g(n)$ is essentially a copy of sequence $f(n)$ with $a - 1$ zeros inserted between each of its samples. We therefore call $g(n)$ an *up-sampled* version of $f(n)$. Because all original samples remain, no information is lost by up-sampling $f(n)$. Once again, we are interested to see what effect this operation has in the frequency domain. By calculating the discrete-time Fourier transform of $g(n)$ using Eq. (4.1), we obtain:

$$g(n) = \begin{cases} f\left(\frac{n}{a}\right), & \frac{n}{a} = \text{integer;} \\ 0, & \text{otherwise.} \end{cases} \Leftrightarrow G\left(e^{j\omega}\right) = F\left(e^{j(a\omega)}\right) \quad (4.46)$$

The spectrum of an up-sampled sequence is compressed by factor a, resulting in a cycles of the periodic spectrum over 2π rad/sec. Equation (4.46) leads to the observation that when a time signal is up-sampled by some factor (stretched in time by inserting zero samples) its spectrum is compressed by the same factor, which shortens the period from 2π rad/sec to $2\pi/a$ rad/sec.

4.1.3.8 *Parseval's Theorem*

Parseval's Theorem[ee] relates the energy of a signal calculated in the time domain to the energy as calculated in the frequency domain. To begin, assume that sequence $f(n)$ may be complex-valued and that $f(n) \Leftrightarrow F\left(e^{j\omega}\right)$. The energy of $f(n)$ is $\sum_{n=-\infty}^{\infty} f(n)f^*(n)$

$$= \sum_{n=-\infty}^{\infty} f(n) \left\{ \frac{1}{2\pi} \int_{2\pi} F\left(e^{j\omega}\right) e^{j\omega n} d\omega \right\}^*.$$

[ee] This theorem is also known in the literature as Rayleigh Theorem.

$$= \sum_{n=-\infty}^{\infty} f(n) \left\{ \frac{1}{2\pi} \int_{2\pi} F^*(e^{j\omega}) e^{-j\omega n} d\omega \right\}$$

$$= \frac{1}{2\pi} \int_{2\pi} F^*(e^{j\omega}) \left\{ \sum_{n=-\infty}^{\infty} f(n) e^{-j\omega n} \right\} d\omega$$

$$= \frac{1}{2\pi} \int_{2\pi} F^*(e^{j\omega}) F(e^{j\omega}) d\omega, \tag{4.47}$$

leading to a statement of the theorem:

$$\boxed{\sum_{n=-\infty}^{\infty} |f(n)|^2 = \frac{1}{2\pi} \int_{2\pi} |F(e^{j\omega})|^2 \, d\omega} \tag{4.48}$$

Next, we summarize the properties of the discrete-time Fourier transform in Table 4.2.

Table 4.2. Table of discrete-time Fourier transform properties.

1. Linearity:	(Eq. (4.34))
$af_1(n) + bf_2(n) \Leftrightarrow aF_1(e^{j\omega}) + bF_2(e^{j\omega})$	
2. Time shift:	(Eq. (4.37))
$f(n \pm n_0) \Leftrightarrow F(e^{j\omega}) e^{\pm j\omega n_0}$	
3. Convolution:	(Eq. (4.38))
$f_1(n) * f_2(n) \Leftrightarrow F_1(e^{j\omega}) \times F_2(e^{j\omega})$	
4. Multiplication: [ff]	(Eq. (4.39))
$f_1(n) \times f_2(n) \Leftrightarrow \frac{1}{2\pi} \{ F_1(e^{j\omega}) \circledast F_2(e^{j\omega}) \}$	

(Continued)

[ff] See Eq. (4.40) for the definition of $F_1(e^{j\omega}) \circledast F_2(e^{j\omega})$.

Table 4.2. (*Continued*)

---

5. Modulation: (Eq. (4.42))

$$f(n)\cos(\omega_0 n) \Leftrightarrow \tfrac{1}{2}F\big(e^{j(\omega-\omega_0)}\big) + \tfrac{1}{2}F\big(e^{j(\omega+\omega_0)}\big)$$

---

6. Modulation (arbitrary carrier phase):   (Eq. (4.43))

$$f(n)\cos(\omega_0 n + \phi) \Leftrightarrow \frac{e^{j\phi}}{2}F\big(e^{j(\omega-\omega_0)}\big) + \frac{e^{-j\phi}}{2}F\big(e^{j(\omega+\omega_0)}\big)$$

---

7. Frequency shift: (Eq. (4.44))

$$f(n)e^{\pm j\omega_0 n} \Leftrightarrow F\big(e^{j(\omega\mp\omega_0)}\big)$$

---

8. Down-sampling by factor $a \in \mathbb{Z}^+$:   (Eq. (4.45))

$$f(an) \Leftrightarrow (1/a)X\big(e^{j\omega/a}\big),$$

$$\text{where } X\big(e^{j\omega}\big) = F\big(e^{j\omega}\big) * \textstyle\sum_{m=0}^{a-1}\delta\big(\omega - \tfrac{m}{a}2\pi\big)$$

---

9. Up-sampling by factor $a \in \mathbb{Z}^+$:   (Eq. (4.46))

$$\left.\begin{array}{l} f(n/a),\ n/a = \text{integer};\\ 0, \quad \text{otherwise.} \end{array}\right\} \Leftrightarrow F\big(e^{j(a\omega)}\big)$$

---

10. Time reversal: (See[gg] below)

$$f(-n) \Leftrightarrow F\big(e^{j(-\omega)}\big)$$

---

11. Conjugation in time: (See[hh] below)

$$f^*(n) \Leftrightarrow F^*\big(e^{j(-\omega)}\big)$$

(*Continued*)

[gg] $\mathcal{F}\{f(-n)\} = \sum_{n=-\infty}^{\infty} f(-n)e^{-j\omega n} = \sum_{k=-\infty}^{\infty} f(k)e^{-j(-\omega)k} = F\big(e^{j(-\omega)}\big)$

[hh] $\mathcal{F}\{f^*(n)\} = \sum_{n=-\infty}^{\infty} f^*(n)e^{-j\omega n} = \big(\sum_{n=-\infty}^{\infty} f(n)e^{j\omega n}\big)^* =$

$\big(\sum_{n=-\infty}^{\infty} f(n)e^{-j(-\omega)n}\big)^* = F^*\big(e^{j(-\omega)}\big)$

Table 4.2. (*Continued*)

---

12. Real, even $f_e(n) = Re\{f_e(-n)\}$: (See footnote on p. 102)

$$f_e(n) \Leftrightarrow \mathcal{F}\{f_e(n)\} = f(0) + 2\sum_{n=1}^{\infty} f_e(n)\cos(\omega n)$$

---

13. Real, odd $f_o(n) = Re\{-f_o(-n)\}$: (See footnote on p. 102)

$$f_o(n) \Leftrightarrow \mathcal{F}\{f_o(n)\} = -2j\sum_{n=1}^{\infty} f_o(n)\sin(\omega n)$$

---

14. Backward difference:

$$f(n) - f(n-1) \Leftrightarrow (1 - e^{-j\omega}) F(e^{j\omega})$$

---

15. Cumulative sum: (See[ii] below)

$$\sum_{k=-\infty}^{n} f(k) = f(n) * u(n) \Leftrightarrow \frac{F(e^{j\omega})}{1-e^{-j\omega}} + \pi F(e^{j0})\delta_{2\pi}(\omega)$$

---

16. Parseval's Theorem: (for energy signals) (Eq. (4.48))

$$E_f = \sum_{n=-\infty}^{\infty} |f(n)|^2 = \frac{1}{2\pi}\int_{2\pi} |F(e^{j\omega})|^2 \, d\omega$$

---

17. Cross-Correlation: (for energy signals): (See[jj] below)

$$\phi_{xy}(n) \Leftrightarrow X^*(e^{j\omega})Y(e^{j\omega})$$

(*Continued*)

[ii] $f(n) * u(n) = \sum_{k=-\infty}^{\infty} f(k)u(n-k) = \sum_{k=-\infty}^{n} f(k)$; $f(n) * u(n) \Leftrightarrow$
$F(e^{j\omega})U(e^{j\omega}) = F(e^{j\omega})\{1/(1-e^{-j\omega}) + \pi\delta_{2\pi}(\omega)\} = F(e^{j\omega})/(1-e^{-j\omega}) +$
$\pi F(e^{j\omega})\delta_{2\pi}(\omega) = F(e^{j\omega})/(1-e^{-j\omega}) + \pi F(e^{j2\pi N})\delta_{2\pi}(\omega) = F(e^{j\omega})/(1-e^{-j\omega}) +$
$\pi F(e^{j0})\delta_{2\pi}(\omega)$, due to the periodicity of $F(e^{j\omega})$.

[jj] $\phi_{xy}(n) = x^*(-n) * y(n)$, therefore $\mathcal{F}\{\phi_{xy}(n)\} = \mathcal{F}\{x^*(-n)\}\mathcal{F}\{y(n)\} =$
$X^*(e^{j\omega})Y(e^{j\omega})$.

Table 4.2. (*Continued*)

18. Autocorrelation: (for energy signals)	(See[kk] below)

$$\phi_{xx}(n) \Leftrightarrow X^*(e^{j\omega})X(e^{j\omega}) = \left|X(e^{j\omega})\right|^2$$

19. Duality: (for periodic signals)	(Eq. (4.57))

$$F_p^*(n) \overset{DFT}{\Longleftrightarrow} N f_p^*(k)$$

4.1.4 *Graphical representation of the Fourier transform*

Engineers and scientists like to have an image of mathematical formulas and functions to facilitate their understanding. Thus far we have presented graphical representations of only real signals in either time or frequency domain. The Fourier transform of a real time-domain signal is generally[ll] a complex function of ω. In this section, we discuss several different ways of graphing such complex-valued waveforms. Such plots are extremely useful in understanding the processing of signals through various devices such as filters, amplifiers, mixers, etc.

Frequency-domain plots are presented with an abscissa (x-axis) scale of ω radians/second in this textbook. Because a discrete-time signal's spectrum is periodic, it is sufficient to plot only one period (e.g., $|\omega| \leq \pi$). As you know, $\omega = 2\pi f$ where the units of f are cycles/second (Hertz). This means that the spectrum's period on the f scale is 1 Hz. Keep in

---

[kk] $\mathcal{F}\{\phi_{xy}(n)\} = X^*(e^{j\omega})Y(e^{j\omega})$, therefore $\mathcal{F}\{\phi_{xx}(n)\} = X^*(e^{j\omega})X(e^{j\omega})$.

[ll] A "real world" signal $x(n)$ will be real, and will almost always have $x_e(n) \neq 0$, $x_o(n) \neq 0$. Therefore, since the Fourier transform $X(e^{j\omega}) = x(0) + 2\sum_{n=1}^{\infty} x_e(n)\cos(\omega n) - 2j\sum_{n=1}^{\infty} x_o(n)\sin(\omega n)$ (see footnote on p. 102), $X(e^{j\omega})$ will almost always be a complex function of ω.

mind that both ω and f are normalized frequency values; the actual frequency value is recovered by multiplying with factor f_s (Hz) / 1 (Hz).[mm]

4.1.4.1 *Rectangular coordinates*

In this form of visualization, a three-dimensional plot is created using the real part of $F(e^{j\omega})$, the imaginary part of $F(e^{j\omega})$, and ω as three axes. Figure 4.6 shows $\mathcal{F}\{\text{rect}_3(n - 100)\} = 7\{\text{sinc}(3.5\omega) * \delta_{2\pi}(\omega)\}e^{-j100\omega}$ plotted this way, for example:

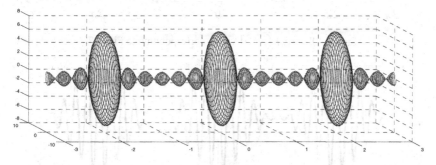

Figure 4.6. A graph of $7\{\text{sinc}(3.5\omega) * \delta_{2\pi}(\omega)\}e^{-j100\omega}$ vs. ω/π.

It may be more instructive to plot two 2-D plots: one of $Re\{F(e^{j\omega})\}$ vs. ω, and another of $Im\{F(e^{j\omega})\}$ vs. ω. For example, Fig. 4.7 shows the 3-D plot of a given $F(e^{j\omega})$:

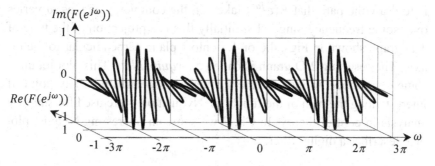

Figure 4.7. A 3-D graph of complex-valued $F(e^{j\omega})$.

[mm] The sampling frequency, in Hz, is $f_s = 1/T_s$ (T_s is the time difference between successive samples of the signal).

Figures 4.8 and 4.9 display the same information, using a pair of 2-D plots, in terms of its real and imaginary components:

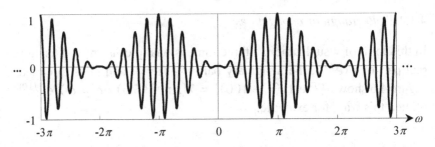

Figure 4.8. $Re\{F(e^{j\omega})\}$ vs. ω, corresponding to Fig. 4.7.

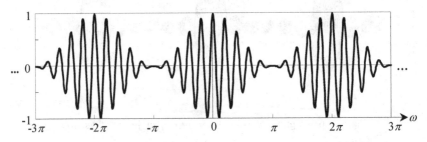

Figure 4.9. $Im\{F(e^{j\omega})\}$ vs. ω, corresponding to Fig. 4.7.

Yet another way of displaying a complex-valued frequency spectrum is to draw the path that $F(e^{j\omega})$ takes, in the complex plane, as ω varies over some frequency range. Essentially this is a projection of the type of 3-D graph shown in Figs. 4.6 or 4.7 onto a plane perpendicular to the ω-axis. The resulting 2-D graph is called a *Nyquist Plot*. This plot requires some extra labels to indicate the frequency associated with any point of interest along the path of the function. Nyquist plots are used for stability analysis in control theory. In Ch. 7 we show an interesting Nyquist plot that describes a digital filter.

4.1.4.2 *Polar coordinates*

As an alternative to the previously-presented graphs of real and imaginary parts of $F(e^{j\omega})$, frequently it is more appropriate to plot magnitude $\left|F(e^{j\omega})\right|$ and phase $\angle F(e^{j\omega})$ vs. ω. Figures 4.10 and 4.11 show these two graphs for a complex frequency-domain function. In general, separately plotting magnitude and phase components is the preferred method for displaying complex frequency-domain waveforms.

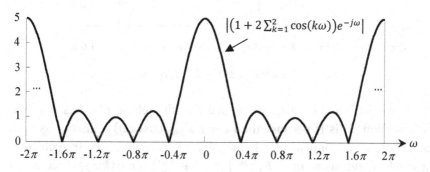

Figure 4.10. A graph of $\left|\left(1 + 2\sum_{k=1}^{2}\cos(k\omega)\right)e^{-j\omega}\right|$ vs. ω.

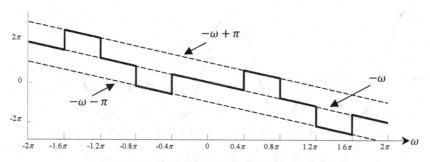

Figure 4.11. A graph of $\angle\left\{\left(1 + 2\sum_{k=1}^{2}\cos(k\omega)\right)e^{-j\omega}\right\}$ vs. ω.

4.1.4.3 *Graphing the amplitude of $F(e^{j\omega})$*

Consider the case where the Fourier transform gives a real, bipolar result: for example, $F(e^{j\omega}) = 1 + 2\sum_{k=1}^{2}\cos(k\omega)$. Plotting its magnitude gives us the same graphs as in Fig. 4.10, and a plot of the phase is shown in Fig. 4.12. Notice that whenever $\left(1 + 2\sum_{k=1}^{2}\cos(k\omega)\right) < 0$, the phase angle is $\pm\pi$ so that $\left|\left(1 + 2\sum_{k=1}^{2}\cos(k\omega)\right)\right|e^{j(\pm\pi)} = -\left|\left(1 + 2\sum_{k=1}^{2}\cos(k\omega)\right)\right|$. [nn]

[nn] Any odd integer value of k will make $e^{jk\pi} = -1$; the values chosen alternate so that the phase angle is an odd function of ω.

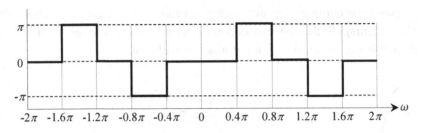

Figure 4.12. A graph of $\angle\left(1 + 2\sum_{k=1}^{2}\cos(k\omega)\right)$ vs. ω.

If we look at Fig. 4.12 we notice that this phase plot has minimal information. This is because the $\left(1 + 2\sum_{k=1}^{2}\cos(k\omega)\right)$ function is real, and phase serves only to specify the polarity of the function. In such case, it is advantageous to plot $F\left(e^{j\omega}\right) = \left(1 + 2\sum_{k=1}^{2}\cos(k\omega)\right)$ directly, as shown in Fig. 4.13:

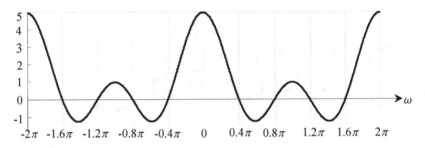

Figure 4.13. A graph of $1 + 2\sum_{k=1}^{2}\cos(k\omega)$ vs. ω.

With access to MATLAB®, all these plots can be readily generated, thus the choice of which one to use should be based on the clarity of the information to be conveyed and how it may be used in a particular application.

4.1.4.4 *Logarithmic scales and Bode plots*

Due to the logarithmic nature of human perception (e.g., we hear frequencies on a logarithmic scale, as reflected by the tuning of a piano keyboard), and to linearize some curves being plotted, it is common to graph one or both axes of a spectrum on a logarithmic scale. For example, one may choose to plot $F(e^{j\omega})$ vs. $\log_{10}(\omega)$ instead of plotting $F(e^{j\omega})$ vs. ω. Compared to a linear frequency scale, the logarithmic scale expands the spacing of frequencies as they approach zero and compresses frequencies together as they approach infinity. Thus, a logarithmic scale never includes zero (d.c.) nor any negative[oo] frequency values.

When plotting the magnitude spectrum $|F(e^{j\omega})|$ it is common to use logarithmic scales for both magnitude[pp] and frequency. The resulting graph is then proportional to $\log|F(e^{j\omega})|$ vs. $\log(\omega)$. A version of the log-log plot, called a Bode magnitude plot, is where the vertical scale is $10\log_{10}|F(e^{j\omega})|^2$ and the horizontal scale is $\log_{10}(\omega)$. The Bode phase plot has vertical scale $\angle F(e^{j\omega})$ and horizontal scale $\log_{10}(\omega)$. Bode magnitude and phase plots are common in system analysis, where they provide a convenient way to combine responses of cascaded systems.

4.1.5 *Fourier transform of periodic sequences*

4.1.5.1 *Comb function*

The comb function, as discussed previously, is another term for an impulse train. This idealized periodic signal plays an important role in our theoretical understanding of analog-to-digital signal conversion, as well as being a convenient tool for modeling periodic signals.

Let us consider the Fourier transform of a comb function $\delta_M(t)$, which is given as Entry #19 in Table 4.1:[qq]

[oo] Negative frequencies do not exist (when defined as #events/second) except as a mathematical tool for efficient notation. This is clearly shown by a cosine wave at frequency ω_0 equal to a sum of complex exponentials at frequencies $\pm\omega_0$: $2\cos(\omega_0 n) = e^{j(\omega_0)n} + e^{j(-\omega_0)n}$

[pp] Phase is always plotted on a linear scale so that zero and negative phase angles may be included.

[qq] We take this as a given, and derive the DFT (Section 4.1.5.3) based on it.

$$\delta_M(t) = \sum_{k=-\infty}^{\infty} \delta(n - kM) \Leftrightarrow \omega_0\, \delta_{\omega_0}(\omega)$$

$$= \omega_0 \sum_{k=-\infty}^{\infty} \delta(\omega - k\omega_0), \text{ where } \omega_0 = 2\pi/M. \qquad (4.49)$$

Clearly both the time domain function and its transform are both comb functions: one has period M samples, while the transformed function has period $\omega_0 = 2\pi/M$ rad/sec. We readily conclude that if one function has closely placed impulses, the other function has widely spaced impulses and vice-versa.[π] Figures 4.14 and 4.15 show the two impulse trains when the period in time domain is $M = 4$ samples:

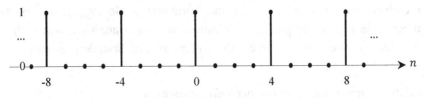

Figure 4.14.   Impulse train $\delta_4(n)$.

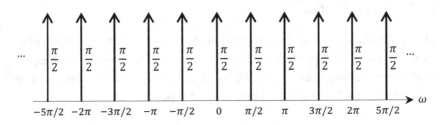

Figure 4.15.   Impulse train $(\pi/2)\delta_{\pi/2}(\omega) = \mathcal{F}\{\delta_4(n)\}$.

4.1.5.2 *Periodic signals as convolution with a comb function*

Periodic power signals play a very important role in linear system analysis. Sinusoidal steady-state analysis of linear electrical circuits is an example of this. Next, we show that an arbitrary periodic signal can be considered as a convolution product of one of its periods with an impulse train.

[π] There is also a corresponding change in areas of the impulses.

Let us consider the convolution of $\text{rect}_5(n)$ and impulse train $\delta_{20}(n)$. These two signals and their convolution product are shown in Figs. 4.16 through 4.18:

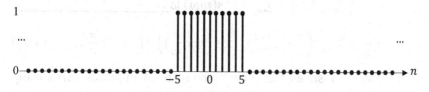

Figure 4.16.   Rectangular pulse $\text{rect}_5(n)$.

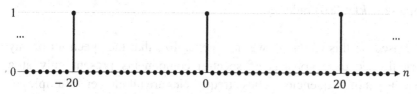

Figure 4.17.   Impulse train $\delta_{20}(n)$.

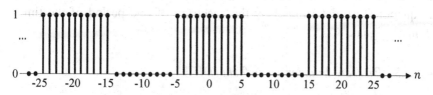

Figure 4.18.   Periodic signal $f_p(n) = \text{rect}_5(n) * \delta_{20}(n)$.

We observe that the resulting signal is a periodic rectangular pulse train having period $N = 20$ samples. This may be written as:

$$f_p(n) = \text{rect}_5(n) * \delta_{20}(n) = \sum_{k=-\infty}^{\infty} \text{rect}_5(n - 20k). \qquad (4.50)$$

Because of this formulation, we can determine the spectrum of any periodic signal in terms of the spectrum of one of its periods. In Fig. 4.18, for example, $\mathcal{F}\{\text{rect}_5(n)\} = 1 + 2\sum_{m=1}^{5} \cos(m\omega)$ and $\mathcal{F}\{\delta_{20}(n)\} = (\pi/10)\delta_{\pi/10}(\omega)$. Therefore, the periodic signal of rectangular pulses of width 11 samples and period 20 samples has a spectrum represented by

$$F(e^{j\omega}) = \mathcal{F}\{rect_5(n) * \delta_{20}(n)\} = \mathcal{F}\{rect_5(n)\} \times \mathcal{F}\{\delta_{20}(n)\}$$

$$= \left(1 + 2\sum_{m=1}^{5}\cos(m\omega)\right) \times \frac{\pi}{10}\delta_{\frac{\pi}{10}}(\omega)$$

$$= \frac{\pi}{10}\sum_{k=-\infty}^{\infty}\left(1 + 2\sum_{m=1}^{5}\cos(m\omega)\right)\delta\left(\omega - k\frac{\pi}{10}\right)$$

$$= \frac{\pi}{10}\sum_{k=-\infty}^{\infty}\left(1 + 2\sum_{m=1}^{5}\cos\left(mk\frac{\pi}{10}\right)\right)\delta\left(\omega - k\frac{\pi}{10}\right). \qquad (4.51)$$

Equation 4.51 states that $F(e^{j\omega})$ consists of impulses periodically spaced in frequency with period $2\pi/20 = \pi/10$ rad/sec, and having areas $(\pi/10)(1 + 2\sum_{m=1}^{5}\cos(mk\pi/10))$. Note that the frequency of the kth impulse is $k(\pi/10)$ rad/sec.

Based on this example, we may generalize that the spectrum of any periodic sequence consists of spectral components present only at a discrete set of frequencies. These frequencies are all integer[ss] multiples of a *fundamental frequency* ω_0,[tt] and members of this set of frequencies are said to be *harmonically-related*. Further, the areas of impulses at these frequencies are determined by the spectrum of one period of the time-domain signal.

4.1.5.3 *Discrete Fourier Transform (DFT)*

In the previous example the periodic signal, rectangular pulses of width 11 that repeated every 20 samples, had frequency spectrum $(e^{j\omega}) = (\pi/10)\sum_{k=-\infty}^{\infty}(1 + 2\sum_{m=1}^{5}\cos(mk\,\pi/10))\delta(\omega - k\,\pi/10)$. Recall, the expression for $F(e^{j\omega}) = \mathcal{F}\{f_p(n)\}$ was obtained by representing periodic signal $f_p(n)$ as a convolution product of one of its periods $rect_5(n)$ with impulse train $\delta_{20}(n)$. In general, any periodic signal $f_p(n)$ may be represented this way:

[ss] including zero and negative integers

[tt] Do not be confused by the notation "ω_0": this represents the period of impulse train $\delta_{\omega_0}(\omega) = \sum_{k=-\infty}^{\infty}(\omega - k\omega_0)$, not the frequency of the impulse corresponding to $k = 0$.

$$f_p(n) = p(n) * \delta_N(n),$$

$$\text{where } p(n) \equiv \begin{cases} f_p(n), & 0 \le n \le N-1; \\ 0, & \text{otherwise.} \end{cases} \tag{4.52}$$

Solving for $F(e^{j\omega}) = \mathcal{F}\{f_p(n)\} = P(e^{j\omega}) \times \omega_0 \delta_{\omega_0}(\omega)$

$$= P(e^{j\omega}) \sum_{k=-\infty}^{\infty} \omega_0 \delta(\omega - k\omega_0)$$

$$= \sum_{k=-\infty}^{\infty} P(e^{j\omega}) \omega_0 \delta(\omega - k\omega_0)$$

$$= \sum_{k=-\infty}^{\infty} P(e^{jk\omega_0}) \omega_0 \delta(\omega - k\omega_0)$$

$$= \sum_{k=-\infty}^{\infty} P(e^{jk\omega_0}) \left(\frac{\omega_0}{2\pi}\right) 2\pi \delta(\omega - k\omega_0)$$

$$= \sum_{k=-\infty}^{\infty} P(e^{jk\omega_0}) \left(\frac{1}{N}\right) 2\pi \delta(\omega - k\omega_0)$$

$$= \frac{1}{N} \sum_{k=0}^{N-1} P(e^{jk\omega_0}) 2\pi \delta_{2\pi}(\omega - k\omega_0)$$

$$= \frac{1}{N} \sum_{k=0}^{N-1} F_p(k) \, 2\pi \delta_{2\pi}(\omega - k\omega_0),$$

$$\text{where } F_p(k) \equiv P(e^{jk\omega_0}) \text{ and } \omega_0 = 2\pi/N. \tag{4.53}$$

Taking the inverse discrete-time Fourier transform of both sides, which is easily done using Fourier transform pair $e^{j\beta n} \Leftrightarrow 2\pi \delta_{2\pi}(\omega - \beta)$, we obtain the following expression for periodic signal $f_p(n)$:

Inverse Discrete Fourier Transform (IDFT)	$f_p(n) = \frac{1}{N} \sum_{k=0}^{N-1} F_p(k) \, e^{j\left(\frac{k2\pi}{N}\right)n}$

$$= IDFT\{F_p(k)\}. \tag{4.54}$$

In Eq. (4.54) we now have a representation of a periodic signal $f_p(n)$ in the form known as the *Inverse Discrete Fourier Transform (IDFT)*. The harmonically-related frequencies $k\omega_0 = k2\pi/N$ (where k is an integer) are called *harmonic frequencies*, or *harmonics* of the fundamental frequency $\omega_0 = 2\pi/N$.[uu] The constants $F_p(k)$, which are functions of harmonic index k, are called the *DFT coefficients*. Just as time-domain signal $f_p(n)$ is discrete and periodic (period = N samples), $F_p(k)$ is also discrete and periodic (period = N samples).

Recall that $F_p(k) \equiv P(e^{jk\omega_0})$, where $P(e^{jk\omega_0})$ is the discrete-time Fourier transform of one period of $f_p(n)$. Thus, we may obtain an expression for $F_p(k)$ in terms of $f_p(n)$:

$$F_p(k) \equiv P(e^{jk\omega_0}) = \sum_{n=-\infty}^{\infty} p(n)e^{-j(k\omega_0)n}$$

$$= \sum_{n=0}^{N-1} p(n)e^{-j(k\omega_0)n} = \sum_{n=0}^{N-1} f_p(n)e^{-jk\omega_0 n}$$

$$= \sum_{n=0}^{N-1} f_p(n)e^{-jk(2\pi/N)n} \quad (k = 0, 1, ..., N-1) \quad (4.55)$$

Therefore:

Discrete Fourier Transform (DFT)	$F_p(k) = \sum_{n=0}^{N-1} f_p(n)e^{-j\left(\frac{k2\pi}{N}\right)n}$
	$= DFT\{f_p(n)\}.$ (4.56)

Equation (4.56), the solution for $F_p(k)$ values using the DFT expression, is called the *Discrete Fourier Transform (DFT)*, and resembles the expression for a discrete-time Fourier transform summation. And Eq. (4.54), the expression for $f_p(n)$ as an inverse DFT, is in fact an inverse discrete-time Fourier transform of a spectrum containing only impulses; hence its integration over a continuum of frequencies simplifies to a summation over discrete frequency values. To summarize, Eqs. (4.56) and (4.54) are the discrete-time Fourier transform and its inverse as applied to a special category of time-domain signals: those discrete-time sequences $f(n)$ that are periodic.

The DFT and its inverse, as described by Eqs. (4.56) and (4.54), may also be derived independently of the discrete-time Fourier transform. This

---

[uu] "Harmonics" usually refers to positive integer multiples of ω_0 (see footnote[oo], p. 123).

approach approximates periodic signal $f_p(n)$ with a weighted sum of harmonically-related basis signals $e^{j(k2\pi/N)n}$. These basis signals are *mutually orthogonal*: one period of ($e^{j(k_1 2\pi/N)n}$)($e^{j(k_2 2\pi/N)n}$)* has zero area when $k_1 \neq k_2$, and nonzero area when $k_1 = k_2$. Mutually-orthogonal basis signals make it possible to easily solve for coefficients $F_p(k)$ due to the disappearance of all but one cross-product term when summing.

The DFT is a mapping of N (usually real, but possibly complex) numbers in the time domain to N (usually complex) numbers in the frequency domain. Both the DFT and its inverse are algorithms that involve summing a finite number of terms, and if all terms are finite-valued then the summations are guaranteed to converge.

The DFT coefficients are plotted like the spectrum sketches described previously, except that now the spectrum will be discrete. Figure 4.19 plots the DFT coefficient spectrum of the periodic rectangular pulse train from Fig. 4.18.

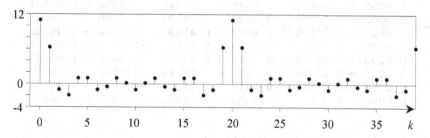

Figure 4.19.   Discrete Fourier Transform spectrum for periodic signal $f_p(n) = \text{rect}_5(n) *$ $\delta_{20}(n) = (1/20) \sum_{k=0}^{19} F_p(k) \, e^{j(k2\pi/20)n}$.

4.1.5.4   *Time-frequency duality of the DFT*

Similarity between summations defining the DFT and the IDFT leads to the following: when $f_p(n) \Leftrightarrow F_p(k)$, it is also true that

$$F_p^*(n) \Leftrightarrow N f_p^*(k).$$ (4.57)[vv]

[vv]$IDFT\{N f_p^*(k)\} = \frac{1}{N}\sum_{k=0}^{N-1}\left(N f_p^*(k)\right)e^{j(k2\pi/N)n} = \left(\sum_{k=0}^{N-1} f_p(k)e^{-j(n2\pi/N)k}\right)^* = F_p^*(n).$

This is the same as saying that the IDFT operation may be done using the DFT:

$$f_p(n) = IDFT\{F_p(k)\} = \frac{1}{N} DFT\{F_p^*(k)\}^*. \qquad (4.58)$$

Example 4.4

Given: periodic pulse train $x_p(n) = rect_2(n - 2) * \delta_{11}(n)$. Write some MATLAB® code to calculate and tabulate its DFT $X_p(k)$ over $0 \le k \le 9$. Then recover one period of $x_p(n)$ from $X_p(k)$ using Eq. (4.58).

```
% Example Ch4_4
% Generate one period of xp(n):
xp = [1 1 1 1 1 0 0 0 0 0 0]'; % 11 samples per period
Xp = fft(xp);
xp1 = conj(fft(conj(Xp)))/11;
disp([xp,Xp,xp1])
```

```
1.0000 + 0.0000i    5.0000 + 0.0000i    1.0000 + 0.0000i
1.0000 + 0.0000i    1.4595 - 3.1958i    1.0000 + 0.0000i
1.0000 + 0.0000i   -0.3413 - 0.3938i    1.0000 + 0.0000i
1.0000 + 0.0000i    1.1549 - 0.3391i    1.0000 + 0.0000i
1.0000 + 0.0000i    0.0846 - 0.5883i    1.0000 + 0.0000i
0.0000 + 0.0000i    0.6423 + 0.4128i   -0.0000 + 0.0000i
0.0000 + 0.0000i    0.6423 - 0.4128i   -0.0000 + 0.0000i
0.0000 + 0.0000i    0.0846 + 0.5883i    0.0000 + 0.0000i
0.0000 + 0.0000i    1.1549 + 0.3391i    0.0000 + 0.0000i
0.0000 + 0.0000i   -0.3413 + 0.3938i   -0.0000 + 0.0000i
0.0000 + 0.0000i    1.4595 + 3.1958i    0.0000 + 0.0000i
```

4.1.5.5   Fast Fourier Transform (FFT)

When efficiently implemented to reduce the total number of multiply-add operations, the DFT is renamed the *FFT (Fast Fourier Transform)* even though both produce the same results. For example, one type of FFT reduces the number of operations from $O(N^2)$ to $O(N\log_2 N)$, which results in ~340 times faster execution when $N = 4096$. In fact, direct implementation of the DFT or the IDFT using the formulas in Eqs. (4.56) and (4.54) is practically no longer done. The FFT algorithm has revolutionized computational digital signal processing, especially in applications where signals need to be processed in real-time, and for this reason some programming languages now include it as a built-in function. In MATLAB®, the FFT (and its inverse, the *IFFT*) are implemented in platform-optimized code with functions **fft** and **ifft**.[ww] We refer those who are interested in the FFT algorithm itself to its description in Leland Jackson's textbook.

4.1.5.6 *Parseval's Theorem*

A periodic signal $f_p(n)$, with period N samples, has power P_f (= average energy per sample) equal to:

$$P_f = \frac{1}{N} \sum_{n=0}^{N-1} |f_p(n)|^2. \tag{4.59}$$

The power of $F_p(k)\, e^{j(k2\pi/N)n}$ is equal to $|F_p(k)|^2$. This is easy to show using the definition given in Eq. (4.59) above.[xx] One can also show[yy] that for two complex exponentials at different frequencies, $F_p(k)e^{jk\omega_0 n}$ and $F_p(l)e^{jl\omega_0 n}$ with ($k \neq l$), the power of their sum is equal to the sum of their powers: $|F_p(k)|^2 + |F_p(l)|^2$.

The result, although derived for the case of superposition of two complex exponential signals, can be readily generalized for multiple sinusoids, as well as for the case when the sinusoidal frequencies are not harmonically related.

[ww] MATLAB® also has versions of these for 2-dimensional signals: **fft2, ifft2**.

[xx] Power $= (1/N) \sum_{n=0}^{N-1} |F_p(k)\, e^{j(k2\pi/N)n}|^2 = (1/N) \sum_{n=0}^{N-1} |F_p(k)|^2 |e^{j(k2\pi/N)n}|^2 =$
$(1/N)|F_p(k)|^2 \sum_{k=0}^{N-1} 1^2 = (1/N)|F_p(k)|^2 N = |F_p(k)|^2.$

[yy] Proof: Let $f_p(n) = F_p(k)e^{jk\omega_0 n} + F_p(l)e^{jl\omega_0 n}$ ($k \neq l$). $P_f = (1/N) \sum_{n=0}^{N-1} |f_p(n)|^2 =$
$(1/N) \sum_{n=0}^{N-1} \left(F_p(k)e^{jk\omega_0 n} + F_p(l)e^{jl\omega_0 n} \right)\left(F_p(k)e^{jk\omega_0 n} + +F_p(l)e^{jl\omega_0 n} \right)^* =$
$(1/N) \sum_{n=0}^{N-1} \left(F_p(k)e^{jk\omega_0 n} + F_p(l)e^{jl\omega_0 n} \right)\left(\left(F_p(k)e^{jk\omega_0 n}\right)^* + \left(F_p(l)e^{jl\omega_0 n}\right)^* \right) =$
$(1/N) \sum_{n=0}^{N-1} \left(F_p(k)e^{jk\omega_0 n} + F_p(l)e^{jl\omega_0 n} \right)\left(F_p^*(k)e^{-jk\omega_0 n} + F_p^*(l)e^{-jl\omega_0 n} \right) =$
$(1/N) \sum_{n=0}^{N-1} \left(F_p(k)e^{jk\omega_0 n} F_p^*(k)e^{-jk\omega_0 n} + F_p(k)e^{jk\omega_0 n} F_p^*(l)e^{-jl\omega_0 n} + \right.$
$\left. F_p(l)e^{jl\omega_0 n} F_p^*(k)e^{-jk\omega_0 n} + F_p(l)e^{jl\omega_0 n} F_p^*(l)e^{-jl\omega_0 n} \right) =$
$(1/N) \sum_{n=0}^{N-1} F_p(k)e^{jk\omega_0 n} F_p^*(k)e^{-jk\omega_0 n} + (1/N) \sum_{n=0}^{N-1} F_p(k)e^{jk\omega_0 n} F_p^*(l)e^{-jl\omega_0 n} +$
$(1/N) \sum_{n=0}^{N-1} F_p(l)e^{jl\omega_0 n} F_p^*(k)e^{-jk\omega_0 n} + (1/N) \sum_{n=0}^{N-1} F_p(l)e^{jl\omega_0 n} F_p^*(l)e^{-jl\omega_0 n} =$
$(1/N) \sum_{n=0}^{N-1} F_p(k)F_p^*(k)e^0 + 0 + 0 + (1/N) \sum_{n=0}^{N-1} F_p(l)F_p^*(l)e^0 =$
$(1/N) \sum_{n=0}^{N-1} |F_p(k)|^2 + (1/N) \sum_{n=0}^{N-1} |F_p(l)|^2 = (1/N)|F_p(k)|^2 \sum_{n=0}^{N-1}(1) +$
$(1/N)|F_p(l)|^2 \sum_{n=0}^{N-1}(1) = (1/N)|F_p(k)|^2 N + (1/N)|F_p(l)|^2 N = |F_p(k)|^2 + |F_p(l)|^2.$

Now we will state Parseval's Theorem in terms of the DFT pair $f_p(n)$ and $F_p(k)$:

Parseval's Theorem for the DFT pair

$$P_f = \frac{1}{N}\sum_{n=0}^{N-1}|f_p(n)|^2$$

$$= \frac{1}{N^2}\sum_{k=0}^{N-1}|F_p(k)|^2. \qquad (4.60)$$

Parseval's Theorem states that power of a periodic signal $f_p(n)$ is equal to the sum of the power in its sinusoidal components.

Example 4.5
Given: $x_p(n) = rect_2(n-2) * \delta_{11}(n)$ (same as in Example 4.4). Using MATLAB®, find power of this signal in both time and frequency domains.

```
% Example Ch4_5
xp = [1 1 1 1 1 1 0 0 0 0 0];
Xp = fft(xp);
P1 = sum(abs(xp).^2)/11;
P2 = sum(abs(Xp).^2)/(11^2);
disp([P1 P2])
```

　　0.4545　　　0.4545

4.1.6 *Summary of Fourier transformations for discrete-time signals*

Table 4.3. Summary of Fourier transformations for discrete-time signals.

Time signal

Frequency spectrum

Discrete in time	Discrete-Time Fourier Transform (DTFT)	Continuous, periodic in frequency
$f(n)$ $\Rightarrow$	$\displaystyle\sum_{n=-\infty}^{\infty} f(n)e^{-j\omega n}$ $\Rightarrow$	$F(e^{j\omega})$
	Inverse Discrete-Time Fourier Transform (IDTFT)	
$f(n)$ $\Leftarrow$	$\displaystyle\frac{1}{2\pi}\int_{2\pi} F(e^{j\omega})e^{j\omega n}d\omega$ $\Leftarrow$	$F(e^{j\omega})$
Discrete, periodic in time	Discrete Fourier Transform (DFT, FFT)	Discrete, periodic in frequency
$f_p(n)$ $\Rightarrow$	$\displaystyle\sum_{n=0}^{N-1} f_p(n)e^{-j(k2\pi/N)n}$ $\Rightarrow$	$F_p(k)$
	Inverse Discrete Fourier Transform (IDFT, IFFT)	
$f_p(n)$ $\Leftarrow$	$\displaystyle\frac{1}{N}\sum_{k=0}^{N-1} F_p(k)\,e^{j(k2\pi/N)n}$ $\Leftarrow$	$F_p(k)$

4.2 Practical Applications

4.2.1 *Spectral analysis using the FFT*

Under certain conditions, the FFT can be used to compute the Fourier transform of a discrete-time sequence $x(n)$. Recall from Section 4.1.5.5: the FFT yields the same results as the DFT, only using fewer calculations. The DFT is a special case of the DTFT when $x(n) = x_p(n)$ is periodic. However, the DFT (and hence FFT) can also efficiently estimate the spectrum of finite-length $x(n)$. We later consider the two cases separately (periodic $x_p(n)$, finite-length $x(n)$).

4.2.1.1 *Frequency resolution*

When we plot $X(e^{j\omega})$, which is periodic with period 2π rad/sec, we assume sequence $x(n)$ has time spacing $T_s = 1$ sec between its samples.[zz] In general, for an arbitrary T_s,

$$\text{DTFT}\{x(n)\} = X(e^{j\omega T_s}), \tag{4.61}$$

whose period $= 2\pi/T_s = \omega_s$ rads/sec.

The DFT (FFT) transforms N samples from the time domain to the frequency domain. The time samples are from one period of periodic signal $x_p(n) = x(n) * \delta_N(n)$, at times nT_s sec $\{n = 0, 1, 2, ..., N - 1\}$. The frequency samples are taken from one period of periodic spectrum $X(e^{j\omega T_s})$, taken at frequencies $k\Delta\omega$ rad/sec $\{k = 0, 1, 2, ..., N - 1\}$. Thus, the value of $\Delta\omega$ (called the *frequency resolution* of spectrum $X(e^{j\omega T_s})$) is equal to:

$$\Delta\omega = \frac{\omega_s}{N} = \frac{2\pi}{NT_s} \text{ rad/sec.} \tag{4.62}$$

Stated in units of cycles per second (Hz):

[zz] See Chapter 6.

$$\Delta f = \frac{f_s}{N} = \frac{1}{NT_s} \text{ Hz} \qquad (4.63)$$

Finer frequency resolution (lower Δf) does not necessarily mean there is more information presented in the graph of the spectrum. For example, it is common to concatenate some zero-valued samples to finite-length $x(n)$ before calculating the FFT for the sole purpose of reducing Δf of the plot for a better appearance. But since those added zeros contain no new information, the information being presented to the viewer remains the same as before.

However, when the data being operated on by the FFT represent a longer time duration, the resulting increase in frequency resolution does indeed provide more information. This is important when trying to detect the presence of sinusoids in a signal that are close to one another in frequency. Since the minimum-discernible frequency difference in a sampled spectral representation is $\Delta f = (NT_s)^{-1}$ Hz, two discernible peaks can appear in a spectral plot only if they are at least $2\Delta f$ Hz apart (peak – dip – peak). To achieve that we may need to decrease Δf by increasing NT_s, which is the time duration of the sampled input signal being analyzed. *Longer analysis time gives higher frequency resolution.*

4.2.1.2 *Periodic sequence*

Per Eq. (4.53), the Fourier transform of periodic sequence $x_p(n)$ is

$$\mathcal{F}\{x_p(n)\} = \frac{1}{N}\sum_{k=0}^{N-1} F_p(k)\, 2\pi\delta_{2\pi}(\omega - k\Delta\omega)$$

$$= \sum_{k=-\infty}^{\infty} \left\{\frac{2\pi}{N} F_p(mod(k,N))\right\} \delta(\omega - k\Delta\omega)$$

$$= \sum_{k=-\infty}^{\infty} \left\{\Delta\omega F_p(mod(k,N))\right\} \delta(\omega - k\Delta\omega), \qquad (4.64)$$

where $\Delta\omega = 2\pi/N$. As we see, the spectrum of $x_p(n)$ contains impulses located at uniformly-spaced frequencies $\omega = k\Delta\omega$ (k integer) and having areas $\Delta\omega F_p(mod(k,N))$. The following MATLAB® code demonstrates using the FFT algorithm to calculate and plot the spectrum of a periodic[aaa] sinusoidal sequence:

```
% generate 4 periods of cosine sequence (10 samples/period)
T0 = 10;
N = 4*T0;
n = 0:(N-1);
x = cos((2*pi/T0)*n);
figure
stem(n,x,'filled')
% calculate spectrum using FFT algorithm
dw = 2*pi/N;
X_impulse_areas = dw*fft(x);
k = 0:(N-1);
figure
stem(k,X_impulse_areas,'filled')
```

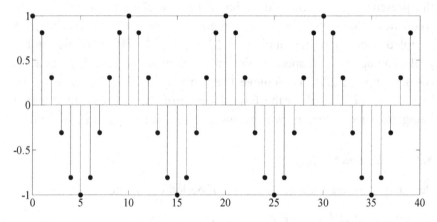

Figure 4.20. A plot of periodic discrete-time sequence $x(n) = \cos(2\pi n/10)$.

The calculated spectrum in Fig. 4.21 is one period of the periodic spectrum given as Entry #12 in Table 4.1,

$$\cos(\omega_0 n) \Leftrightarrow \pi\delta_{2\pi}(\omega + \omega_0) + \pi\delta_{2\pi}(\omega - \omega_0), \qquad (4.65)$$

where the cosine frequency $\omega_0 = 2\pi/T_0 = 2\pi/10$ rad/sec, and impulse at $k = 4$ represents the frequency $\omega = k\Delta\omega = k(2\pi/N) = 4(2\pi/40) =$

[aaa] Recall from Section 2.1.2.3 that not every sinusoidal sequence is periodic.

$2\pi/10$ rad/sec, which is the same as ω_0.[bbb] The impulse areas are correctly calculated to be π.

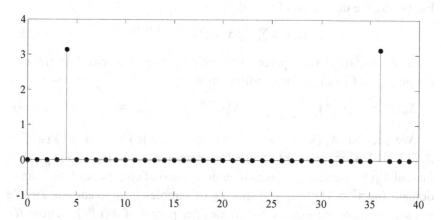

Figure 4.21. A plot of the spectrum of $x(n) = \cos(2\pi n/10)$, calculated using the Fast Fourier Transform (FFT).

4.2.1.3 *Finite-length sequence*

The frequency spectrum of sequence $x(n)$ is found using the discrete-time Fourier transform, given by Eq. (4.1):

$$X(e^{j\omega}) = \sum_{n=-\infty}^{\infty} x(n)e^{-j\omega n}. \qquad (4.66)$$

Practically all sequences that we analyze are of finite length (equal to zero outside of some range of sample index values n). Without loss of generality, assume that $x(n)$ is N samples in length and spans index range $0 \le n \le N - 1$. Then the DTFT expression simplifies to contain a sum of only N terms:

$$X(e^{j\omega}) = \sum_{n=0}^{N-1} x(n)e^{-j\omega n}. \qquad (4.67)$$

Note that since $x(n)$ is finite-length it cannot be periodic. How then may the FFT, which assumes that the time sequence being transformed is

---

[bbb] Impulses occur throughout the periodic spectrum at $m2\pi \pm \omega_0$, m integer (corresponding to index values $k = mN \pm 4$, m integer), hence explaining the presence of another impulse at frequency index $k = 36$.

periodic, be useful for us to efficiently calculate $X(e^{j\omega})$? The answer lies in the similarity between the Fourier transform expressions. Recall, from Eq. (4.56), the definition of the DFT:

$$X_p(k) = \sum_{n=0}^{N-1} x_p(n)e^{-j(k2\pi/N)n}. \tag{4.68}$$

In Eq. (4.66), if we replace $x(n)$ with $x_p(n)$ and ω with $k2\pi/N$, we obtain Eq. (4.67) under the condition that:

$$X_p(k) = X(e^{j\omega})\big|_{\omega=k2\pi/N} = X(e^{jk\Delta\omega}) \qquad (\Delta\omega = 2\pi/N). \tag{4.69}$$

We see that $X_p(k) = FFT_N\{x(n)\}$ is a sampled version of $X(e^{j\omega})$. $X_p(k)$ no longer represents {impulse areas $\div \Delta\omega$} as for $FFT\{x_p(n)\}$; instead $X_p(k)$ are samples taken over one period of continuous frequency-domain function $X(e^{j\omega})$. The spacing of samples in frequency is $2\pi/N$ rad/sec, so there are exactly N samples per period of $X(e^{j\omega})$. These N samples contain all the information contained in the continuous function $X(e^{j\omega})$, which means that $X(e^{j\omega})$ can be exactly reconstructed from the N samples that $X_p(k)$ represents $(k = 0, 1, ..., N-1)$.[ccc]

To produce nicer-looking graphs we may calculate samples of $X(e^{j\omega})$ at finer frequency resolution. This is done by choosing FFT size N greater than the number of nonzero samples in $x(n)$. The technique is called *zero-padding* the sequence $x(n)$ prior to taking its FFT (concatenating extra time-domain samples that are all zeros). No new information is gained by doing this.

The following two figures show an extreme example of how zero-padding may improve the display of a spectrum based on its samples. In this example $x(n) = 0$ except for samples in the range $0 \le n \le 4$, where the sample values are: {0.0975, 0.2785, 0.5469, 0.9575, 0.9649}. Taking a 5-point FFT yields $X_p(k) = \{2.8453, -0.7353 + 0.8942j, -0.4435 +$

[ccc] The IFFT (or IDFT) recovers N samples in time $(x(n), 0 \le n \le N-1)$ from the N spectral samples $(X_p(k), 0 \le k \le N-1)$. Once $x(n)$ is found this way, then the DTFT gives us $X(e^{j\omega}) = \sum_{n=0}^{N-1} x(n)e^{-j\omega n}$. Here we are sampling in the frequency domain; the same concepts apply to sampling in the time-domain, which is the topic of Ch. 6.

0.0129j, −0.4435 − 0.0129j, −0.7353 − 0.8942j}. The magnitudes of these five spectral samples are plotted in Fig. 4.22.

```
x = [0.0975, 0.2785, 05.469, 0.9575, 0.9649];
N = length(x);
% second argument N is optional when equal to length(x):
Xp = fft(x,N);
w = linspace(0,2*pi,N); w(end) = [];
plot(w,abs(Xp))
```

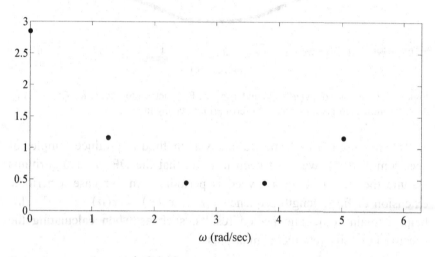

ω (rad/sec)

Figure 4.22. Samples of $|X(e^{j\omega})| = |\mathcal{F}\{x(n)\}|$ found using the FFT method, when $x(n) = \{0.0975, 0.2785, 0.5469, 0.9575, 0.9649\}$ for $0 \le n \le 4$ and $x(n) = 0$ elsewhere.

Next, the same five samples together with 100 zeros are processed using a 105-pt. FFT, and the magnitude of its results shown in Fig. 4.23. Theoretically both Figs. 4.22 and 4.23 contain the same information, but the latter better visually represents magnitude spectrum $|X(e^{j\omega})|$.

```
x = [0.0975, 0.2785, 05.469, 0.9575, 0.9649];
% zero-pad 5 samples to a total length of N = 105:
N = 150;
Xp = fft(x,N);
w = linspace(0,2*pi,N); w(end) = [];
plot(w,abs(Xp))
```

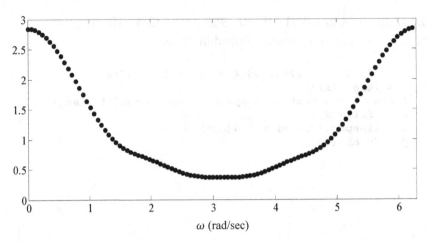

Figure 4.23. Samples of $|X(e^{j\omega})|$ found using the FFT method for the same $x(n)$ as in Fig. 4.22, this time zero-padding with 100 zeros prior to taking the FFT.

When using the FFT spectral analysis method to produce samples of spectrum $X(e^{j\omega})$, we must keep in mind that the DFT/FFT algorithms assume the signal being analyzed is periodic – in our case a periodic extension of finite-length sequence $x(n)$: $x_p(n) = x(n) * \delta_N(n)$. This helps explain some unexpected results, such as when calculating the spectrum of a discrete rectangular pulse:

```
% let x(n) be a delayed rect pulse:
% x(n) = 1 for n = [0,4], x(n) = 0 elsewhere.
w = linspace(0,2*pi,6); w(6) = [];
plot(w,abs(fft(ones(1,5))),'o')
axis([0 2*pi -2 6])
```

The resulting plot is shown in Fig. 4.24. It seems there is only energy at $\omega = 0$ (and other integer multiples of 2π rad/sec), and this is indeed true when one considers that the periodic extension of the sequence $[1,1,1,1,1]$ is the constant sequence $x_p(n) = 1 \ \forall n$, whose discrete-time Fourier transform is $2\pi\delta_{2\pi}(\omega)$.[ddd]

[ddd] Figure 4.24 shows $X_p(0) = 5$ instead of 2π because $X_p(k) = (N/2\pi) \times$ impulse areas in $\mathcal{F}\{x_p(n)\}$.

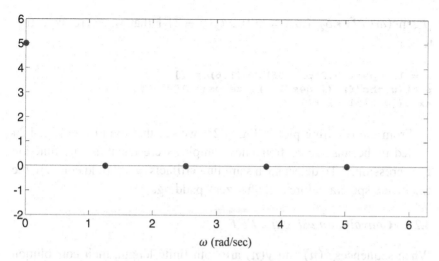

Figure 4.24. Samples of $|X(e^{j\omega})| = |\mathcal{F}\{x(n)\}|$ found using the FFT method, when $x(n) = \{1, 1, 1, 1, 1\}$ for $0 \leq n \leq 4$ and $x(n) = 0$ elsewhere.

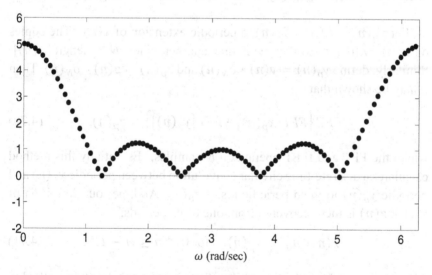

Figure 4.25. Samples of $|X(e^{j\omega})|$ found using the FFT method for the same $x(n)$ as in Fig. 4.24, this time zero-padding with 100 zeros prior to taking the FFT.

Repeating the same example, only this time padding the sequence $[1, 1, 1, 1, 1]$ with 100 zeros, shows us samples of magnitude spectrum

$|T \sin c(\omega T/2) * \delta_{2\pi}(\omega)| = |\mathcal{F}\{rect_2(n-2)\}|$ that we were expecting to see:

```
w = linspace(0,2*pi,106); w(106) = [];
plot(w,abs(fft([ones(1,5), zeros(1,100)])),'o')
axis([0 2*pi -2 6])
```

From the resulting plot in Fig. 4.25, we see that the plot in Fig. 4.24 misled us because some frequency samples were taken at sinc function zero-crossings! To detect such sampling artifacts one should compare the calculated spectra before and after zero-padding.

4.2.2 Convolution using the FFT

When sequences $x(n)$ and $y(n)$ are both finite-length, their convolution product $z(n) = x(n) * y(n)$ will also be finite-length. As discussed in Section 2.1.5.2, length$\{z(n)\} = N = $ length$\{x(n)\} + $ length$\{y(n)\} - 1$.

Let $x_p(n) = x(n) * \delta_N(n)$, a periodic extension of $x(n)$. The copies of $x(n)$ will not overlap with one another since $N > $ length$\{x(n)\}$. Similarly, define $y_p(n) = y(n) * \delta_N(n)$ and $z_p(n) = z(n) * \delta_N(n)$. Then it may be shown that

$$IFFT\left\{FFT\{x_p(n)\} \cdot FFT\{y_p(n)\}\right\} = z_p(n), \qquad (4.70)$$

where the FFT and IFFT operate on N samples. Essentially this method calculates a discrete-time circular convolution between periodic $x_p(n)$ and periodic $y_p(n)$ to yield periodic result $z_p(n)$. And, per our definition of $z_p(n)$, $z(n)$ is then recovered from one of its periods:[eee]

$$z(n + n_0) = z_p(n), \quad \text{for } 0 \le n \le N - 1, \qquad (4.71)$$

where n_0 depends on the time shifts of $x(n)$ and $y(n)$. For this method to work, both $x(n)$ and $y(n)$ must be zero-padded to at least N points (the sum of their lengths -1) before calculating their FFT's, so that one period of $z_p(n)$ is large enough to store sequence $z(n)$.

[eee] This is what MATLAB® does in its discrete convolution function **conv**.

Example 4.6

Given: $x(n) = [1, 1, 1, 1, 1]$ over the range $-2 \leq n \leq 2$ and zero elsewhere; and $y(n) = [1, 2, 3]$ over the range $2 \leq n \leq 4$ and zero elsewhere. Write some MATLAB® code to calculate convolution product $z(n) = x(n) * y(n)$ using the FFT method.

```
% Example Ch4_6
x = [1 1 1 1 1];
y = [1 2 3];
N = length(x)+length(y)-1;
z = ifft(fft(x,N).*fft(y,N))    % same as z=conv(x,y)

z = 1.0000  3.0000  6.0000  6.0000  6.0000  5.0000  3.0000
```

Note that, per Section 2.1.5.2, the calculated samples of $z(n)$ fall in the time index range $0 \leq n \leq 6$ ($n_{xmin} + n_{ymin} \leq n \leq n_{xmax} + n_{ymax}$).

4.2.3 *Autocorrelation using the FFT*

From Eq. (2.49) on p. 36, the autocorrelation of sequence $x(n)$ may be expressed as a convolution product:

$$\phi_{xx}(n) = x^*(-n) * x(n). \tag{4.72}$$

When $x(n)$ has length M and lies in the range $m_0 \leq n \leq m_0 + M - 1$, $x^*(-n)$ will lie in the range $-m_0 - M + 1 \leq n \leq -m_0$ and have length M. Convolution product of $x^*(-n)$ and $x(n)$ will have length $2M - 1$ (always an odd number) and lie in index range $-(M - 1) \leq n \leq (M - 1)$. Therefore, the autocorrelation of finite-length $x(n)$ will be a finite-length sequence that is centered at $n = 0$.

Since $x^*(-n) \Leftrightarrow X^*(e^{j\omega})$ from Table 4.2 Properties #10 and #11, we see that the FFT-based convolution described in the previous section may be adapted to calculating autocorrelation of finite-length sequence $x(n)$ as

$$IFFT\left\{\left(FFT\{x_p(n)\}\right)^* \cdot FFT\{x_p(n)\}\right\}, \tag{4.73}$$

which is the same as:

$$IFFT\left\{\left|FFT\{x_p(n)\}\right|^2\right\} \tag{4.74}$$

Example 4.7
Given: $x(n) = [1, 2, 3, 4, 5]$ over the range $0 \le n \le 4$ and zero elsewhere. Write some MATLAB® code to calculate $\phi_{xx}(n)$ using the FFT method.

```
% Example Ch4_7
x = [1 2 3 4 5];
N = 2*length(x)-1;
phi_xx = ifft(abs(fft(x,N)).^2)

phi_xx = 55.0000 40.0000 26.0000 14.0000 5.0000 5.0000
14.0000 26.0000 40.0000
```

How do we interpret this result? As stated above, we know that the time range of $\phi_{xx}(n)$ in this case will be $-4 \le n \le 4$ ($M = 5$). The first value 55.000 corresponds to $\phi_{xx}(0)$, which is the *zero-lag* autocorrelation value that is the energy of sequence $x(n)$.[fff] Considering that the IFFT output is one period of a periodic sequence, we conclude that the corresponding time index values are $n = [0, 1, 2, 3, 4, -4, -3, -2, -1]$. Because this is a bit confusing, MATLAB® provides a function to circularly shift the output of IFFT (or FFT) operation so that the presumptive index values are in proper sequence: `fftshift`.

Here is the revised line of code to calculate autocorrelation of $x(n)$, having length M samples, and present the results sequentially between $-(M-1)$ to $+(M-1)$ time index values:

```
phi_xx = fftshift(ifft(abs(fft(x,N)).^2))

phi_xx = 5.0000 14.0000 26.0000 40.0000 55.0000 40.0000
26.0000 14.0000 5.0000
```

This autocorrelation result is over time index range $-4 \le n \le 4$.

[fff] See p. 37.

4.2.4 *Discrete Cosine Transform (DCT)*

Consider periodic sequence $x_p(n) = x_p(n + N)$, which is also even: $x_p(n) = x_p(-n)$. Then the DFT of this sequence is:

$$DFT\{x_p(n)\} = X_p(k) = \sum_{n=0}^{N-1} x_p(n)e^{-j(k2\pi/N)n}$$

$$= \sum_{n=0}^{N-1} x_p(n)\{\cos(k2\pi n/N) - j\sin(k2\pi n/N)\}$$

$$= \sum_{n=0}^{N-1} x_p(n)\cos(k2\pi n/N)$$

$$-j\sum_{m=0}^{N-1} x_p(n)\sin(k2\pi n/N)$$

$$= \sum_{n=0}^{N-1} x_p(n)\cos(k2\pi n/N), \quad 0 \le k \le N - 1. \quad (4.75)^{ggg}$$

When $N = \#$ samples/period is an even number, all but two of these N samples come in pairs since $x_p(n) = x_p(-n)$. We may therefore use this transform to map $M = 2 + (N - 2)/2 = (N + 2)/2$ unique samples in time to the same number of unique samples in frequency.[hhh] When period N is an odd number, all but one of these N samples come in pairs and we may map $M = 1 + (N - 1)/2 = (N + 1)/2$ unique samples in time to that many unique samples in frequency. And because the DFT is invertible, the frequency-domain samples may be processed to recover the original time-domain samples.

This algorithm and its variants are called the "Discrete Cosine Transform" (DCT), since the transform formula includes only the cosine function. The DCT is used to transform M real coefficients in time to M real coefficients in frequency without requiring any complex math when $x_p(n)$ is real. All the properties of the discrete-time Fourier transform also apply to the DCT. The DCT has a 2-dimensional form that is widely used for data compression of digital images.[iii]

[ggg] $x_p(n)\sin(k2\pi n/N)$ is the product of even and odd periodic sequences, and therefore it must be odd and periodic (sum of samples in one period = 0).

[hhh] since information content in both domains must be equal.

[iii] JPEG image compression algorithm.

4.3 Useful MATLAB® Code

4.3.1 *Plotting the spectrum of a discrete-time signal*

The spectrum of a discrete-time signal, $\mathcal{F}\{h(n)\} = H(e^{j\omega})$, is periodic. There is nothing to be gained by plotting more than one period other than to point out the periodicity. In fact, when the time sequence $h(n)$ is real (usually the case), then $|H(e^{j\omega})|$ is even and $\angle H(e^{j\omega})$ is odd, it is unnecessary to plot more than half of a period of $H(e^{j\omega})$. Thus, we normally plot only the positive frequency side of the period centered at $\omega = 0$, which on a normalized frequency scale corresponds to frequency range $0 \leq \omega \leq \pi$ rad/sec.

Example 4.8
Given: $H(e^{j\omega}) = 0.634(1 + e^{-j\omega})/(1 + 0.268e^{-j\omega})$. Plot its magnitude and phase of over frequency range $0 \leq \omega \leq \pi$ rad/sec.

```
% Example_Ch4_8
Npts = 1e4; % specify # samples of H(e^jw) to be plotted
w = linspace(0,pi,Npts);
H = 0.634*(1 + exp(-j*w))./(1 + 0.268*exp(-j*w));
figure  % open new figure window
plot(w,abs(H))
xlabel('frequency in rad/sec')
ylabel('magnitude of H(e^j^\omega)')
title(['H(e^j^\omega) = ', ...
       '0.634(1+e^-^j^\omega)/(1+0.268e^-^j^\omega)'])
grid on

figure  % open new figure window
plot(w,angle(H))
xlabel('frequency in rad/sec')
ylabel('angle of H(e^j^\omega) in radians')
title(['H(e^j^\omega) = ', ...
       '0.634(1+e^-^j^\omega)/(1+0.268e^-^j^\omega)'])
grid on
```

The resulting graphs of magnitude spectrum $|H(e^{j\omega})|$ and phase spectrum $\angle H(e^{j\omega})$ are shown in Figs. 4.26 and 4.27, respectively.

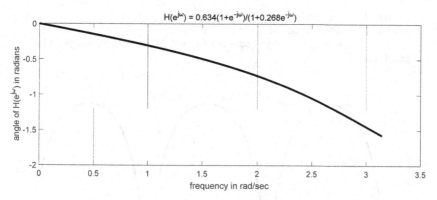

Figure 4.26. Plot of phase spectrum $\angle H(e^{j\omega})$ from Example 4.8.

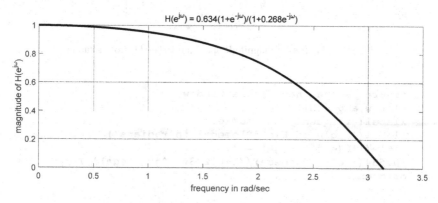

Figure 4.27. Plot of magnitude spectrum $|H(e^{j\omega})|$ from Example 4.8.

Example 4.9

As in Example 4.8, $H(e^{j\omega}) = 0.634(1 + e^{-j\omega})/(1 + 0.268e^{-j\omega})$. Plot its magnitude and phase, this time over three periods (frequency range $-3\pi \le \omega \le 3\pi$ rad/sec):

```
% Example_Ch4_9
Npts = 1e4;  % specify # samples of H(e^jw) to be plotted
w = linspace(-3*pi,3*pi,Npts);
H = 0.634*(1 + exp(-j*w))./(1 + 0.268*exp(-j*w));
figure  % open new figure window
plot(w,abs(H))
xlabel('frequency in rad/sec')
```

```
ylabel('magnitude of H(e^j^\omega)')
title(['H(e^j^\omega) = ', ...
       '0.634(1+e^-^j^\omega)/(1+0.268e^-^j^\omega)'])
grid on
```

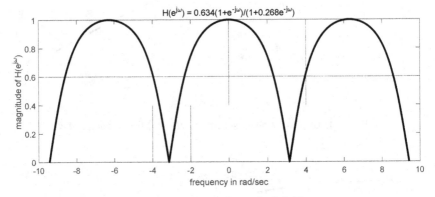

Figure 4.28. Plot of magnitude spectrum $\left|H\left(e^{j\omega}\right)\right|$ from Example 4.9.

```
figure  % open new figure window
plot(w,angle(H))
xlabel('frequency in rad/sec')
ylabel('angle of H(e^j^\omega) in radians')
title(['H(e^j^\omega) = ', ...
'0.634(1+e^-^j^\omega)/(1+0.268e^-^j^\omega)'])
grid on
```

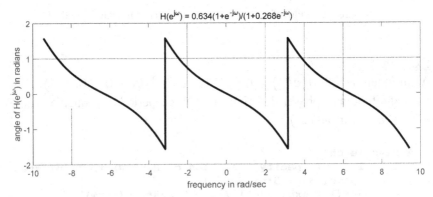

Figure 4.29. Plot of phase spectrum $\angle H\left(e^{j\omega}\right)$ from Example 4.9.

When plotting in MATLAB it may be useful to change the labels of plot coordinates so that the plot is easier to understand. For example, instead of displaying frequency in radians/sec over the range $[0, \pi]$ we can display normalized fractional frequency as ω/ω_s over the range $[0, 1/2]$.[iii] This is shown in the following example.

Example 4.10

Given: $H(e^{j\omega}) = 0.18(1 - e^{-j2\omega})/(1 - 0.50e^{-j\omega} + 0.64e^{-j2\omega})$. Plot its magnitude & phase over normalized frequency range $0 \leq \omega/\omega_s \leq 0.5$:

```
% Example_Ch4_10
Npts = 1e4;   % specify # samples of H(e^jw) to be plotted
wn = linspace(0,0.5,Npts);
w = 2*pi*wn;
H = 0.18*(1 - exp(-j*2*w))./(1 - 0.50*exp(-j*w) ...
+ 0.64*exp(-j*2*w));
figure   % open new figure window
plot(wn,abs(H))
xlabel('normalized frequency \omega/\omega_s')
ylabel('magnitude of H(e^j^\omega)')
title(['H(e^j^\omega) = 0.18(1-e^-^j^2^\omega)', ...
   '/(1-0.50e^-^j^\omega+0.64e^-^j^2^\omega)'])
grid on
```

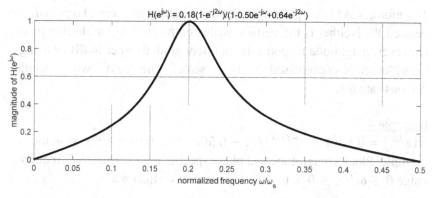

Figure 4.30. Plot of magnitude spectrum $|H(e^{j\omega})|$ from Example 4.10.

---

iii Constant ω_s is the "sampling frequency" that is explained in Chapter 6. Using label ω/ω_s makes the graph's frequency axis markings independent of the value of ω_s.

```
figure  % open new figure window
plot(wn,angle(H)*180/pi)
xlabel('normalized frequency \omega/\omega_s')
ylabel('phase of H(e^j^\omega) in degrees')
title(['H(e^j^\omega) = 0.18(1-e^-^j^2^\omega)', ...
    '/(1-0.504e^-^j^\omega+0.64e^-^j^2^\omega)'])
grid on
```

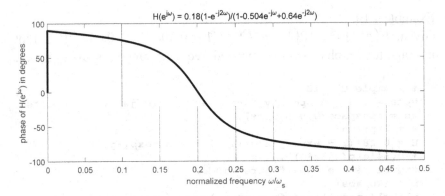

Figure 4.31.   Plot of phase spectrum $\angle H\left(e^{j\omega}\right)$ from Example 4.10.

Plotting magnitude and phase of periodic function $H(e^{j\omega})$ is normally not done on a logarithmically-compressed horizontal frequency scale. Therefore, MATLAB plotting functions **loglog** and **semilogx** will not be used.[kkk] Neither is the vertical scale compressed when plotting phase. However, magnitude response is routinely plotted either in dB or using a logarithmically-compressed vertical scale. The next two examples demonstrate this.

Example 4.11

$H\left(e^{j\omega}\right) = 0.18\left(1 - e^{-j2\omega}\right)/(1 - 0.50e^{-j\omega} + 0.64e^{-j2\omega})$, as in the last example. Plot its magnitude and phase spectra, over normalized frequency range $0 \le \omega/\omega_s \le 0.5$, using a logarithmic vertical scale:

```
% Example_Ch4_11
Npts = 1e4;  % specify # samples of H(e^jw) to be plotted
wn = linspace(0,0.5,Npts);
```

[kkk] As they often are to plot frequency spectra of continuous-time signals (see MATLAB examples at the end of Chapter 5).

```
w = 2*pi*wn;
H = 0.18*(1 - exp(-j*2*w))./(1 - 0.50*exp(-j*w) ...
+ 0.64*exp(-j*2*w));
figure  % open new figure window
semilogy(wn,abs(H).^2)
axis([0 0.5 1e-2 1e0])  % change vertical axis limits
                        % for better-looking plot
xlabel('normalized frequency \omega/\omega_s')
ylabel('|H(e^j^\omega)|^2')
title(['H(e^j^\omega) = 0.18(1-e^-^j^2^\omega)', ...
       '/(1-0.50e^-^j^\omega+0.64e^-^j^2^\omega)'])
grid on
```

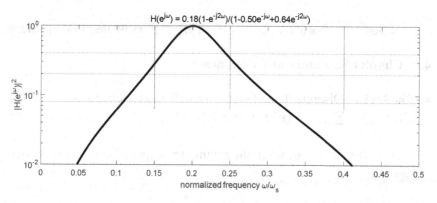

Figure 4.32. Plot of magnitude spectrum $\left|H\left(e^{j\omega}\right)\right|$ from Example 4.11.

Example 4.12

Plot the magnitude-squared value of $H\left(e^{j\omega}\right) = 0.5 + 0.5e^{-j5\omega}$, over normalized frequency range $0 \le \omega/\omega_s \le 0.5$, using a dB vertical scale:

```
% Example_Ch4_12
Npts = 1e4;  % specify # samples of H(e^jw) to be plotted
wn = linspace(0,0.5,Npts);
w = 2*pi*wn;
H = 0.5 + 0.5*exp(-j*5*w);
figure; plot(wn,20*log10(abs(H)))
axis([0 0.5 -50 10])  % change vertical axis limits
                      % for better-looking plot
xlabel('normalized frequency \omega/\omega_s')
ylabel('|H(e^j^\omega)|^2 (dB)')
title('H(e^j^\omega) = 0.5 + 0.5e^-^j^5^\omega')
grid on
```

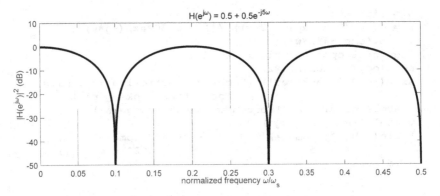

Figure 4.33. Plot of magnitude spectrum $\left|H\left(e^{j\omega}\right)\right|^2$ in dB, from Example 4.12.

4.4 Chapter Summary and Comments

- The DTFT (Discrete-Time Fourier Transform) is defined as:
$\mathcal{F}\{f(n)\} = \sum_{n=-\infty}^{\infty} f(n)e^{-j\omega n} = F(e^{j\omega})$

- $F(e^{j\omega})$ does not exist when the summation expression defining it does not converge.

- $F(e^{j\omega})$ is periodic, having period 2π rad/sec.

- When $F(e^{j\omega})$ exists, it contains the same information as does $f(n)$ only in a different domain.

- The spectra of basic discrete-time signals are periodically-extended versions of their continuous-time counterparts. For example, $rect_k(n) \Leftrightarrow T sinc(\omega T/2) * \delta_{2\pi}(\omega)$, where $T = 2k + 1$.

- The Inverse DTFT is defined as $f(n) = \frac{1}{2\pi}\int_{2\pi} F(e^{j\omega})e^{j\omega n}d\omega$, where the integral is over any period of $F(e^{j\omega})$.

- Tables of Fourier transforms and their properties make it possible to transform basic signals from time domain to frequency domain (and from frequency domain to time domain) using substitutions instead of summations and integrations.

- DTFT properties parallel those of the CTFT (continuous-time Fourier Transform) except for that multiplication of two sequences in time results in the circular convolution of their spectra in frequency: $f_1(n)f_2(n) \Leftrightarrow \frac{1}{2\pi}\int_{2\pi} F_1(e^{j\phi})F_2(e^{j(\phi-\omega)})d\phi$. When this convolution product is calculated in the frequency domain using the normal convolution integral, the result does not converge due to the periodicity of both $F_1(e^{j\omega})$ and $F_2(e^{j\omega})$.

- When $f(n) = f_p(n)$ is periodic, then $F(e^{j\omega})$ periodic and discrete (as if multiplied by an impulse train in frequency). The transformation between the N samples that form one period of $f_p(n)$ to the N samples that form one period of $F_p(k)$ (= list of impulse areas $\times 2\pi/N$ in DTFT$\{f_p(n)\}$) is called the Discrete Fourier Transform (DFT).

- DFT: $F_p(k) = \sum_{n=0}^{N-1} f_p(n)e^{-jk2\pi n/N}$, which needs to be calculated only over $k = 0, 1, \ldots, N-1$ (one period of the spectrum).

- IDFT: $f_p(n) = \frac{1}{N}\sum_{k=0}^{N-1} F_p(k)e^{jk2\pi n/N}$, which needs to be calculated only over $n = 0, 1, \ldots, N-1$ (one period of the time-domain signal).

- A very efficient implementation of the DFT (IDFT) is the FFT (IFFT).

- The FFT (Fast Fourier Transform) speeds up processing time by factor $N/log_2(N)$ as compared to a direct implementation of the DFT. For example, the FFT is 50,000 times faster than the DFT when $N = 1$ million samples.

- It is usually faster to perform discrete-time convolution between two long finite-length sequences by using the FFT to convert to the frequency domain, multiply spectra, and then convert back to the time domain with the IFFT. The speed of the FFT algorithm has revolutionized digital signal processing. The FFT is often a built-in function on specialized digital signal processing chips (DSP processors).

4.5 Homework Problems

P4.1 Calculate the DTFT of each $f(n)$ using the definition
$\mathcal{F}\{f(n)\} = \sum_{n=-\infty}^{\infty} f(n)e^{-j\omega n}$:
a) $f(n) = 3\delta(n)$
b) $f(n) = \delta(n-1) + \delta(n+1)$
c) $f(n) = \text{rect}_1(n)$
d) $f(n) = \Delta_3(n)$

P4.2 Calculate the IDTFT of each $F(e^{j\omega})$ using the definition
$\mathcal{F}^{-1}\{F(e^{j\omega})\} = (1/2\pi) \int_{2\pi} F(e^{j\omega})e^{j\omega n} d\omega$:
a) $F(e^{j\omega}) = \delta_{2\pi}(\omega)$
b) $F(e^{j\omega}) = \delta_\pi(\omega - \pi/2)$
c) $F(e^{j\omega}) = (\delta(\omega-1) + \delta(\omega+1)) * \delta_{2\pi}(\omega)$
d) $F(e^{j\omega}) = \text{rect}(\omega) * \delta_{2\pi}(\omega)$

P4.3 Given: $f(n) \Leftrightarrow F(e^{j\omega})$, $g(n) \Leftrightarrow G(e^{j\omega})$. Using the property of linear superposition, find the Fourier transforms of each of the following, expressed in terms of $F(e^{j\omega})$ and $G(e^{j\omega})$:
a) $f(n) + g(n)$
b) $3f(n) - 2g(n)$
c) $f(n) + \delta(n)g(n)$
d) $f(n)\{1 + g(n)\delta(n-1)\}$

P4.4 By applying linear superposition to the DTFT pairs given in Table 4.1, find Fourier transforms of each of the following:
a) $1 - u(n) + \delta(n)$
b) $2u(n) - \delta(n) - 1$
c) $0.5^n u(n) + 0.5^{-n} u(-n)$
d) $0.5\,\delta(n-10) + 0.5\,\delta(n+10)$
e) $j\cos(10n - \pi/2)$
f) $\sin(2n) - j\cos(2n)$
g) $(\text{rect}_5(n) - \Delta_6(n))/2$
h) $2\text{sinc}(n\pi) + \text{sinc}^2(n\pi/2)$
i) $2\delta(n) + \text{sinc}^2(n\pi/2)$
j) $\delta_4(n) - \delta_2(n)$

P4.5 By applying properties of the DTFT given in Table 4.2, find Fourier transforms of each of the following:

a) $0.5^{n-1}u(n-1)$

b) $\delta(n+1) * \text{sgn}(n)$

c) $u(3n)$

d) $\cos(n)\cos(2n)$

e) $-2je^{j2(n-1)}$

f) $(1/2\pi)\,\text{sinc}(n/4)\cos(n/2)$

g) $\text{rect}_2(n)e^{-j2n}$

h) $\text{sinc}(\pi n) * \text{sinc}(\pi n/2)$

i) $\text{sinc}(\pi n/2)\,\text{sinc}(\pi n/2)$

j) $\text{sinc}^2(\pi n/4)$

k) $u(-n)$

l) $\cos(n\pi/4)\,e^{j\pi/4}$

m) $\text{sgn}(n) - \text{sgn}(n-1)$

n) $\sum_{k=-\infty}^{n}\Delta_{100}(k)$

P4.6 By applying the DTFT pairs and DTFT properties given in Tables 4.1 and 4.2, find the inverse Fourier transforms of each of the following:

a) $\delta_{2\pi}(\omega)$

b) $\delta_{2\pi}(\omega - \pi)$

c) $(\pi/4)\text{rect}(2\omega/\pi) * \delta_{2\pi}(\omega)$

d) $\text{sinc}(2.5\omega) * \delta_{2\pi}(\omega)$

e) $e^{-j5\omega}$

f) $e^{j5\omega} + e^{-j5\omega}$

g) $\cos(5\omega)$

h) $(1/\omega) * \delta_{2\pi}(\omega)$

i) $(1 + e^{j\omega})/(1 - e^{j\omega})$

j) $\delta_{2\pi/3}(\omega)$

P4.7 Given: $\sum_{n=-\infty}^{\infty}|f(n)|^2 = 1$. What must be the area of $|F(e^{j\omega})|^2$ over $0 \le \omega \le 2\pi$?

P4.8 What is the energy of sequence $\text{sinc}(\pi n/2)$?

P4.9 What is the energy of sequence $\mathrm{sinc}(\pi n/2) + \mathrm{sinc}(\pi n/4)$?

P4.10 What is the power of periodic sequence $\mathrm{rect}_4(n) * \delta_{20}(n)$?

P4.11 What is the power of periodic sequence $\Delta_4(n) * \delta_{20}(n)$?

P4.12 Calculate the DFT of each $f_p(n)$ given below using definition
$F_p(k) = \sum_{n=0}^{N-1} f_p(n)e^{-j(2\pi kn/N)}$ (specify $F_p(k)$ over its period
$0 \le k < N$; your first task will be to determine N):
a) $f_p(n) = \delta_4(n)$
b) $f_p(n) = \delta_4(n-2)$
c) $f_p(n) = \delta(n-3) * \delta_4(n)$
d) $f_p(n) = \mathrm{rect}_1(n) * \delta_4(n)$

P4.13 Calculate the IDFT of each $F_p(k)$ given below using definition
$f_p(n) = (1/N) \sum_{n=0}^{N-1} F_p(k)e^{j(2\pi kn/N)}$ (specify $f_p(n)$ over its
period $0 \le n < N$; your first task will be to determine N):
a) $F_p(k) = \delta_3(k)$
b) $F_p(k) = \delta_3(k+1)$
c) $F_p(k) = \mathrm{rect}_1(n) * \delta_3(n)$
d) $F_p(k) = (-1)^k$

P4.14 Plot the magnitude-squared value of $H(e^{j\omega}) =$
$2(e^{j\omega} - 0.95)/(e^{j\omega} - 0.90)$, over normalized frequency range
$0 \le \omega/\omega_s \le 0.5$, using a dB vertical scale.

Chapter 5

Frequency Analysis of Continuous-Time Signals

5.1 Theory

5.1.1 *Fourier Transform*

Frequency analysis may be used to better understand, compare, and manipulate signals in time. It is done decomposing a time-varying signal into a summation of sinusoids. With that representation, systems that process time-varying signals may be developed based on the understanding of how they affect a signal's individual sinusoidal components.

Frequency analysis is based on the theory developed by Joseph Fourier (1768–1830) in connection with his research on heat. We will view the Fourier transform as a mapping of a signal in the time domain to its uniquely associated components in the frequency domain. The mathematical relationship between these two functions $f(t)$ and $F(\omega)$ is known at the Continuous-Time Fourier Transform (CTFT):

Fourier Transform (CTFT)

$$F(\omega) = \int_{-\infty}^{\infty} f(t)e^{-j\omega t}dt, \qquad (5.1)$$

Inverse Fourier Transform

$$f(t) = \frac{1}{2\pi}\int_{-\infty}^{\infty} F(\omega)e^{j\omega t}d\omega. \qquad (5.2)$$

Or, as written in shorthand notation:

$$F(\omega) = \mathcal{F}\{f(t)\}, \qquad (5.3)$$

$$f(t) = \mathcal{F}^{-1}\{F(\omega)\}. \qquad (5.4)$$

$F(\omega)$ is called a *frequency spectrum* because it describes $f(t)$ as a summation of sinusoids having frequency values over the entire spectrum of possibilities. Likewise, $|F(\omega)|$ and $\angle F(\omega)$ are called the *magnitude spectrum* and *phase spectrum*, respectively.

We are interested in signals whose Fourier transforms are invertible, meaning that both the Fourier transform integral and its inverse exist to give a unique mapping between $f(t)$ and $F(\omega)$. When that is true, the following notation is used:

$$f(t) \Leftrightarrow F(\omega) \qquad (5.5)$$

Practically speaking, signals that may be measured and recorded are those having invertible Fourier transforms.[a] Engineers often call these *real-world* signals in the sense that they are continuous, have finite amplitudes, and have finite energy. But many signals of interest for theoretical analysis do not fall in this category. For example: a waveform that is constant with respect to time, such as that corresponding to a d.c. voltage, has a Fourier integral (Eq. (5.1)) that does not converge at $\omega = 0$. Likewise, the Fourier integral of a periodic signal does not converge for at least one value of ω. Fortunately, in most cases of interest the problem of Fourier integral non-convergence may be overcome, and the Fourier transform made invertible, by permitting the use of generalized functions[b] such as $\delta(x)$ in both time and frequency domains.

Note that, although we discuss Fourier analysis in terms of time variation and frequency content, t and ω are two independent variables that may have different interpretations in other fields of science, engineering and mathematics where Fourier analysis is used. As a matter of fact, in some applications such as digital image processing, Fourier

---

[a] Identifying the class of signals whose Fourier transforms exist and are invertible is an advanced mathematical topic.

[b] The Fourier transform of all generalized functions has been shown to exist (Lighthill).

analysis is even applied to multidimensional functions (e.g., a function of $\{x, y\}$ is mapped to a function of $\{\omega_x, \omega_y\}$).

5.1.2 *Fourier transforms of basic signals*

Let us now derive, step-by-step, the Fourier transforms of a few of the basic functions introduced previously in Ch. 3. For many other functions it is straightforward to find the Fourier transform by evaluating Eq. (5.1), which may be simplified with the help of a table of integrals.[c] In other instances, application of Fourier transform properties is an excellent tool available to us. Examples worked out in the following section will provide experience in such techniques.

5.1.2.1 *Exponentially decaying signal*

Let $x(t) = u(t)e^{-bt}$ (assuming real $b > 0$). Its Fourier transform $X(\omega)$ is then found according to the definition in Eq. (5.1):[d]

$$X(\omega) = \mathcal{F}\{u(t)e^{-bt}\} = \int_0^\infty e^{-bt}e^{-j\omega t}\,dt$$

$$= \int_0^\infty e^{-(b+j\omega)t}\,dt = \left.\frac{e^{-(b+j\omega)t}}{-(b+j\omega)}\right|_0^\infty = 0 - \frac{1}{-(b+j\omega)} = \frac{1}{b+j\omega}, \quad (5.6)$$

or $u(t)e^{-bt}$ and $1/(b+j\omega)$ comprise a "Fourier transform pair":

$$\boxed{u(t)e^{-bt} \Leftrightarrow \frac{1}{b+j\omega} \quad (b > 0).} \quad (5.7)$$

This causal signal, $x(t) = u(t)e^{-bt}$, when flipped about $t = 0$ becomes the anticausal $x(-t) = u(-t)e^{bt}$, whose Fourier transform is similarly shown to be:[e]

---

[c] Abramowitz *et al.*; Dwight.

[d] In this derivation, note that $\lim\limits_{t\to\infty} e^{-(b+j\omega)t} = \lim\limits_{t\to\infty} e^{-bt}e^{-j\omega t} = \lim\limits_{t\to\infty} e^{-bt} \cdot \lim\limits_{t\to\infty} e^{-j\omega t} = 0 \cdot \lim\limits_{t\to\infty} e^{-j\omega t} = 0$, since $b > 0$ and $\left|e^{-j\omega t}\right| = 1 < \infty \;\; \forall t$.

[e] $\mathcal{F}\{u(-t)e^{bt}\} = \int_{-\infty}^{0} e^{bt}e^{-j\omega t}\,dt = \int_{-\infty}^{0} e^{(b-j\omega)t}\,dt = \left.\frac{e^{(b-j\omega)t}}{b-j\omega}\right|_{-\infty}^{0} = \frac{1}{b-j\omega} - 0 = \frac{1}{b-j\omega}$ (when $b > 0$).

$$u(-t)e^{bt} \Leftrightarrow \frac{1}{b-j\omega} \qquad (b > 0) \qquad (5.8)$$

In preparation for the next derivation, we will calculate the Fourier transform of the following signal that is symmetric about $t = 0$:

$$\mathcal{F}\{u(t)e^{-bt} + u(-t)e^{bt}\} = \mathcal{F}\{u(t)e^{-bt}\} + \mathcal{F}\{u(-t)e^{bt}\}$$

$$= \frac{1}{b+j\omega} + \frac{1}{b-j\omega} = \frac{2b}{b^2+\omega^2} \qquad (b > 0). \qquad (5.9)$$

Note that $u(t)e^{-bt} + u(-t)e^{bt} = e^{-b|t|}$. We now have another useful Fourier transform pair:

$$e^{-b|t|} \Leftrightarrow \frac{2b}{b^2+\omega^2} \qquad (b > 0) \qquad (5.10)$$

5.1.2.2 Constant value

As mentioned previously, a constant waveform such as $x(t) = 1$ presents us with the problem that its Fourier transform integral does not converge at $\omega = 0$. This integral may be solved with the help of generalized functions, however. First, consider the frequency-domain function $2\pi\delta(\omega)$. What is its inverse Fourier transform? The sifting property makes Eq. (5.2) easy to solve:

$$\mathcal{F}^{-1}\{2\pi\delta(\omega)\} = \frac{1}{2\pi}\int_{-\infty}^{\infty} 2\pi\delta(\omega)e^{j\omega t}\,d\omega$$

$$= \int_{-\infty}^{\infty} \delta(\omega)e^{j0t}\,d\omega = \int_{-\infty}^{\infty} \delta(\omega)\,d\omega = 1. \qquad (5.11)$$

Therefore, if the Fourier transform is invertible then $\mathcal{F}\{1\} = 2\pi\delta(\omega)$. Let us verify this with the help of generalized function $x(t) = \lim_{b\to 0} e^{-b|t|} = 1$:

$$X(\omega) = \mathcal{F}\left\{\lim_{b\to 0} e^{-b|t|}\right\} = \lim_{b\to 0}\left(\frac{2b}{b^2+\omega^2}\right) = \begin{cases} 0, & \omega \neq 0; \\ \infty, & \omega = 0. \end{cases} \qquad (5.12)^{\text{f}}$$

What kind of function $X(\omega)$ is zero everywhere except at a single point $\omega = 0$? One candidate is the Dirac delta function $\delta(\omega)$ times a constant

---

[f] $\mathbb{Z}$ is the symbol that represents the set of integers (Zahlen is a German word for numbers).

(having unspecified area). To find the area of this impulse, note that the area of $2b/(b^2 + \omega^2)$ is independent of b:

$$\int_{-\infty}^{\infty} \frac{2b}{b^2+\omega^2} \, d\omega = 2\, \text{Tan}^{-1}\left(\frac{\omega}{b}\right)\Big|_{\omega=-\infty}^{\omega=\infty} = 2\pi. \qquad (5.13)$$

We confirm, therefore, that $x(t) = 1$ and $X(\omega) = 2\pi\, \delta(\omega)$ are a Fourier transform pair:

$$\boxed{1 \Leftrightarrow 2\pi\delta(\omega)} \qquad (5.14)$$

5.1.2.3 *Impulse function*

By substituting $\delta(t)$ for $f(t)$ in Eq. (5.1), the Fourier transform definition, we obtain:

$$F(\omega) = \int_{-\infty}^{\infty} \delta(t)\, e^{-j\omega t} \, dt = \int_{-\infty}^{\infty} \delta(t)\, e^{-j\omega 0} \, dt$$

$$= \int_{-\infty}^{\infty} \delta(t)(1)\, dt = 1. \qquad (5.15)$$

Using similar arguments as before, we may also show that the inverse Fourier transform of 1 is $\delta(t)$. We now have another Fourier transform pair:

$$\boxed{\delta(t) \Leftrightarrow 1} \qquad (5.16)$$

5.1.2.4 *Delayed impulse function*

Similarly, and just as easily, we may find the Fourier transform of $\delta(t - t_0)$:

$$\mathcal{F}\{\delta(t - t_0)\} = \int_{-\infty}^{\infty} \delta(t - t_0)e^{-j\omega t} \, dt = \int_{-\infty}^{\infty} \delta(t - t_0)e^{-j\omega t_0} \, dt$$

$$= e^{-j\omega t_0} \int_{-\infty}^{\infty} \delta(t - t_0) \, dt = e^{-j\omega t_0}. \qquad (5.17)$$

This gives us the Fourier transform pair:

$$\boxed{\delta(t - t_0) \Leftrightarrow e^{-j\omega t_0}} \qquad (5.18)$$

5.1.2.5 *Signum function*

The signum function is defined in Chapter 3 as

$$\text{sgn}(t) \equiv \begin{cases} -1, & t < 0; \\ 0, & t = 0; \\ 1, & t > 0. \end{cases} \tag{5.19}$$

Because the Fourier integral of $\text{sgn}(t)$ does not converge in the ordinary sense, we must consider using functions that approach $\text{sgn}(t)$ in the limit. One such generalized function, among many that may be considered, is:

$$\text{sgn}(t) = \lim_{b \to 0} \{u(t)e^{-bt} - u(-t)e^{bt}\} \tag{5.20}$$

Solving for $\mathcal{F}\{\text{sgn}(t)\}$ in this formulation:

$$\mathcal{F}\{\text{sgn}(t)\} = \lim_{b \to 0}\left(\int_{-\infty}^{\infty} u(t)e^{-bt}e^{-j\omega t}\, dt - \int_{-\infty}^{\infty} u(-t)e^{bt}e^{-j\omega t}\, dt\right)$$

$$= \lim_{b \to 0}\left(\int_{0}^{\infty} e^{-bt}e^{-j\omega t}\, dt - \int_{-\infty}^{0} e^{bt}e^{-j\omega t}\, dt\right)$$

$$= \lim_{b \to 0}\left(\int_{0}^{\infty} e^{-(b+j\omega)t}\, dt - \int_{-\infty}^{0} e^{(b-j\omega)t}\, dt\right)$$

$$= \lim_{b \to 0}\left(\frac{e^{-(b+j\omega)t}}{-(b+j\omega)}\Big|_{0}^{\infty} - \frac{e^{(b-j\omega)t}}{(b-j\omega)}\Big|_{-\infty}^{0}\right)$$

$$= \lim_{b \to 0}\left(\frac{1}{b+j\omega} + \frac{1}{-b+j\omega}\right) = \frac{2}{j\omega}. \tag{5.21}$$

One can also show[g] that $\mathcal{F}^{-1}\{2/j\omega\} = \text{sgn}(t)$. Therefore, we obtain the Fourier transform pair:

---

[g] $\mathcal{F}^{-1}\left\{\frac{2}{j\omega}\right\} = \frac{1}{2\pi}\int_{-\infty}^{\infty}\left(\frac{2}{j\omega}\right)e^{j\omega t}\, d\omega = \frac{1}{2\pi}\int_{-\infty}^{\infty}\left(\frac{2}{j\omega}\right)(\cos(\omega t) + j\sin(\omega t))\, d\omega =$

$\frac{1}{2\pi}\int_{-\infty}^{\infty}\left(\frac{2}{j\omega}\right)j\sin(\omega t)\, d\omega = \frac{1}{\pi}\int_{-\infty}^{\infty}|t|\,\text{sinc}(\omega t)d\omega = \begin{cases} \frac{t}{\pi}\int_{-\infty}^{\infty}\text{sinc}(\omega t)d\omega, & t > 0; \\ 0, & t = 0; \\ \frac{-t}{\pi}\int_{-\infty}^{\infty}\text{sinc}(\omega t)d\omega, & t < 0. \end{cases} =$

$\begin{cases} \frac{t}{\pi}\left(\frac{\pi}{t}\right), & t > 0; \\ 0, & t = 0; \\ \frac{-t}{\pi}\left(\frac{\pi}{t}\right), & t < 0. \end{cases} = \begin{cases} 1, & t > 0; \\ 0, & t = 0; \\ -1, & t < 0. \end{cases} = \text{sgn}(t).$

$$\boxed{\text{sgn}(t) \Leftrightarrow \frac{2}{j\omega}} \qquad (5.22)$$

5.1.2.6 *Unit step function*

As with $\text{sgn}(t)$, the Fourier transform of $u(t)$ can be found by expressing it in the form of a generalized function. Alternatively, now that $\mathcal{F}\{\text{sgn}(t)\}$ and $\mathcal{F}\{1\}$ are known, we may express $u(t)$ in terms of these functions and quickly find its Fourier transform using the linear superposition property[h]:

$$u(t) = \tfrac{1}{2}(\text{sgn}(t) + 1), \qquad (5.23)$$

$$\mathcal{F}\{u(t)\} = \mathcal{F}\{\tfrac{1}{2}(\text{sgn}(t) + 1)\} = \tfrac{1}{2}\mathcal{F}\{\text{sgn}(t)\} + \tfrac{1}{2}\mathcal{F}\{1\},$$

$$= \tfrac{1}{2}\left(\frac{2}{j\omega}\right) + \tfrac{1}{2}\left(2\pi\,\delta(\omega)\right) = \frac{1}{j\omega} + \pi\delta(\omega). \qquad (5.24)$$

Therefore, because this Fourier transform may be shown to be invertible, we have the Fourier transform pair[i]:

$$\boxed{u(t) \Leftrightarrow \pi\delta(\omega) + \frac{1}{j\omega}} \qquad (5.25)$$

5.1.2.7 *Complex exponential function*

To find the Fourier transform of complex exponential signal $e^{j\omega_0 t}$, which is essentially a phasor at frequency ω_0, we begin by solving for the inverse Fourier transform of frequency-shifted impulse function $\delta(\omega - \omega_0)$ using Eq. (5.2):

$$\mathcal{F}^{-1}\{\delta(\omega - \omega_0)\} = \frac{1}{2\pi}\int_{-\infty}^{\infty}\delta(\omega - \omega_0)\,e^{j\omega t}d\omega$$

$$= \frac{1}{2\pi}\int_{-\infty}^{\infty}\delta(\omega - \omega_0)\,e^{j\omega_0 t}d\omega$$

$$= \frac{e^{j\omega_0 n}}{2\pi}\int_{-\pi}^{\pi}\delta(\omega - \omega_0)\,d\omega = \frac{e^{j\omega_0 n}}{2\pi} \qquad (5.26)$$

---

[h] Linear superposition and other properties of the Fourier transform are discussed in Section 5.1.3.1 on page 168.

[i] $\mathcal{F}^{-1}\left\{\frac{1}{j\omega} + \pi\delta(\omega)\right\} = \mathcal{F}^{-1}\left\{\frac{1}{j\omega}\right\} + \mathcal{F}^{-1}\{\pi\delta(\omega)\} = \frac{1}{2}\text{sgn}(t) + \frac{1}{2} = u(t)$.

After multiplying both sides by 2π, we get the desired result: $2\pi\,\mathcal{F}^{-1}\{\delta(\omega-\omega_0)\} = \mathcal{F}^{-1}\{2\pi\,\delta(\omega-\omega_0)\} = e^{j\omega_0 t}$, or

$$\boxed{e^{j\omega_0 t} \Leftrightarrow 2\pi\,\delta(\omega-\omega_0)} \qquad (5.27)$$

5.1.2.8 *Sinusoid*

The Fourier transform of $A\cos(\omega_0 t + \theta)$ is derived by expressing it in terms of complex exponential functions using Euler's identity:

$$A\cos(\omega_0 t + \theta) = A\left(\tfrac{1}{2}e^{j(\omega_0 t+\theta)} + \tfrac{1}{2}e^{-j(\omega_0 t+\theta)}\right). \qquad (5.28)$$

Then, taking advantage of $e^{j\omega_0 t} \Leftrightarrow 2\pi\,\delta(\omega-\omega_0)$:

$$\mathcal{F}\{A\cos(\omega_0 t + \theta)\} = \tfrac{A}{2}\mathcal{F}\{e^{j(\omega_0 t+\theta)} + e^{-j(\omega_0 t+\theta)}\}$$

$$= \tfrac{A}{2}\mathcal{F}\{e^{j\omega_0 t}e^{j\theta} + e^{-j\omega_0 t}e^{-j\theta}\} = \tfrac{A}{2}e^{j\theta}\,\mathcal{F}\{e^{j\omega_0 t}\} + \tfrac{A}{2}e^{-j\theta}\mathcal{F}\{e^{-j\omega_0 t}\}$$

$$= Ae^{j\theta}\pi\,\delta(\omega-\omega_0) + Ae^{-j\theta}\pi\delta(\omega+\omega_0), \qquad (5.29)$$

or

$$\boxed{\begin{array}{c} \cos(\omega_0 t + \theta) \Leftrightarrow \\[4pt] e^{-j\theta}\pi\delta(\omega+\omega_0) + e^{j\theta}\pi\delta(\omega-\omega_0). \end{array}} \qquad (5.30)$$

Setting $\theta = \{0, -\pi/2\}$ gives us these two Fourier transform pairs:

$$\boxed{\cos(\omega_0 t) \Leftrightarrow \pi\delta(\omega+\omega_0) + \pi\delta(\omega-\omega_0)} \qquad (5.31)$$

$$\boxed{\sin(\omega_0 t) \Leftrightarrow j\pi\delta(\omega+\omega_0) - j\pi\delta(\omega-\omega_0)} \qquad (5.32)$$

Note that, as demonstrated above, the Fourier transform of a real even signal is real, and the Fourier transform of a real odd signal is imaginary.[j]

---

[j] With signals: even $\times$ even = even, even $\times$ odd = odd, odd $\times$ odd = even. Also, $\int_{-\infty}^{\infty} f_e(t)dt = 2\int_0^{\infty} f_e(t)dt$, and $\int_{-\infty}^{\infty} f_o(t)dt = 0$. Therefore $\mathcal{F}\{x(t)\} = \int_{-\infty}^{\infty} x(t)e^{-j\omega t}dt = \int_{-\infty}^{\infty}\left(x_e(t) + x_o(t)\right)(\cos(\omega t) - j\sin(\omega t))dt = \int_{-\infty}^{\infty} x_e(t)\cos(\omega t)\,dt + \int_{-\infty}^{\infty} x_o(t)\cos(\omega t)\,dt - j\int_{-\infty}^{\infty} x_e(t)\sin(\omega t)\,dt - j\int_{-\infty}^{\infty} x_o(t)\sin(\omega t)\,dt = 2\int_0^{\infty} x_e(t)\cos(\omega t)\,dt + 0 - j0 - 2j\int_0^{\infty} x_o(t)\sin(\omega t)\,dt.$ $\therefore \mathcal{F}\{x(t)\} = 2\int_0^{\infty}(x_e(t)\cos(\omega t) - jx_o(t)\sin(\omega t))dt,$

5.1.2.9 *Rectangular pulse function*

The Fourier transform of a rectangular pulse function is derived by direct integration, applying the definition given by Eq. (5.1):

$$F(\omega) = \int_{-\infty}^{\infty} \text{rect}(t/\tau)e^{-j\omega t}dt = \int_{-\tau/2}^{\tau/2}(1)e^{-j\omega t}dt$$

$$= \frac{e^{-j\omega t}}{-j\omega}\Big|_{t=-\frac{\tau}{2}}^{t=\frac{\tau}{2}} = \frac{e^{-j\omega\tau/2}}{-j\omega} - \frac{e^{j\omega\tau/2}}{-j\omega} = \frac{e^{j\omega\tau/2}-e^{-j\omega\tau/2}}{j\omega}$$

$$= \frac{2j\sin(\omega\tau/2)}{j\omega} = \tau\,\frac{\sin(\omega\tau/2)}{(\omega\tau/2)} = \tau\,\text{sinc}(\omega\tau/2). \tag{5.33}$$

We now have the Fourier transform pair:

$$\boxed{\text{rect}\left(\frac{t}{\tau}\right) \Leftrightarrow \tau\,\text{sinc}\left(\frac{\omega\tau}{2}\right)} \tag{5.34}$$

These and some other continuous-time Fourier transform pairs are summarized in Table 5.1.

Table 5.1. Table of continuous-time Fourier transform pairs.

Entry #	$f(t)$	$F(\omega)$	Refer to Eq. #
1.	$\frac{1}{2\pi}\int_{-\infty}^{\infty}F(\omega)e^{j\omega t}d\omega$	$\int_{-\infty}^{\infty}f(t)e^{-j\omega t}dt$	(5.1), (5.2)
2.	$e^{-bt}u(t)$	$\frac{1}{b+j\omega}; \ b>0$	(5.7)
3.	$e^{bt}u(-t)$	$\frac{1}{b-j\omega}; \ b>0$	(5.8)

(*Continued*)

which clearly shows why real and even $x(t)$ has real $\mathcal{F}\{x(t)\}$, and real and odd $x(t)$ has imaginary $\mathcal{F}\{x(t)\}$.

Table 5.1. (*Continued*)

Entry #	$f(t)$	$F(\omega)$	Refer to Eq. #
4.	$e^{-b\lvert t \rvert}$	$\dfrac{2b}{b^2 + \omega^2};\quad b > 0$	(5.10)
5.	1	$2\pi\delta(\omega)$	(5.14)
6.	$\delta(t)$	1	(5.16)
7.	$\delta(t - t_0)$	$e^{-j\omega t_0}$	(5.18)
8.	$\text{sgn}(t)$	$\dfrac{2}{j\omega}$	(5.22)
9.	$u(t)$	$\pi\delta(\omega) + \dfrac{1}{j\omega}$	(5.25)
10.	$e^{j\omega_0 t}$	$2\pi\,\delta(\omega - \omega_0)$	(5.27)
11.	$\cos(\omega_0 t + \theta)$	$e^{-j\theta}\pi\delta(\omega + \omega_0)$ $+ e^{j\theta}\pi\delta(\omega - \omega_0)$	(5.30)
12.	$\cos(\omega_0 t)$	$\pi\delta(\omega + \omega_0)$ $+ \pi\delta(\omega - \omega_0)$	(5.31)
13.	$\sin(\omega_0 t)$	$j\pi\delta(\omega + \omega_0)$ $- j\pi\delta(\omega - \omega_0)$	(5.32)
14.	$\text{rect}(t/T)$ pulse width $= T$ sec	$T\,\text{sinc}(\omega T/2)$	(5.34)

(*Continued*)

Table 5.1. (*Continued*)

Entry #	$f(t)$	$F(\omega)$	Refer to Eq. #
15.	$\dfrac{W}{\pi}\operatorname{sinc}(Wt)$	$\operatorname{rect}\left(\dfrac{\omega}{2W}\right)$ bandwidth = W rad/sec	(5.40)
16.	$\Delta(t/T)$ pulse width = T sec	$\dfrac{T}{2}\operatorname{sinc}^2(\omega T/4)$	Lathi, p.702
17.	$\dfrac{W}{2\pi}\operatorname{sinc}^2(Wt/2)$	$\Delta\left(\dfrac{\omega}{2W}\right)$ bandwidth = W rad/sec	Lathi, p.702
18.	$e^{-t^2/(2\sigma^2)}$	$\sigma\sqrt{2\pi}\,e^{-\sigma^2\omega^2/2}$	Lathi, p.702
19.	$\delta_{T_0}(t) =$ $\sum_{n=-\infty}^{\infty}\delta(t-nT_0)$ Impulse train in time, period = T_0 sec	$\omega_0\delta_{\omega_0}(\omega) =$ $\omega_0\sum_{n=-\infty}^{\infty}\delta(\omega-n\omega_0)$ Impulse train in frequency, period $\omega_0 = 2\pi/T_0$ rad/sec	(5.50) Lathi, p.702

5.1.3 *Fourier transform properties*

In most applications, we will need to derive Fourier transforms of many different functions. Frequently it is convenient to use some of the mathematical properties of the Fourier integral given in Eq. (5.1). Therefore, in this section we will discuss some of the important properties and then show in a table all other properties that may be useful. Once again you will find that it is straightforward to verify these entries in Table 5.2.

5.1.3.1 *Linearity*

Let us consider two different functions of time, $f_1(t)$ and $f_2(t)$, whose Fourier transforms are known to be $F_1(\omega)$ and $F_2(\omega)$, respectively: $f_1(t) \Leftrightarrow F_1(\omega)$ and $f_2(t) \Leftrightarrow F_2(\omega)$. Then from the CTFT definition in Eq. (5.1), for two arbitrary constants a and b, the following relationship[k] holds true:

$$af_1(t) + bf_2(t) \Leftrightarrow aF_1(\omega) + bF_2(\omega) \qquad (5.35)$$

With this property, we may find Fourier transforms of signals that are linear combinations of signals having known transforms. The following example illustrates this property and shows the plot of the function and its transform.

Example 5.1

Find the Fourier transform of $x(t)$ that is shown in Fig. 5.1.

To solve this problem, express $x(t)$ as the sum of two rectangular pulses:

$$x(t) = 0.5 \, \text{rect}(t/4) + 1.5 \, \text{rect}(t/2). \qquad (5.36)$$

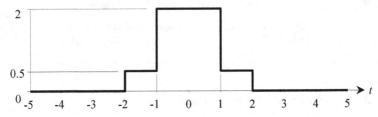

Figure 5.1. Signal $x(t)$ to be transformed to the frequency domain in Example 5.1.

We see that $x(t) = af_1(t) + bf_2(t)$: $a = 0.5$, $f_1(n) = \text{rect}(t/4)$, $b = 1.5$ and $f_2(n) = \text{rect}(t/2)$. Therefore, applying the linearity property, we can say that $\mathcal{F}\{x(t)\} = a\mathcal{F}\{f_1(t)\} + b\mathcal{F}\{f_2(t)\}$:

---

[k] This is called the "Linear Superposition" property. It combines the scaling property ($\mathcal{F}\{af(t)\} = a\mathcal{F}\{f(t)\}$) and the additive property ($\mathcal{F}\{f_1(t) + f_2(t)\} = \mathcal{F}\{f_1(t)\} + \mathcal{F}\{f_2(t)\}$) into a single expression.

$$\mathcal{F}\{x(n)\} = 0.5\,\mathcal{F}\{\text{rect}(t/4)\} + 1.5\,\mathcal{F}\{\text{rect}(t/2)\}$$

$$= 0.5\big(4\,\text{sinc}(2\omega)\big) + 1.5\big(2\,\text{sinc}(\omega)\big)$$

$$= 2\text{sinc}(2\omega) + 3\text{sinc}(\omega). \tag{5.37}$$

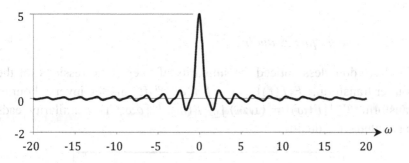

Figure 5.2. From Example 5.1: $\mathcal{F}\{x(t)\} = X(\omega) = 2\text{sinc}(2\omega) + 3\text{sinc}(\omega)$.

5.1.3.2 *Time shifting*

In practice, quite often we encounter a signal whose Fourier transform is known but for the fact that it is delayed (or advanced) in time. Here we ask a question – is there a simple relationship between the two Fourier transforms?

Consider our experience with sinusoidal signal analysis in a basic circuit theory course. We know that a time shifted sinusoid at a frequency ω_0 rad/sec is a sinusoid at the same frequency and amplitude, but now has a phase offset from its original value. Since the Fourier transform of an arbitrary time function is a linear combination of various frequencies,[1]

---

[1] The inverse Fourier transform integral may be written as: $f(t) = (1/2\pi) \int_{-\infty}^{\infty} F(\omega)(\cos(\omega t) + j\sin(\omega t))d\omega = (1/2\pi)\int_{-\infty}^{\infty} F(\omega)\cos(\omega t)\,d\omega + (j/2\pi)\int_{-\infty}^{\infty} F(\omega)\sin(\omega t)d\omega = (1/2\pi)\int_{-\infty}^{\infty}(F_e(\omega) + F_o(\omega))\cos(\omega t)\,d\omega + (j/2\pi)\int_{-\infty}^{\infty}(F_e(\omega) + F_o(\omega))\sin(\omega t)\,d\omega = (1/2\pi)\int_{-\infty}^{\infty} F_e(\omega)\cos(\omega t)\,d\omega + (j/2\pi)\int_{-\infty}^{\infty} F_o(\omega)\sin(\omega t)\,d\omega = (1/\pi)\int_{0}^{\infty} F_e(\omega)\cos(\omega t)\,d\omega + (j/\pi)\int_{0}^{\infty} F_o(\omega)\sin(\omega t)\,d\omega$. In other words, $f(t)$ is a weighted sum of sines and cosines at all possible frequencies (where the weights for sines and cosines are $(j/\pi)F_o(\omega)$ and $(1/\pi)F_e(\omega)$, respectively).

each of these frequency components will now have a corresponding phase offset associated with it:

$$\boxed{f(t \pm t_0) \Leftrightarrow F(\omega)e^{\pm j\omega t_0}}$$ (5.38)[m]

5.1.3.3 *Time/frequency duality*

You have doubtless noticed the similarity of integral expressions for the Fourier transform, $\mathcal{F}\{f(t)\} = \int_{-\infty}^{\infty} f(t)e^{-j\omega t}dt$, and the inverse Fourier transform, $\mathcal{F}^{-1}\{F(\omega)\} = (1/2\pi)\int_{-\infty}^{\infty} F(\omega)e^{j\omega t}d\omega$.[n] This similarity leads to an interesting duality:

$$\boxed{\text{If } f(t) \Leftrightarrow F(\omega) \text{ then } F(t) \Leftrightarrow 2\pi f(-\omega)}$$ (5.39)

Example 5.2
Find the Fourier transform of sinc(Wt) by applying the time/frequency duality property.

Equation 5.34 states that $\text{rect}(t/\tau) \Leftrightarrow \tau \text{ sinc}(\omega\tau/2)$. Therefore, by duality, $\tau \text{ sinc}(t\tau/2) \Leftrightarrow 2\pi \text{ rect}(-\omega/\tau) = 2\pi \text{ rect}(\omega/\tau)$. After substituting $\tau = 2W$ we obtain $2W \text{ sinc}(tW) \Leftrightarrow 2\pi \text{ rect}(\omega/2W)$. Thus, we have another useful Fourier transform pair:

$$\boxed{\text{sinc}(Wt) \Leftrightarrow \frac{\pi}{W} \text{ rect}\left(\frac{\omega}{2W}\right)}$$ (5.40)

---

[m] Proof: $\mathcal{F}\{f(t \pm t_0)\} = \int_{-\infty}^{\infty} f(t \pm t_0)e^{-j\omega t}dt = \int_{-\infty}^{\infty} f(\beta)e^{-j\omega(\beta \mp t_0)}d\beta = \int_{-\infty}^{\infty} f(\beta)e^{-j\omega\beta}e^{\pm j\omega t_0}d\beta = e^{\pm j\omega t_0}\int_{-\infty}^{\infty} f(\beta)e^{-j\omega\beta}d\beta = e^{\pm j\omega t_0}\mathcal{F}\{f(t)\} = e^{\pm j\omega t_0}F(\omega)$.

[n] The two look even more similar when stated in terms of non-normalized frequency variable $f = \frac{\omega}{2\pi}$: $\mathcal{F}\{f(t)\} = \int_{-\infty}^{\infty} f(t)e^{-j2\pi ft}dt$, $\mathcal{F}^{-1}\{F(f)\} = \int_{-\infty}^{\infty} F(f)e^{j2\pi ft}df$.

5.1.3.4 *Convolution*

Many signal processing applications require either multiplying or convolving together two signals. As we shall now see, multiplication and convolution are closely related through the Fourier transform. What may be a difficult-to-simplify convolution of two signals in one domain is simply expressed as a multiplicative product in the other domain. Here we show the correspondence between multiplication and convolution in time and frequency domains.

If $f_1(t) \Leftrightarrow F_1(\omega)$ and $f_2(t) \Leftrightarrow F_2(\omega)$, then:

$$f_1(t) * f_2(t) \Leftrightarrow F_1(\omega)F_2(\omega)$$

(5.41)°

$$f_1(t)f_2(t) \Leftrightarrow \frac{1}{2\pi} F_1(\omega) * F_2(\omega)$$

(5.42)ᵖ

<u>Example 5.3</u>

Find the Fourier transform of $x(t) = \text{rect}(t/2) * \text{sinc}(\pi t)$:

Since $\text{rect}(t/2) \Leftrightarrow 2\,\text{sinc}(\omega)$ and $\text{sinc}(\pi t) \Leftrightarrow \text{rect}(\omega/2\pi)$, then
$\mathcal{F}\{\text{rect}(t/2) * \text{sinc}(\pi t)\} = \mathcal{F}\{\text{rect}(t/2)\}\mathcal{F}\{\text{sinc}(\pi t)\} = 2\,\text{sinc}(\omega) \times \text{rect}(\omega/2\pi)$:

° Proof: $\mathcal{F}\{f_1(t) * f_2(t)\} = \int_{-\infty}^{\infty} \left(\int_{-\infty}^{\infty} f_1(\tau)f_2(t-\tau)d\tau \right)e^{-j\omega t}dt =$
$\int_{-\infty}^{\infty} f_1(\tau)\left(\int_{-\infty}^{\infty} f_2(t-\tau)e^{-j\omega t}dt \right)d\tau = \int_{-\infty}^{\infty} f_1(\tau)\mathcal{F}\{f_2(t-\tau)\}d\tau =$
$\int_{-\infty}^{\infty} f_1(\tau)\left(e^{-j\omega \tau}F_2(\omega) \right)d\tau = \int_{-\infty}^{\infty} f_1(\tau)e^{-j\omega \tau}d\tau \times F_2(\omega) = F_1(\omega) \times F_2(\omega)$.

ᵖ Proof: $\mathcal{F}\{f_1(t)f_2(t)\} = \int_{-\infty}^{\infty} f_1(t)f_2(t)e^{-j\omega t}dt = \int_{-\infty}^{\infty} \mathcal{F}^{-1}\{F_1(\omega)\}f_2(t)e^{-j\omega t}dt =$
$\int_{-\infty}^{\infty} \left((1/2\pi) \int_{-\infty}^{\infty} F_1(\beta)e^{j\beta t}d\beta \right) f_2(t)e^{-j\omega t}dt =$
$(1/2\pi) \int_{-\infty}^{\infty} F_1(\beta) \int_{-\infty}^{\infty} e^{j\beta t} f_2(t)e^{-j\omega t}dt\, d\beta =$
$(1/2\pi) \int_{-\infty}^{\infty} F_1(\beta) \int_{-\infty}^{\infty} f_2(t)e^{-j(\omega-\beta)t}dt\, d\beta = (1/2\pi) \int_{-\infty}^{\infty} F_1(\beta)F_2(\omega - \beta)\, d\beta =$
$(1/2\pi)F_1(\omega) * F_2(\omega)$.

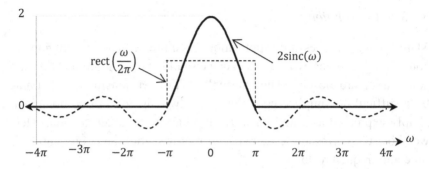

Figure 5.3. The spectrum $X(\omega) = 2\,\text{sinc}(\omega) \times \text{rect}(\omega/2\pi)$ in Example 5.3.

5.1.3.5 *Modulation*

In communication applications we find that an information signal $f(t)$, such as speech, may be wirelessly transmitted over radio waves by first multiplying it with a sinusoid at a relatively high frequency ω_0 (called the carrier frequency). Therefore, it is important to understand the frequency-domain effects of multiplication with a sinusoid in time. We find, as shown below, that the modulation process shifts $F(\omega)$ by $\pm\omega_0$ in frequency. Consider a signal $m(t) = f(t)\cos(\omega_0 t)$. From Eq. (5.1), we have:

$$\mathcal{F}\{m(t)\} = M(\omega) = \int_{-\infty}^{\infty} f(t)\cos(\omega_0 t)e^{-j\omega t}\,dt$$

$$= \int_{-\infty}^{\infty} f(t)\left(\tfrac{1}{2}e^{j\omega_0 t} + \tfrac{1}{2}e^{-j\omega_0 t}\right)e^{-j\omega t}\,dt$$

$$= \tfrac{1}{2}\int_{-\infty}^{\infty} f(t)\left(e^{-j(\omega-\omega_0)t} + e^{-j(\omega+\omega_0)t}\right)dt$$

$$= \tfrac{1}{2}\int_{-\infty}^{\infty} f(t)e^{-j(\omega-\omega_0)t}\,dt + \tfrac{1}{2}\int_{-\infty}^{\infty} f(t)e^{-j(\omega+\omega_0)t}\,dt$$

$$= \tfrac{1}{2}F(\omega - \omega_0) + \tfrac{1}{2}F(\omega + \omega_0). \tag{5.43}$$

This is called the modulation property of the Fourier transform:

$$\boxed{f(t)\cos(\omega_0 t) \Leftrightarrow \tfrac{1}{2}F(\omega - \omega_0) + \tfrac{1}{2}F(\omega + \omega_0)} \tag{5.44}$$

When allowing for an arbitrary cosine phase shift ϕ, the modulation property is:

$$f(t)\cos(\omega_0 t + \phi) \Leftrightarrow$$

$$\tfrac{1}{2}\left(e^{j\phi}F(\omega - \omega_0) + e^{-j\phi}F(\omega + \omega_0)\right) \qquad (5.45)^q$$

Example 5.4

Plot $\mathcal{F}\{\text{rect}(t/3\pi)\cos(4\pi t)\}$.

To begin, $\text{rect}(t/3\pi) \Leftrightarrow 3\pi\,\text{sinc}(\omega 3\pi/2)$ from Entry #14 in Table 5.1. This spectrum is plotted in Fig. 5.4:

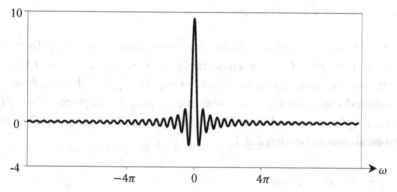

Figure 5.4. Plot of $\mathcal{F}\{\text{rect}(t/3\pi)\} = 3\pi\,\text{sinc}(\omega 3\pi/2)$, from Example 5.4.

The modulation property (Eq. (5.44)) tells us that

$$\mathcal{F}\{f(t)\cos(4\pi t)\} = \tfrac{1}{2}F(\omega - 4\pi) + \tfrac{1}{2}F(\omega - 4\pi).$$

Therefore, $\mathcal{F}\{\text{rect}(t/3\pi)\cos(4\pi t)\} =$

$$(3\pi/2)\text{sinc}((\omega - 4\pi)3\pi/2) + (3\pi/2)\text{sinc}((\omega + 4\pi)3\pi/2),$$

as shown in Fig. 5.5.

---

^{q} So, for example, when $\phi = -\pi/2$: $f(t)\sin(\omega_0 t) \Leftrightarrow \tfrac{1}{2j}F(\omega - \omega_0) - \tfrac{1}{2j}F(\omega + \omega_0)$

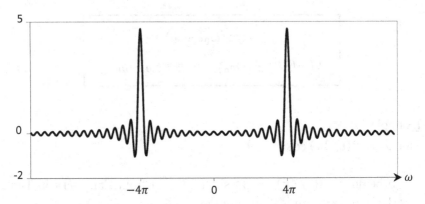

Figure 5.5.   Plot of $\mathcal{F}\{\text{rect}(t/3\pi)\cos(4\pi t)\} = (3\pi/2)\text{sinc}((\omega - 4\pi)3\pi/2) + (3\pi/2)$
$\text{sinc}((\omega + 4\pi)3\pi/2)$, from Example 5.4.

Modulation, therefore, shifts[r] the energy/power of $f(t)$ from the original ("baseband") frequency range that is occupied by $F(\omega)$ to the frequency range that is in vicinity of ω_0. If ω_0 is high enough so that the shifted copies of $F(\omega)$ are non-overlapping in frequency, then $f(t)$ may easily be recovered from the modulated signal $f(t)\cos(\omega_0(t))$. This will be shown in Section 5.2.4.

5.1.3.6 *Frequency shift*

We saw in the previous section that modulation is the multiplication of $f(t)$ by a sinusoid at frequency ω_0, and it results in the shifting of $F(\omega)$ by $\pm\omega_0$ in the frequency domain. To shift $F(\omega)$ only in one direction, multiply $f(t)$ by $e^{\pm j\omega_0 t}$:

$$\boxed{f(t)e^{\pm j\omega_0 t} \Leftrightarrow F(\omega \mp \omega_0)} \tag{5.46}$$

5.1.3.7 *Time scaling*

Another interesting property of the Fourier transform is that of scaling in time. We consider a signal $f(t)$ and scale independent variable t by

---

[r] Modulation, or the frequency shifting of $F(\omega)$, is also commonly referred to as *frequency heterodyning*.

constant factor $a > 0$, resulting in signal $f(at)$. Depending on the numerical value of a, signal $f(t)$ is either expanded or compressed in time.[s] Finding the Fourier transform of $f(at)$ when $a > 0$:

$$\mathcal{F}\{f(at)\} = \int_{-\infty}^{\infty} f(at)e^{-j\omega t}dt = \int_{\beta/a=-\infty}^{\beta/a=\infty} f(\beta)e^{-j\omega(\beta/a)}d(\beta/a)$$

$$= \int_{\beta=-\infty}^{\beta=\infty} f(\beta)e^{-j\omega(\beta/a)}\frac{1}{a}d\beta = \frac{1}{a}\int_{-\infty}^{\infty} f(\beta)e^{-j\beta(\omega/a)}d\beta = \frac{1}{a}F\left(\frac{\omega}{a}\right).$$

For $a < 0$, $\mathcal{F}\{f(at)\} = \frac{1}{-a}F\left(\frac{\omega}{a}\right)$. Combining these results:

$$\boxed{f(at) \Leftrightarrow \frac{1}{|a|}F\left(\frac{\omega}{a}\right)} \tag{5.47}$$

Equation (5.47) leads to the observation: if a time signal is expanded by factor a then its spectrum is contracted by factor a.

Example 5.5

Find the Fourier transform of $x(t) = \text{rect}(2t - 2)$.

Method 1 – consider $x(t)$ to be the result of first time-shifting, then time-scaling, a rectangular pulse function:

$\text{rect}(t) \Leftrightarrow \text{sinc}(\omega/2)$	Table 5.1, #14
$\text{rect}(t - 2) \Leftrightarrow \text{sinc}(\omega/2)e^{-j2\omega}$	Table 5.2, #2
$\text{rect}(2t - 2) \Leftrightarrow (1/|2|)\text{sinc}((\omega/2)/2)e^{-j2(\omega/2)}$	Table 5.2, #9
$\text{rect}(2t - 2) \Leftrightarrow 0.5\,\text{sinc}(\omega/4)e^{-j\omega}$	simplification

Method 2 – consider $x(t)$ to be the result of first time-scaling, then time-shifting, a rectangular pulse function:

$\text{rect}(t) \Leftrightarrow \text{sinc}(\omega/2)$	Table 5.1, #14
$\text{rect}(2t) \Leftrightarrow (1/|2|)\text{sinc}((\omega/2)/2)$	Table 5.2, #9
$\text{rect}(2t) \Leftrightarrow 0.5\,\text{sinc}(\omega/4)$	simplification
$\text{rect}(2(t - 1)) \Leftrightarrow 0.5\,\text{sinc}(\omega/4)e^{-j\omega}$	Table 5.2, #2
$\text{rect}(2t - 2) \Leftrightarrow 0.5\,\text{sinc}(\omega/4)e^{-j\omega}$	simplification

---

[s] $f(at)$ with $0 < a < 1$ is an expansion of $f(t)$, by factor a, away from $t = 0$; $f(at)$ with $a > 1$ is a compression of $f(t)$ in time, by factor a, toward $t = 0$.

Note that the order of operations does not matter, although Method 1 seems more logical due to the stated argument of the rectangular pulse in $x(t)$: $(2t - 2) \rightarrow$ implies first shift, then scaling.

5.1.3.8 *Parseval's Theorem*

Parseval's Theorem[t] relates the energy of a signal calculated in the time domain to the energy as calculated in the frequency domain. To begin, assume that signal $f(t)$ may be complex-valued and that $f(t) \Leftrightarrow F(\omega)$. The energy of $f(t)$ is $\int_{-\infty}^{\infty} f(t) f^*(t)\, dt$

$$= \int_{-\infty}^{\infty} f(t) \left\{ \frac{1}{2\pi} \int_{-\infty}^{\infty} F^*(\omega) e^{-j\omega t}\, d\omega \right\} dt$$

$$= \frac{1}{2\pi} \int_{-\infty}^{\infty} F^*(\omega) \{ \int_{-\infty}^{\infty} f(t) e^{-j\omega t}\, dt \} d\omega$$

$$= \frac{1}{2\pi} \int_{-\infty}^{\infty} F^*(\omega)\, F(\omega)\, d\omega, \qquad (5.48)$$

leading to a statement of the theorem:

$$\boxed{\int_{-\infty}^{\infty} |f(t)|^2\, dt = \frac{1}{2\pi} \int_{-\infty}^{\infty} |F(\omega)|^2\, d\omega} \qquad (5.49)$$

Example 5.6
Find the energy of $\mathrm{sinc}(t)$.

$$\text{Energy}\{\mathrm{sinc}(t)\} = \int_{-\infty}^{\infty} |\mathrm{sinc}(t)|^2\, dt = \frac{1}{2\pi} \int_{-\infty}^{\infty} |\mathcal{F}\{\mathrm{sinc}(t)\}|^2\, d\omega$$

$$= \frac{1}{2\pi} \int_{-\infty}^{\infty} \left| \pi\, \mathrm{rect}\left(\frac{\omega}{2}\right) \right|^2 d\omega = \frac{1}{2\pi} \int_{-1}^{1} |\pi\,(1)|^2\, d\omega$$

$$= \frac{1}{2\pi} \int_{-1}^{1} \pi^2\, d\omega = \frac{1}{2\pi}(2\pi^2) = \pi.$$

Next, we summarize the properties of the continuous-time Fourier transform in Table 5.2.

---

[t] This theorem is also known in the literature as Rayleigh Theorem.

Table 5.2. Table of continuous-time Fourier transform properties.

1.	Linearity:	(Eq. (5.35))
	$af_1(t) + bf_2(t) \Leftrightarrow aF_1(\omega) + bF_2(\omega)$	
2.	Time shift:	(Eq. (5.38))
	$f(t \pm t_0) \Leftrightarrow F(\omega)e^{\pm j\omega t_0}$	
3.	Convolution:	(Eq. (5.41))
	$f_1(t) * f_2(t) \Leftrightarrow F_1(\omega)F_2(\omega)$	
4.	Multiplication:	(Eq. (5.42))
	$f_1(t)f_2(t) \Leftrightarrow \frac{1}{2\pi}\{F_1(\omega) * F_2(\omega)\}$	
5.	Modulation:	(Eq. (5.44))
	$f(t)\cos(\omega_0 t) \Leftrightarrow \frac{1}{2}F(\omega - \omega_0) + \frac{1}{2}F(\omega + \omega_0)$	
6.	Modulation (arbitrary carrier phase) :	(Eq. (5.45))
	$f(t)\cos(\omega_0 t + \phi) \Leftrightarrow \frac{e^{j\phi}}{2}F(\omega - \omega_0) + \frac{e^{-j\phi}}{2}F(\omega + \omega_0)$	
7.	Frequency shift:	(Eq. (5.46))
	$f(t)e^{\pm j\omega_0 t} \Leftrightarrow F(\omega \mp \omega_0)$	
8.	Duality:	(Eq. (5.39))
	$F(t) \Leftrightarrow 2\pi f(-\omega)$	
9.	Time scaling:	(Eq. (5.47))
	$f(at) \Leftrightarrow \dfrac{F(\omega/a)}{|a|}$	

(Continued)

Table 5.2. (*Continued*)

---

10. Time reversal: (See [u] below)

$$f(-t) \Leftrightarrow F(-\omega)$$

---

11. Conjugation in time: (See [v] below)

$$f^*(t) \Leftrightarrow F^*(-\omega)$$

---

12. Real, even $f_e(t) = Re\{f_e(-t)\}$: (See footnote on p. 164)

$$f_e(t) \Leftrightarrow \mathcal{F}\{f_e(t)\} = 2 \int_0^\infty f(t) \cos(\omega t)\, dt$$

---

13. Real, odd $f_o(t) = Re\{-f_o(-t)\}$: (See footnote on p. 164)

$$f_o(t) \Leftrightarrow \mathcal{F}\{f_o(t)\} = -2j \int_0^\infty f(t) \sin(\omega t)\, dt$$

---

14. Differential: (See [w] below)

$$\frac{df(t)}{dt} \Leftrightarrow j\omega\, F(\omega)$$

---

15. Cumulative integral: (See [x] below)

$$\int_{-\infty}^t f(\beta)d\beta = f(t) * u(t) \Leftrightarrow F(\omega)/j\omega + \pi F(0)\delta(\omega)$$

(*Continued*)

---

[u] $\mathcal{F}\{f(-t)\} = \int_{-\infty}^\infty f(-t)e^{-j\omega t}dt = \int_{-\infty}^\infty f(\beta)e^{-j(-\omega)\beta}d\beta = F(-\omega)$.

[v] $\mathcal{F}\{f^*(t)\} = \int_{-\infty}^\infty f^*(t)e^{-j\omega t}dt = \left(\int_{-\infty}^\infty f(t)e^{j\omega t}dt\right)^* = \left(\int_{-\infty}^\infty f(t)e^{-j(-\omega)t}dt\right)^* = F^*(-\omega)$.

[w] $\frac{d}{dt}f(t) = \frac{d}{dt}\left(\frac{1}{2\pi}\int_{-\infty}^\infty F(\omega)e^{j\omega t}d\omega\right) = (1/2\pi)\int_{-\infty}^\infty F(\omega)\frac{d}{dt}\left(e^{j\omega t}\right)d\omega = (1/2\pi)\int_{-\infty}^\infty F(\omega) j\omega\, e^{j\omega t}d\omega = (1/2\pi)\int_{-\infty}^\infty \{j\omega F(\omega)\}e^{j\omega t}d\omega = \mathcal{F}^{-1}\{j\omega F(\omega)\}$, or: $\frac{d}{dt}f(t) \Leftrightarrow j\omega F(\omega)$.

[x] $f(t) * u(t) = \int_{-\infty}^\infty f(\beta)u(t-\beta)d\beta = \int_{-\infty}^t f(\beta)d\beta; f(t) * u(t) \Leftrightarrow F(\omega)\{1/j\omega + \pi\delta(\omega)\} = F(\omega)/j\omega + \pi F(\omega)\delta(\omega) = F(\omega)/j\omega + \pi F(0)\delta(\omega)$.

Table 5.2. (*Continued*)

16. Parseval's Theorem: (for energy signals)   (Eq. (5.49)) $$E_f = \int_{-\infty}^{\infty} |f(t)|^2 \, dt = \frac{1}{2\pi} \int_{-\infty}^{\infty} |F(\omega)|^2 \, d\omega$$
17. Cross-Correlation: (for energy signals):   (See [y] below) $$\phi_{xy}(t) \Leftrightarrow X^*(\omega)Y(\omega)$$
18. Autocorrelation: (for energy signals)   (See [z] below) $$\phi_{xx}(t) \Leftrightarrow X^*(\omega)X(\omega) = |X(\omega)|^2$$

5.1.4 *Graphical representation of the Fourier transform*

Engineers and scientists like to have an image of mathematical formulas and functions to facilitate their understanding. So far we have presented graphical representations of only real signals in either time or frequency domain. The Fourier transform of a real time-domain signal is generally[aa] a complex function of ω. In this section, we discuss several different ways of graphing such complex-valued waveforms. Such plots are extremely useful in understanding the processing of signals through various devices such as filters, amplifiers, mixers, etc.

In this textbook, plots are presented with an abscissa (x-axis) scale of ω radians/second. As you know, $\omega = 2\pi f$ where f has units cycles per

---

[y] $\phi_{xy}(t) = x^*(-t) * y(t)$, therefore $\mathcal{F}\{\phi_{xy}(t)\} = \mathcal{F}\{x^*(-t)\}\mathcal{F}\{y(t)\} = X^*(\omega)Y(\omega)$.

[z] $\mathcal{F}\{\phi_{xy}(t)\} = X^*(\omega)Y(\omega)$, therefore $\mathcal{F}\{\phi_{xx}(t)\} = X^*(\omega)X(\omega)$.

[aa] A "real world" signal $x(t)$ will be real, and almost always have $x_e(t) \neq 0$, $x_o(t) \neq 0$. Thus, since the Fourier transform $X(\omega) = 2\int_0^{\infty}(x_e(t)\cos(\omega t) - jx_o(t)\sin(\omega t))dt$ (from footnote on p. 164), $X(\omega)$ will almost always be a complex function of ω.

second (Hertz). In some applications, such as communication engineering, the preferred abscissa scale is frequency f in Hz. This amounts to a change in scale by factor 2π, which can be readily accomplished.

5.1.4.1 *Rectangular coordinates*

In this form of visualization, a three-dimensional plot is created using the real part of $F(\omega)$, the imaginary part of $F(\omega)$, and ω as three axes. For

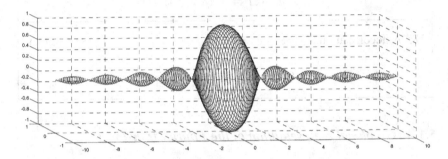

Figure 5.6. A graph of $\text{sinc}(\omega/2)e^{-j20\omega}$ vs. ω/π.

example, Fig. 5.6 shows $\mathcal{F}\{rect(t-20)\} = \text{sinc}(\omega/2)e^{-j20\omega}$ plotted this way. Instead of a 3-D plot, frequently it is more instructive to plot two 2-D plots; one of $Re\{F(\omega)\}$ vs. ω, and another of $Im\{F(\omega)\}$ vs. ω. Consider Fig. 5.7, which shows the 3-D plot of a given $F(\omega)$ in rectangular coordinates (as in Fig. 5.6):

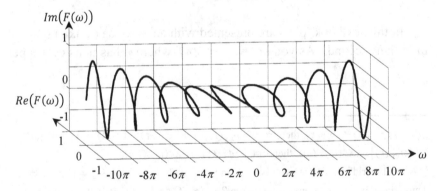

Figure 5.7. A 3-D graph of complex-valued $F(\omega)$.

Figures 5.8 and 5.9 display the same information using a pair of 2-D plots, one for the real part vs. frequency and the other for the imaginary part vs. frequency:

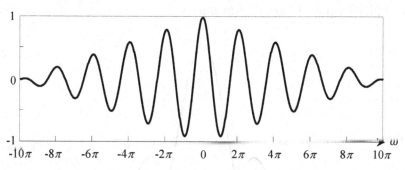

Figure 5.8. $Re\{F(\omega)\}$ vs. ω, corresponding to Fig. 5.7.

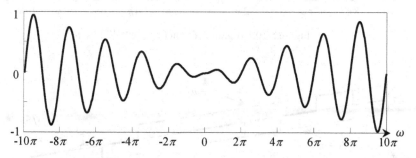

Figure 5.9. $Im\{F(\omega)\}$ vs. ω, corresponding to Fig. 5.7.

Yet another way of displaying a complex-valued frequency spectrum is to draw the path that $F(\omega)$ takes, in the complex plane, as ω varies over some frequency range. Essentially this is a projection of the type of 3-D graph shown in Figs. 5.6 or 5.7 onto a plane perpendicular to the ω-axis. The resulting 2-D graph is called a *Nyquist Plot*. This plot requires some extra labels to indicate the frequency associated with any point of interest along the path of the function. Nyquist plots are used for stability analysis in control theory.

5.1.4.2 *Polar coordinates*

As an alternative to the previously-presented graphs of real and imaginary parts of $F(\omega)$, frequently it is more appropriate to plot magnitude $|F(\omega)|$

and phase $\angle F(\omega)$ vs. ω. Figures 5.10 and 5.11 show these two graphs for a complex frequency-domain function.[bb] In general, separately plotting magnitude and phase components is the preferred method for displaying complex frequency-domain waveforms.

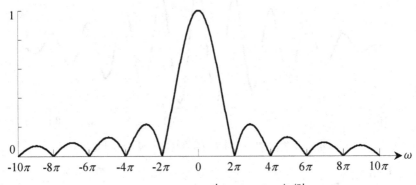

Figure 5.10.   A graph of $|sinc(\omega/2)e^{-j\omega/5}|$ vs. ω.

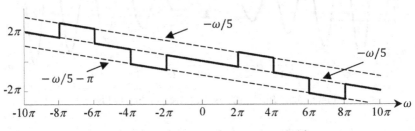

Figure 5.11.   A graph of $\angle\{sinc(\omega/2)e^{-j\omega/5}\}$ vs. ω.

5.1.4.3 *Graphing the amplitude of F(ω)*

Consider the case where the Fourier transform gives a real, bipolar result: for example, $F(\omega) = sinc(\omega/2)$. Plotting its magnitude gives us the same graphs as in Fig. 5.10, and a plot of the phase is shown next:

[bb] Note that the magnitude plot is the same for all θ in $F(\omega) = sinc(\omega/2)e^{-j\theta}$; hence Fig. 5.10 also pertains to the signal plotted in the previous example.

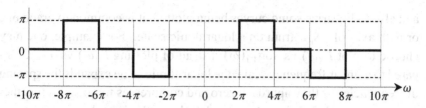

Figure 5.12. A graph of $\angle\text{sinc}(\omega/2)$ vs. ω.

You will notice that whenever $\text{sinc}(\omega/2) < 0$, the phase angle is chosen[cc] to be $\pm\pi$ so that $|\text{sinc}(\omega/2)|e^{j(\pm\pi)} = -|\text{sinc}(\omega/2)|$.

If we look at Fig. 5.12 we notice that this phase plot has minimal information. This is because the $\text{sinc}(\omega/2)$ function is real, and phase serves only to specify the polarity of the function. In such case, it is advantageous to plot $F(\omega) = \text{sinc}(\omega/2)$ directly, as shown in Fig. 5.13 below:

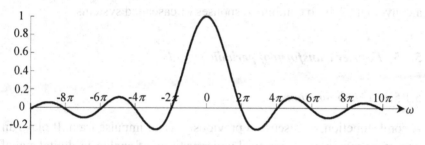

Figure 5.13. A graph of $\text{sinc}(\omega/2)$ vs. ω.

With access to MATLAB®, all these plots can be readily generated, thus the choice of which one to use should be based on the clarity of the information to be conveyed and how it will be applied.

5.1.4.4 *Logarithmic scales and Bode plots*

Due to the logarithmic nature of human perception (e.g., we hear frequencies on a logarithmic scale, as reflected by the tuning of a piano keyboard),

---

[cc] Any odd integer value of k will make $e^{jk\pi} = -1$; the values chosen alternate so that the phase angle is an odd function of ω.

and also to linearize some curves being plotted, it is common to graph one or both axes of a spectrum on a logarithmic scale. For example, one may choose to plot $F(\omega)$ vs. $\log_{10}(\omega)$ instead of plotting $F(\omega)$ vs. ω. Compared to a linear frequency scale, the logarithmic scale expands the spacing of frequencies as they approach zero and compresses frequencies together as they approach infinity. As a result, a logarithmic scale never includes zero (d.c.) nor any negative[dd] frequency values.

When plotting the magnitude spectrum $|F(\omega)|$ it is common to use logarithmic scales for both magnitude[ee] and frequency. The resulting graph is then proportional to $\log|F(\omega)|$ vs. $\log(\omega)$. A version of the log-log plot, called a Bode magnitude plot, is where the vertical scale is $10\log_{10}|F(\omega)|^2$ and the horizontal scale is $\log_{10}(\omega)$. The Bode phase plot has vertical scale $\angle F(\omega)$ and horizontal scale $\log_{10}(\omega)$. Bode magnitude and phase plots are common in system analysis, where they provide a convenient way to combine responses of cascaded systems.

5.1.5 *Fourier transform of periodic signals*

5.1.5.1 *Comb function*

A comb function, as discussed previously, is an impulse train. It plays an important role in our theoretical understanding of analog-to-digital signal conversion, as well as being a convenient tool for modeling periodic signals.

Let us consider the Fourier transform of a comb function $\delta_{T_0}(t)$, which is Entry #19 in Table 5.1:

[dd] Negative frequencies do not exist (when defined as #events/second) except as a mathematical tool for efficient notation. This is clearly shown by a cosine wave at frequency ω_0 equal to a sum of complex exponentials at frequencies $\pm\omega_0$: $\cos(\omega_0 t) = (1/2)e^{j(\omega_0)t} + (1/2)e^{j(-\omega_0)t}$.

[ee] Phase is always plotted on a linear scale so that zero and negative phase angles may be included.

$$\delta_{T_0}(t) = \sum_{k=-\infty}^{\infty} \delta(t - kT_0) \Leftrightarrow \omega_0 \delta_{\omega_0}(\omega)$$

$$= \omega_0 \sum_{k=-\infty}^{\infty} \delta(\omega - k\omega_0), \text{ where } \omega_0 = 2\pi/T_0. \quad (5.50)$$

Clearly both the time domain function and its transform are both comb functions: one has period T_0 seconds, while the transformed function has period $\omega_0 = 2\pi/T_0$ rad/sec. We readily conclude that if one function has closely placed impulses, the other function has widely spaced impulses and vice-versa.[ff] Figures 5.14 and 5.15 show the two impulse trains when the period in time domain is $T_0 = 4$ sec:

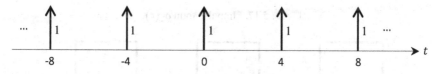

Figure 5.14.   Impulse train $\delta_4(t)$.

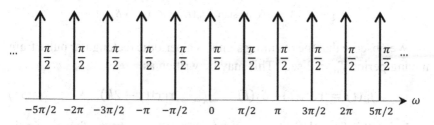

Figure 5.15.   Impulse train $(\pi/2)\delta_{\pi/2}(\omega) = \mathcal{F}\{\delta_4(t)\}$.

5.1.5.2 *Periodic signals as convolution with a comb function*

Periodic power signals play a very important role in linear system analysis. Sinusoidal steady-state analysis of linear electrical circuits is an example of this. Next we show that an arbitrary periodic signal can be considered as a convolution product of one its periods with an impulse train.

Let us consider the convolution of rect(t) and impulse train $\delta_2(t)$. These two signals and their convolution product are shown in Figs. 5.16 through 5.18:

[ff] There is also a corresponding change in areas of the impulses.

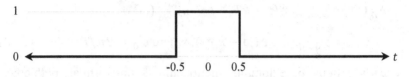

Figure 5.16. Rectangular pulse rect(t).

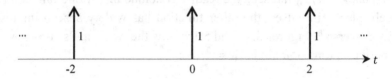

Figure 5.17. Impulse train $\delta_2(t)$.

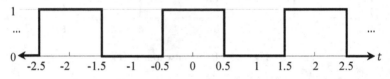

Figure 5.18. Periodic signal $f_p(t) = \text{rect}(t) * \delta_2(t)$.

We observe that the resulting signal is a periodic rectangular pulse train having period $T_0 = 2$ sec. This may be written as:

$$f_p(t) = \text{rect}(t) * \delta_2(t) = \sum_{k=-\infty}^{\infty} \text{rect}(t - 2k). \qquad (5.51)$$

Using this formulation, we can determine the spectrum of any periodic signal in terms of the spectrum of one of its periods. For example, $\mathcal{F}\{\text{rect}(t)\} = \text{sinc}(\omega/2)$ and $\mathcal{F}\{\delta_2(t)\} = \pi \sum_{k=-\infty}^{\infty} \delta(\omega - n\pi)$ in Fig. 5.18. Therefore, the periodic signal of rectangular pulses of width 1 and period 2 has a spectrum represented by

$$F(\omega) = \mathcal{F}\{\text{rect}(t) * \delta_2(t)\} = \mathcal{F}\{\text{rect}(t)\} \times \mathcal{F}\{\delta_2(t)\}$$

$$= \text{sinc}\left(\frac{\omega}{2}\right) \times \pi \sum_{k=-\infty}^{\infty} \delta(\omega - k\pi)$$

$$= \sum_{k=-\infty}^{\infty} \pi \, \text{sinc}\left(\frac{\omega}{2}\right) \delta(\omega - k\pi)$$

$$= \sum_{k=-\infty}^{\infty} \pi \, \text{sinc}\left(\frac{k\pi}{2}\right) \delta(\omega - k\pi). \qquad (5.52)$$

Equation 5.52 states that $F(\omega)$ consists of impulses periodically spaced in frequency with period π rad/sec, and having areas $\pi\mathrm{sinc}(k\pi/2)$. Note that the frequency of the kth impulse is $k\pi$ rad/sec.

Based on this example, we may generalize that the spectrum of any periodic signal consists of spectral components present only at a discrete set of frequencies. These frequencies are all integer[gg] multiples of a *fundamental frequency* ω_0,[hh] and members of this set of frequencies are said to be *harmonically-related*. Further, the areas of impulses at these frequencies are determined by the spectrum of one period of the time-domain signal.

5.1.5.3 *Exponential Fourier Series*

In the previous example the periodic signal, rectangular pulses of width 1 second that repeated every 2 seconds, had frequency spectrum $F(\omega) = \sum_{n=-\infty}^{\infty} \pi \, sinc(n\pi/2)\delta(\omega - n\pi)$. Recall, the expression for $F(\omega) = \mathcal{F}\{f_p(t)\}$ was obtained by representing periodic signal $f_p(t)$ as a convolution product of one of its periods $\mathrm{rect}(t)$ with impulse train $\delta_2(t)$. In general, any periodic signal $f_p(t)$ may be represented this way:

$$f_p(t) = p(t) * \delta_{T_0}(t),$$

$$\text{where } p(t) \equiv \begin{cases} f_p(t), & 0 \le t < T_0; \\ 0, & \text{otherwise.} \end{cases} \quad (5.53)$$

Solving for $F(\omega) = \mathcal{F}\{f_p(t)\} = P(\omega) \times \omega_0 \delta_{\omega_0}(\omega)$

$$= P(\omega) \sum_{n=-\infty}^{\infty} \omega_0 \delta(\omega - n\omega_0)$$

$$= \sum_{n=-\infty}^{\infty} P(\omega)\omega_0 \delta(\omega - n\omega_0)$$

$$= \sum_{n=-\infty}^{\infty} P(n\omega_0)\omega_0 \delta(\omega - n\omega_0)$$

[gg] including zero and negative integers.

[hh] Do not be confused by the notation "ω_0": this represents the period of impulse trainδ_{ω_0} $(\omega) = \sum_{k=-\infty}^{\infty}(\omega - k\omega_0)$ and not the frequency of the impulse corresponding to $k = 0$.

$$= \sum_{n=-\infty}^{\infty} P(n\omega_0)(\omega_0/2\pi)2\pi\delta(\omega - n\omega_0)$$

$$= \sum_{n=-\infty}^{\infty} P(n\omega_0)(1/T_0)2\pi\delta(\omega - n\omega_0)$$

$$= \sum_{n=-\infty}^{\infty} D_n \, 2\pi\delta(\omega - n\omega_0),$$

where $D_n \equiv P(n\omega_0)/T_0$ and $\omega_0 = 2\pi/T_0$. \hfill (5.54)

Taking the inverse Fourier transform of both sides, easily done using Fourier transform pair $e^{j\beta t} \Leftrightarrow 2\pi\delta(\omega - \beta)$, we obtain the following expression for periodic signal $f_p(t)$:

Exponential Fourier Series	$f_p(t) = \sum_{n=-\infty}^{\infty} D_n \, e^{jn\omega_0 t}.$ \hfill (5.55)

In Eq. (5.55) we now have a representation of a periodic signal $f_p(t)$ in the form known as the *Exponential Fourier Series*. Each term in this summation is of the form $D_n \, e^{jn\omega_0 t} = D_n\{\cos(n\omega_0 t) + j\sin(n\omega_0 t)\}$, having frequency $n\omega_0$ (n integer). Thus terms in the Fourier Series summation are harmonically-related and are harmonics of fundamental frequency $\omega_0 = 2\pi/T_0$.[ii] The constants D_n, which are functions of harmonic index n, are called the *Exponential Fourier Series coefficients*.

Recall that $D_n = P(n\omega_0)/T_0$, where $P(\omega)$ is the Fourier transform of one period of $f_p(t)$. From this we may obtain an expression for D_n in terms of $f_p(t)$:

$$D_n = \frac{1}{T_0} P(n\omega_0) = \frac{1}{T_0} \int_{-\infty}^{\infty} p(t)e^{-j(n\omega_0)t} dt$$

$$= \frac{1}{T_0} \int_0^{T_0} f_p(t)e^{-j(n\omega_0)t} dt. \hfill (5.56)$$

The same result is obtained when the integration is done over interval $[t_1, t_1 + T_0) \, \forall t_1$. Therefore:

Solution for Exponential Fourier Series Coefficients	$D_n = \frac{1}{T_0} \int_{T_0} f_p(t)e^{-jn\omega_0 t} dt. \hfill (5.57)$

[ii] "Harmonics" usually refers to positive integer multiples of ω_0. See the footnote on page 184 for more about negative frequencies.

Equation (5.57), the solution for D_n values in an Exponential Fourier series, resembles the expression for a Fourier transform integral. And Eq. (5.55), the expression for $f_p(t)$ as a Fourier series, is in fact an inverse Fourier transform of a spectrum containing only impulses; hence its integration over a continuum of frequencies simplifies to a summation over discrete frequency values. To summarize, Eqs. (5.55) and (5.57) are the continuous-time Fourier transform and its inverse as applied to a special category of time-domain signals: those continuous-time signals $f(t)$ that are periodic.

The Exponential Fourier Series, as described by Eqs. (5.55) and (5.57), may also be derived independently of the Fourier transform.[jj] This approach approximates periodic signal $f_p(t)$ with a weighted sum of harmonically-related basis signals $e^{jn\omega_0 t}$. These basis signals are *mutually orthogonal*: one period of $e^{jn_1\omega_0 t}(e^{jn_2\omega_0 t})^*$ has zero area when $n_1 \neq n_2$, and nonzero area when $n_1 = n_2$. Mutually-orthogonal basis signals make it possible to easily solve for optimal[kk] coefficients D_n due to the disappearance of most cross-product terms when integrating. This method of deriving the Fourier Series also better explains how $f_p(t)$ is approximated at time instants where it is not continuous.[ll]

The Exponential Fourier Series coefficients are plotted in a manner similar to the spectrum sketches described previously, except that now the spectrum will be discrete. Figure 5.19 shows the Exponential Fourier

---

[jj] In the majority of text books the concept of Fourier series is introduced first, and then the Fourier transform is introduced as a limiting case of the Fourier series: e.g., $f(t) = \lim_{T_0 \to \infty} f_p(t)$.

[kk] D_n's are chosen to minimize the power of error signal $e_p(t) = f_p(t) - \left(\sum_{n=-\infty}^{\infty} D_n e^{jn\omega_0 t}\right)$.

[ll] An optimal Fourier Series approximation to periodic signal $f_p(t)$ is one whose error signal has zero power, but this does not guarantee that $\sum_{n=-\infty}^{\infty} D_n e^{jn\omega_0 t}$ converges to exactly equal $f_p(t)$ at all instants of time: at discontinuities in $f_p(t)$ its Fourier Series converges to the mean of amplitudes on either side of the jump. (No wonder that, at the time of a discontinuity, functions such as $\text{rect}(t)$, $\text{sgn}(t)$ and $u(t)$ are defined to equal the mean of amplitudes on either side of the jump.)

Series coefficient spectrum of the periodic rectangular pulse train from Fig. 5.18.

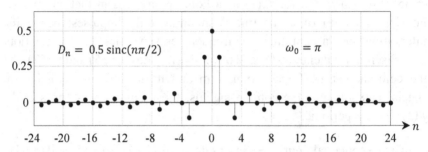

Figure 5.19. Exponential Fourier Series spectrum for periodic signal $f_p(t) = \text{rect}(t) *$ $\delta_2(t) = \sum_{n=-\infty}^{\infty} D_n \, e^{jn(\pi)t}$.

5.1.5.4 *Trigonometric Fourier Series*

Two other Fourier series formulations are commonly discussed in text-books. They are the Trigonometric Fourier Series (using sine and cosine functions), and a compact version of it. We will present these next. As you will notice, once the Fourier coefficients in one formulation are determined, the coefficients in all other formulations can be easily found from them.

Let us consider Eq. (5.29), the complex exponential form of the Fourier series:

$$f_p(t) = \sum_{n=-\infty}^{\infty} D_n \, e^{jn\omega_0 t}$$

$$= D_0 + \sum_{n=1}^{\infty} \{ D_n e^{jn\omega_0 t} + D_{-n} e^{-jn\omega_0 t} \}$$

$$= D_0 + \sum_{n=1}^{\infty} \left\{ \begin{array}{l} D_n(\cos(n\omega_0 t) + j\sin(n\omega_0 t)) \\ + D_{-n}(\cos(n\omega_0 t) - j\sin(n\omega_0 t)) \end{array} \right\}$$

$$= D_0 + \sum_{n=1}^{\infty} \left\{ \begin{array}{l} (D_n + D_{-n})\cos(n\omega_0 t) \\ + j(D_n - D_{-n})\sin(n\omega_0 t) \end{array} \right\}. \qquad (5.58)$$

Define:
$$a_0 = D_0$$
$$a_n = (D_n + D_{-n}) \qquad (n = 1,2,3,\cdots)$$
$$b_n = j(D_n - D_{-n}) \qquad (n = 1,2,3,\cdots). \qquad (5.59)$$

Then we obtain the expression for the Trigonometric Fourier Series:

Trigonometric Fourier Series

$$f_p(t) = a_0 + \sum_{n=1}^{\infty} \begin{Bmatrix} a_n \cos(n\omega_0 t) \\ +b_n \sin(n\omega_0 t) \end{Bmatrix}. \qquad (5.60)$$

Coefficient a_0 is the average value of $f_p(t)$, and b_0 does not appear in the expression ($b_0 = 0$). Equation (5.60) separates the periodic signal $f_p(t)$ into its even (cosine terms and constant) and odd (sine terms) components. Also, by replacing complex exponentials with sines and cosines, there is no need to have any negative frequency values.

If one wishes to find the Trigonometric Fourier Series coefficients directly from $f_p(t)$, then the following expressions may be used (as found from Eqs. (5.57) and (5.59)):

$$a_0 = \frac{1}{T_0} \int_{T_0} f_p(t)\, dt, \qquad (5.61)$$

$$a_n = \frac{2}{T_0} \int_{T_0} f_p(t) \cos(n\omega_0 t)\, dt \quad (n = 1,2,3,\cdots), \qquad (5.62)$$

$$b_n = \frac{2}{T_0} \int_{T_0} f_p(t) \sin(n\omega_0 t)\, dt \quad (n = 1,2,3,\cdots). \qquad (5.63)$$

In most cases we deal with time-domain periodic signals that are real, in which case the Trigonometric Fourier Series coefficients a_n's and b_n's will be real. The Exponential Fourier Series coefficients are usually complex, even when $f_p(t)$ is real. These properties are shown below.

Always true: $\qquad D_n = \frac{1}{2}(a_n - jb_n),$

$$D_{-n} = \frac{1}{2}(a_n + jb_n) \text{ for } n \geq 0, \qquad (5.64)$$

When $f_p(t)$ is real: $\qquad a_n = 2\, Re\{D_n\} = 2\, Re\{D_{-n}\},$

$$b_n = -2\, Im\{D_n\}. \qquad (5.65)$$

<u>Example 5.7</u>
Find the trigonometric Fourier series coefficients of periodic function $f_p(t) = \sum_{k=-\infty}^{\infty} \text{rect}(t - 4k)$.

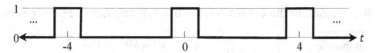

Figure 5.20. Periodic signal $f_p(t) = \text{rect}(t) * \delta_4(t)$, in Example 5.7.

The period of this periodic pulse train, shown in Fig. 5.20, is $T_0 = 4$. Over the period centered at $= 0$, the function $f_p(t) = \text{rect}(t)$. Also, $\omega_0 = 2\pi/T_0 = \pi/2$. We may calculate the trigonometric Fourier series coefficients using Eqs. 5.61–5.63:

$$a_0 = \frac{1}{4}\int_{-2}^{2} f_p(t)\,dt = \frac{1}{4}\int_{-1/2}^{1/2}(1)\,dt = \frac{1}{4}(1) = 0.25, \tag{5.66}$$

$$a_n\ (n = 1,2,3,\cdots) = \frac{2}{4}\int_{-1/2}^{1/2}(1)\cos\left(\frac{n\pi}{2}t\right)dt = 0.5\,\frac{\sin\left(\frac{n\pi}{2}t\right)}{\frac{n\pi}{2}}\Bigg|_{-1/2}^{1/2}$$

$$= \frac{1}{n\pi}\left\{\sin\left(\frac{n\pi}{4}\right) - \sin\left(-\frac{n\pi}{4}\right)\right\} = \frac{2}{n\pi}\sin\left(\frac{n\pi}{4}\right), \tag{5.67}$$

$$b_n\ (n = 1,2,3,\cdots) = \frac{2}{4}\int_{-\frac{1}{2}}^{\frac{1}{2}}(1)\sin\left(\frac{n\pi}{2}t\right)dt = 0. \tag{5.68}$$

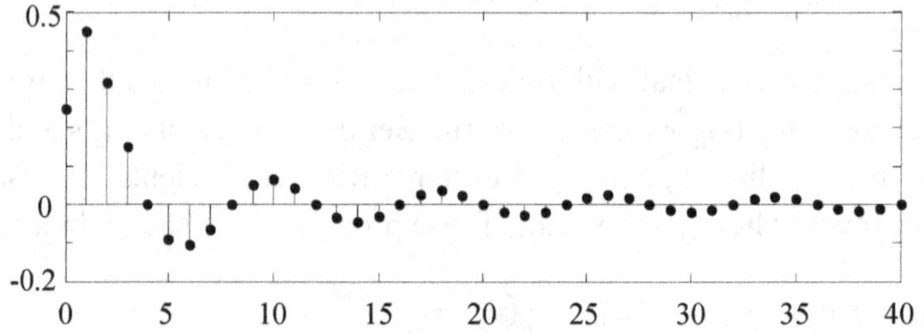

Figure 5.21. Plot of Trigonometric Fourier Series coefficients a_0-a_{40} for periodic signal $f_p(t) = \text{rect}(t) * \delta_4(t)$ in Example 5.7.

The b_n's are all zero because $f(t)$ is an even function.[mm] The spectrum a_n vs. n is shown in Fig. 5.21.

[mm] The Trigonometric Fourier Series of an even signal contains only even (cosine) terms, and hence all sine terms are missing because $b_1, b_2, b_3 \ldots = 0$. Similarly, an odd signal will have $a_0, a_1, a_2, a_3 \ldots = 0$ in its Trigonometric Fourier Series.

5.1.5.5 *Compact Trigonometric Fourier Series*

A compact form of the trigonometric Fourier series is readily obtained by combining the sine and cosine terms for each frequency index n via Euler's identity. We obtain:

$$a_n \cos(n\omega_0 t) + b_n \sin(n\omega_0 t) = C_n \cos(n\omega_0 t + \theta_n). \qquad (5.69)$$

This gives us the expression for the *Compact Trigonometric Fourier Series*:

Compact Trigonometric Fourier Series

$$f_p(t) = C_0 + \sum_{n=1}^{\infty} C_n \cos(n\omega_0 t + \theta_n) \qquad (5.70)$$

Where, when $f_p(t)$ is real:

$$C_0 = a_0 = D_0, \qquad (5.71)$$

$$C_n = \sqrt{a_n^2 + b_n^2} = 2|D_n| \qquad (n = 1,2,3,\cdots), \qquad (5.72)$$

$$\theta_n = Tan^{-1}\left(-\frac{b_n}{a_n}\right) = \angle D_n \qquad (n = 1,2,3,\cdots). \qquad (5.73)$$

As shown above, coefficients $\{C_n$ and $\theta_n\}$ can be determined from either $\{a_n$ and $b_n\}$ or D_n. With C_n's and θ_n's determined, one would plot the magnitude spectrum (C_n vs. n) and the phase spectrum (θ_n vs. n) as is done with the Fourier transform of a signal.

5.1.5.6 *Parseval's Theorem*

A periodic signal $f_p(t)$, with period $T_0 = 2\pi/\omega_0$, has power $P_{f_p} =$ energy per unit time:

$$P_{f_p} = \frac{1}{T_0} \int_{T_0} |f_p(t)|^2 \, dt. \qquad (5.74)$$

The power of signal $D_n e^{jn\omega_0 t}$ is equal to $|D_n|^2$. This is easy to show using above definition, and is something that you would have encountered in your basic circuit theory course.[nn] One can also show that for two complex exponentials at different frequencies, $D_n e^{jn\omega_0 t}$ and $D_m e^{jm\omega_0 t}$ ($n \neq m$), the power of their sum is equal to the sum of their powers: [oo]

$$\text{Power}\{D_n e^{jn\omega_0 t} + D_m e^{jm\omega_0 t}\} = |D_n|^2 + |D_m|^2. \qquad (5.75)$$

The result, although derived for the case of superposition of two complex exponential signals, can be readily generalized for multiple sinusoids, as well as for the case when the sinusoidal frequencies are not harmonically related. Now we will state Parseval's Theorem in terms of Fourier Series coefficients D_n:

Parseval's Theorem for the Fourier Series	$P_f = \frac{1}{T_0}\int_{T_0}\left|f_p(t)\right|^2 dt = \sum_{n=-\infty}^{\infty}|D_n|^2 \qquad (5.76)$

Or equivalently, when $f_p(t)$ is real:

$$P_f = \frac{1}{T_0}\int_{T_0} f_p^2(t)\, dt = C_0^2 + \frac{1}{2}\sum_{n=1}^{\infty} C_n^2 \qquad (5.77)$$

Parseval's Theorem states that power of a periodic signal $f_p(t)$ is equal to the sum of the power in its sinusoidal components.

[nn] $Power = \frac{1}{T_0}\int_{T_0}\left|D_n e^{jn\omega_0 t}\right|^2 dt = \frac{1}{T_0}\int_{T_0}|D_n|^2\left|e^{jn\omega_0 t}\right|^2 dt = |D_n|^2 \frac{1}{T_0}\int_{T_0}\left|e^{jn\omega_0 t}\right|^2 dt = |D_n|^2 \frac{1}{T_0}\int_{T_0}(1)dt = |D_n|^2 \frac{1}{T_0}(T_0) = |D_n|^2.$

[oo] Proof: Let $f_p(t) = D_n e^{jn\omega_0 t} + D_m e^{jm\omega_0 t}$ ($n \neq m$). $P_f = \frac{1}{T_0}\int_{T_0}(D_n e^{jn\omega_0 t} + D_m e^{jm\omega_0 t})(D_n e^{jn\omega_0 t} + D_m e^{jm\omega_0 t})^* dt = \frac{1}{T_0}\int_{T_0}(D_n e^{jn\omega_0 t} + D_m e^{jm\omega_0 t})(\{D_n e^{jn\omega_0 t}\}^* + \{D_m e^{jm\omega_0 t}\}^*)dt = \frac{1}{T_0}\int_{T_0}(D_n e^{jn\omega_0 t} + D_m e^{jm\omega_0 t})(D_n^* e^{-jn\omega_0 t} + D_m^* e^{-jm\omega_0 t})dt = \frac{1}{T_0}\int_{T_0}(D_n e^{jn\omega_0 t}D_n^* e^{-jn\omega_0 t} + D_m e^{jm\omega_0 t}D_n^* e^{-jn\omega_0 t} + D_n e^{jn\omega_0 t}D_m^* e^{-jm\omega_0 t} + D_m e^{jm\omega_0 t}D_m^* e^{-jm\omega_0 t})\, dt = \frac{1}{T_0}\int_{T_0}(D_n e^{jn\omega_0 t}D_n^* e^{-jn\omega_0 t} + D_m e^{jm\omega_0 t}D_m^* e^{-jm\omega_0 t})\, dt = \frac{1}{T_0}\int_{T_0}|D_n|^2 e^{jn\omega_0 t}e^{-jn\omega_0 t}\, dt + \frac{1}{T_0}\int_{T_0}|D_m|^2 e^{jm\omega_0 t}e^{-jm\omega_0 t}\, dt = |D_n|^2 \frac{1}{T_0}\int_{T_0} e^{jn\omega_0 t}e^{-jn\omega_0 t}\, dt + |D_m|^2 \frac{1}{T_0}\int_{T_0} e^{jm\omega_0 t}e^{-jm\omega_0 t}\, dt = |D_n|^2 \frac{1}{T_0}\int_{T_0}\left|e^{jn\omega_0 t}\right|^2 dt + |D_m|^2 \frac{1}{T_0}\int_{T_0}\left|e^{jm\omega_0 t}\right|^2 dt = |D_n|^2 \frac{1}{T_0}\int_{T_0}(1)\, dt + |D_m|^2 \frac{1}{T_0}\int_{T_0}(1)\, dt = |D_n|^2 \frac{1}{T_0}(T_0) + |D_m|^2 \frac{1}{T_0}(T_0) = |D_n|^2 + |D_m|^2.$

5.1.6 *Summary of Fourier transformations for continuous-time signals*

Table 5.3. Summary of Fourier transformations for continuous-time signals.

Time signal		Frequency spectrum
Continuous in time	Continuous-Time Fourier Transform (CTFT)	Continuous in frequency
$f(t)$ $\Rightarrow$	$\displaystyle \int_{-\infty}^{\infty} f(t)e^{-j\omega t}dt$ $\Rightarrow$	$F(\omega)$
	Inverse Continuous-Time Fourier Transform (ICTFT)	
$f(t)$ $\Leftarrow$	$\displaystyle \frac{1}{2\pi}\int_{-\infty}^{\infty} F(\omega)e^{j\omega t}d\omega$ $\Leftarrow$	$F(\omega)$
Continuous, periodic in time	Fourier Series (calculation of D_k)	Discrete in frequency
$f_p(t)$ $\Rightarrow$	$\displaystyle \frac{1}{T_0}\int_{T_0} f_p(t)e^{-j(k\omega_0)t}dt$ $\Rightarrow$	D_k
	Fourier Series (reconstruction of $f_p(t)$)	
$f_p(t)$ $\Leftarrow$	$\displaystyle \sum_{k=-\infty}^{\infty} D_k e^{j(k\omega_0)t}$ $\Leftarrow$	D_k

5.2 Practical Applications

5.2.1 *Frequency scale of a piano keyboard*

The frequency domain may seem to be, upon first consideration, a strange place to analyze signals. After all, each of us can sense the passing of time and the amplitude-vs.-time waveform description of a signal makes perfect sense. The piano keyboard, however, has such direct connection to the frequency domain that we must mention it here as a practical application of frequency-domain signal representation.

Pressing a key on the piano keyboard activates a mechanism to produce a sound whose energy is concentrated at a specific frequency. Comparing the frequencies corresponding to the piano keys, one will notice that they are logarithmically spaced (with respect to key index) in frequency. That is, the nth key frequency is a constant value times the $(n-1)^{\text{th}}$ key frequency:

$$f_{n+1} = c f_n. \tag{5.78}$$

This means that shifting to the right by m keys will increase the frequency by factor c^m:

$$f_{n+m} = c^m f_n. \tag{5.79}$$

On a standard piano keyboard the frequency doubles after moving to the right by 12 keys (including both black and white):

$$f_{n+12} = c^{12} f_n = 2 f_n. \tag{5.80}$$

From this relationship, we may solve for c and arrive at the expression:

$$f_{n+1} = 2^{1/12} f_n \cong 1.0595 \, f_n. \tag{5.81}$$

Why are the frequencies of keys on a piano keyboard uniformly spaced on a log scale? Why not uniform spacing on a linear scale? The answer is that the human sense of hearing perceives frequency of sound on a log scale. It is no wonder that shifting a melody by 12 notes (one octave) results in the same-sounding "key."

In addition to doubling of frequency with each octave, there exist other mathematical relationships between frequency ratios of common musical intervals in the 12-tone equal temperament tuning method (uniform spacing on a log frequency scale):

$2^{1/12} \cong 1.0595$: minor 2nd

$2^{2/12} \cong 1.1225$: Major 2nd

$2^{3/12} \cong 1.1892$: minor 3rd

$2^{4/12} \cong 1.2599$: Major 3rd

$2^{5/12} \cong 1.3348$: perfect 4th

$2^{6/12} \cong 1.4142$: augmented 4th, diminished 5th

$2^{7/12} \cong 1.4983$: perfect 5th

$2^{8/12} \cong 1.5874$: minor 6th

$2^{9/12} \cong 1.6818$: Major 6th

$2^{10/12} \cong 1.7818$: minor 7th

$2^{11/12} \cong 1.8877$: Major 7th

Amazingly, the human mind perceives some frequency steps as sounding sad and others as sounding happy (minor and major intervals, respectively). Therefore, there may be emotional information conveyed by a pair of frequencies!

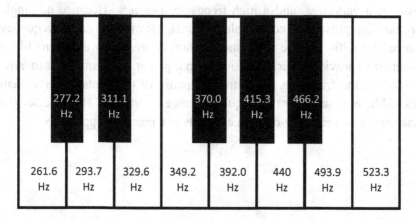

Figure 5.22. Frequencies of piano keys over the middle octave, with A-440 tuning. Note that each key frequency value increases in frequency by factor 1.0595.

5.2.2 *Frequency-domain loudspeaker measurement*

Aside from power-handling capability, the performance of a loudspeaker is usually characterized by a plot of its frequency magnitude response. How is this response measured? First, the loudspeaker is placed in an anechoic chamber (essentially a room without echoes, due to sound absorptive materials on the walls, floor and ceiling) and hooked up to an amplifier. A high-quality microphone, one having nearly flat frequency response in the audio range, is placed in front and at a specific distance from the speaker. Then the amplifier is driven by a pure sinusoid at a fixed amplitude and frequency. The resulting microphone output signal's amplitude is measured and recorded. The measurement is repeated for different frequencies in the audio range, and a curve is then plotted to include the resulting data points. This curve is plotted on a dB scale (mentioned later in this chapter). Because phase response is usually not measured,[pp] the loudspeaker is not fully characterized by this measurement. Even so, the information provided by a magnitude frequency plot is useful for rating the quality of a loudspeaker and judging its suitability for a specific application.

Loudspeaker systems often have three speakers: a low-frequency *woofer*, a *midrange*, and a high-frequency *tweeter*. Because no single speaker can provide an acceptable output signal over the audio frequency range 20 to 20k Hz, the three speaker output pressure waveforms blend together to provide a fairly-uniform output power vs. frequency. Ideally, of course, the frequency magnitude response of the loudspeaker system should be constant over the frequency range of interest. Here we see the usefulness of frequency-domain concepts in a practical application.

[pp] Some argue that the human ear is more sensitive to magnitude than to phase, but it depends on the situation.

5.2.3 *Effects of various time-domain operations on frequency magnitude and phase*

In this section, we consider how some basic time-domain operations affect the magnitude and phase components of the frequency spectrum. We base our conclusions on the Fourier transform properties given in Table 5.2.

(a) Multiplication by a constant:

Since $af(t) \Leftrightarrow aF(\omega)$, we see that multiplying by a constant in the time domain results in multiplication by the same constant in the frequency domain. This constant may be called the *gain factor*. When the constant a is positive real, then magnitude scales by a and phase is unchanged: $|aF(\omega)| = |a||F(\omega)| = a|F(\omega)|$; $\angle aF(\omega) = \angle a + \angle F(\omega) = \angle F(\omega)$. When gain factor a is complex, then both magnitude and phase are affected: $|aF(\omega)| = |a||F(\omega)|$, $\angle aF(\omega) = \angle a + \angle F(\omega)$.

(b) Adding a constant:

Since $f_1(t) + f_2(t) \Leftrightarrow F_1(\omega) + F_2(\omega)$, we see that adding a constant to $f(t)$ results in the addition of an impulse function at frequency $\omega = 0$ in the frequency domain: $f(t) + a \Leftrightarrow F(\omega) + a \cdot 2\pi\delta(\omega)$. Hence magnitude and phase are unaffected at all nonzero frequencies. [qq]

(c) Time delay:

Since $f(t - t_0) \Leftrightarrow F(\omega)e^{-j\omega t_0}$, delaying $f(t)$ by t_0 seconds results in no change in the magnitude spectrum: $|F(\omega)e^{-j\omega t_0}| = |F(\omega)||e^{-j\omega t_0}| = |F(\omega)|$. The phase, however, is modified by an

---

[qq] At zero frequency, magnitude becomes ∞ and phase is meaningless.

additive frequency-dependent linear offset: $\angle\{F(\omega)e^{-j\omega t_0}\} = \angle F(\omega) + \angle e^{-j\omega t_0} = \angle F(\omega) - \omega t_0$. Note that this linear offset will have negative slope for a time delay, and positive slope for a time advance.

(d) Time reversal:

We will consider the case that time-domain signal $f(t)$ is real. Since $f(-t) \Leftrightarrow F(-\omega)$ and $f^*(t) \Leftrightarrow F^*(-\omega)$, the effects of time-reversal may be determined by both conjugating and time-reversing $f(t)$: $f(-t) = f^*(-t) \Leftrightarrow F(-\omega)$. Since we know that $|F(\omega)|$ is an even function of frequency when $f(t)$ is real, we conclude that the magnitude spectrum of real $f(t)$ will not change from time-reversal. On the other hand, $\angle F(\omega)$ is an odd function of frequency when $f(t)$ is real so that time reversal results in a negation of the phase spectrum.

(e) Compression or expansion in time:

From Property #9 in Table 5.2, $f(at) = F(\omega/a)/|a|$. Assume constant a is positive real. When $a > 1$ we are compressing $f(t)$ in time (towards $t = 0$) causing both magnitude and phase of $F(\omega)$ to stretch out in frequency (away from $\omega = 0$). When $0 < a < 1$ we are stretching $f(t)$ in time, causing both magnitude and phase of $F(\omega)$ to compress in frequency.

(f) Adding a delayed version of the signal to itself:

Let $g(t) = f(t) + f(t - t_0)$. How do the magnitude and phase spectra of $G(\omega)$ compare to those of $F(\omega)$? To answer this question, we rewrite $g(t) = f(t) * \{\delta(t) + \delta(t - t_0)\}$, or $G(\omega) = F(\omega) \times \mathcal{F}\{\delta(t) + \delta(t - t_0)\} = F(\omega)\{1 + e^{-j\omega t_0}\}$. The magnitude spectrum is $|G(\omega)| = |F(\omega)||1 + e^{-j\omega t_0}| = 2|F(\omega)||\cos(\omega t_0/2)|$, and we see that magnitude response is multiplied by a comb-like function having zeros at $\omega = \pi/t_0, 3\pi/t_0, 5\pi/t_0, \dots$ ᵀᵀ The phase spectrum $\angle G(\omega) = \angle\{F(\omega)(1 + e^{-j\omega t_0})\} = \angle F(\omega) + \angle(1 + e^{-j\omega t_0}) = \angle F(\omega) + Arg\{-\omega t_0\}/2$. ˢˢ

(g) Subtracting a delayed version from itself:

Let $g(t) = f(t) - f(t - t_0)$. How do the magnitude and phase spectra of $G(\omega)$ compare to those of $F(\omega)$? To answer this question, we rewrite $g(t) = f(t) * \{\delta(t) - \delta(t - t_0)\}$, or $G(\omega) = F(\omega) \times \mathcal{F}\{\delta(t) - \delta(t - t_0)\} = F(\omega)\{1 - e^{-j\omega t_0}\}$. The magnitude spectrum is $|G(\omega)| = |F(\omega)||1 - e^{-j\omega t_0}| = 2|F(\omega)||\sin(\omega t_0/2)|$, and we see that the magnitude response is multiplied by a comb-like function having zeros at $\omega = 0, 2\pi/t_0, 4\pi/t_0, \dots$ ᵗᵗ Phase spectrum $\angle G(\omega) = \angle\{F(\omega)(1 - e^{-j\omega t_0})\} = \angle F(\omega) + \angle(1 - e^{-j\omega t_0}) = \angle F(\omega) + Arg\{\pi - \omega t_0\}/2$. ᵘᵘ

5.2.4 *Communication by frequency shifting*

One of the most important applications of signal processing is communications, which is the transfer of information over distance. This is especially true in wireless communications, because typically the message signal is converted to a higher frequency range for efficient transmission using electromagnetic waves.

ᵀᵀ $|1 + e^{-j\omega t_0}| = \sqrt{(1 + \cos(\omega t_0))^2 + (-\sin(\omega t_0))^2} = \sqrt{1 + 2\cos(\omega t_0) + \cos^2(\omega t_0) + \sin^2(\omega t_0)} = \sqrt{2 + 2\cos(\omega t_0)} = \sqrt{4\cos^2(\omega t_0/2)} = 2|\cos(\omega t_0/2)|$.
ˢˢ $\angle(1 + e^{-j\omega t_0})$ can be shown to equal $-\pi/2 \le Arg\{-\omega t_0\}/2 \le \pi/2$, which is half the principal value of $-\omega t_0$.
ᵗᵗ $|1 - e^{-j\omega t_0}| = \sqrt{(1 - \cos(\omega t_0))^2 + (\sin(\omega t_0))^2} = \sqrt{1 - 2\cos(\omega t_0) + \cos^2(\omega t_0) + \sin^2(\omega t_0)} = \sqrt{2 - 2\cos(\omega t_0)} = \sqrt{4\sin^2(\omega t_0/2)} = 2|\sin(\omega t_0/2)|$.
ᵘᵘ $\angle(1 - e^{-j\omega t_0})$ can be shown to equal $-\pi/2 \le Arg\{\pi - \omega t_0\}/2 \le \pi/2$, which is half the principal value of $\pi - \omega t_0$.

Consider the magnitude spectrum $|M(\omega)| = \Delta(\omega/2W)$, pictured in Fig. 5.23, which we will use to model a message signal (e.g. audio or video). This is called a *baseband signal* since its spectrum represents the signal in its original form. The energy of $M(\omega)$ is distributed over frequency range (bandwidth) $|\omega| < W$ rad/sec. Is it possible to shift the information that $M(\omega)$ represents to a higher frequency range, the *transmission band*, and then back to baseband, without any loss of information? The answer is yes, when the bandwidth of $M(\omega)$ is finite as we have assumed.

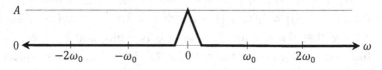

Figure 5.23.   Sample baseband spectrum $M(\omega)$.

In the time domain, represent the message as $m(t) = \mathcal{F}^{-1}\{M(\omega)\}$. Since our goal is to shift $M(\omega)$ to a higher frequency range, we notice the frequency shift property of the Fourier transform (#7 in Table 5.2):

$$m(t)e^{j\omega_0 t} \Leftrightarrow M(\omega - \omega_0).$$

This does exactly what we need, but the problem is that in the real world we cannot generate $e^{j\omega_0 t}$. Instead, some real signal must be used. The solution is to instead multiply $m(t)$ by $\cos(\omega_0 t) = \frac{1}{2}e^{j\omega_0 t} + \frac{1}{2}e^{-j\omega_0 t}$. This is the modulation property (#5 in Table 5.2):

$$m(t)\cos(\omega_0 t) \Leftrightarrow \frac{1}{2}M(\omega - \omega_0) + \frac{1}{2}M(\omega + \omega_0). \qquad (5.82)$$

Thus, by multiplying message signal $m(t)$ with a high-frequency sinusoid $\cos(\omega_0 t)$ we shift a scaled version of baseband spectrum $M(\omega)$ up and down in frequency, as shown in Fig. 5.24. The closeness of transmission band center frequency ω_0 to the baseband is exaggerated here; normally $\omega_0 \gg$ bandwidth of $m(t)$. Note that we have achieved our goal of shifting an exact copy of the information in signal $m(t)$ to a higher

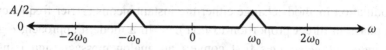

Figure 5.24. Spectrum of $m(t)\cos(\omega_0 t)$, which is $\frac{1}{2}M(\omega - \omega_0) + \frac{1}{2}M(\omega + \omega_0)$.

frequency range! The cosine used here is called the *carrier* wave, because its amplitude now reflects changes in $m(t)$ and thus carries its information over the transmission frequency band. In general, multiplying $m(t)$ by a high-frequency carrier is called *Amplitude Modulation (AM)*.

To recover $m(t)$ from the signal sent to the receiver, $m(t)\cos(\omega_0 t)$, one performs another modulation operation (that is called *demodulation*):

$$\{m(t)\cos(\omega_0 t)\}\{2\cos(\omega_0 t)\} = m(t) \cdot 2\cos^2(\omega_0 t)$$

$$= m(t)(1 + \cos(2\omega_0 t))$$

$$= m(t) + m(t)\cos(2\omega_0 t), \qquad (5.83)$$

whose frequency domain spectrum is

$$= M(\omega) + \frac{1}{2}M(\omega - 2\omega_0) + \frac{1}{2}M(\omega + 2\omega_0), \qquad (5.84)$$

as shown in Fig. 5.25:

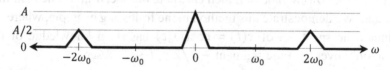

Figure 5.25. Sample baseband spectrum $M(\omega)$.

The spectral copies at $\pm 2\omega_0$ are easily eliminated using a *lowpass filter*.[vv] What remains is the original baseband spectrum $M(\omega)$, and our amplitude modulation system has succeeded in recovering message signal $m(t)$ from the amplitude-modulated carrier signal $m(t)\cos(\omega_0 t)$. Other

---

[vv] Analog lowpass filtering is covered in Ch. 8.

modulation methods change either the instantaneous carrier frequency or its phase in direct proportion to $m(t)$.[ww] In a modern approach, *digital communication systems* first convert an analog message signal into a stream of 1's and 0's using *analog-to-digital conversion* (see Ch. 6) before modulating a carrier with binary-coded information.

5.2.5 Spectral analysis using time windowing

Assume signal $x(t)$ is only known over some finite-width time interval $t_1 \leq t \leq t_2$. How may we best estimate $X(\omega)$ from only what is known about $x(t)$? The answer is: it depends on the signal, and what is meant by "best". For example, when $x(t)$ is periodic having period T_0, then all we need is one period of this signal, $x(t_0 \leq t \leq t_0 + T_0)$ for any t_0, to exactly determine $X(\omega)$. However, when nothing is known about $x(t)$ outside of the interval $t_1 \leq t \leq t_2$ all we can do is hope that the spectrum estimated from $x(t_1 \leq t \leq t_2)$ closely resembles the spectrum of $x(-\infty \leq t \leq \infty)$. Therefore, when signal $x(t)$ has a high degree of self-similarity vs. time our estimate can be more accurate. The assumptions we make, and how they closely they agree with reality, affect the accuracy of our spectral estimate $\tilde{X}(\omega) \cong X(\omega)$.

A common way to estimate the spectrum of $x(t)$ is to assume that it equals zero outside of our observation interval, multiply what we know of $x(t)$ by a *windowing function*, then calculate the spectrum of the resulting product. We demonstrate this method in the following example, where we estimate the spectrum of $x(t) = \cos(\omega_0 t)$ based on knowledge of the signal only over the time segment $-T/2 \leq t \leq T/2$ sec.

Example 5.8
Estimate the spectrum of $x(t)$, based on the information that $x(t) = \cos(\omega_0 t)$ over time interval $-T/2 \leq t \leq T/2$. At other times $x(t)$ is not known.

[ww] Resulting in *Frequency Modulation* (*FM*) and *Phase Modulation* (*PM*).

By assuming $x(t) = 0$ outside of $-T/2 \le t \le T/2$ and calculating the Fourier transform over all time, we are in effect estimating $X(\omega)$ as $\tilde{X}(\omega) = \mathcal{F}\{\cos(\omega_0 t)\,\text{rect}(t/T)\}$. From Tables 5.1 and 5.2:

$$\cos(\omega_0 t) \Leftrightarrow \pi\delta(\omega + \omega_0) + \pi\delta(\omega - \omega_0) = \pi\delta(\omega \pm \omega_0)$$

$$\text{rect}(t/T) \Leftrightarrow T\text{sinc}(\omega T/2)$$

$$\therefore \cos(\omega_0 t)\,\text{rect}(t/T) \Leftrightarrow \frac{T}{2}\text{sinc}(\omega T/2) * \delta(\omega \pm \omega_0) = \tilde{X}(\omega),$$

$$\text{or } \tilde{X}(\omega) = \frac{T}{2}\text{sinc}((\omega + \omega_0)T/2) + \frac{T}{2}\text{sinc}((\omega - \omega_0)T/2). \quad (5.85)$$

Noting that $X(\omega) = \pi\delta(\omega \pm \omega_0)$ has its energy at frequencies $\pm\omega_0$, estimated spectrum $\tilde{X}(\omega)$ differs in that its energy is smeared to either side of $\pm\omega_0$ (see Fig. 5.26). This is called the *time windowing effect* in spectral estimation: instead of $x(t)$ we analyze $x(t)w(t)$, which distorts our spectral estimate because $X(\omega)$ is convolved with $W(\omega)$: [xx]

$$x(t)w(t) \Leftrightarrow \frac{1}{2\pi} X(\omega) * W(\omega) \quad (5.86)\text{[yy]}$$

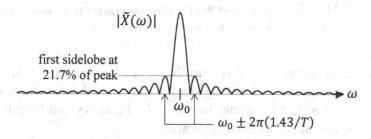

Figure 5.26. Magnitude spectrum of $\tilde{X}(\omega) = \mathcal{F}\{\cos(\omega_0 t)\,\text{rect}(t/T)\}$ (sinusoid multiplied by a rectangular time window) shown near $\omega = \omega_0$, from Example 5.8.

The windowing effect may mislead us to believe that multiple sinusoidal components are present, despite $x(t)$ having energy only at ω_0, due to

---

[xx] In this example, $x(t) = \cos(\omega_0 t)$ and $w(t) = \text{rect}(t/T)$.

[yy] Property #4 in Table 5.2 (p. 177)

the *sidelobes* of energy that appear by convolving with a sinc function that is $W(\omega)$ in this case.

The distortion caused by the windowing effect depends on the shape and width of time window $w(t)$, and on the nature of $x(t)$ whose spectrum is being estimated.[zz] In general, the wider $w(t)$ is in time – the better.

In most applications, another desirable characteristic of windowing function $w(t)$ is that it tapers gradually to zero at both ends. We demonstrate this concept by replacing the rectangular time window with a triangular one when estimating the spectrum of $x(t) = \cos(\omega_0 t)$:

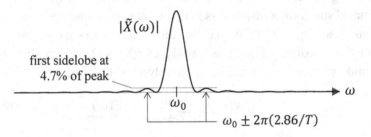

Figure 5.27. Magnitude spectrum of $\tilde{X}(\omega) = \mathcal{F}\{\cos(\omega_0 t)\,\Delta(t/T)\}$ (sinusoid multiplied by a triangular time window) shown near $\omega = \omega_0$, from Example 5.8.

Different windowing functions have been proposed for spectral estimation purposes, each having specific frequency-domain characteristics (such as main lobe width, first sidelobe frequency and height) that should be considered for each application.[aaa]

[zz] In some cases, the windowing effect is negligible (when $x(t)w(t) \cong x(t)$).

[aaa] These windows are typically defined in the discrete-time domain so that computer-based methods may be used to calculate the Fourier transform; our example is given in the continuous time domain to simplify the explanations.

5.2.6 Representing an analog signal with frequency-domain samples

Chapter 6 introduces the theory and applications of sampling a continuous-time signal $x(t)$ to obtain a list of numbers, and then, if certain conditions are met, recovering $x(t)$ from that list. This amazing concept can also be seen from a frequency domain perspective, which we examine here.

Consider finite-length signal $x(t)$ that equals zero outside of the time interval $[-T_0/2, T_0/2]$.[bbb] Thus $x(t) = x(t)\text{rect}(t/T_0)$. Create periodic signal $x_p(t)$ by convolving $x(t)$ with impulse train $\delta_{T_0}(t)$:

$$x_p(t) = x(t) * \delta_{T_0}(t). \qquad (5.87)$$

Next, because of its periodicity, we may decompose $x_p(t)$ into a weighted sum of complex exponentials (Exponential Fourier Series),

$$x_p(t) = \sum_{n=-\infty}^{\infty} D_n e^{jn\omega_0 t}, \qquad (5.88)$$

where $\omega_0 = 2\pi/T_0$. A solution for the Fourier series coefficients D_n was derived in Section 5.17 (Eq. (5.57)):

$$D_n = \frac{1}{T_0} \int_{-T_0/2}^{T_0/2} x_p(t) e^{-jn\omega_0 t} dt. \qquad (5.89)$$

Although exact solutions for coefficients D_n via integration may not be possible, numerical approximation methods can make this a practical approach. Noting that

$$x(t) = x_p(t)\text{rect}(t/T_0) = \left\{ \sum_{n=-\infty}^{\infty} D_n e^{jn\omega_0 t} \right\}\text{rect}(t/T_0), \qquad (5.90)$$

we see that the original finite-length signal $x(t)$ is exactly recovered from the frequency-domain coefficients D_n, which themselves are derived from $x_p(t)$ and hence $x(t)$. Waveform $x(t)$ and values D_n are therefore interchangeable when representing the signal's information.

[bbb] Assume there are no impulse functions located at the endpoints of this interval.

Consider a special case: $x(t)$ is such that $D_n = 0$ for $\forall n$ such that $|n| >$ M. This means that a *finite* number of values can be used to represent the *infinite* number of signal amplitudes $x(t_k)$: $t_k \in [-T_0/2, T_0/2]$. How is this possible? The answer is: in this case $x(t)$ does not contain the maximum information possible for a pulse having width T_0 seconds (there is redundancy). Therefore, only $2M + 1$ coefficients (D_{-M} through D_{+M}) represent its information in the frequency domain.[ccc]

This example has demonstrated the concept of storing a continuous-time signal's information in a discrete format. It is a form of *frequency-domain sampling*. This is seen from Eq. (5.56), rewritten here in terms of $x(t)$:

$$D_n = \frac{1}{T_0} X(n\omega_0) = \frac{1}{T_0} \mathcal{F}\{x(t)\}|_{\omega = n\omega_0}. \qquad (5.91)$$

We see that D_n is spectrum $X(\omega)$ sampled at frequency $\omega = n\omega_0$ and scaled by amplitude factor $1/T_0$.

5.3 Useful MATLAB® Code

In this section, we will concentrate on the practical task of plotting the spectrum of signal $h(t)$. The phase and magnitude components of frequency this spectrum $\mathcal{F}\{h(t)\} = H(\omega) = |H(\omega)|e^{j\angle H(\omega)}$ are usually plotted separately. When $h(t)$ is real (as is typically the case), then $|H(\omega)|$ is even and $\angle H(\omega)$ is odd, so that no additional information is gained by plotting these components over both positive and negative frequencies. Thus, we normally plot only the positive frequency sides of the magnitude and phase spectra.

Example 5.9
Plot the magnitude and phase of $H(\omega) = j\omega/(5 + j\omega)$ over frequency range $0 \leq \omega \leq 20$ rad/sec.

[ccc] When $x(t)$ is real then $D_{-k} = D_k^*$, which means that only D_k for $k \geq 0$ need to be stored.

```
% Example_Ch5_9
Npts = 1e4;
w = linspace(0,20,Npts);
H = (j*w)./(5 + j*w);
figure; plot(w,abs(H))
xlabel('frequency in rad/sec')
ylabel('magnitude of H(\omega)')
title('H(\omega) = j\omega/(5+j\omega)')
grid on
```

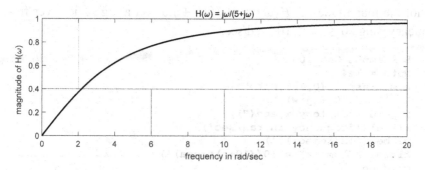

Figure 5.28. Plot of $|H(\omega)| = |j\omega/(5 + j\omega)|$ vs. ω, in Example 5.9.

```
figure; plot(w,angle(H))
xlabel('frequency in rad/sec')
ylabel('angle of H(\omega) in radians')
title('H(\omega) = j\omega/(5+j\omega)')
grid on
```

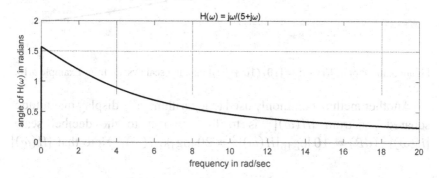

Figure 5.29. Plot of $\angle H(\omega) = \angle(j\omega/(5 + j\omega))$ vs. ω, in Example 5.9.

We often plot magnitude $|H(\omega)|$ on a logarithmic scale.[ddd] Because $\log(0) = -\infty$, at frequencies where $|H(\omega)| = 0$ this type of plot is meaningless. At frequencies where $|H(\omega)| > 0$, however, a logarithmic compression makes it possible to display both very large and very small magnitudes together on the same plot. In MATLAB®, the y-axis is displayed logarithmically using function **semilogy**.

<u>Example 5.10</u>

Plot the magnitude of $H(\omega) = 10/(10 + j\omega)$ on a log scale, over frequency range $0 \le \omega \le 100$ rad/sec:

```
% Example_Ch5_10
Npts = 1e4;
w = linspace(0,100,Npts);
H = 10./(10 + j*w);
figure; semilogy(w,abs(H))
xlabel('frequency in rad/sec')
ylabel('|H(\omega)|')
title('H(\omega) = 10/(10+j\omega)')
grid on
```

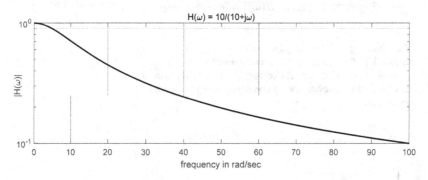

Figure 5.30. Plot of $|H(\omega)| = |10/(10 + j\omega)|$ on a log scale vs. ω, from Example 5.10.

Another method commonly used to logarithmically display magnitude-squared spectrum $|H(\omega)|^2$ is to first map it to the decibel scale: $|H(\omega)|^2$ $(dB) = 10 \log_{10}|H(\omega)|^2 = 20 \log_{10}|H(\omega)|$. Note that $|H(\omega)|$

[ddd] Phase angle is always plotted using a linear scale, typically over the principal value range $-\pi \le \angle H(\omega) \le \pi$ radians (or $-180° \le \angle H(\omega) \le 180°$ degrees).

and $|H(\omega)|^2$ are usually dimensionless, and thus the decibel scale also represents a dimensionless value. Decibels are not units; instead, the *dB* label is only used to remind us of the logarithmic mapping that was done.

<u>Example 5.11</u>

Plot the magnitude-squared value of $H(\omega) = 10/(10 + j\omega)$ in dB, over the frequency range $0 \leq \omega \leq 100$ rad/sec:

```
% Example_Ch5_11
Npts = 1e4;
w = linspace(0,100,Npts);
H = 10./(10 + j*w);
figure; plot(w,20*log10(abs(H)))
xlabel('frequency in rad/sec')
ylabel('|H(\omega)|^2 (dB)')
title('H(\omega) = 10/(10+j\omega)')
grid on
```

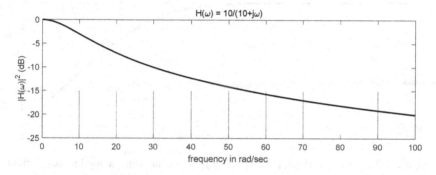

Figure 5.31. Plot of $|H(\omega)|^2 = |10/(10 + j\omega)|^2$ in dB vs. ω, from Example 5.11.

Another common display method is the log-log plot: both the magnitude scale (y-axis) and the frequency scale (x-axis) are compressed logarithmically when plotting. This has the effect of linearizing curves such as $1/\omega$ vs. ω and thus making the plot easier to use for extrapolating data. (For this plot to be useful we cannot include points where either $|H(\omega)| = 0$ or $\omega = 0$, however.)

In MATLAB®, both x and y axes are displayed logarithmically using the function **loglog**. The example below also uses function **logspace** instead of `linspace` when initializing frequency values so that the plot has sample points that are uniformly spaced along the log-frequency axis.

Example 5.12

Plot the magnitude of $H(\omega) = 10/(10 + j\omega)$ on a log-log scale, over the frequency range $0.1 \leq \omega \leq 100$ rad/sec:

```
% Example_Ch5_12
Npts = 1e4;
w = logspace(log10(0.1),log10(100),Npts);
H = 10./(10 + j*w);
figure; loglog(w,abs(H))
xlabel('frequency in rad/sec')
ylabel('|H(\omega)|')
title('H(\omega) = 10/(10+j\omega)')
grid on
```

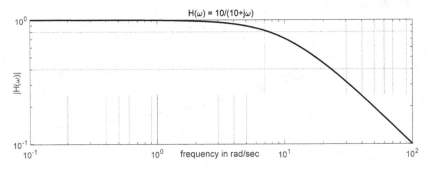

Figure 5.32.   Plot of $|H(\omega)| = |10/(10 + j\omega)|$ vs. ω using a log-log scale, from Example 5.12.

The same result is achieved by plotting $|H(\omega)|^2$ in dB vs. ω on a log scale. In Example 5.13, we take advantage of the MATLAB® `semilogx` graphing function.

Example 5.13

Plot the magnitude-squared value of $H(\omega) = 10/(10 + j\omega)$, in dB, vs. ω on a log scale, over frequency range $0.1 \leq \omega \leq 100$ rad/sec:

```
% Example_Ch5_13
Npts = 1e4;
w = logspace(log10(0.1),log10(100),Npts);
```

```
H = 10./(10 + j*w);
figure; semilogx(w,20*log10(abs(H)))
xlabel('frequency in rad/sec')
ylabel('|H(\omega)|^2 (dB)')
title('H(\omega) = 10/(10+j\omega)')
grid on
```

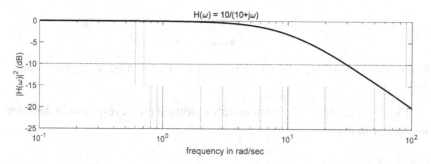

Figure 5.33. Plot of $|H(\omega)|^2 = |10/(10 + j\omega)|^2$ in dB, vs. ω on a log scale, from Example 5.13.

In practical applications engineers rarely specify frequency in rad/sec; f (Hz) $= \omega/2\pi$ is the frequency unit of choice. We repeat the previous example on a Hz frequency scale.

Example 5.14

Plot the magnitude-squared value of $H(f) = 10/(10 + j2\pi f)$, in dB, vs. f on a log scale, over frequency range $0.01 \le f \le 20$ rad/sec:

```
% Example_Ch5_14
Npts = 1e4;
f = logspace(log10(0.01),log10(20),Npts);
w = 2*pi*f;
H = 10./(10 + j*w);
figure; semilogx(f,20*log10(abs(H)))
xlabel('frequency in Hz')
ylabel('|H(f)|^2 (dB)')
title('H(f) = 10/(10+j2\pif)')
grid on
```

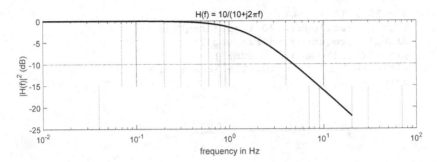

Figure 5.34. Plot of $H(f) = 10/(10 + j2\pi f)$ in dB, vs. f on a log scale, from Example 5.14.

Finally, Example 5.15 demonstrates plotting phase response in degrees instead of in radians.

Example 5.15

Plot the phase of $H(f) = 10/(10 + j2\pi f)$, in degrees, vs. f on a log scale, over frequency range $0.01 \leq f \leq 10$ KHz:

```
% Example_Ch5_15
Npts = 1e4;
f = logspace(log10(0.01),log10(10e3),Npts);
w = 2*pi*f;
H = 10./(10 + j*w);
figure; semilogx(f,angle(H)*180/pi)
xlabel('frequency in Hz')
ylabel('phase in degrees')
title('H(f) = 10/(10+j2\pif)')
grid on
```

Figure 5.35. Plot of $\angle H(f) = \angle(10/(10 + j2\pi f))$ in degrees vs. f on a log scale, in Example 5.15.

5.4 Chapter Summary and Comments

- The CTFT (Continuous-Time Fourier Transform) is defined as $\mathcal{F}\{f(t)\} = \int_{-\infty}^{\infty} f(t)e^{-j\omega t}dt = F(\omega)$.

- Strictly speaking, $F(\omega)$ does not exist when the integral expression defining it does not converge at one or more frequencies. In some cases, we can get around this problem by including Dirac delta functions in $F(\omega)$ at frequencies of non-convergence.

- When $F(\omega)$ exists, it contains the same information as does $f(t)$ only in a different domain.

- The Inverse CTFT is defined as $f(t) = \frac{1}{2\pi}\int_{-\infty}^{\infty} F(\omega)e^{j\omega t}d\omega$.

- The CTFT of a convolution product in time may be calculated as a multiplicative product of spectra.

- The CTFT of a multiplicative product in time may be calculated as a convolution product of spectra.

- Tables of Fourier transforms and their properties make it possible to transform basic signals from time domain to frequency domain (and from frequency domain to time domain) using substitutions instead of integrations.

- When $f(t) = f_p(t)$ is periodic, then $F(\omega)$ is discrete – that is, composed of impulses that are uniformly spaced in frequency. The areas of these impulses in $F(\omega)$ convey the information of $f_p(t)$.

- The transformation between one period of $f_p(t)$ to the Fourier Series coefficients D_k (= impulse areas in CTFT$\{f_p(t)\} \times 2\pi$) is called the Exponential Fourier Series.

- The Trigonometric Fourier Series is a different form of the Exponential Fourier Series.

- When $f_p(t)$ is real, the Trigonometric Fourier Series expression may be written using only real numbers and non-negative frequency values.

- The Compact Fourier Series is a special form of the Trigonometric Fourier Series; it is used only when $f_p(t)$ is real.

5.5 Homework Problems

P5.1 Calculate the CTFT of each $f(t)$ via integration – using the definition $\mathcal{F}\{f(t)\} = \int_{-\infty}^{\infty} f(t)e^{-j\omega t} dt$:
  a) $f(t) = 3\delta(t)$
  b) $f(t) = \delta(t - 1) + \delta(t + 1)$
  c) $f(t) = \text{rect}(t)$
  d) $f(t) = \Delta(t)$

P5.2 Calculate the ICTFT of each $F(\omega)$ via integration – using the definition $\mathcal{F}^{-1}\{F(\omega)\} = (1/2\pi) \int_{-\infty}^{\infty} F(\omega)e^{j\omega t} d\omega$:
  a) $F(\omega) = \delta(\omega)$
  b) $F(\omega) = \delta(\omega - \pi/2)$
  c) $F(\omega) = \delta(\omega - 1) + \delta(\omega + 1)$
  d) $F(\omega) = \text{rect}(\omega)$

P5.3 Given: $f(t) \Leftrightarrow F(\omega)$, $g(t) \Leftrightarrow G(\omega)$. Using the property of linear superposition, find the Fourier transforms of each of the following, expressed in terms of $F(\omega)$ and $G(\omega)$:
  a) $f(t) + g(t)$
  b) $3f(t) - 2g(t)$
  c) $f(t) + \delta(t)g(t)$
  d) $f(t)\{1 + g(t)\delta(t - 1)\}$

P5.4 By applying linear superposition to the CTFT pairs given in Table 5.1, find Fourier transforms of each of the following:
  a) $1 - u(t) + \delta(t)$
  b) $2u(t) - \delta(t) - 1$
  c) $e^{-2t}u(t) + e^{2t}u(-t)$

d) $0.5\,\delta(t - 10) + 0.5\,\delta(t + 10)$

e) $j\cos(10t - \pi/2)$

f) $\sin(2t) - j\cos(2t)$

g) $(\mathrm{rect}(t/5) - \Delta(t/6))/2$

h) $2\mathrm{sinc}(\pi t) + \mathrm{sinc}^2(\pi t/2)$

i) $2\delta(t) + \mathrm{sinc}^2(\pi t/2)$

j) $\delta_4(t) - \delta_2(t)$

P5.5   By applying properties of the CTFT given in Table 5.2, find
Fourier transforms of each of the following:

a) $e^{-3(t-1)}u(t - 1)$

h) $\delta(t + 1) * \mathrm{sgn}(t)$

c) $u(3t)$

d) $\cos(t)\cos(2t)$

e) $-2je^{j2(t-1)}$

f) $(1/2\pi)\,\mathrm{sinc}(t/4)\cos(t/2)$

g) $\mathrm{rect}(t/2)e^{-j2t}$

h) $\mathrm{sinc}(\pi t) * \mathrm{sinc}(\pi t/2)$

i) $\mathrm{sinc}(\pi t/2)\,\mathrm{sinc}(\pi t/2)$

j) $\mathrm{sinc}^2(\pi t/4)$

k) $u(-t)$

l) $\cos(\pi t/4)\,e^{j\pi/4}$

m) $\mathrm{sgn}(t) - \mathrm{sgn}(t - 1)$

n) $\int_{\beta=-\infty}^{t} \Delta(\beta/100)\,d\beta$

P5.6   By applying the CTFT pairs and CTFT properties given in
Tables 5.1 and 5.2, find the inverse Fourier transforms of each
of the following:

a) $\delta(\omega)$

b) $\delta(\omega - \pi)$

c) $(\pi/4)\mathrm{rect}(2\omega/\pi)$

d) $\mathrm{sinc}(2.5\omega)$

e) $e^{-j5\omega}$

f) $e^{j5\omega} + e^{-j5\omega}$

g) $\cos(5\omega)$

h) $1/\omega$

i) $(2/j\omega) * \delta_{2\pi}(\omega)$

j) $\delta_{2\pi/3}(\omega)$

P5.7 Given: $\int_{\beta=-\infty}^{t} |f(t)|^2 \, d\beta = 1$. What must be the area of $|F(\omega)|^2$?

P5.8 What is the energy of sequence $\text{sinc}(\pi t/2)$?

P5.9 What is the energy of sequence $\text{sinc}(\pi t/2) + \text{sinc}(\pi t/4)$?

P5.10 What is the power of periodic sequence $\text{rect}(t/4) * \delta_{20}(t)$?

P5.11 What is the power of periodic sequence $\Delta(t/4) * \delta_{20}(t)$?

P5.12 Calculate the Exponential Fourier Series coefficients D_n of each $f_p(t)$ given below using $D_n = (1/T_0) \int_{T_0} f_p(t) e^{-jn\omega_0 t} dt$ (your first task will be to determine T_0):

a) $f_p(t) = \delta_4(t)$

b) $f_p(t) = \delta_4(t - 2)$

c) $f_p(t) = \delta(t - 3) * \delta_4(t)$

d) $f_p(t) = \text{rect}(t) * \delta_4(t)$

P5.13 Calculate the $f_p(t)$ of each set of Fourier Series coefficients D_n given below using definition $f_p(t) = \sum_{n=-\infty}^{\infty} D_n e^{jn\omega_0 t}$, where $T_0 = 3$ seconds:

a) $D_{\pm 1} = 1$, all other $D_n = 0$

b) $D_n = \text{sinc}(2\pi n/3)$

c) $D_n = \text{sinc}^2(\pi n/4)$

d) $D_n = 1/2$, $\forall n$

P5.14 Given: $H(\omega) = 2\left(e^{j\omega} - 0.95\right)/\left(e^{j\omega} - 0.90\right)$. Plot $|H(\omega)|^2$ over frequency range $0 \le \omega \le 2\pi$, using a dB vertical scale.

Chapter 6

Sampling Theory and Practice

6.1 Theory

6.1.1 Sampling a continuous-time signal

A list of numbers specifying signal values measured at specific times defines "samples" of the signal, and obtaining these samples from the continuous-time signal is called "sampling." Sampling theory is the mathematical description of the effects of sampling, in both the time and frequency domains. Thus, the relationship between continuous-time and discrete-time signals may be described using sampling theory. Converting an analog signal to a digital signal not only involves sampling in time, but also rounding amplitudes to one of a finite number of levels; this last step is called amplitude *quantization*, which we discuss later in this chapter. Sampling and quantization operations, taken together, form the basis for *Analog-to-Digital Conversion*, which is usually implemented using specialized electronic circuits.

It may seem odd that we would want to convert a continuous-time waveform, which specifies a signal's values at *all* times, to a list of samples taken at *discrete* times. But the reason we do this is to make possible more accurate storage/recovery, transmission and processing of the information conveyed by that signal. And, when certain conditions hold, sampling theory states that we can exactly reconstruct the signal's continuous-time waveform from only the knowledge of its samples. Therefore, under those conditions no information is lost because of sampling.

To begin our theoretical discussion, let us assume that samples are taken at uniformly-spaced instants of time and that one of those times is $t = 0$. Under these assumptions it is convenient to model the act of sampling as multiplication with a train of impulses having spacing T_s sec on the time scale:

$$x(t) \times \delta_{T_s}(t) = x(t) \sum_{n=-\infty}^{\infty} \delta(t - nT_s)$$

$$= \sum_{n=-\infty}^{\infty} x(nT_s)\delta(t - nT_s). \qquad (6.1)$$

We see that the area of impulse located at time nT_s sec is $x(nT_s)$. Clearly the list of values corresponding to $x(t)|_{t=nT_s}$, $n \in \mathbb{Z}$, is the list of areas of the impulses in $x(t)\delta_{T_s}(t)$.[a] Therefore the sifting (also called *sampling*) property of the Dirac impulse function makes it possible to extract the values of $x(t)$ at specific times and preserve them as areas of impulses. It should be noted that sampling a signal by multiplying with an impulse train is only a convenient mathematical model for us to use, for impulse functions and impulse trains do not exist in real life.

Instead of writing "$x(nT_s)$" to describe the list of samples, we may instead write "$x(n)$." This does not necessarily mean that $T_s = 1$ sec, just that we now refer to each sample using its index value n and the time that sample was taken is implied to be nT_s sec. The entire sequence of sampled amplitude values of $x(t)$ is labelled $x(n)$, which is a list of values and each value has a specific time association.

Let us consider the frequency-domain effects of sampling. Recall from Chapter 5 that the Fourier Transform of an impulse train is another impulse train:

$$\delta_{T_s}(t) \Leftrightarrow \omega_s \delta_{\omega_s}(\omega) \qquad (\omega_s = 2\pi/T_s). \qquad (6.2)$$

Also, recall that multiplication in time gives convolution in the frequency domain:

---

[a] The areas of impulses in $\delta_{T_s}(t)$ are all 1.

$$x(t)y(t) \Leftrightarrow \frac{1}{2\pi}\left(X(\omega) * Y(\omega)\right). \tag{6.3}$$

It then follows that

$$\mathcal{F}\{x(t) \times \delta_{T_s}(t)\} = \frac{1}{2\pi}\left(\mathcal{F}\{x(t)\} * \mathcal{F}\{\delta_{T_s}(t)\}\right)$$

$$= \frac{1}{2\pi}\left(X(\omega) * \omega_s\delta_{\omega_s}(\omega)\right)$$

$$= \frac{\omega_s}{2\pi}\left(X(\omega) * \delta_{\omega_s}(\omega)\right) = \frac{1}{T_s}\left(X(\omega) * \sum_{k=-\infty}^{\infty}\delta(\omega - k\omega_s)\right)$$

$$= \frac{1}{T_s}\sum_{k=-\infty}^{\infty}X(\omega - k\omega_s), \text{ or}$$

$$x(t) \times \delta_{T_s}(t) \Leftrightarrow \frac{1}{T_s}\sum_{k=-\infty}^{\infty}X(\omega - k\omega_s). \quad (\omega_s = 2\pi/T_s). \tag{6.4}$$

The infinitely many frequency-shifted copies of $X(\omega)$ add together to give a periodic spectrum, having period ω_s rad/sec. Thus, sampling in time produces periodicity in frequency.

6.1.2 *Relation between CTFT and DTFT based on sampling*

We now have sufficient theoretical background to mathematically relate continuous-time and discrete-time Fourier transforms using sampling theory. Given a continuous-time Fourier transform pair $f(t) \Leftrightarrow F(\omega)$, sample $f(t)$ at a rate of one sample per second ($T_s = 1$ sec):

$$f(t)\delta_1(t) = f(t)\sum_{n=-\infty}^{\infty}\delta(t - n(1))$$

$$= \sum_{n=-\infty}^{\infty}f(t)\,\delta(t - n) = \sum_{n=-\infty}^{\infty}f(n)\delta(t - n). \tag{6.5}$$

Normally at this point we would begin calling the list of impulse areas the "sequence $f(n)$" and switch over to discrete signal analysis. However, for the purpose at hand, we choose to view $\sum_{n=-\infty}^{\infty}f(n)\delta(t - n)$ as a continuous-time signal and find its Fourier transform:

$$\mathcal{F}\{\sum_{n=-\infty}^{\infty}f(n)\delta(t - n)\} = \sum_{n=-\infty}^{\infty}\mathcal{F}\{f(n)\delta(t - n)\}$$

$$= \sum_{n=-\infty}^{\infty} f(n) \, \mathcal{F}\{\delta(t-n)\} = \sum_{n=-\infty}^{\infty} f(n) \, e^{-j\omega n}, \quad \text{or}$$

$$f(t)\delta_1(t) \Leftrightarrow \sum_{n=-\infty}^{\infty} f(n) \, e^{-j\omega n}. \tag{6.6}$$

You will notice that the summation expression $\sum_{n=-\infty}^{\infty} f(n) \, e^{-j\omega n}$ is nothing else but the discrete-time Fourier transform, as defined in Ch. 4! In other words, the spectrum of sequence $f(n)$ may be thought of as the spectrum of $f(t)\delta_1(t)$, where $f(t)$ is a continuous-time signal that when sampled once per second gives us the samples of sequence $f(n)$.

Because sampling in time was shown to result in a periodic spectrum, in this case the periodicity must also exist. The period in frequency is $\omega_s = 2\pi \text{ rad}/T_s \text{ sec} = 2\pi \text{ rad/sec} \ (T_s = 1 \text{ sec})$, which explains why in Ch. 4 the discrete-time Fourier Transform was shown to always give a periodic spectrum $F(e^{j\omega})$ having period 2π. The discrete-time Fourier transform, or $DTFT$, may be thought of as the continuous-time Fourier transform of an analog waveform that had been sampled at a rate of one sample per second. Conversely, the discrete-time Fourier transform summation may be used to calculate the spectra of sampled continuous-time signals, because the integral of a weighted impulse train simplifies to a summation of terms.

If we sample $f(t)$ using $T_s \neq 1$ sec, may we still find $\mathcal{F}\{f(t)\delta_{T_s}(t)\}$ using the $DTFT$? Yes – it is easily shown:

$$\mathcal{F}\{f(t)\delta_{T_s}(t)\} = \sum_{n=-\infty}^{\infty} f(nT_s) \, e^{-j\omega n T_s}$$

$$= \sum_{n=-\infty}^{\infty} g(n) \, e^{-j\omega n T_s} = G\left(e^{j(\omega T_s)}\right), \tag{6.7}$$

where sequence $g(n)$ is the result of sampling $f(t)$ every T_s seconds. The resulting spectrum is $DTFT\{g(n)\}$ scaled in frequency so that its period is $2\pi \cdot (1/T_s) = 2\pi \cdot (\omega_s/2\pi) = \omega_s$ rad/sec. To accomplish this, we simply re-label the frequency axis accordingly.

Figure 6.1 shows relationships between the continuous-time and discrete-time Fourier transforms that are based on sampling. Further, note that sampling theory also plays a role when the Fourier transform of a periodic signal is calculated. This is because a periodic signal may be modeled as a pulse that is convolved with an impulse train, and the corresponding frequency-domain effect is a multiplication of the pulse spectrum by an impulse train: this is sampling in the frequency domain!

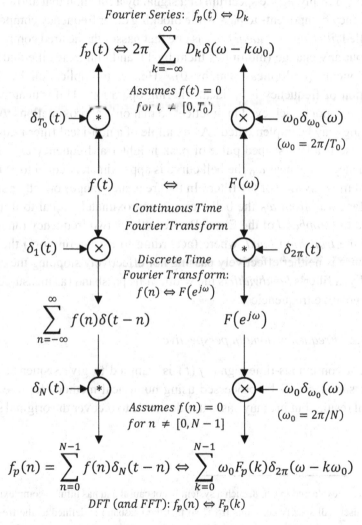

Figure 6.1. Sampling-based Fourier transform relationships.

6.1.3   *Recovering a continuous-time signal from its samples*

6.1.3.1   *Filtering basics*

Chapter 8 will formally introduce this concept, but for now here are the basics: multiplying the spectrum of a signal by a function that allows some frequency components to pass while blocking other frequency components is called *filtering*. An *ideal filter* is one that passes the desired components without any change (multiplies them by 1), and eliminates the undesired components (multiplies them by 0). Hence, multiplication by a rect function of frequency is an ideal filtering operation. Unfortunately, this ideal may not be achieved in practice[b] so that only approximations to ideal filtering may be implemented. An example of a non-ideal filter frequency response is a bell-shaped pulse of peak height 1 at frequency ω_0. In the frequency range near ω_0 the bell curve is approximately equal to 1; this is called the *passband* of the filter. In the frequency ranges on either side of and far away from ω_0 the bell curve is approximately equal to 0; this is called the *stopband* of the filter. Between those two frequency ranges we have the *transition band*, where (according to user definition[c]) the filter response is neither effectively passing nor effectively stopping the signal. Finally, a filter's *bandwidth* is the width of its passband (as measured only over positive frequencies).

6.1.3.2   *Frequency domain perspective*

Once a continuous-time signal $f(t)$ is sampled to give sequence $f(n)$, these samples may be processed using numerical methods to give some useful result,[d] but in many cases it is desirable to recover the original signal

---

[b] Since, as explained in Ch. 8, such a system is not causal and has infinite complexity.

[c] The user will specify constants $\{a, b\}$ so that the passband is defined as the frequency range over which $|H(\omega)| \geq a$ (a is just less than 1), and that the stopband is defined as the frequency range over which $|H(\omega)| \leq b$ (b is just greater than 0).

[d] As an example, to identify a speaker's identity from an analog recording of his/her speech.

by reconstructing it from its samples.[e] Let us see under what conditions the original analog signal may be reconstructed from its uniformly-spaced samples in time.

Assume that $F(\omega) = \mathcal{F}\{f(t)\}$ equals zero outside frequency range $-\omega_{max} < \omega < \omega_{max}$. Even if this assumption is not exactly true, when $|F(\omega)|^2$ is negligible when $|\omega| > \omega_{max}$ we say that signal $f(t)$ is *band-limited* to frequencies below ω_{max} (referring to positive frequencies) then the following is, for all practical purposes, true. When $f(t)$ is sampled in time to give $f_s(t) = f(t)\delta_{T_s}(t)$, the resulting spectrum is made periodic:

$$f_s(t) \Leftrightarrow F_s(\omega) = \frac{1}{T_s} F(\omega) * \delta_{\omega_s}(\omega)$$

$$= \frac{1}{T_s}\sum_{k=-\infty}^{\infty} F(\omega - k\omega_s) \quad (\omega_s = 2\pi/T_s). \quad (6.8)$$

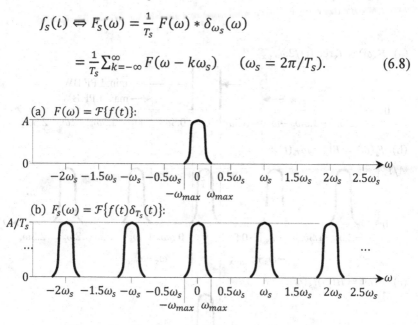

Figure 6.2. Signal spectrum before and after sampling.

In other words, sampling $f(t)$ every T_s seconds creates copies of the original (*baseband*) spectrum $F(\omega)$ and places them at every integer multiple of ω_s in the frequency domain (and scales these copies by $1/T_s$,

---

[e] This is the case in mobile telephony, where the original analog speech signal is sampled, processed to reduce the number of bits per second needed to be transmitted, and then reconstructed to once again be an analog speech signal at the receiver.

but that is not so important). Figure 6.2 shows the spectrum of $f(t)$ before and after sampling (where $\omega_s > 2\omega_{max}$). Note that, in this plot of spectrum $F_s(\omega)$, there is no overlap between copies of the original spectrum $F(\omega)$ after they were shifted to multiples of ω_s and added together. This makes it easy to recover the original signal $f(t)$ from its sampled version $f_s(t)$: simply pass $f_s(t)$ through a "lowpass" filter that blocks all but the copy of $F(\omega)$ that is centered at $\omega = 0$ (that is, the original *baseband* spectrum[f]). If ideal, the lowpass filter would need to have a bandwidth of anywhere between ω_{max} and $\omega_s - \omega_{max}$ (and a gain factor of T_s) to exactly recover $f(t)$ from $f_s(t)$, as shown in Fig. 6.3.

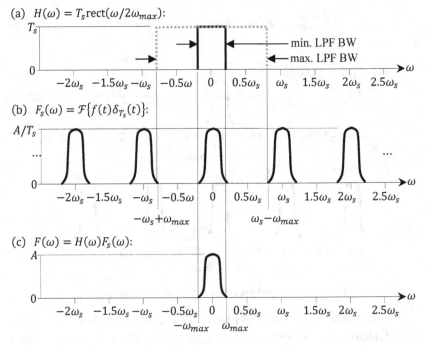

Figure 6.3.   Ideal lowpass filtering to recover $F(\omega)$ from $F_s(\omega)$.

---

[f] *Baseband* is a name given to the frequency range of the original spectrum before it was processed (in this case, by sampling).

This lowpass filter is called the *reconstruction* filter because it reconstructs the original $f(t)$ from its sampled version $f_s(t)$.[g] In Fig. 6.3 the reconstruction filter is assumed to be ideal. Because there is a gap between adjacent shifted copies of $F(\omega)$ within $F_s(\omega)$, the reconstruction lowpass filter frequency response need not be ideal so long as its passband gain is constant and its stopband gain is zero. An example of this is shown in Fig. 6.4:

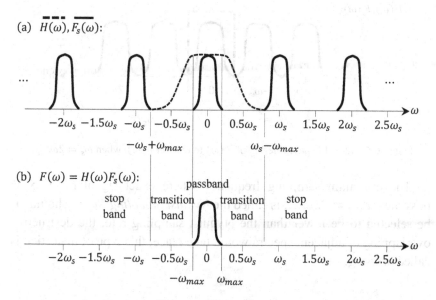

Figure 6.4. Non-ideal lowpass filtering to recover $F(\omega)$ from $F_s(\omega)$.

Consider the case that the sampling frequency ω_s value is adjustable. When using an ideal reconstruction LPF, what is the minimum value of ω_s for which we could still recover the original signal $f(t)$ from its

---

[g] Recall that $f_s(t)$ is an impulse train – the area of each impulse specifies the value of $f(t)$ at that impulse location in time. This is our mathematical way of representing what happens when the information of $f(t)$ is retained at only the sampling time instants. In practice the sample values are stored as binary numerical codes. So, to reconstruct $f(t)$, what needs to be done is to first synthesize an approximation to the impulse train signal $f_s(t)$ from the list of binary numerical codes, and then pass $f_s(t)$ through the reconstruction lowpass filter.

sampled version $f_s(t)$? The answer depends on the bandwidth of $f(t)$. When the highest frequency component within $f(t)$ is ω_{max}, then it must be sampled at a frequency of at least $\omega_s = 2\omega_{max}$, or otherwise the destructive overlapping of spectra will occur within $F_s(\omega)$. The spectrum of $f_s(t)$ in the case that ω_s is at its lowest possible value to permit exact recovery of $f(t)$ is shown in Fig. 6.5.

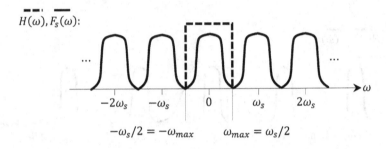

Figure 6.5. Ideal lowpass filtering of $F_s(\omega)$ to recover $F(\omega)$, when $\omega_s = 2\omega_{max}$.

The minimum sampling frequency where exact signal recovery is possible, or $\omega_s = 2\omega_{max}$, is called the *Nyquist Sampling Rate*. Should ω_s be selected to be lower than the Nyquist sampling rate, the destructive overlapping of adjacent spectra would take place; this type of distortion is called *aliasing*.

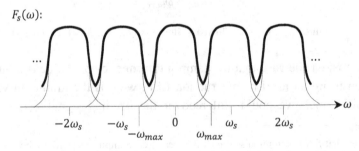

Figure 6.6. The case where $\omega_s < 2\omega_{max}$, producing aliasing distortion.

So, we should take the bandwidth of $f(t)$ $(=\omega_{max})$ into account when selecting ω_s. Choosing $\omega_s < 2\omega_{max}$ results in aliasing distortion and this makes it impossible to recover $f(t)$ from $f_s(t)$.[h] Too few samples per second $(= \omega_s/2\pi)$ are taken, which is called *undersampling* the signal $f(t)$. On the other hand, choosing $\omega_s > 2\omega_{max}$ will prevent aliasing but will also produce a sequence of samples that contain redundancy. This is called the *oversampled* case (as shown in Figs. 6.2 and 6.3). The cost of storing and transmitting these samples grows proportionally to their number, so in that sense oversampling a signal is wasteful. Sampling at the Nyquist rate $\omega_s = 2\omega_{max}$ lies on the border between undersampling and oversampling, and is often referred to as *critical sampling*. Note that when critical sampling is used the spectral copies of $F(\omega)$ have no gaps between them, and an ideal lowpass filter must be used (because of its zero-width transition band) to recover the original signal $f(t)$.

Finally, it is worth explaining the significance of the term "aliasing." When sampling sinusoids at higher than the Nyquist rate there is never any ambiguity as to the frequency of the original waveform based on its sample values: the sinusoid may be uniquely and exactly reconstructed from its samples. But when a sinusoid is sampled below the Nyquist rate we can no longer be exactly certain as to its original frequency. This is demonstrated in Fig. 6.7: two sinusoids at different frequencies have the same sample values. One of these sinusoids is undersampled and the other is properly sampled:

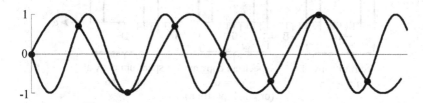

Figure 6.7. Sine waves at different frequencies can give identical samples if at least one of them is undersampled.

The frequency domain explains this phenomenon best. Figure 6.8 shows: (a) the spectrum of $x(t) = \cos(3t)$ after sampling at $\omega_s = 8$ rad/sec (oversampling), and (b) the spectrum of $y(t) = \cos(5t)$ after

---

[h] Except in special circumstances, which we discuss later in Section 6.1.6.

sampling at $\omega_s = 8$ rad/sec (undersampling). The two spectra are identical! Due to aliasing, which means that because of sampling there was overlapping of adjacent frequency-shifted copies of $Y(\omega)$ to give $Y_s(\omega)$, the illusion of a 3 rad/sec sinusoid was created even though the original frequency was 5 rad/sec. The sinusoid at frequency 5 takes on the *alias* of being at 3, and there is no way to distinguish between the two cases based on their sample values (or spectra).

(a) Spectrum of $x_s(t) = \cos(3t)\, \delta_{2\pi/8}(t)$:

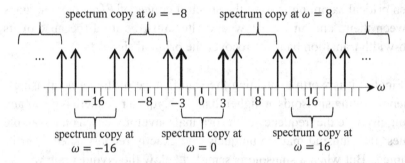

(b) Spectrum of $x_s(t) = \cos(5t)\, \delta_{2\pi/8}(t)$:

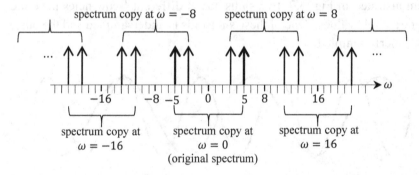

Figure 6.8. Identical spectra result when $\cos(3t)$ and $\cos(5t)$ are sampled at $\omega_s = 8$ rad/sec ($T_s = 2\pi/8$ sec), which demonstrates "aliasing."

6.1.3.3 *Time domain perspective*

It is also instructive to analyze signal sampling and reconstruction in the time domain. Take, for example, the case of sampling a pure sinusoid.

How many samples per period correspond to the Nyquist rate? The answer is 2 samples/period: $T_0/T_s = \omega_s/\omega_0 = 2\omega_0/\omega_0 = 2.$[i] Another time-domain item of interest is the impulse response of the reconstruction filter that is required to reconstruct an analog signal from its samples. That is the topic of the following paragraphs.

Assume we are sampling signal $x(t)$ at exactly the Nyquist rate ω_s. That is, the sampling frequency is twice the highest frequency component in $X(\omega)$. Our mathematical model for the result of sampling is the impulse train signal $x_s(t)$. To recover $x(t)$ from $x_s(t)$ we pass the latter through an ideal lowpass filter; this LPF must have bandwidth $\omega_s/2$ (as is customary, we measure bandwidth over positive frequencies only), passband gain T_s, and stopband gain 0. The impulse response of this ideal lowpass reconstruction filter is therefore

$$h(t) = \text{sinc}\left(\frac{\omega_s t}{2}\right) \Leftrightarrow T_s \, \text{rect}\left(\frac{\omega}{\omega_s}\right) = H(\omega). \qquad (6.9)$$

The reconstructed signal in the frequency domain is determined by the multiplicative product $X(\omega) = X_s(\omega)H(\omega)$, and the same information in the time domain may be calculated using the convolution product $x(t) = x_s(t) * h(t) = x_s(t) * \text{sinc}(\omega_s t/2)$. Expressing $x_s(t) = x(t)\delta_{T_s}(t)$, an expression for the reconstructed signal $x(t)$ in terms of its own samples is:

$$x(t) = x(t)\delta_{T_s}(t) * \text{sinc}\left(\frac{\omega_s t}{2}\right).$$

---

[i] When sinusoid $A\cos(\omega_0 t + \theta)$ is sampled at the Nyquist rate of 2 samples per period ($\omega_s = 2\omega_0$), there is still some ambiguity that prevents us from reconstructing the original waveform from its samples. All samples will have the same magnitude (in the range between 0 and A, depending on the value of θ) and toggle in polarity from one sample to the next. In either case, there are infinitely many combinations of $\{A, \theta\}$ that could have produced such results. Although this dilemma occurs anytime we sample a signal whose spectrum has an impulse at the highest frequency component, it is of theoretical interest only. In practice one would sample the signal above the Nyquist rate to simplify reconstruction lowpass filtering.

$$= \{\textstyle\sum_{n=-\infty}^{\infty} x(nT_s)\delta(t - nT_s)\} * \text{sinc}\left(\frac{\omega_s t}{2}\right)$$

$$= \textstyle\sum_{n=-\infty}^{\infty} x(nT_s)\text{sinc}\left(\frac{\omega_s(t-nT_s)}{2}\right). \qquad (6.10)$$

This equation describes the sum of infinitely many sinc functions delayed by integer multiples of T_s sec, scaled by the sampled values of $x(t)$ at those delay times, and added together to reconstruct the original waveform $x(t)$. The sinc functions may be thought of as building-block waveforms whose weighted sum is the original signal (the weights being samples of the original signal). In the case described here where Nyquist sampling is used, each of the sinc functions used to reconstruct $x(t)$ has zero-crossings at every sampling time except for the one where it is located. For this reason, as shown in Fig. 6.9: only one sinc function determines the value of $x(t)$ at $t = kT_s$, but infinitely many sinc functions add together to determine the value of $x(t)$ at $t \neq kT_s$.[j]

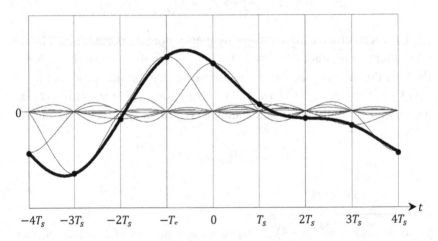

Figure 6.9. Reconstructing $x(t)$ as the sum of weighted sinc functions.

If we were sampling above the Nyquist rate and reconstructing with an ideal LPF then the zero-crossings of the sinc pulses would not line up as

---

[j] Stated another way: in this example, to exactly recover $x(t)$ at $t \neq nT_s$ we need the knowledge of *all* sample values – both forward and backward in time.

nicely with the sample times. In any case, it is important to note that between one and infinitely many samples are needed to reconstruct the waveform at any time.

There are some aspects of the previous discussion that are of academic interest only. For example, the ideal lowpass cannot be practically implemented because it is not causal (sinc $\neq 0 \; \forall t < 0$). Also, because the ideal lowpass filter has a rect pulse as its frequency response, and this rect pulse has no transition band between passband and stopband, realizing it using analog circuitry would require infinite cost.[k] One way to approximate the frequency-domain filtering characteristics of an ideal lowpass filter response for accurately recovering a signal from its samples is to use a filter having the following impulse response:

$$\tilde{h}_{LPF}(t) = \text{sinc}\left(\frac{\omega_s(t-kT_s)}{2}\right) \text{rect}\left(\frac{t-kT_s}{2kT_s}\right) \quad (k \in \mathbb{Z}). \quad (6.11)$$

The ideal $h_{LPF}(t)$ and its approximation $\tilde{h}_{LPF}(t)$ in Figs. 6.10 and 6.11 clearly show us that the latter is a truncated and delayed version of the former. Also shown are their magnitude frequency responses. Truncating the sinc pulse at a zero crossing by forcing the rectangular pulse width to be an integer multiple of T_s helps minimize spectral distortion due to the truncation.[l] The kT_s sec delay was introduced to make $\tilde{h}_{LPF}(t)$ causal and therefore possible to realize; delay has no effect on magnitude frequency response.[m]

---

[k] A rational $H(s)$ would need to have infinite order (Ch. 8).

[l] Setting a waveform to zero outside of some range by multiplying it with a finite-width waveform is called *windowing*. Here the rect pulse is called the *time window* function.

[m] Although introducing extra delay is undesirable in some applications, e.g., two-way communications.

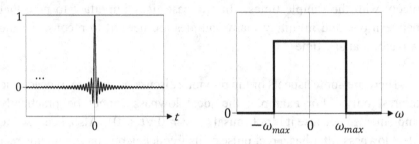

Figure 6.10. Ideal LPF impulse response (sinc), and its magnitude spectrum (rect).

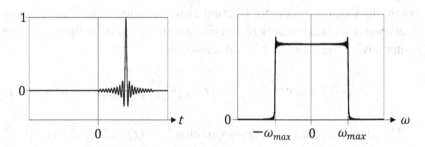

Figure 6.11. Non-ideal LPF impulse response (truncated, delayed sinc), and its magnitude spectrum (rect pulse with overshoot).

6.1.4 *Oversampling to simplify reconstruction filtering*

We have seen that Nyquist-rate sampling requires an ideal lowpass filter to separate the baseband image of the original analog signal spectrum from its copies. Because a lowpass filter cannot be realized in practice, what is to be done? If we insist on sampling at the Nyquist rate, which is the lowest possible frequency for which aliasing does not occur and the original spectral copies have no gaps between them, then reconstruction lowpass filtering using a practical filter will usually have two consequences: (1) the filter will attenuate some desired signal components at the high-frequency end of the passband, and (2) the filter will not fully eliminate some undesired adjacent spectrum components at the low-frequency end of the stopband. This situation is depicted in Fig. 6.12.

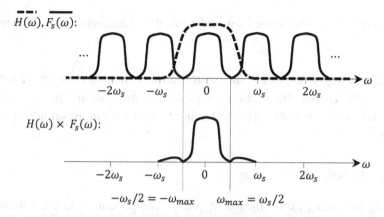

$H(\omega), F_s(\omega)$:

$H(\omega) \times F_s(\omega)$:

$-\omega_s/2 = -\omega_{max}$ $\omega_{max} = \omega_s/2$

Figure 6.12. The spectral consequences of reconstructing a Nyquist-rate-sampled signal using a non-ideal lowpass filter.

To avoid the distortion due to imperfect filtering as shown in the previous figure, there is a simple solution: sample the signal above the Nyquist rate. This situation was previously depicted in Fig. 6.4. By putting a gap between adjacent spectra, we make room for the non-ideal lowpass filter transition band.

6.1.5 *Eliminating aliasing distortion*

As defined previously, aliasing distortion is the overlapping of adjacent spectral copies when a continuous-time signal is undersampled ($\omega_s <$ $2\omega_{max}$). Typically, this distortion irrecoverably corrupts the baseband spectrum so that it is impossible to recover the original signal from its samples.

When the signal energy at $\omega_s/2$ or above is negligible, then in some applications the aliased components may not cause too much trouble. For example, human speech signals have nearly all of their energy below 10^4 Hz ($\omega = 2\pi \cdot 10^4$rad/sec) during voiced (vowel) sounds, so choosing $\omega_s = 2(2\pi \cdot 10^4$rad/sec) results in very little aliasing. Higher frequency energy does occur during unvoiced speech such as the "sh" sound; however, even when aliasing occurs it is not very objectionable due to the random nature of the waveform and how it is perceived by a human listener.[n]

The best way to avoid aliasing distortion is to sample at or above the Nyquist rate (twice the highest frequency component in the signal), although this comes at the price of having to deal with more samples per second.

6.1.5.1 *Anti-alias post-filtering*

One way of getting rid of aliased frequency components is to filter them out during the reconstruction filtering step. This is shown in Fig. 6.13 for the case that an ideal lowpass filter is used. When the original signal has highest frequency component ω_{max} and is sampled at ω_s, then the aliased components will fall as low as $\omega_s - \omega_{max}$ in frequency. So, if it is necessary to undersample the original signal that results in aliasing, aliased

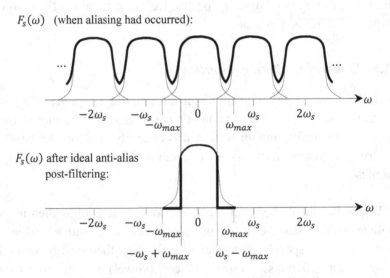

Figure 6.13. Post-filtering to remove aliasing distortion.

---

[n] We must be careful, though, since aliasing can bring high-energy sounds at frequencies above the human hearing range into a frequency range that is audible.

components can be filtered out when reconstructing the signal from its samples using an ideal lowpass filter having bandwidth $\omega_s - \omega_{max}$ instead of the usual ω_{max}. This blocks the aliased frequency components, but it also blocks the original signal energy in the frequency range $[\omega_s - \omega_{max}, \omega_{max}]$. If the reconstruction lowpass filter used is not ideal and has a wide transition band, then this method loses its effectiveness.

6.1.5.2 *Anti-alias pre-filtering*

A better way to deal with aliasing is to eliminate it before it occurs. That is, before sampling a signal at frequency ω_s the signal is first lowpass filtered to have a highest frequency component of $\omega_s/2$. This eliminates some of the original signal frequencies, but not to the extent as in the previously mentioned post-filtering technique since $\omega_s/2 > \omega_s - \omega_{max}$

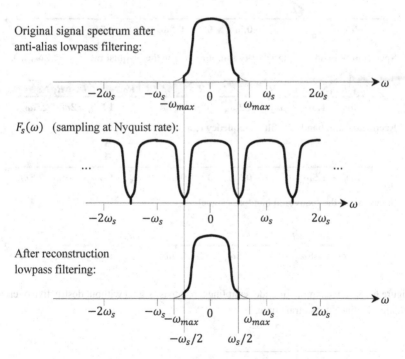

Figure 6.14. Pre-filtering to prevent aliasing distortion.

when undersampling.[°] Figure 6.14 demonstrates the effects of anti-aliasing pre-filtering. The pre-filtering technique comes at the cost of requiring that two lowpass filters are used in the sampling/reconstruction process instead of just one.

6.1.6 *Sampling bandpass signals*

When a signal is bandpass in nature (only contains frequency components over some narrow frequency range $[\omega_1, \omega_2]$) then it is possible to undersample it and still recover the signal from its samples (using a bandpass filter instead of a lowpass filter), as shown in Fig. 6.15. This is relevant for software-defined radios where the goal is to have a minimum

Original bandpass signal:

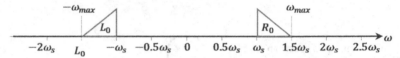

Spectrum of bandpass signal after sampling below the Nyquist rate ($\omega_s < 2\omega_{max}$):

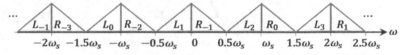

Reconstruction bandpass filter frequency response:

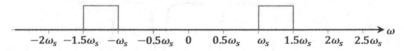

Spectrum of the recovered bandpass signal:

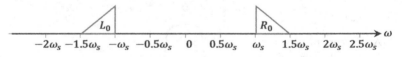

Figure 6.15. An example of undersampling a bandpass signal without destructive overlap-adding of adjacent spectral copies.

[°] When undersampling: $\omega_s < 2\omega_{max}$, or $\omega_s/2 < \omega_{max}$. Therefore $-\omega_s/2 > -\omega_{max}$, which gives $\omega_s/2 > \omega_s - \omega_{max}$ after ω_s is added to both sides of the inequality.

of analog circuitry prior to sampling a high-frequency bandpass radio signal, so that radio functionality may easily be adapted to signal and interference conditions by executing one of many different software algorithms. In this scheme, because no destructive overlapping of the frequency-shifted spectra occurs, it is possible to recover a message signal from the sampled bandpass signal using digital signal processing algorithms.[p]

6.1.7 *Approximate reconstruction of a continuous-time signal from its samples*

To this point we mathematically modeled the process of sampling and reconstructing a continuous-time signal using an impulse train function. Impulse trains do not exist in nature, so in this section we describe more realistic alternative methods.

6.1.7.1 *Zero-order hold method*

Although a Dirac impulse function cannot exist in the real world[q] its convolution product with another signal *can* exist.[r] Consider the waveform shown in Fig. 6.16. Shown are the original analog signal $x(t)$, and the result of passing $x_s(t) = x(t)\delta_{T_s}(t)$ through what is called a "zero-order hold" filter to produce a step-like waveform. The zero-order hold filter gets its name because it replaces each impulse at the input with a

---

[p] In this example, even though we undersample the signal (the Nyquist criterion $\omega_s \geq 2\omega_{max}$ is not satisfied) and adjacent spectra overlap one another to produce "aliasing", the overlapping spectral copies fit into gaps that are inherent to the bandpass nature of the signal. Thus, there is no *destructive* overlap-adding of spectra. Care must be taken to make sure that this indeed is the case when sampling bandpass signals. For example, if the bandpass signal has bandwidth $BW = \omega_{max} - \omega_{min}$ then the theoretically minimum possible sampling rate ω_s that may work in some cases (depending on the values of ω_{min} and ω_{max}) is $2 \cdot BW$. This is the case in Fig. 6.15.

[q] Instantaneously switching amplitude from zero to infinity, and then instantaneously switching it back to zero, requires an infinite amount of energy to accomplish.

[r] For example: if $x(t)$ exists then $x(t) * \delta(t)$ also exists.

pulse whose height is a zero-order polynomial (a constant value) at the output. It can be said that the output of this filter is equal to the area of the nearest impulse in the input,[s] which is the value of $x(t)$ at the nearest sampling time. Therefore, the impulse response of a zero-order hold filter is $\text{rect}(t/T_s)$ and its output signal is equal to $\text{rect}(t/T_s)$ convolved with $x_s(t)$. Importantly, the step-like waveform at the zero-hold filter output can be produced with a high degree of accuracy in the real world, and the accuracy of this approximation to $x(t)$ improves as T_s decreases (the step widths get smaller). Thus, we can represent the sampled values of $x(t)$ using a waveform that does not include any impulses. In practice the step-

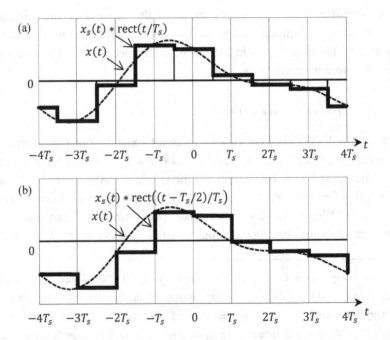

Figure 6.16. Reconstructing $x(t)$ as the sum of weighted rect functions: (a) $x_s(t) *$ $\text{rect}(t/T_s)$; (b) $x_s(t) * \text{rect}((t - T_s/2)/T_s)$.

---

[s] This requires a non-causal filter because the nearest impulse may be found almost $T_s/2$ sec into the future. However, a causal filter results when the output signal is delayed by $T_s/2$ sec.

like waveform is easily created by sampling $x(t)$ using a sample-and-hold circuit, as shown in Fig. 6.17.

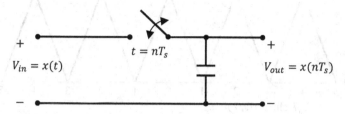

Figure 6.17. Sample-and-hold circuit to obtain $x_s(t) * \text{rect}((t - T_s/2)/T_s)$ from $x(t)$.

Assume that the switch closes at time $t = nT_s$ sec to instantaneously charge the capacitor to voltage $x(nT_s)$, and then immediately opens so that the capacitor holds that charge for T_s sec. The resulting output waveform will appear as shown in Fig. 6.16(b). Note that this is the same waveform as results from passing the impulse train sampled signal $x_s(t)$ through a zero-order hold filter that includes $T_s/2$ sec delay. Using this simple circuit, we have achieved a sampling of $x(t)$ to give the convolution product

$$x_s(t) * \text{rect}\left(\frac{t - T_s/2}{T_s}\right)$$

$$= x(t)\delta_{T_s}(t) * \text{rect}((t - T_s/2)/T_s) \approx x(t). \qquad (6.12)$$

Let's compare the magnitude spectrum of $x_s(t)$ to the magnitude spectrum of $\text{rect}(t/T_s) * x_s(t)$. (We are ignoring the $T_s/2$ sec delay, because delay affects phase but not magnitude frequency response.) Since $\mathcal{F}\{\text{rect}(t/T_s)\} = T_s\text{sinc}(\omega T_s/2)$, and because convolution in time is multiplication in frequency, we obtain the following result:

$$\mathcal{F}\{\text{rect}(t/T_s) * x_s(t)\} = T_s\text{sinc}(\omega T_s/2) X_s(\omega). \qquad (6.13)$$

Thus, the only difference between spectrum $X_s(\omega)$ and that at the output of a zero-hold filter is a multiplication by waveform $T_s\text{sinc}(\omega T_s/2)$ in the latter. Figure 6.18 shows three spectra: (a) $|X_s(\omega)|$, (b) $|T_s\text{sinc}(\omega T_s/2)|$, and (c) $|T_s\text{sinc}(\omega T_s/2) X_s(\omega)|$. In the plot of (c) we see that the baseband image of $X(\omega)$ is distorted because of multiplication by

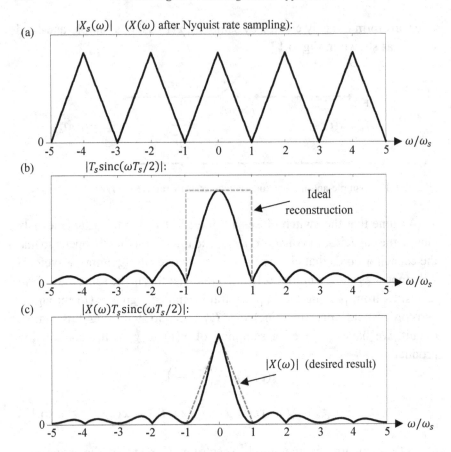

Figure 6.18. (a) Original spectrum, (b) spectral distortion due to sample/hold process, and (c) the resulting product of these: $|T_s \text{sinc}(\omega T_s/2) X_s(\omega)|$.

the sinc function. Fortunately, we can compensate for this distortion of the baseband spectrum by post-multiplying the spectrum with inverse sinc function $1/(T_s \text{sinc}(\omega T_s/2))$, which is finite at all frequencies that are not exact integer multiples of ω_s. Then, an ideal lowpass filter will be able to exactly recover the original baseband spectrum $X(\omega)$.

6.1.7.2 *First-order hold method*

In the previous section, we learned that although it is practically impossible to obtain sampled signal $x_s(t) = x(t)\delta_{T_s}(t)$, it is simple to obtain

$\text{rect}((t - T_s/2)/T_s) * x_s(t)$ using a sample-an-hold analog circuit. The distortion produced by this convolution can be compensated for using inverse-sinc filtering.[t] The sample-and-hold unit output signal, which is the same as passing $x_s(t)$ though a zero-hold filter, is characterized by its amplitudes between any two sampling times: they are polynomials of the 0th order (constants). Alternatively, we may convolve $x_s(t)$ with a triangular pulse: $y(t) = x_s(t) * \Delta(t/2T_s)$. As shown in Fig. 6.19, this is the same as "connecting the dots" between adjacent sampled values of $x(t)$. The resulting waveform $y(t)$ is composed of first-order polynomials (sloped line segments) between sample points, and thus this convolution is named "first-order hold" filtering. The name is a bit of a misnomer, as the waveform is not "held" at a constant level while waiting for the next sample time; instead, the waveform anticipates the arrival of the next sample amplitude by linearly heading toward that value.[u]

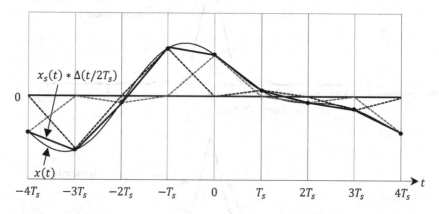

Figure 6.19. Reconstructing $x(t)$ as the sum of weighted triangular pulse functions (first-order hold filter).

Finally, we consider the frequency-domain effects of first-order hold filtering. Since $\Delta(t/2T_s) \Leftrightarrow T_s\text{sinc}^2(\omega T_s/2)$, the output spectrum is:

---

[t] Of course, we cannot compensate for the $T_s/2$ sec delay since advance (opposite of delay) units do not exist.

[u] To be implemented as a causal filter, the first-order hold impulse response needs to be delayed by T_s seconds.

$$Y(\omega) = \mathcal{F}\{\Delta(t/2T_s) * x_s(t)\} = T_s \text{sinc}^2(\omega T_s/2) \, X_s(\omega). \qquad (6.14)$$

The effects of first-order hold filtering on the reconstructed signal spectrum are demonstrated next in Fig. 6.20.

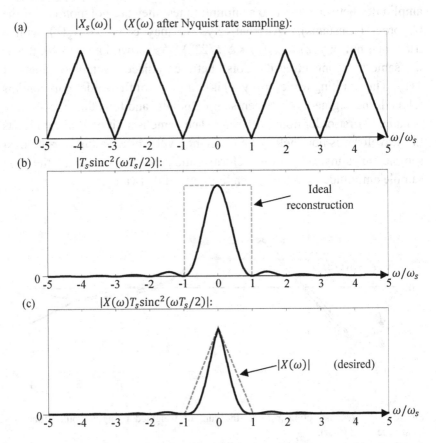

Figure 6.20. (a) Original spectrum, (b) spectral distortion due to 1st-order hold process, and (c) the resulting product of these: $|T_s \text{sinc}^2(\omega T_s/2) \, X_s(\omega)|$.

A causal version of the first-order hold filter may be simulated using a programmable-slope ramp generator (integrator circuit) to produce a piecewise linear waveform in response to input signal $x_s(t)$.

6.1.8 *Digital-to-analog conversion*

The process of converting a binary code representing a sampled amplitude value to a corresponding voltage is termed "digital-to-analog conversion" (DAC). An *N-bit digital-to-analog converter* is a hybrid of digital and analog circuitry that performs the calculation

$$x = \text{constant} \times \sum_{n=0}^{N-1} 2^n b_n, \qquad (6.15)$$

where b_n is a bit in the binary codeword representing a sampled value whose amplitude is x volts. Note that b_0 is the least-significant bit (carries the lowest weight) and b_{N-1} is the most-significant bit (carries the highest weight). The number of different voltage levels that this circuit can produce is equal to the number of permutations of N bits: 2^N. So, an 8-bit DAC can produce 256 different output voltages while a 16-bit DAC can produce 65,536. For every extra bit in the binary code the number of possible output amplitude levels doubles.

N digital inputs → N-bit DAC → Single-wire analog output

Figure 6.21. Circuit diagram symbol for a digital-to-analog converter (DAC).

6.1.9 *Analog-to-digital conversion*

The process of quantizing a voltage and assigning to it a binary code is termed "analog-to-digital conversion", or *ADC* for short. An *N-bit analog-to-digital converter* accepts a single analog signal and outputs *N* digital signals (one representing each bit in the binary code assigned to the quantized amplitude level). As with the DAC, an ADC utilizes a hybrid of both analog and digital circuitry and is fabricated on a single integrated circuit. The circuits needed to accomplish analog-to-digital conversion combine sampling, quantization (described in the following section) and binary coding; they are more complicated than digital-to-analog converters designed for the equivalent number of amplitude levels. In fact, some ADC's have a DAC as one of their internal subsystems!

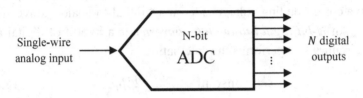

Figure 6.22.　Circuit diagram symbol for the analog-to-digital converter (ADC).

6.1.10 *Amplitude quantization*

In this section, we briefly discuss the technique and effects of amplitude quantization, which makes it possible to represent each of the samples of a signal using a finite-length code. This operation is normally bundled together with sampling in an analog-to-digital converter. When describing the probabilistic aspects of amplitude quantization in this section we assume the reader has some knowledge of probability theory.

6.1.10.1 *Definition*

Amplitude quantization, normally referred to as simply "*quantization*", is the act of mapping a continuously-distributed input value to one in a discrete set of possible output values. If this operation is performed on a scalar, it is called *scalar quantization*. If performed on a group of scalar values (called a vector), it is called *vector quantization*. For example: given a scalar amplitude value that is known to be anywhere in the range 0–5 volts, we may round it to the nearest tenth of a volt to produce an output value. This rounding to the nearest available output level is a type of scalar quantization. Other, more complicated ways may be used to choose which output level is selected based on the input signal vs. time. For example, when dealing with audio signals, a perceptually-based criterion is often used instead of rounding-to-nearest.

A system that performs amplitude quantization on its input signal to instantaneously produce the resulting output signal is called a *quantizer*. Figure 6.23 shows a possible input-output relation for a quantizer. The quantizer characteristic shown is called *uniform quantization* because of the uniformly-spaced output levels:

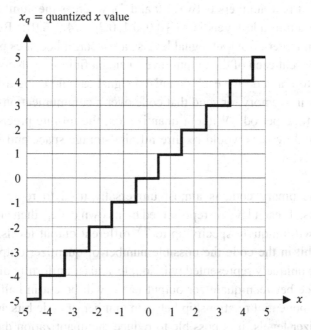

Figure 6.23. Input-output description of a uniform quantizer that rounds input value x to the nearest integer value.

Even though we model a quantizer as a system having an input and output, measuring its impulse response is meaningless because this system is nonlinear. The many-to-one mapping of amplitude values caused by passing a signal through a quantizer (or *quantizing* that signal) results in an irreversible loss of information. However, this behavior is predictable and it is also controllable: by designing a quantizer to have many closely-spaced output levels, its action will not change the input waveform very much.

6.1.10.2 *Why quantize?*

Quantization would not be used if not for that, in practice, a finite number of quantizer output levels are produced. Returning to the previous example of rounding a voltage in the range 0–5 volts to the nearest tenth of a volt: the number of input signal amplitude possibilities is infinite (a count of real numbers between 0 and 5), whereas the number of possible output amplitude levels is 51 $\{0.0, 0.1, 0.2, \dots, 4.9, 5.0\}$. Because only a finite number of output signal levels can occur, it becomes possible for us to represent each of the output levels using a finite-length code.[v] Then we may store a code for each quantized signal sample using a finite number of bits in memory, or send that code over a communications channel in a finite time period. Without quantization, the infinite precision of signal amplitude values would require infinite storage space and infinite transmission time.

The binary code is almost universally used to represent quantizer outputs. If each level is represented by its own code,[w] then an N-bit binary code will uniquely specify up to 2^N different output levels. With every extra bit in the code the possible number of quantizer output levels that can be uniquely represented will double, and (if uniformly distributed) the distance between quantizer output levels will be cut in half. Because of this exponential relationship between number of code bits and number of quantizer levels, it is possible to reduce the quantization distortion to an acceptable level and still have reasonably short binary code words. Next, we examine a quantitative measure of the distortion caused by rounding-to-nearest quantizers.

---

[v] For example, we could use a two-digit decimal code (00, 01, ... , 49, 50) or a 6-bit binary code (000000 thru 110010) to represent each of the 51 outcomes.

[w] This is not always the case. For example, pairs or triplets of output levels may be represented using a single binary code word. Such grouping of quantizer outputs prior to binary code assignment is common when variable-length codes are used, taking advantage of the fact that combinations do not occur with equal probability (e.g., Huffman codes).

6.1.10.3 *Signal to quantization noise power ratio* (SNR_Q)

Consider a scalar quantizer that instantaneously rounds the input amplitude value x to the nearest one of L available output levels, producing output amplitude value x_q. Assume that these L output levels are uniformly distributed over range $[-x_{max}, +x_{max}]$, and that the input amplitude value x is equally likely to be anywhere in this range. Using the tools of probability theory, we will express the input to the quantizer as a sample of random variable X, where its probability density function (PDF) is:

$$f_X(x) = \frac{1}{2x_{max}} \text{rect}\left(\frac{x}{2x_{max}}\right) \qquad (6.16)$$

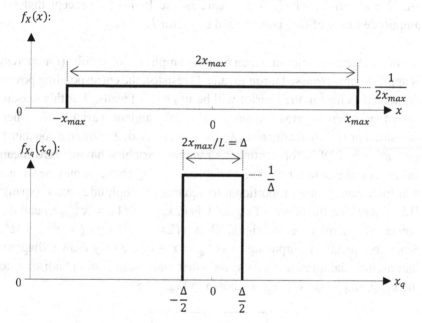

Figure 6.24. Probability density functions of input and output signals to a uniform quantizer, as described above.

Chopping the quantizer range into L contiguous intervals of width $\Delta = 2x_{max}/L$, and placing an output amplitude level in the middle of each interval will give the result that each amplitude level is selected with equal probability ($=1/L$). This is due to the input signal amplitude distribution (uniform over the quantizer input range) in combination with the specific placement of quantizer output levels. Let random variable X_q describe the quantizer output signal. Figure 6.24 shows the probability density functions of random variables X and X_q.

Define *quantization error* $q \equiv x_q - x$. This error is probabilistically described by random variable $Q = X_q - X$. The amplitude range of quantization error will be $[-x_{max}/L, +x_{max}/L] = [-\Delta/2, +\Delta/2]$ since the input signal amplitude is never more than half-a-step-size Δ away from the nearest quantizer output level to which it is rounded. This means that the probability density function of the random variable Q is uniform between $-\Delta/2$ and $+\Delta/2$, which is the same as the PDF of X except that the amplitude range of Q is compressed by factor L.

In an electronic circuit, when a signal amplitude of v volts (i amperes) is applied across (passed through) an r Ω resistor, the corresponding power dissipated as heat in the resistor will be v^2/r ($i^2 r$) watts. For this reason, we often refer to mean-squared value of random variable X, when measured probabilistically as $E[X^2]$, as the *power* of X (within a constant). For uniform PDF's for continuous random variables having zero mean value, as is the case for random variables X and Q above, it may be shown that their powers are proportional to squares of amplitude spans: $(\text{span})^2/12$. Therefore, the power of signal X is $(2x_{max})^2/12 = x_{max}^2/3$ and the power of quantizer error signal Q is $((2x_{max})/L)^2/12 = x_{max}^2/3L^2$. Since the quantizer output signal is $x_q = x + q$, we may draw a diagram that models the quantizer as a system that adds quantization "noise" q to the input amplitude value x as shown in Fig. 6.25:

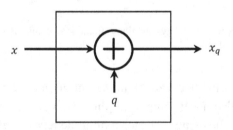

Figure 6.25.   Noise-additive model for a quantizer.

A measure of quantizer distortion is commonly chosen to be signal power divided by noise power at the output. This measure, commonly used in electrical engineering, is known as Signal-to-Noise Ratio (SNR). In this application, it is known as SNR_Q – the ratio of signal power to quantization noise power. In the example discussed above where a signal is uniformly distributed over amplitude range $[-x_{max}, +x_{max}]$, and is matched by the range of a uniform L-level quantizer to which it is inputted, the value of SNR_Q is:

$$SNR_Q = \frac{E[X^2]}{E[Q^2]} = \frac{x_{max}^2/3}{x_{max}^2/3L^2} = L^2 \qquad (6.17)$$

If, as is usually the case, the quantizer output is represented with a fixed-length binary code of length N bits, then the number of quantizer output levels $L = 2^N$. Substituting, the expression for SNR_Q becomes 2^{2N}. When stated using a dB scale,

$$SNR_Q \ (dB) = 10 \log_{10}(2^{2N}) = 20N \log_{10}(2)$$

$$= 20N \ (0.301) = 6.02N$$

$$\approx 6N \ dB. \qquad (6.18)$$
$$\text{(uniformly-distributed signal)}$$

The 6 dB per quantized signal code bit rule-of-thumb only holds for signals whose amplitudes are uniformly distributed over the entire uniform quantizer range. Let's look at some simple examples of signal waveforms having non-uniformly distributed amplitudes. Consider a sinusoid whose amplitude spans the quantizer range $-x_{max}$ to $+x_{max}$. A sinusoid having amplitude C has power value $C^2/2$ regardless of its frequency or phase shift; therefore, the quantizer input signal power is $x_{max}^2/2$. After uniform quantization with step size $\Delta = x_{max}/L$, the error due to rounding has the same power as for any other input signal [x]; this gives a quantization error

---

[x] We assume that the input signal amplitudes are continuously distributed so that the quantizer error signal still has a uniform distribution over $[-\Delta/2, +\Delta/2]$. Without this assumption, it is even possible for quantization error to equal zero (on the second pass of a signal through the same rounding-to-nearest quantizer).

power $= x_{max}^2/3L^2$. Calculating SNR_Q when binary encoding is used so that $L = 2^N$ gives us:

$$SNR_Q \ (dB) = 10 \log_{10} \left(\frac{x_{max}^2/2}{x_{max}^2/(3 \cdot 2^{2N})} \right)$$

$$\approx 6N + 1.76 \ dB \qquad\qquad (6.19)$$
$$\text{(sinusoidal signal)}$$

The highest possible input signal power occurs for waveforms having amplitudes $\pm x_{max}$, such as a square wave. The power of these signals is x_{max}^2, or double the power of sinusoids having the same amplitude span. Of course, one would never quantize a signal whose waveform is already discrete ($|x(t)| = x_{max}$), but we shall use this special case to find the upper limit of SNR_Q:

$$SNR_Q \ (dB) = 10 \log_{10} \left(\frac{x_{max}^2}{x_{max}^2/(3 \cdot 2^{2N})} \right)$$

$$\approx 6N + 4.77 \ dB \qquad\qquad (6.20)$$
$$\text{(square wave)}$$

In practice, signals of interest are more likely to have amplitudes near zero value than at their extremes. Thus, SNR_Q will be lower because input signal power is less than $(\text{span})^2/12$, while the quantization error stays constant at $\Delta^2/12$ (where Δ, as defined previously, is the quantizer step size $= \text{span}/L$). Take for example a random waveform whose amplitude distribution is Gaussian, such that 99% of the time it falls in the range $[-x_{max}, +x_{max}]$. From probability theory, we can determine that the power of this signal (for which variance $\sigma^2 =$ mean squared-value, since mean value $= 0$) is $x_{max}^2/6.63$. After N-bit uniform quantization (and ignoring the 1% of the time when input amplitude exceeds quantizer input range) the quantization error power is, as in the previous examples, $x_{max}^2/(3 \cdot 2^{2N})$. Hence SNR_Q for this more-realistic situation is:

$$SNR_Q \ (dB) = 10 \log_{10} \left(\frac{x_{max}^2/6.63}{x_{max}^2/(3 \cdot 2^{2N})} \right)$$

$$\approx 6N - 3.44 \ dB \qquad\qquad (6.21)$$
$$\text{(Gaussian PDF, 99\% in range)}$$

To summarize, it is the amplitude distribution of the quantizer's input signal that determines the signal-to-noise power ratio at the quantizer output. When binary codes are used to represent which quantizer output level was chosen, the number of output levels is chosen to be a power of 2. For an N-bit uniform quantizer, the output SNR_Q is $\approx 6N + C$ (dB), where typically the constant C is a few decibels in the negative direction (but has a theoretically maximum value $+4.77$ dB).

One final concept: what happens if the input signal amplitude range does not take full advantage of the uniform quantizer's input range? For example, if only half of the quantizer input range is spanned by signal amplitudes then only half of the available L quantization levels will ever appear at the output. Hence, it is as if the quantizer had one fewer bit in its output code (since each bit corresponds to a doubling of output levels). Then the *effective* number of quantizer bits is $N_e = N - 1$, and we may use the previously-derived formulas with N_e substituted for N. A uniformly-distributed signal spanning only half the input range of a 10-bit quantizer, for example, gives $SNR_Q \approx 6N_e = 6(10 - 1)$ $dB = 54$ dB. Calculating the effective number of quantizer bits N_e for other situations is just as simple: $2^{N_e} = L \times$ (fraction of output levels used) $= L_e$, or $N_e = \log_2 L_e$, where L_e is the effective number of output levels (number of output levels that may ever appear at the quantizer output).

6.1.10.4 *Non-uniform quantization*

A non-uniform quantizer will have output levels that are non-uniformly distributed over its amplitude range. The input-output description of a non-uniform quantizer is shown in Fig. 6.26. A justification for non-uniformly distributed output levels is that, in certain applications, rounding errors are less objectionable at higher amplitude levels than at lower amplitude levels. For example, psychoacoustics research has shown that a human listener is less likely to notice quantization noise when it is masked by higher levels of signal power. Such perceptual properties of hearing permit larger step sizes to be used as the signal amplitude increases

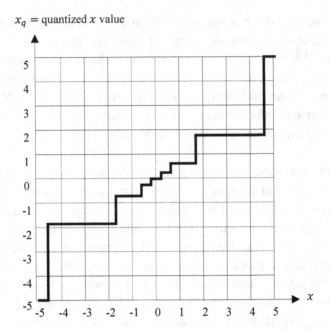

Figure 6.26.   Input-output description of a non-uniform quantizer.

without noticeably higher distortion as detected by the listener. Taking advantage of the nonlinear spacing we can reduce the number of levels L needed to cover the amplitude range of the quantizer, and hence fewer bits are required to represent the quantized signal. Another example of this perceptual phenomenon is in image intensity quantization. In a given set of lighting conditions, the human visual system is such that only 20 or so non-uniformly spaced intensity levels will fool the eye to "see" a continuous distribution. Non-uniform quantization, therefore, takes advantage of the application-specific *effects* of quantization noise.

It is much simpler to implement a uniform quantizer than a non-uniform one. For this reason, non-uniform quantization is often simulated using a series of nonlinear amplifiers called a *compressor* and *expander* (or *compander* when referring to both). Figure 6.27 shows a cascade of a compressor, uniform quantizer, and expander that together realize a non-

uniform quantizer characteristic. The compressor's function is to nonlinearly process the amplitudes of input signal before they reach the quantizer,
in a way that compresses the higher amplitudes (at both positive and
negative polarities) towards zero. The compressed signal is passed
through a uniform quantizer, and then expanded to counteract the
compressor characteristic.[y] Overall, the effect is non-uniform quantization.

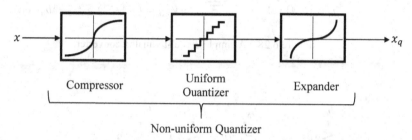

Figure 6.27. Simulating a non-uniform quantization characteristic.

6.2 Practical Applications

6.2.1 *Practical digital-to-analog conversion*

Assume that each bit value b_n is represented by one of two voltages:
$\{0, A\}$ volts. Given input binary code word bits $\{b_0, b_1, b_2, ..., b_{N-1}\}$, the
inverting op-amp circuit in Fig. 6.28 implements a simple digital-to-
analog converter having $V_{out} = (-A/2^N) \sum_{n=0}^{N-1} 2^n b_n$ volts. Another
implementation of a simple digital-to-analog converter takes advantage of
a resistive network called the "R-2R Ladder". Only two different resistor
values are used, as shown in Fig. 6.29.

---

[y] Normally the uniform quantizer will produce a binary-coded output signal that is
transmitted or stored. Later, upon recovery, these binary codes are used to synthesize
compressed & quantized amplitude values that are passed through an expander. Thus, the
compressor is located at the signal source, and the expander is located at the receiving end
of the overall system.

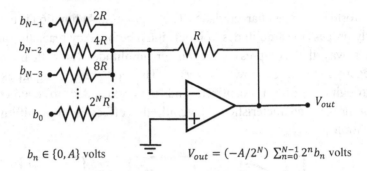

$b_n \in \{0, A\}$ volts $V_{out} = (-A/2^N) \sum_{n=0}^{N-1} 2^n b_n$ volts

Figure 6.28.   A simple digital-to-analog converter.

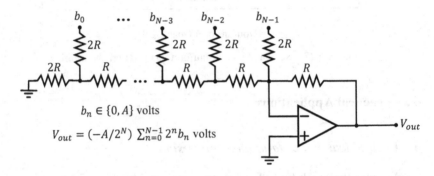

$b_n \in \{0, A\}$ volts

$V_{out} = (-A/2^N) \sum_{n=0}^{N-1} 2^n b_n$ volts

Figure 6.29.   A digital-to-analog converter using an R-2R ladder network.

6.2.2 *Practical analog-to-digital conversion*

6.2.2.1 *Successive approximation ADC*

Assume we have an N-bit DAC that accurately translates a digital word into the corresponding voltage. Place this DAC in a feedback loop as in Fig. 6.30. The input and DAC output signals feed a voltage comparator, which outputs digital '1' when {input signal voltage > DAC output voltage} and '0' when {input signal voltage < DAC output voltage}. Based on this voltage comparator output logic level, an N-bit binary counter is either incremented or decremented at every clock pulse from a high-frequency clock. Then, at equally-spaced increments of time the up/down binary counter outputs are sampled and latched to provide an ADC output code. If the up/down counter can keep up with changes in the input signal voltage, its binary count value will be within one of the desired value.[z] This ADC method suffers from being slow in responding to wide swings in the input voltage, and for some waveforms may not keep up with the amplitude changes. This is because up to $2^N - 1$ high-frequency clock intervals may be necessary for the N-bit binary counter to reach the correct count.[aa] However, this is the simplest ADC architecture to understand and construct.

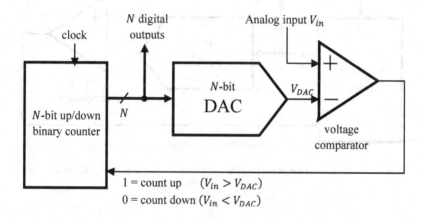

Figure 6.30. Successive approximation analog-to-digital converter.

---

[z] Accuracy may be increased by outputting only the most-significant N bits of an $(N + k)$-bit up/down counter that drives an $(N + k)$-bit DAC, but then the clock frequency must by higher by factor 2^k.

[aa] A 16-bit counter may require $2^{16} - 1 = 65,535$ high-frequency clock pulses to bring it to the correct output state; if we wish to sample and accurately quantize a signal every $1/44,100$ sec (sampling frequency $f_s = 44.1$ kHz), then the high-frequency clock must have a frequency of at least $65,535 \times 44,100 = 2.89$ GHz!

6.2.2.2 *Logarithmic successive approximation ADC*

A modified form of the successive approximation ADC is shown in Fig. 6.31. There a sequential logic circuit drives the DAC instead of an up/down binary counter. The action of this circuit is to determine the values of the N codeword bits, in the order from most to least significant, based on a series of N comparisons. With each comparison, the search for correct output level code is reduced to half of the available choices, so that this method is called a *logarithmic successive approximation* ADC.

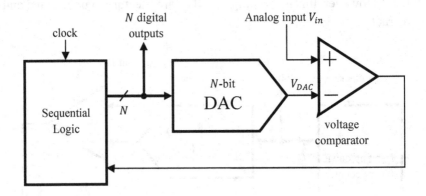

Figure 6.31.   Logarithmic successive approximation analog-to-digital converter.

Here is the algorithm used by the combinational logic circuit to home in on the correct binary code word using just N comparison steps:

(a) Initialize bit counter: $m \leftarrow N$ (number of DAC input bits)
(b) Decrement the bit counter: $m \leftarrow m - 1$
(c) Set DAC input bit b_m to 1, clear all the lower-significant bits (b_{m-1} through b_0) to 0
(d) If $V_{DAC} > V_{in}$ then clear bit b_m: $b_m \leftarrow 0$
(e) If $m > 0$ then repeat from Step (b)
(f) $m = 0$: conversion is complete (ADC output bits = DAC input bits).

6.2.2.3 *Flash ADC*

Although the number of internal clock cycles required for the logarithmic successive approximation ADC to quantize and assign a binary code to input voltage v_{in} is only N (one per output bit), by adding additional hardware complexity it is indeed possible to almost instantly convert an input voltage to a binary code word. Consider the circuit shown in Fig. 6.32:

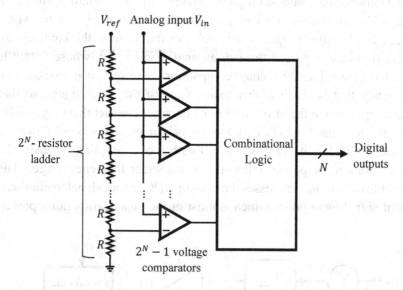

Figure 6.32. Flash analog-to-digital converter.

What we see is a resistive voltage divider composed of 2^N resistors, whose 2^N tapped outputs produce $2^N - 1$ voltages $V_n = V_{ref} \times n/2^N$, for each tap index value n, where $1 \leq n \leq 2^N - 1$. V_{in} is compared to each tap voltage to produce voltage comparator output logic level 1 ($V_{in} > V_n$) or logic level 0 ($V_{in} < V_n$). Note that when input voltage V_{in} falls in the range $0 < V_{in} < V_{ref}$, comparators below a certain spot on the ladder will output 1 and those above them will all output 0. A combinational logic circuit analyzes all $2^N - 1$ comparator outputs to calculate the N-bit index of the lowest comparator in the ladder whose output level is logic 0 (in case all comparators output 1, the output word is N ones). These N bits serve as the Flash ADC output binary codeword. There is no clock involved; the output bits are ready for output as soon as their values have stabilized. The results are available "in a flash" compared to the successive approximation methods, hence the name given to this hardware-intensive circuit is the *Flash ADC*.

6.2.2.4 *Delta-Sigma (ΔΣ) ADC*

The Delta-Sigma analog-to-digital converter block diagram is shown in Fig. 6.33. At its core is a voltage comparator (essentially a 1-bit ADC) whose level-shifted output is fed back to subtract from the input signal. This difference is passed through an analog integrator, whose output is then fed forward to the voltage comparator. The system is clocked at a frequency that is much higher than $2f_{max}$ of the input signal, so that oversampling by a factor of 10 or more occurs. The fact that only one bit per sample is used, which introduces much quantization noise power, is compensated for in two ways. Firstly, oversampling spreads the quantization noise power uniformly over a wider frequency range. The digital lowpass filter processes the one-bit ADC output signal to eliminate out-of-signal-band noise, which is most of the quantization noise power.

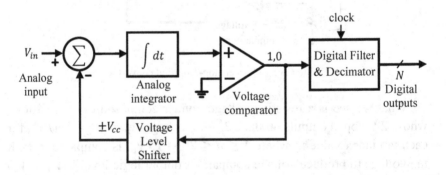

Figure 6.33. Delta-Sigma analog-to-digital converter.

Secondly, the integrator within the feedback system has the effect of lowpass filtering the signal component while shaping the quantization noise power distribution so that it is shifted to higher frequencies and away from the signal band. Taken together, these two effects can increase SNR_Q

the levels typical of 10-16 bit uniform ADC's. Another advantage of the Delta-Sigma ADC is that anti-alias pre-filtering may be very simple, such as a basic RC ladder network, due to the high input sampling rate.

6.2.3 *Useful MATLAB® Code*

6.2.3.1 *Amplitude quantization*

Although amplitude quantization is normally a by-product of ADC hardware, we may simulate its effects using MATLAB's **round** instruction. The following function accepts a matrix of arbitrary size and rounds each matrix coefficient to the nearest of L specified amplitude levels:

```
function Xq = Quantize(X,r)
% Quantize data in X to the nearest levels as specified
%   by r.
%-----------------------------------------------------
% Input:   data matrix/array X (may be any size)
%          output levels in array r (ascending order)
% Output:  quantized data in Xq (same size as X)
    [rows,cols] = size(X);
    X = X(:);
    Xq = X;
    L = length(r);
    r = reshape(r,1,L);
    % Array t contains L+1 transition levels:
    t = [min(X)-eps,mean([r(1:L-1);r(2:end)]),max(X)+eps];
    for k = 1:L
        indexes = find( X>=t(k) & X<t(k+1) );
        Xq(indexes) = r(k);
    end
    Xq = reshape(Xq,rows,cols);
end
```

The quantization function listed above is applied to quantizing a sine wave to 9 amplitude levels that are nonuniformly distributed:

```
% Example_Ch6_1
% Display a sinusoid with and without nonuniform
%   quantization
```

```
%-----------------------------------------------------------
    t = linspace(0,10,1e4);
    x = sin((2*pi/5)*t);
    r = [-0.9,-0.5,-0.2,-0.1,0,0.1,0.2,0.5,0.9];
    xq = Quantize(x,r);
    plot(t,x); hold on
    plot(t,xq)
    axis([0,10,-1.1,1.1])
    title('Nonuniform Quantization of a Sinusoid')
    xlabel('time in seconds')
    ylabel('amplitude, original and quantized')
                             end
```

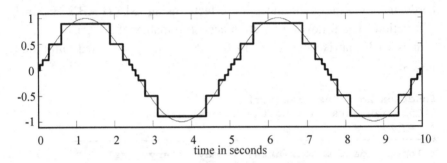

Figure 6.34.   Non-uniform quantization of a sinusoid (from Example 6-1).

Another application of quantization is to convert a black-and-white image to binary format: pixels below threshold level 127.5 are rounded to 0 (black intensity), and pixels above amplitude level 127.5 are rounded to 255 (white intensity).[bb]

```
% Example_Ch6_2
% Display image of a flower and its quantized version
    A = imread('flower_photo.jpg');
    figure; imshow(A); \
    title('Original B&W photo')
```

[bb] Since the interior pages of this textbook are printed using only one color of ink (black), the "un-quantized" image on the left is actually quantized to two levels: presence of ink, and absence of ink. To create the optical illusion that gray-level image intensities are present, the publisher used a digital halftoning algorithm (itself a signal processing operation) to vary the density of ink deposited per unit area on a fine scale.

```
% specify only two output levels (black,white)
r = [0 255];
Aq = uint8(Quantize(double(A),r));
figure; imshow(Aq);
title('Photo quantized to 2 levels')
end
```

Original B&W photo   Photo quantized to 2 levels

Figure 6.35.   Gray-level image before and after 2-level quantization.

6.3   Chapter Summary and Comments

- Sequence $x(n)$ may be obtained by sampling continuous-time signal $x(t)$.

- Signal $x(t)$ is sampled by multiplying it with impulse train $\delta_{T_s}(t)$.

- Sequence $x(n)$ is a list of impulse areas in $x(t)\delta_{T_s}(t)$.

- In practice, $x(t)$ is not sampled by multiplying with an impulse train: this is only a mathematical tool for describing and better understanding the sampling process.

- The Fourier Transform of $x(t)\delta_{T_s}(t)$ is $(1/T_s)\,X(\omega) * \delta_{\omega_s}(\omega)$: amplitude-scaled copies of $X(\omega)$ are frequency-shifted to every multiple of $\omega_s = 2\pi/T_s$ and added together to form a periodic spectrum.

- $\mathcal{F}\{f(n)\}$ may be thought of as $\mathcal{F}\{f(t)\delta_1(t)\}$, which is the continuous-time Fourier Transform of analog signal $f(t)$ after it had been sampled every $T_s = 1$ sec. From sampling theory, therefore, we see why $F(e^{j\omega})$ is periodic with period $= 2\pi$ rad/sec.

- Reconstruction filtering, which is the recovery of $x(t)$ from $x(t)\delta_{T_s}(t)$, is done by filtering out all copies of $X(\omega)$ except the one that is centered at $\omega = 0$.

- The minimum sampling frequency before periodically-repeated copies of $X(\omega)$ begin to overlap is called the "Nyquist rate": $\omega_{s_{min}} = 2\omega_{max}$, where ω_{max} is the highest frequency component in $X(\omega)$.

- Sampling above the Nyquist rate $(\omega_s > 2\omega_{max})$ is called oversampling; in this case recovery of $x(t)$ from $x(t)\delta_{T_s}(t)$ is possible but more samples than necessary are taken.

- Sampling below the Nyquist rate $(\omega_s < 2\omega_{max})$ is called undersampling; in this case recovery of $x(t)$ from $x(t)\delta_{T_s}(t)$ is not possible due to aliasing (destructive overlap-adding of adjacent copies of $X(\omega)$ in the periodic spectrum).

- Aliasing can make one frequency component look like another, hence its name. For example, sampling both $\cos(2t)$ and $\cos(3t)$ at $\omega_s = 5$ rad/sec gives the same sequence.

- The best way to eliminate aliasing distortion is not to have any aliasing in the first place: choose $\omega_s \geq 2\omega_{max}$. This, however, adds the cost of having to deal with more samples of the signal per unit time (e.g., requires more memory to store them, it takes longer to transmit them at a fixed data rate, etc.).

- The next best way to eliminate aliasing distortion is to pre-filter $x(t)$ so that it has zero energy above $\omega_s/2$ before sampling. This reduces the bandwidth of the reconstructed signal so that it no longer is exactly equal to the original $x(t)$, but in certain applications the results may be acceptable. Pre-filtering to reduce the bandwidth of $x(t)$ is called "Anti-alias filtering", and it adds the cost of an additional filtering operation.

- The least desirable way to eliminate aliasing distortion is to reduce the bandwidth of the reconstruction lowpass filter from $\omega_s/2$ to $\omega_s - \omega_{max}$. This filters out any leakage from adjacent spectra but significantly reduces the bandwidth of the reconstructed signal.

- Analog-to-digital conversion is the combination of sampling, quantization, and binary coding operations. It is more complex than digital-to-analog conversion; in fact, some ADC's have a DAC as part of their internal circuitry.

- ADC's may be rated according to number of their output bits, speed of conversion, and accuracy.

6.4 Homework Problems

Problems P6.1-P6.23: Write an expression for the Fourier Transform of the specified signal, and then plot that spectrum over frequency range $-10 \leq \omega \leq 10$ rad/sec.

P6.1 $x(t) = \frac{1}{\pi}\cos(3t)$

P6.2 $x_s(t) = \frac{1}{\pi}\cos(3t) \cdot \frac{2\pi}{5}\delta_{2\pi/5}(t)$

P6.3 $y(t) = \frac{2}{\pi}\text{sinc}(2t)$

P6.4 $y_s(t) = \frac{2}{\pi}\text{sinc}(2t) \cdot \frac{2\pi}{6}\delta_{2\pi/6}(t)$

P6.5 $w(t) = \frac{1}{2\pi}\text{sinc}^2(t/2)$

P6.6 $w_s(t) = \frac{1}{2\pi}\text{sinc}^2\left(\frac{t}{2}\right) \cdot \pi\delta_\pi(t)$

P6.7　$f(t) = \frac{1}{\pi}\cos(7t)$

P6.8　$f_s(t) = \frac{1}{\pi}\cos(7t) \cdot \frac{2\pi}{5}\delta_{\frac{2\pi}{5}}(t)$

P6.9　$x(t) = \frac{3}{\pi}\text{sinc}^2(3t)$

P6.10　$x_{s1}(t) = \frac{3}{\pi}\text{sinc}^2(3t) \cdot \frac{\pi}{6}\delta_{\pi/6}(t)$

P6.11　$x_{r1}(t) = $ Ideal LPF$\left\{\frac{3}{\pi}\text{sinc}^2(3t) \cdot \frac{\pi}{6}\delta_{\pi/6}(t)\right\}$ having bandwidth = 6 rad/sec

P6.12　$x_{s2}(t) = \frac{3}{\pi}\text{sinc}^2(3t) \cdot \frac{\pi}{5}\delta_{\pi/5}(t)$

P6.13　$x_{r2}(t) = $ Ideal LPF$\left\{\frac{3}{\pi}\text{sinc}^2(3t) \cdot \frac{\pi}{5}\delta_{\pi/5}(t)\right\}$ having bandwidth = 6 rad/sec

P6.14　$x_{r3}(t) = $ Ideal LPF$\left\{\frac{3}{\pi}\text{sinc}^2(3t) \cdot \frac{\pi}{5}\delta_{\pi/5}(t)\right\}$ having bandwidth = 4 rad/sec

P6.15　$y(t) = $ Ideal LPF$\left\{\frac{3}{\pi}\text{sinc}^2(3t)\right\}$ having bandwidth = 5 rad/sec

P6.16　$y_s(t) = y(t) \cdot \frac{\pi}{5}\delta_{\pi/5}(t)$　(Assume $y(t)$ from Problem P6.15)

P6.17　$x_{r4}(t) = $ Ideal LPF$\{y_s(t)\}$, bandwidth = 5 rad/sec. (Assume $y_s(t)$ from Problem P6.16)

P6.18　Which of the following is the most accurate reconstruction of $x(t)$ in Problem P6.9: $x_{r2}(t)$ in Problem 6.13, $x_{r3}(t)$ in Problem 6.14, or $x_{r4}(t)$ in Problem 6.17? Explain.

P6.19　$w_s(t) = \frac{1}{2\pi}\text{sinc}^2(t/2) \cdot \frac{2\pi}{9}\delta_{2\pi/9}(t)$

P6.20　$h(t) \Leftrightarrow H(\omega) = \Delta\left(\frac{\omega+4}{8}\right) + \Delta\left(\frac{\omega}{8}\right) + \Delta\left(\frac{\omega-4}{8}\right)$

P6.21 $w_r(t) = w_s(t) * h(t) \Leftrightarrow W_s(\omega)H(\omega)$ (Assume $w_s(t)$ from Problem P6.19, $h(t)$ from Problem P6.20.)

P6.22   What is the advantage of using $H(\omega)$ to recover $w(t)$ instead of using an ideal LPF?   (Assume $w_s(t)$ from Problem P6.19, $h(t)$ from Problem P6.20.)

P6.23   What is the disadvantage of using $H(\omega)$ to recover $w(t)$ instead of using an ideal LPF?   (Assume $w_s(t)$ from Problem P6.19, $h(t)$ from Problem P6.20.)

Chapter 7

Frequency Analysis of
Discrete-Time Systems

7.1 Theory

7.1.1 *Introduction*

In previous chapters, we have introduced several useful time-domain signals, and considered how each may be represented as weighted sum of sinusoids at different frequencies. This is what is called a signal's frequency-domain representation, or *frequency spectrum*. In the engineering field a practical system or device is studied using a mathematical model. Typically, such a model accepts one input signal and in response produces one output signal. In the discrete-time case, to reduce mathematical complexity, we prefer working with system models that are completely described using constant-coefficient linear difference equations.[a] Such models fall in the category of *Linear Shift-Invariant (LSI) Systems*. Linear systems are those whose input/output relationship follows the principle of linear superposition, and shift-invariance means that the system parameters do not vary with time. However, most real-world systems or devices exhibit behavior that is both time-varying and nonlinear. To simplify the representation and analysis of such systems one may develop difference equations that accurately approximate nonlinear behavior within a limited amplitude range, or that accurately approximate time-varying behavior during a limited time span. Therefore, in this text

---

[a] Linear difference equations are introduced in Section 7.5.2.

269

we will concern ourselves only with systems assumed to be linear and shift-invariant.[b]

This chapter considers how a discrete-time signal is modified as it is processed by a constant-coefficient linear difference equation. Specifically, we will examine how the frequency spectrum of the output signal compares to that of the input signal. First we consider a general linear system that is defined only mathematically (as if the processor was hidden inside a "black box"), and after that we present specific networks that may be used to implement some useful system responses.

7.1.2 *Linear shift-invariant discrete-time system*

7.1.2.1 *Impulse response*

Let us represent a linear shift-invariant discrete-time system as a black box, having one input and one output. A very important characteristic is how this system responds to an impulse input signal when the system is at zero initial conditions.[c] Figure 7.1 depicts this situation. When, in response to an impulse $\delta(n)$ at the input, the output signal is $h(n)$, we call $h(n)$ the *impulse response* of the system:

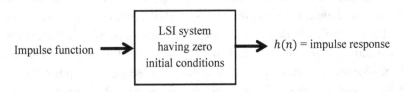

Figure 7.1. Measuring the impulse response of a discrete-time system.

---

[b] The terms "shift-invariant" and "time-invariant" are used interchangeably for discrete-time systems.

[c] This is called a *zero-state response* of the system. Throughout this chapter it is assumed that all systems are initially at zero state (no component of the output is due to anything other than the signal being applied at the input).

The system's impulse response $h(n)$ may also be determined from the difference equation modeling the system. We will not be concerned with that aspect here. For the time being we will assume that we have knowledge of $h(n)$ by performing the experiment depicted in Fig. 7.1.

The behavior of a LSI discrete-time system is completely described by its impulse response. Why is this so? Let us first support the notion that it may be possible. We begin by showing that unless the system is linear, it is impossible for the system's impulse response $h(n)$ – or the system's response to any other input signal, for that matter – to completely describe its behavior.

Assume that the input signal to our LSI system is some arbitrary sequence $f(n)$ and the resulting output signal is $g(n)$, as shown in Fig. 7.2:

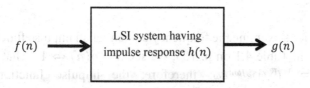

$$f(n) \longrightarrow \boxed{\begin{array}{c} \text{LSI system having} \\ \text{impulse response } h(n) \end{array}} \longrightarrow g(n)$$

Figure 7.2. The relationship between input and output signals of a linear, shift invariant system is *completely* described by the system's impulse response.

Since input signal $f(n)$ results in output signal $g(n)$, the scaling property of linearity dictates that inverting the input signal will result in an inverted output signal. If this system were nonlinear, however, then it would be possible for both $f(n)$ and $-f(n)$ to result in the same output $g(n)$.[d] Such reasoning also leads us to realize that two different nonlinear systems may have the same impulse responses.[e] But if a system is to be completely categorized by its impulse response, then two different systems

---

[d] For example, this is true for a nonlinear system whose input/output relationship is described by $g(n) = f^2(n)$.

[e] For example, the two nonlinear systems defined by $g(n) = f^2(n)$ and $g(n) = |f(n)|$ have identical impulse responses.

may not have the same impulse response. Therefore, it is only possible that a *linear* system may be completely described by its impulse response.

Next, we show that only a shift-invariant system may be completely described by its response to a given signal (such as the impulse). "Shift-invariance", meaning that a system's behavior is the same regardless of *when* the input is applied, is a characteristic described by the following: if applying input $f(n)$ results in $g(n)$ at the output, then input $f(n - n_0)$ will result in $g(n - n_0)$ at the output. That is, time-shifting the input will produce a time-shifted but otherwise identical output signal as before. But for a system that is not shift-invariant, its impulse response will be different at different times, and a single impulse response cannot possibly completely describe this system's behavior.[f] Thus we see that system linearity and shift invariance are prerequisites for the system's full description being conveyed by a single sequence, such as its impulse response.

Now let's examine the frequency content of an impulse function. From Entry #6 in Table 4.1 on p. 103, we see that $\delta(n) \Leftrightarrow 1$. That is, $\delta(n) = \mathcal{F}^{-1}\{1\} = \frac{1}{2\pi}\int_{-\pi}^{\pi}(1)e^{j\omega n}d\omega$; therefore, the impulse function contains equally-weighted terms $e^{j\omega n}$ at all frequencies (over period $-\pi \leq \omega \leq \pi$, and hence over all periods of its periodic spectrum). As will be shown later in this chapter, $e^{j\omega n}$ is an eigenfunction of an LSI discrete-time system: when it is applied as an input, the output is the same except for a constant scaling factor:

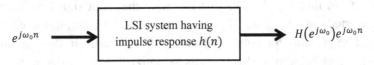

Figure 7.3.　The effect of passing eigenfunction $e^{j\omega_0 n}$ through a linear, shift-invariant discrete-time system ($H(e^{j\omega_0})$ is a complex constant).

---

[f] It is possible to completely categorize a time-varying linear system using a time-varying impulse response, but this greatly complicates matters.

Eigenfunction $e^{j\omega_0 n}$ is at frequency ω_0. The impulse function $\delta(n)$ is composed of eigenfunctions at all possible frequencies. It makes sense that the spectrum of an impulse has all frequencies present, and that these terms are at the same amplitude, if what we want to do is measure the response of our system over the entire spectrum of frequencies. The system's response at frequency ω_0 may be described by the constant $H(e^{j\omega_0})$, and the frequency response of the system over the entire spectrum of frequencies is exactly specified by $H(e^{j\omega})$.[g] Since the system is linear, when $\delta(n) = (1/2\pi) \int_{-\pi}^{\pi} (e^{j\omega n}) d\omega$ is applied to the input the system will produce $h(n) = (1/2\pi) \int_{-\pi}^{\pi} (H(e^{j\omega}) e^{j\omega n}) d\omega$ at the output. We see from this expression for $h(n)$, therefore, that $h(n)$ and $H(e^{j\omega})$ are a Fourier Transform pair. $H(e^{j\omega})$ is called the *frequency response function* of the LSI system. Since $H(e^{j\omega})$ completely describes the system in the frequency domain, then $h(n)$ must also completely describe the system in the time domain. This concludes our conceptual argument.

7.1.2.2 *Input/output relations*

Recall the sifting property of the impulse function, as given in Eq. (2.2) on page 19:

$$x(n_0) = \sum_{n=-\infty}^{\infty} x(n)\delta(n - n_0). \qquad (7.1)$$

Note that $x(n_0)$ is a scalar constant that is obtained by sampling the sequence $x(n)$: that is, $x(n_0) = x(n)|_{n=n_0}$. With a change in variables, we may use the sifting property to obtain a different result:

$$x(n) = \sum_{n_0=-\infty}^{\infty} x(n_0)\delta(n - n_0). \qquad (7.2)$$

Here $x(n)$ represents an entire sequence. This expression, previously introduced as Eq. (2.41) on page 33, shows that an arbitrary discrete-time signal $x(n)$ may be decomposed into a summation of delayed, scaled impulse functions.

---

[g] Eigenfunction $e^{j\omega n}$ is periodic in ω, with period equal to 2π rad/sec. Proof: $e^{j(\omega + k2\pi)n} = e^{j\omega n} e^{j2\pi(kn)} = e^{j\omega n}(1) = e^{j\omega n}$. Hence $H(e^{j\omega})$ is also periodic in ω, with period equal to 2π rad/sec.

We will use "$x(n) \rightarrow y(n)$" to denote the following input/output relationship: when LSI discrete-time system having zero initial conditions is driven by input signal $x(n)$, then the resulting output signal is $y(n)$. Following this notation, the following is a step-by-step derivation of the relationship between input signal $x(n)$, system impulse response $h(n)$, and the output signal $y(n)$:

input $\rightarrow$ output

$\delta(n) \rightarrow h(n)$ $\qquad\qquad$ (system impulse response)

$\delta(n - k) \rightarrow h(n - k)$ $\qquad\qquad$ (shift-invariance)

$x(k)\delta(n - k) \rightarrow x(k)h(n - k)$ $\qquad\qquad$ (linearity)

$x(n)\delta(n - k) \rightarrow x(k)h(n - k)$ $\qquad\qquad$ (sifting property)

$\sum_{k=-\infty}^{\infty} x(n)\delta(n - k) \rightarrow \sum_{k=-\infty}^{\infty} x(k)h(n - k)$ $\qquad\qquad$ (linearity)

$x(n) \sum_{k=-\infty}^{\infty} \delta(n - k) \rightarrow \sum_{k=-\infty}^{\infty} x(k)h(n - k)$ $\qquad\qquad$ (simplified)

$x(n) \rightarrow \sum_{k=-\infty}^{\infty} x(k)h(n - k)$ $\qquad\qquad$ (simplified)

$x(n) \rightarrow x(n) * h(n) = y(n)$ $\qquad\qquad$ (definition of convolution)

$X(e^{j\omega}) \rightarrow X(e^{j\omega}) \times H(e^{j\omega}) = Y(e^{j\omega})$ $\qquad\qquad$ (Fourier Transform)

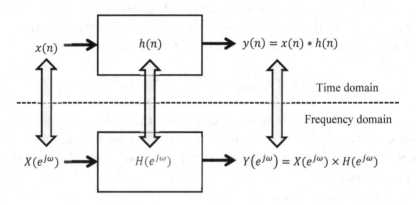

Figure 7.4. The input/output relationships of an LSI system, shown in both time and frequency domains.

Figure 7.4 represents the process of determining the output of a linear, shift-invariant discrete-time system having impulse response $h(n)$ to any input $x(n)$. With the knowledge of impulse response one can determine the output of the system via convolution in the time domain, *OR*, if it is advantageous, transform all signals to the frequency domain, perform multiplication, and then convert the product to the time domain using the inverse Fourier Transform.

We stated previously that $H(e^{j\omega})$ is called the *frequency response function* of the LSI system. It is also referred to as the *transfer function*, since it determines how the eigenfunction components of the input are scaled when they transfer to the output. Frequently, the system behavior may be specified more conveniently through its transfer function than with its impulse response, as we will observe in the following section on filters.

We will next verify that, as assumed previously, $e^{j\omega_0 n}$ is an eigenfunction of an LSI discrete-time system. When $e^{j\omega_0 n}$ is applied at the input, an LSI system having impulse response $h(n)$ will produce the following output:

$$e^{j\omega_0 n} * h(n) = \sum_{k=-\infty}^{\infty} e^{j\omega_0 k} h(n-k)$$

$$= \sum_{m=-\infty}^{\infty} e^{j\omega_0(n-m)} h(m) = \sum_{m=-\infty}^{\infty} e^{j\omega_0 n} e^{-j\omega_0 m} h(m)$$

$$= e^{j\omega_0 n} \sum_{m=-\infty}^{\infty} e^{-j\omega_0 m} h(m) = e^{j\omega_0 n} H(e^{j\omega_0}), \qquad (7.3)$$

where constant $H(e^{j\omega_0})$ is $\mathcal{F}\{h(n)\}$ evaluated at $\omega = \omega_0$. The same result can be obtained via the frequency-domain:

$$\mathcal{F}\{e^{j\omega_0 n} * h(n)\} = \mathcal{F}\{e^{j\omega_0 n}\}\mathcal{F}\{h(n)\}$$

$$= 2\pi\delta_{2\pi}(\omega - \omega_0)H(e^{j\omega}) = 2\pi\delta_{2\pi}(\omega - \omega_0)H(e^{j\omega_0})$$

$$= \mathcal{F}\{e^{j\omega_0 n}\}H(e^{j\omega_0}) = \mathcal{F}\{e^{j\omega_0 n} H(e^{j\omega_0})\}, \text{ or }$$

$$e^{j\omega_0 n} * h(n) = e^{j\omega_0 n} H(e^{j\omega_0}). \qquad (7.4)$$

It is more practical to consider what output results when sinusoid $\cos(\omega_0 n)$ is the input signal to an LSI system. This is because $\cos(\omega_0 n)$

is a real sequence that is a weighted sum of eigenfunctions at frequencies $\pm\omega_0$ $(= 0.5e^{j\omega_0 n} + 0.5e^{-j\omega_0 n})$, whereas $e^{j\omega_0 n}$ is not real and therefore cannot be represented with a single sequence of real numbers. When inputted to a linear system having transfer function $H(e^{j\omega})$, each of these eigenfunction components produces the following output components:

$$0.5e^{j\omega_0 n} \Rightarrow 0.5H(e^{j\omega_0}), \tag{7.5}$$

$$0.5e^{j(-\omega_0)n} \Rightarrow 0.5H(e^{j(-\omega_0)}). \tag{7.6}$$

By linearity, therefore,

$$0.5\,e^{j\omega_0 n} + 0.5\,e^{-j\omega_0 n} = \cos(\omega_0 n)$$

$$\Rightarrow 0.5e^{j\omega_0 n} H(e^{j\omega_0}) + 0.5e^{-j\omega_0 n} H(e^{-j\omega_0}). \tag{7.7}$$

Next, because a real-world system will have a real impulse response, assume that $h(n)$ is real. That results in magnitude response $|H(e^{j\omega})|$ being an even function of ω, and phase response $\angle H(e^{j\omega})$ being an odd function of ω. Since $H(e^{j\omega_0}) = |H(e^{j\omega_0})|e^{j\angle H(e^{j\omega_0})}$:

$$\cos(\omega_0 n) \Rightarrow \tfrac{1}{2}e^{j\omega_0 n}\,|H(e^{j\omega_0})|e^{j\angle H(e^{j\omega_0})},$$

$$+\tfrac{1}{2}e^{-j\omega_0 n}\,|H(e^{-j\omega_0})|e^{j\angle H(e^{-j\omega_0})}. \tag{7.8}$$

$$\cos(\omega_0 n) \Rightarrow \tfrac{1}{2}e^{j\omega_0 n}\,|H(e^{j\omega_0})|e^{j\angle H(e^{j\omega_0})},$$

$$+\tfrac{1}{2}e^{-j\omega_0 n}\,|H(e^{j\omega_0})|e^{-j\angle H(e^{j\omega_0})}. \tag{7.9}$$

$$\cos(\omega_0 n) \Rightarrow \tfrac{1}{2}|H(e^{j\omega_0})|\,e^{j(\omega_0 n + \angle H(e^{j\omega_0}))},$$

$$+\tfrac{1}{2}|H(e^{j\omega_0})|\,e^{-j(\omega_0 n + \angle H(e^{j\omega_0}))}. \tag{7.10}$$

From this we see that an input sinusoid will produce an output sinusoid at the same frequency, but with amplitude multiplied by $|H(e^{j\omega_0})|$ and with phase additively offset by $\angle H(e^{j\omega_0})$:

$$\cos(\omega_0 n) \Rightarrow |H(e^{j\omega_0})| \cos(\omega_0 n + \angle H(e^{j\omega_0})), \qquad (7.11)$$

when $h(n)$ is real. This relation gives us a practical alternative to system characterization via impulse response measurement: apply a sinusoid at frequency ω_0 to the input of the system, measure the changes in amplitude and phase of the output sinusoid with respect to that of the input sinusoid, and repeat the measurement at sufficiently many other frequencies until $H(e^{j\omega})$, at all values of ω, is deduced from the measured data points.

7.1.3 Digital filtering concepts

Filters are an important class of digital signal processing algorithms. Specifically, they play a major role in communication systems. *Filtering*, the function of filters, is a frequency-domain concept that is best described by a linear system's transfer function $H(e^{j\omega})$ as opposed to its impulse response $h(n)$. Here we discuss filters having idealized frequency response functions. An ideal filter passes some frequencies while stopping others from passing. Its transfer function, $H(e^{j\omega})$, multiplies the input signal's spectrum by either 1 or 0. The range of frequencies where $H(e^{j\omega}) = 1$ is called the *passband* and the range of frequencies where $H(e^{j\omega}) = 0$ is called the *stopband*. There are four different types of ideal digital filters: *lowpass, highpass, bandpass*, and *band-elimination* filters.[h] Figure 7.5 shows the transfer functions for these filters.

Since $H(e^{j\omega})$ is periodic, digital filters have their passband/stopband pattern repeating infinitely many times along the frequency scale. Considering only frequency range $-\pi \le \omega \le \pi$, which is the period of $H(e^{j\omega})$ centered at $\omega = 0$, the names of digital filters are quite representative of their frequency response characteristics.

---

[h] Given that an ideal filter's transfer function $H(e^{j\omega})$ is either zero or one at any frequency, two other trivial filter types may be defined: the *all-stop* filter where $H(e^{j\omega}) = 0$, and the *all-pass* filter where $H(e^{j\omega}) = 1$. We should note that another, less restrictive filter commonly bears the name *allpass* filter; it is characterized by magnitude $|H(e^{j\omega})| = 1$ and arbitrary phase.

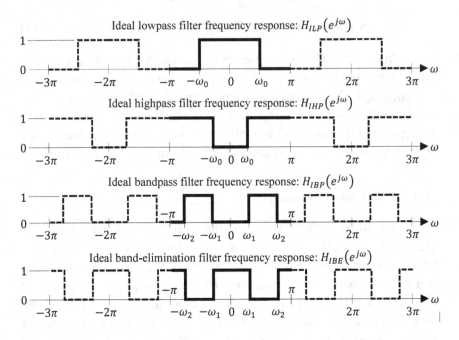

Figure 7.5. Frequency responses of the four ideal digital filter types (the period centered at $\omega = 0$ is highlighted).

7.1.3.1 Ideal lowpass filter

The ideal lowpass filter passes all frequencies below a certain specified *cutoff frequency* ω_0. If the input to a lowpass filter has frequency components at $|\omega| > \omega_0$, then those components are absent at the output of the filter.[i] The ideal lowpass filter's frequency response $H_{ILP}(e^{j\omega})$ is defined as a rectangular pulse (continuous in frequency) of width $2\omega_0$ ($\omega_0 < \pi$) that is periodically extended to repeat every 2π rad/sec on the ω scale:

$$H_{ILP}(e^{j\omega}) = \text{rect}\left(\frac{\omega}{2\omega_0}\right) * \delta_{2\pi}(\omega) \qquad (7.12)$$

---

[i] The ideal lowpass filter passes all frequencies $|\omega| < \omega_0$ when considering only frequency range $-\pi \leq \omega \leq \pi$. This may be extended to all periods of $H(e^{j\omega})$ by using principal value of ω in the expression: $|\text{pv } \omega| < \omega_0$.

From Table 4.1, the inverse Fourier Transform of $H_{ILP}(e^{j\omega})$ is the impulse response of an ideal lowpass filter:

$$h_{ILP}(n) = \frac{\omega_0}{\pi} \text{sinc}(\omega_0 n). \tag{7.13}$$

We note that $h_{ILP}(n)$ is noncausal,[j] and therefore an ideal lowpass filter cannot be realized to operate in real-time (because present output is partly a function of future inputs).

7.1.3.2 *Ideal highpass filter*

The ideal highpass filter transfer function is *complementary* to that of the ideal lowpass filter, which is to say $H_{IHP}(e^{j\omega}) = 1 - H_{ILP}(e^{j\omega})$:

$$H_{IHP}(e^{j\omega}) = 1 - \text{rect}\left(\frac{\omega}{2\omega_0}\right) * \delta_{2\pi}(\omega). \tag{7.14}$$

Multiplication by transfer function $H_{IHP}(e^{j\omega})$ passes all frequencies above a certain specified frequency ω_0 (over the period of $H_{IHP}(e^{j\omega})$ that is centered at 0). Just as for the lowpass filter, the impulse response of an ideal highpass filter is noncausal:

$$h_{IHP}(n) = \delta(n) - \frac{\omega_0}{\pi} \text{sinc}(\omega_0 n). \tag{7.15}$$

7.1.3.3 *Ideal bandpass filter*

The ideal bandpass filter passes all frequency components in the range $\omega_1 < |\omega| < \omega_2$ (when looking only at the period of $H_{IBP}(e^{j\omega})$ that is centered at zero). An ideal bandpass filter can be realized by placing ideal lowpass and highpass filters in series (where ω_2 is the cutoff frequency of the lowpass filter, and ω_1 is the cutoff frequency of the highpass filter):

$$H_{IBP}(e^{j\omega}) = H_{ILP}(e^{j\omega})H_{IHP}(e^{j\omega}), \tag{7.16}$$

---

[j] The sinc function is not zero for all $n < 0$. As a matter of fact, it has nonzero values (or "has support") over time index range $(-\infty, \infty)$.

$$= \left\{ \text{rect}\left(\frac{\omega}{2\omega_2}\right) * \delta_{2\pi}(\omega) \right\} \left\{ 1 - \text{rect}\left(\frac{\omega}{2\omega_1}\right) * \delta_{2\pi}(\omega) \right\}$$

$$= \text{rect}\left(\frac{\omega}{2\omega_2}\right) * \delta_{2\pi}(\omega) - \text{rect}\left(\frac{\omega}{2\omega_2}\right) \text{rect}\left(\frac{\omega}{2\omega_1}\right) * \delta_{2\pi}(\omega)$$

$$= \left\{ \text{rect}\left(\frac{\omega}{2\omega_2}\right) - \text{rect}\left(\frac{\omega}{2\omega_2}\right) \text{rect}\left(\frac{\omega}{2\omega_1}\right) \right\} * \delta_{2\pi}(\omega). \qquad (7.17)$$

$$(0 < \omega_1 < \omega_2 < \pi)$$

Therefore,

$$H_{IBP}\left(e^{j\omega}\right) = \left\{ \text{rect}\left(\frac{\omega}{2\omega_2}\right) - \text{rect}\left(\frac{\omega}{2\omega_1}\right) \right\} * \delta_{2\pi}(\omega), \qquad (7.18)$$

because $\text{rect}(\omega/2\omega_2)\,\text{rect}(\omega/2\omega_1) = \text{rect}(\omega/2\omega_1)$ when $\omega_2 > \omega_1$. The corresponding impulse response is the inverse Fourier transform of Eq. (7.18):

$$h_{IBP}(n) = \frac{\omega_2}{\pi}\text{sinc}(\omega_2 n) - \frac{\omega_1}{\pi}\text{sinc}(\omega_1 n). \qquad (7.19)$$

An equally valid formulation for $H_{IBP}\left(e^{j\omega}\right)$ is a sum of frequency-shifted rectangular pulses (periodically extended):

$$H_{IBP}\left(e^{j\omega}\right) = \left\{ \text{rect}\left(\frac{\omega+\omega_c}{W}\right) + \text{rect}\left(\frac{\omega-\omega_c}{W}\right) \right\} * \delta_{2\pi}(\omega) \qquad (7.20)$$

$$\left(W = \omega_2 - \omega_1, \ \omega_c = \frac{\omega_1+\omega_2}{2} \right).$$

The corresponding impulse response is the inverse Fourier transform of Eq. (7.20):

$$h_{IBP}(n) = \frac{W}{\pi}\text{sinc}\left(\frac{Wn}{2}\right)\cos(\omega_c n) \qquad (7.21)$$

$$\left(W = \omega_2 - \omega_1, \ \omega_c = \frac{\omega_1+\omega_2}{2} \right).$$

7.1.3.4 *Ideal band-elimination filter*

The band-elimination filter is the complement of the bandpass filter, making the stopband range $\omega_1 < |\omega| < \omega_2$ (when looking only at the period of $H_{IBE}(e^{j\omega})$ that is centered at zero). Referring to the ideal bandpass filter expressions in Section 7.1.3.3, the corresponding results for band-elimination filter are therefore:

$$H_{IBE}\left(e^{j\omega}\right) = 1 - \left\{\text{rect}\left(\frac{\omega}{2\omega_2}\right) - \text{rect}\left(\frac{\omega}{2\omega_1}\right)\right\} * \delta_{2\pi}(\omega), \quad (7.22)$$

Alternatively:

$$H_{IBE}\left(e^{j\omega}\right) = 1 - \left\{\text{rect}\left(\frac{\omega+\omega_c}{W}\right) + \text{rect}\left(\frac{\omega-\omega_c}{W}\right)\right\} * \delta_{2\pi}(\omega) \quad (7.23)$$

$$\left(W = \omega_2 - \omega_1, \ \omega_c = \frac{\omega_1+\omega_2}{2}\right).$$

The corresponding time-domain impulse response expressions are

$$h_{IBE}(n) = \delta(n) - \left\{\frac{\omega_2}{\pi}\text{sinc}(\omega_2 n) - \frac{\omega_1}{\pi}\text{sinc}(\omega_1 n)\right\} \quad (7.24)$$

and

$$h_{IBE}(n) = \delta(n) - \frac{W}{\pi}\text{sinc}\left(\frac{Wn}{2}\right)\cos(\omega_c n). \quad (7.25)$$

$$\left(W = \omega_2 - \omega_1, \ \omega_c = \frac{\omega_1+\omega_2}{2}\right)$$

All the ideal filters that we have discussed are not causal and thus cannot be implemented in practice. Another reason they cannot be realized is that infinite filter complexity is required to achieve the sudden transitions between passband and stopband in frequency response.[k] Later in this chapter we will consider some low-order digital filters that are practical to realize, but whose frequency transfer functions only approximate the ideal case.

---

[k] The presence of an instantaneous jump in the magnitude frequency response characteristic (at passband-to-stopband transition) requires infinitely many multiply-add operations to be exactly realized as a digital filter.

7.1.4 *Discrete-time filter networks*

7.1.4.1 *Digital filter building blocks*

Linear digital filtering may be implemented in hardware using either dedicated digital electronic circuits that perform specific operations repeatedly and quickly, or less efficiently on a general-purpose computer that is programmed to do the same task. When viewed at the lowest level, the primary operations of digital signal processing are memory storage and retrieval, addition, and multiplication of binary codes representing real numbers.[1] These basic discrete-time filter blocks may be interconnected to build very complicated networks.[m] The digital networks introduced in this chapter are of the simplest type, accomplishing simple filtering functions with a minimum of operations.

The simplest discrete-time operation is the *sample delay*. It is implemented in hardware by storing a numerical code in memory for one clock cycle, then retrieving it from memory during the next clock cycle. The sample delay may be described in the time domain as a linear, shift-invariant operation:

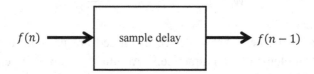

Figure 7.6. Time-domain representation of the sample delay function.

Note that $f(n-1) = f(n) * \delta(n-1)$, so the system's impulse response $h(n)$ is identified to be $\delta(n-1)$. The transfer function of this discrete-time system is then

---

[1] Complex numbers are handled by processing their real and imaginary parts separately.

[m] An example of a complicated and specialized network is a processor that accepts N samples representing one period of sequence $f_p(n)$, performs the DFT using its computationally-efficient FFT implementation, and then returns N samples corresponding to one period of its frequency-domain transformed version $F_p(k)$.

$$H(e^{j\omega}) = \mathcal{F}\{\delta(n-1)\} = \sum_{n=-\infty}^{\infty} \delta(n-1)e^{-j\omega n} = e^{-j\omega}. \quad (7.26)$$

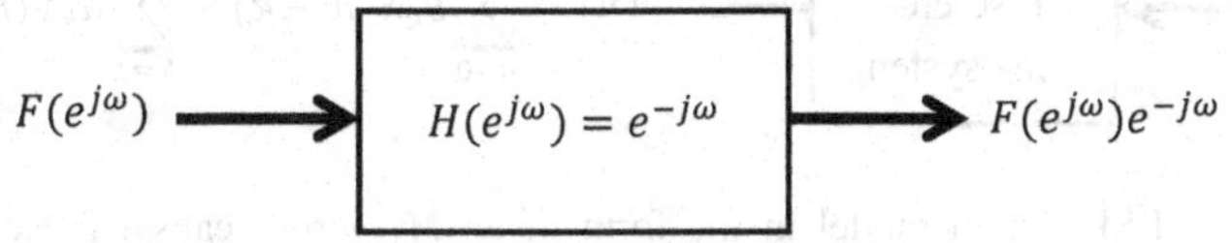

Figure 7.7. Frequency-domain representation of the sample delay function.

The effect of multiplying $F(e^{j\omega})$ by transfer function $e^{-j\omega}$ is that the magnitude of its frequency spectrum is unchanged, and that its phase spectrum is additively offset by $-\omega$:

$$\left|F(e^{j\omega})e^{-j\omega}\right| = \left|F(e^{j\omega})\right|\left|e^{-j\omega}\right| = \left|F(e^{j\omega})\right|, \quad (7.27)$$

$$\angle\{F(e^{j\omega})e^{-j\omega}\} = \angle F(e^{j\omega}) + \angle e^{-j\omega} = \angle F(e^{j\omega}) - \omega. \quad (7.28)$$

7.1.4.2 *Linear difference equations*

The discrete-time equivalent of a linear differential equation is a linear difference equation. Linear difference equations express the output sample of a discrete-time linear system as a weighted sum of input samples and other output samples. These samples may be at past, present and future times. However, if it is possible for a difference equation to be implemented for real-time operations (without introducing any extra delay[n]) then the difference equation must be causal. The general form of a causal difference equation is shown in Fig. 7.8.

---

[n] We may simulate looking into the future by introducing a delay and have a *pipeline* memory unit to fill with signal sample values during that time. Then, processing commences as usual except that we operate in a delayed time frame of reference; our *now* is actually in the past by amount of delay introduced by the memory pipeline. According to our delayed time frame of reference, any *future* signal values that appear in pipeline memory will be available for processing. This method, of course, is nothing more than introducing a time delay and relabeling the time scale to make it appear that we are looking a short time into the future. Certain applications, such as 2-way voice communications, cannot tolerate excessive processing delays. Other applications, such as enhancing the audio quality of a recording made some 50 years ago for the purpose of re-issuing a

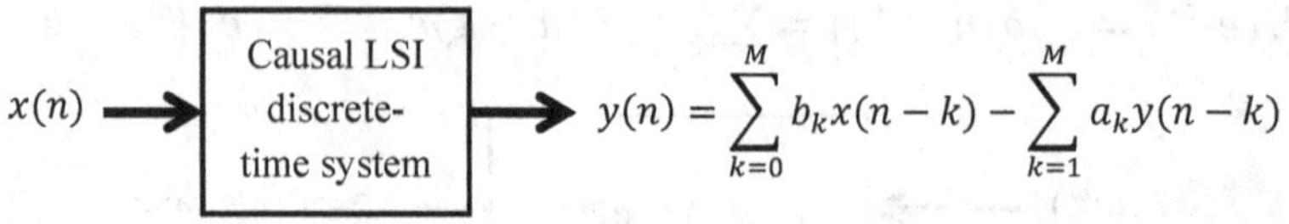

$$x(n) \rightarrow \boxed{\begin{array}{c} \text{Causal LSI} \\ \text{discrete-} \\ \text{time system} \end{array}} \rightarrow y(n) = \sum_{k=0}^{M} b_k x(n-k) - \sum_{k=1}^{M} a_k y(n-k)$$

Figure 7.8.   LSI system model in the form of an Mth-order causal linear difference equation.

     Output sequence $y(n)$ is a function of $M+1$ past and present input signal samples $(x(n), x(n-1), x(n-2), \ldots, x(n-M))$ and M past output signal samples $(y(n-1), y(n-2), \ldots, y(n-M))$. Because the calculation for present output value $y(n)$ does not involve future samples of either input or output signals, the system itself is also said to be causal.[o] Let's calculate the impulse response of this system. When $x(n) = \delta(n)$,

$$y(n) = \sum_{k=0}^{M} b_k \delta(n-k) - \sum_{k=1}^{M} a_k y(n-k) \quad (= h(n)). \quad (7.29)$$

     Taking the discrete-time Fourier Transform of all terms in the equation, we obtain an expression for spectrum $Y(e^{j\omega})$:

$$Y(e^{j\omega}) = \sum_{k=0}^{M} b_k e^{-j\omega k} - \sum_{k=1}^{M} a_k e^{-j\omega k} \, Y(e^{j\omega}). \quad (7.30)$$

Solving for $H(e^{j\omega}) = Y(e^{j\omega})$ (since $X(e^{j\omega}) = 1$) we obtain:

$$H(e^{j\omega}) = \frac{\sum_{k=0}^{M} b_k e^{-j\omega k}}{1 + \sum_{k=1}^{M} a_k e^{-j\omega k}}. \quad (7.31)$$

     This is called a *rational* form for $H(e^{j\omega})$, since it is a ratio of polynomials in terms of complex variable $z = e^{j\omega}$:

$$H(z) = \frac{\sum_{k=0}^{M} b_k z^{-k}}{1 + \sum_{k=1}^{M} a_k z^{-k}} = \frac{b_0 z^M + b_1 z^{M-1} + b_2 z^{M-2} + \cdots + b_{M-1} z + b_M}{z^M + a_1 z^{M-1} + a_2 z^{M-2} + \cdots + a_{M-1} z + a_M}. \quad (7.32)$$

(The generalization of $e^{j\omega}$ to become any complex value z leads to the Z-Transform, which is covered in Ch. 9.)

---

recording, may be able to tolerate even an hour of processing delay without much of a problem.

[o] A causal system's impulse response is a causal signal.

Example 7.1

Let $H(e^{j\omega}) = 1/(1 - 0.5e^{-j\omega})$, which corresponds to the general case in Eq. (7.21) with $M = 1$, $b_0 = 1$, $b_1 = 0$, and $a_1 = -0.5$. Find $\left|H(e^{j\omega})\right|^2$:

$$\left|H(e^{j\omega})\right|^2 = \left|\frac{1}{1-0.5e^{-j\omega}}\right|^2 = \frac{|1|^2}{|1-0.5e^{-j\omega}|^2} = \frac{1}{|1-0.5\{\cos(\omega)-j\sin(\omega)\}|^2}$$

$$= \frac{1}{|(1-0.5\cos(\omega))+j(0.5\sin(\omega))|^2} = \frac{1}{(1-0.5\cos(\omega))^2+(0.5\sin(\omega))^2}. \qquad (7.33)$$

A plot of $\left|H(e^{j\omega})\right|^2$ vs. ω over frequency range $|\omega| \le 2\pi$ shows, as expected, that this magnitude-squared spectrum is periodic in ω with period $= 2\pi$ rad/sec:

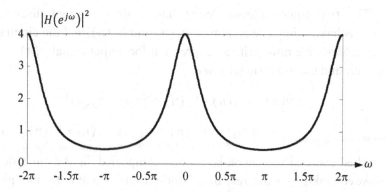

Figure 7.9. Magnitude-squared vs. frequency plot of $H(e^{j\omega}) = 1/(1 - 0.5e^{-j\omega})$.

Note that, from Entry #2 in Table 4.1 on page 103, this transfer function is the Fourier transform of the causal signal $h(n) = 0.5^n u(n)$. Therefore, a linear system having this impulse response will have a magnitude-squared frequency response as shown in Fig. 7.9.

7.1.4.3 *Basic feedback network*

Now let us see how a causal difference equation may be implemented in practice. Consider the network shown in Fig. 7.10:

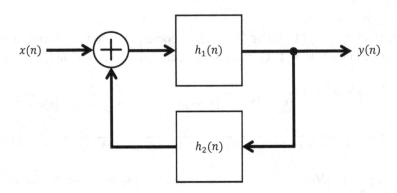

Figure 7.10. Time domain description of a basic discrete-time feedback network.

The two square blocks depict linear, shift-invariant discrete-time systems having impulse responses $h_1(n)$ and $h_2(n)$, and the circular unit is an adder. We now write an expression for output signal $y(n)$ in terms of itself and the input signal $x(n)$:

$$y(n) = h_1(n) * \left(x(n) + y(n) * h_2(n)\right)$$

$$= h_1(n) * x(n) + h_1(n) * y(n) * h_2(n). \qquad (7.34)$$

This expression cannot be further simplified in the time domain. However, taking the discrete-time Fourier transform of both sides gives us a frequency-domain expression that is easier to manipulate because in that domain convolutions are replaced by multiplications:

$$Y\left(e^{j\omega}\right) = H_1\left(e^{j\omega}\right)X\left(e^{j\omega}\right) + H_1\left(e^{j\omega}\right)Y\left(e^{j\omega}\right)H_2\left(e^{j\omega}\right). \qquad (7.35)$$

This simplifies to:

$$\frac{Y(e^{j\omega})}{X(e^{j\omega})} = \frac{H_1(e^{j\omega})}{1 - H_1(e^{j\omega})H_2(e^{j\omega})}. \qquad (7.36)$$

Shown in Fig. 7.11 is the same feedback system, only labeled in terms of its frequency-domain[p] parameters:

---

[p] Due to linearity of the transform, addition in time results in addition in frequency; that is why the adder block remains unchanged.

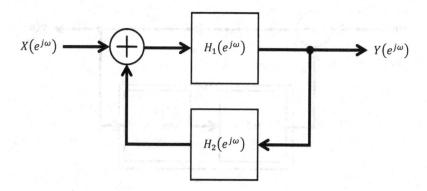

Figure 7.11. Frequency domain description of the basic feedback network in Fig. 7.10, having transfer function $Y(e^{j\omega})/X(e^{j\omega}) = H_1(e^{j\omega})/(1 - H_1(e^{j\omega})H_2(e^{j\omega}))$.

Example 7.2

Referring to Fig. 7.11, let $H_1(e^{j\omega}) = 1$ and $H_2(e^{j\omega}) = 0.5e^{-j\omega}$. Draw a network diagram to realize this system.

From Eq. (7.36) we obtain the transfer function

$$\frac{H_1(e^{j\omega})}{1 - H_1(e^{j\omega})H_2(e^{j\omega})} = \frac{1}{1 - 0.5e^{-j\omega}}, \tag{7.37}$$

which means that $Y(e^{j\omega})/X(e^{j\omega}) = 1/(1 - 0.5e^{-j\omega})$, or the same frequency response function as in Example 7.1. To physically realize this system, we must find linear systems whose transfer functions are $H_1(e^{j\omega}) = 1$ and $H_2(e^{j\omega}) = 0.5e^{-j\omega}$:

$$h_1(n) = \mathcal{F}^{-1}\{H_1(e^{j\omega})\} = \mathcal{F}^{-1}\{1\} = \delta(n); \tag{7.38}$$

$$h_2(n) = \mathcal{F}^{-1}\{H_2(e^{j\omega})\} = \mathcal{F}^{-1}\{0.5e^{-j\omega}\} = 0.5\,\delta(n - 1). \tag{7.39}$$

Since $x(n) * \delta(n) = x(n)$, we can replace the system having impulse response $h_1(n)$ with no operation.[q] The other system is implemented with a one-sample delay unit, combined with scaling by factor 0.5 as shown in Fig. 7.12:

---

[q] In electrical engineering jargon, we replace the system with "a straight wire."

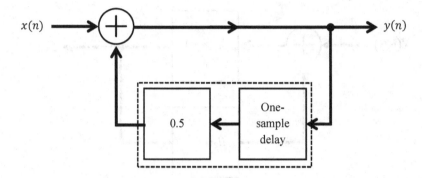

Figure 7.12.   Network used to implement the transfer function $H(e^{j\omega}) = 1/(1 - 0.5e^{-j\omega})$.

Whether it is implemented in hardware, or as a computer program, this transfer function is easily realizable. This was an example of a simple digital lowpass filter.

7.1.4.4 *Generalized feedback network*

The network described in Fig. 7.12 belongs to a class of networks where delay unit outputs are used for feedback, as shown in Fig. 7.13.

These networks contain only delay units, multipliers and adders, and realize difference equations of the form:

$$y(n) = x(n) - \sum_{k=1}^{M} a_k y(n - k) \qquad (7.40)$$

Taking the Fourier transform of both sides and then solving for the frequency response function $H(e^{j\omega}) = Y(e^{j\omega})/X(e^{j\omega})$ gives us:

$$H(e^{j\omega}) = \frac{1}{1 + \sum_{k=1}^{M} a_k e^{-j\omega k}} \qquad (7.41)$$

The impulse response of this feedback network continues for all time because, as shown in Fig. 7.13, the present output value $y(n)$ is fed back to produce future outputs even after the single nonzero sample of $x(n) = \delta(n)$ at $n = 0$ has long since passed. Therefore, this system is called an *Infinite (-duration) Impulse Response* (IIR) network. Because the output

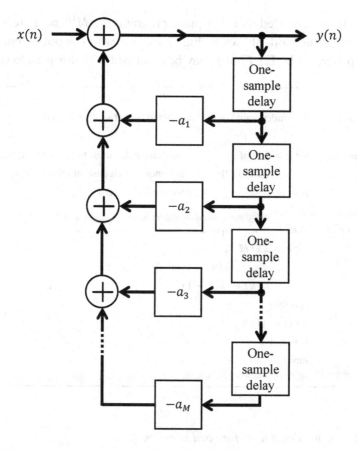

Figure 7.13.   General M^{th}-order discrete-time feedback network.

feeds on itself, it is quite possible that the impulse response sample values of this network will continue to grow without bound as $n \to \infty$. That would make the network *unstable* and not of much use for signal processing purposes. Chapter 11 will describe methods for determining whether a given rational function $H(e^{j\omega}) = H(z)|_{z=e^{j\omega}}$ describes a stable or unstable network.

When implemented as a computer program, the Mth order feedback discrete-time network shown in Fig. 7.13, and whose input/output relationship is stated by Eq. (7.30), may be described with this pseudo code:[r]

Initialize1: $n \leftarrow$ index value of first calculated output sample $y(n)$

Initialize2:   for $k = 1$ to M / Initialize delay units to zero output; this
                    $w(k) \leftarrow 0$ / assumes that all past samples of $y(n)$
              end for   / were equal to zero.

Loop: $y(n) \leftarrow x(n) - a_1 w(1)$
        for $k = 2$ to M
                $y(n) \leftarrow y(n) - a_k w(k)$
                $w(k) \leftarrow w(k-1)$
        end for
        $w(1) \leftarrow y(n)$
        $n \leftarrow n + 1$
        jump to Loop

7.1.4.5 Generalized feed-forward network

Figure 7.14 describes a class of networks where only feed-forward connections are present. These networks contain only delay units, multipliers and adders, and can simulate difference equations of the form:

$$y(n) = \sum_{k=0}^{M} b_k x(n-k). \tag{7.42}$$

Taking the Fourier transform of both sides of Eq. (7.42) and then solving for the frequency response function $H(e^{j\omega}) = Y(e^{j\omega})/X(e^{j\omega})$ gives us:

---

[r] "Pseudo code" is used to describe an algorithm without assuming the syntax of any particular programming language.

$$H(e^{j\omega}) = \sum_{k=0}^{M} b_k e^{-j\omega k}. \tag{7.43}$$

The impulse response of this feed-forward network has at most $M + 1$ nonzero samples in time because, as shown in Fig. 7.14, the single nonzero sample corresponding tò $n = 0$ in $x(n) = \delta(n)$ is flushed from the delay-line memory of the network after M sample delays have passed. Therefore, this system is called a *Finite (-duration) Impulse Response* (FIR) network.

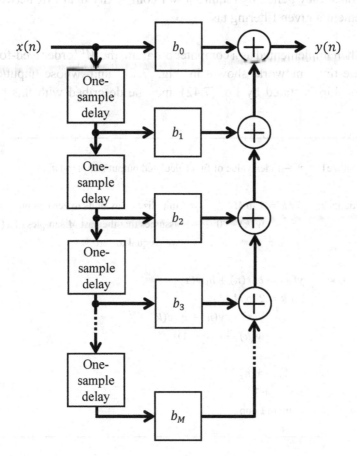

Figure 7.14. General M^{th} order discrete-time feed-forward network.

Because there are no feedback connections this network is guaranteed to be stable[s] and therefore desirable for signal processing purposes. Note, however, that the FIR network complexity is proportional to the number of nonzero samples in its impulse response $h(n)$. For this reason, the ideal lowpass (or highpass, bandpass, band-elimination) filter cannot be practically realized with a FIR network because its impulse response has infinite duration. IIR networks, despite their potential instability, are often used since they generally require lower complexity than FIR networks to implement a given filtering task.

When implemented as a computer program, the Mth order feed-forward discrete-time network shown in Fig. 7.14, and whose input/output relationship is stated by Eq. (7.42), may be described with this pseudo code:

Initialize1: $n \leftarrow$ index value of first calculated output sample $y(n)$

Initialize2:   for $k = 1$ to M / Initialize delay units to zero output; this
       $w(k) \leftarrow 0$ / assumes that the past M samples of $x(n)$
     end for   / were equal to zero.

  Loop: $y(n) \leftarrow b_0 x(n) + b_1 w(1)$
      for $k = 2$ to M
         $y(n) \leftarrow y(n) + b_k w(k)$
         $w(k) \leftarrow w(k - 1)$
      end for
      $w(1) \leftarrow x(n)$
      $n \leftarrow n + 1$
      jump to Loop

---

[s] When the network coefficients b_k $(k = 0,1,2, ..., M)$ are finite, and input sequence $x(n)$ is finite-valued, the output sequence $y(n)$ is guaranteed to be finite-valued.

7.1.4.6 *Combined feedback and feed-forward network*

A cascade of the two networks presented in Figs 7.13 and 7.14 is shown in Fig. 7.15. The new network will have a transfer function that is the product of the two:

$$H(e^{j\omega}) = \frac{\sum_{k=0}^{M} b_k e^{-j\omega k}}{1 + \sum_{k=1}^{M} a_k e^{-j\omega k}}. \tag{7.44}$$

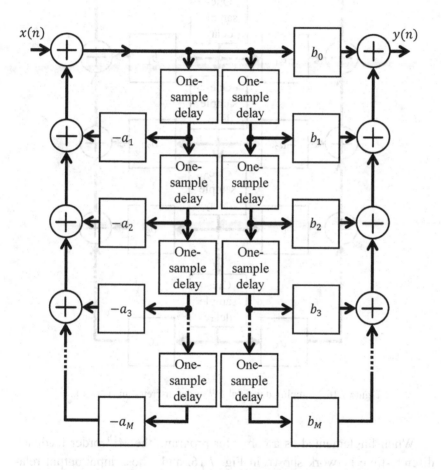

Figure 7.15. General M^{th} order discrete-time network as a cascade of feedback and feed-forward networks.

By combining the delay lines in Fig. 7.15 one may achieve significant savings in complexity, which is important when implemented in dedicated hardware. The simplified network is shown in Fig. 7.16.

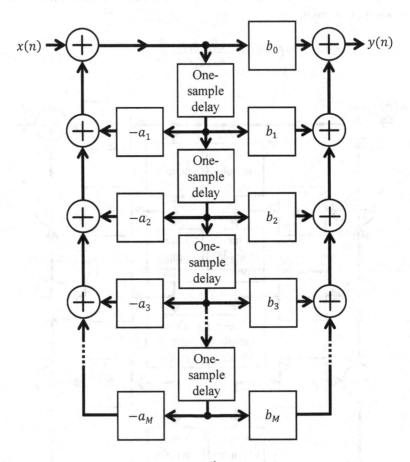

Figure 7.16. Simplified general M^{th} order discrete-time network.

When implemented as a computer program, the Mth order feedback discrete-time network shown in Fig. 7.16, and whose input/output relationship is stated by Eq. (7.44), may be described with this pseudo code:

Define: $c_k \equiv b_k/b_0$ $(k = 1, 2, ..., M)$

Initialize1: $n \leftarrow$ index value of first calculated output sample $y(n)$

Initialize2: for $k = 1$ to M / Initialize delay units to zero output; this
$\qquad\qquad\qquad w(k) \leftarrow 0$ / assumes all past samples of $y(n)$ and the
$\qquad$ end for / past M samples of $x(n)$ were equal to zero.

Loop: $T \leftarrow x(n) - a_1 w(1)$
$\qquad y(n) \leftarrow c_1 w(1)$
$\qquad$ for $k = 2$ to M
$\qquad\qquad T \leftarrow T - a_k w(k)$
$\qquad\qquad y(n) \leftarrow y(n) + c_k w(k)$
$\qquad\qquad w(k) \leftarrow w(k-1)$
$\qquad$ end for
$\qquad w(1) \leftarrow T$
$\qquad y(n) \leftarrow b_0(y(n) + T)$
$\qquad n \leftarrow n + 1$
$\qquad$ jump to Loop

7.2 Practical Applications

7.2.1 First-order digital filters

Just as many useful analog filter circuits may be designed using only one capacitor (or inductor) and one resistor, which have first-order differential equations describing their time-domain operation, first-order discrete-time networks are useful for the same reason: they can implement simple low-pass and highpass frequency responses. Substituting $M = 1$ into Eq. (7.44), we see that the frequency-domain transfer function of the first-order discrete-time network is:

$$H(e^{j\omega}) = Y(e^{j\omega})/X(e^{j\omega}) = \frac{b_0 + b_1 e^{-j\omega}}{1 + a_1 e^{-j\omega}}. \qquad (7.45)$$

The general discrete-time network for $M = 1$ (composed of one sample delay unit, 3 multipliers and 2 adders) is shown in Fig. 7.17:

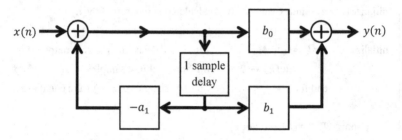

Figure 7.17. General 1st order discrete-time network.

From Eq. (7.45) the difference equation is found using some algebraic manipulation and the inverse Fourier transform:

$$Y(e^{j\omega})(1 + a_1 e^{-j\omega}) = X(e^{j\omega})(b_0 + b_1 e^{-j\omega})$$

$$Y(e^{j\omega}) + a_1 e^{-j\omega} Y(e^{j\omega}) = b_0 X(e^{j\omega}) + b_1 e^{-j\omega} X(e^{j\omega})$$

$$Y(e^{j\omega}) = b_0 X(e^{j\omega}) + b_1 e^{-j\omega} X(e^{j\omega}) - a_1 e^{-j\omega} Y(e^{j\omega})$$

$$y(n) = b_0 x(n) + b_1 x(n-1) - a_1 y(n-1). \tag{7.46}$$

When $x(n) = \delta(n)$ then $y(n)$ is the impulse response $h(n)$: [t]

$$h(n) = b_0 \delta(n) + b_1 \delta(n-1) - a_1 h(n-1). \tag{7.47}$$

Because $\delta(n) = 1$ only at $n = 0$, and $\delta(n-1) = 1$ only at $n = 1$, we can write:

$$h(n) = \begin{cases} 0, & n < 0; \\ b_0, & n = 0; \\ b_1 - a_1 b_0, & n = 1; \\ -a_1 h(n-1), & n > 1. \end{cases} \tag{7.48}$$

Unfortunately, this expression for impulse response $h(n)$ is recursively stated for all sample indexes $n > 1$. Let's take a different approach to find the closed-form solution for impulse response $h(n)$:

---

[t] When initial conditions are zero, which means that $y(n) = 0$, $\forall n < 0$.

$$H(e^{j\omega}) = \frac{b_0 + b_1 e^{-j\omega}}{1 + a_1 e^{-j\omega}} = b_0 \left\{ \frac{1}{1 + a_1 e^{-j\omega}} \right\} + b_1 e^{-j\omega} \left\{ \frac{1}{1 + a_1 e^{-j\omega}} \right\}. \quad (7.49)$$

Then, using the Fourier transform pair Entry #2 in Table 4.1 (under the constraint that $|a_1| < 1$), and the time shift Property #2 in Table 4.2, we find $h(n)$ to be:

$$h(n) = b_0 (-a_1)^n u(n) + b_1 (-a_1)^{n-1} u(n - 1). \quad (7.50)$$

This is the impulse response of a first-order digital filter, which can now be expressed as:

$$h(n) = \begin{cases} 0, & n < 0; \\ b_0, & n = 0; \\ b_0 (-a_1)^n + b_1 (-a_1)^{n-1}, & n \geq 1. \end{cases} \quad (7.51)$$

7.2.1.1 Lowpass filter

The following transfer function realizes a practical digital lowpass filter:

$$H_{LPF}(e^{j\omega}) = Y(e^{j\omega})/X(e^{j\omega}) = \frac{C_1 + C_1 e^{-j\omega}}{1 - \alpha e^{-j\omega}}. \quad (7.52)$$

Note this matches the form of a first-order digital filter in Eq. (7.45) when $M = 1$, $b_0 = C_1$, $b_1 = C_1$ and $a_1 = -\alpha$. Constants C_1 and α are found from desired half-power frequency ω_0 as follows:

$$C_1 = \frac{\sin(\omega_0/2)}{\cos(\omega_0/2) + \sin(\omega_0/2)},$$

$$\alpha = \frac{\cos(\omega_0/2) - \sin(\omega_0/2)}{\cos(\omega_0/2) + \sin(\omega_0/2)}; \qquad 0 < \omega_0 < \pi. \quad (7.53)$$

In terms of ω_0 the filter's magnitude-squared transfer function is

$$|H_{LPF}(\omega)|^2 = \frac{\sin^2(\omega_0/2)(1 + \cos(\omega))}{1 - \cos(\omega_0)\cos(\omega)}. \quad (7.54)$$

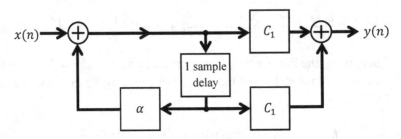

Figure 7.18. First-order lowpass digital filter network, having transfer function $H_{LPF}(e^{j\omega}) = Y(e^{j\omega})/X(e^{j\omega}) = (C_1 + C_1 e^{-j\omega})/(1 - \alpha e^{-j\omega})$.

We can simplify this network by eliminating one multiply operation as shown here:

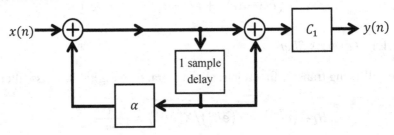

Figure 7.19. First-order lowpass digital filter network, simplified.

Referring to Eq. (7.50), we see that the impulse response of this first-order digital lowpass filter is:

$$h_{LPF}(n) = C_1 \alpha^n u(n) + C_1 \alpha^{n-1} u(n - 1). \qquad (7.55)$$

<u>Example 7.3</u>
Design a first-order digital lowpass filter having half-power frequency $\omega_0 = \pi/11$ rad/sec.

From Eq. (7.53), $C_1 \cong 1/8$ and $\alpha \cong 3/4$. Using these parameters in Eq. (7.52) gives the desired transfer function:

$$H_{LPF}(e^{j\omega}) = \left(\frac{1}{8} + \frac{1}{8}e^{-j\omega}\right)/\left(1 - \frac{3}{4}e^{-j\omega}\right). \qquad (7.56)$$

When evaluating Eq. (7.56) at frequencies $\omega = 0$ and $\omega = \omega_0$, we confirm that $\left|H_{LPF}(e^{j\omega_0})\right|^2 / \left|H_{LPF}(e^{j0})\right|^2 \cong 0.5/1 = 1/2$, which confirms that ω_0 is the half-power (-3 dB) frequency as desired.

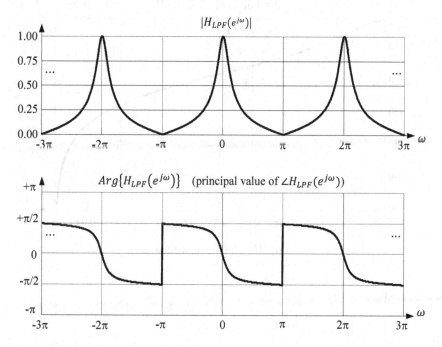

Figure 7.20. Magnitude and phase of $H(e^{j\omega})$ in Example 7.3 (first-order digital lowpass filter).

Recall that every $H(e^{j\omega})$ is periodic, having period $= 2\pi$ radians/sec. Also, in this case[u], $\left|H_{LPF}(e^{j\omega})\right|$ is even and $\angle H_{LPF}(e^{j\omega})$ is odd (with respect to ω) as shown in Fig. 7.20.

Since magnitude function $\left|H_{LPF}(e^{j\omega})\right|$ is even and periodic, and since phase function $\angle H_{LPF}(e^{j\omega})$ is odd and periodic, it is sufficient to plot these waveforms over half of one period: $0 \leq \omega \leq \pi$ rad/sec. The

---

[u] In Chapters 9 and 11 we shall see that $h(n)$ is real whenever rational $H(z)$ has poles and zeroes that are real and/or come in complex-conjugate pairs, which is the case here. Also, recall that when $h(n)$ is real then $\left|H(e^{j\omega})\right|$ is even and $\angle H(e^{j\omega})$ is odd.

graphs in Fig. 7.21 show $|H_{LPF}(e^{j\omega})|^2$ in dB, and $\angle H_{LPF}(e^{j\omega})$ in radians, over that frequency range.

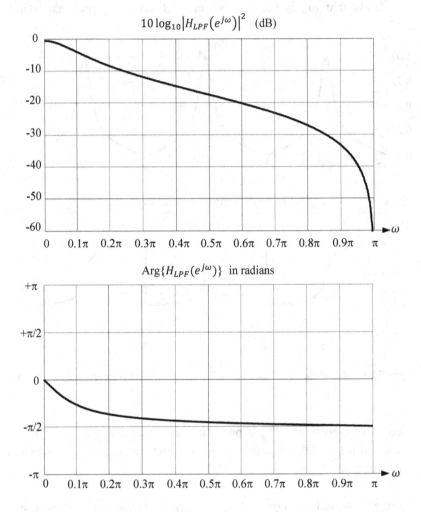

Figure 7.21.   Magnitude-squared and phase of $H_{LPF}(e^{j\omega})$ in Example 7.3 (first-order digital lowpass filter).

Finally, we know from Eq. (7.50) that the impulse response of the lowpass filter in this example is

$$h_{LPF}(n) = \frac{1}{8}\left(\frac{3}{4}\right)^n u(n) + \frac{1}{8}\left(\frac{3}{4}\right)^{n-1} u(n-1). \qquad (7.57)$$

7.2.1.2 Highpass filter

The following transfer function realizes a practical digital highpass filter:

$$H_{HPF}\left(e^{j\omega}\right) = Y\left(e^{j\omega}\right)/X\left(e^{j\omega}\right) = \frac{C_2 - C_2 e^{-j\omega}}{1 - \alpha e^{-j\omega}}. \qquad (7.58)$$

Note this matches the form of a first-order digital filter in Eq. (7.45) when $M = 1$, $b_0 = C_2$, $b_1 = -C_2$ and $a_1 = -\alpha$. Constants C_2 and α are found from desired half-power frequency ω_0 as follows:

$$C_2 = \frac{\cos(\omega_0/2)}{\cos(\omega_0/2)+\sin(\omega_0/2)}, \quad \alpha = \frac{\cos(\omega_0/2)-\sin(\omega_0/2)}{\cos(\omega_0/2)+\sin(\omega_0/2)}; \quad 0 < \omega_0 < \pi \qquad (7.59)$$

In terms of ω_0 the filter's magnitude-squared transfer function is

$$|H_{HPF}(\omega)|^2 = \frac{\cos^2(\omega_0/2)(1-\cos(\omega))}{1-\cos(\omega_0)\cos(\omega)}. \qquad (7.60)$$

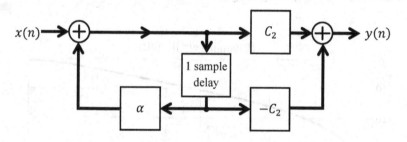

Figure 7.22. First-order highpass digital filter network, having transfer function $H_{HPF}\left(e^{j\omega}\right) = Y\left(e^{j\omega}\right)/X\left(e^{j\omega}\right) = (C_2 - C_2 e^{-j\omega})/(1 - \alpha e^{-j\omega})$.

We can simplify this network by replacing one multiply operation with a negation, as shown in Fig. 7.23. Referring to Eq. (7.50), we see that the impulse response of this first-order digital highpass filter is:

$$h_{LPF}(n) = C_2 \alpha^n u(n) - C_2 \alpha^{n-1} u(n-1). \qquad (7.61)$$

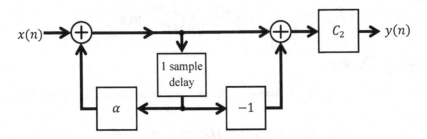

Figure 7.23. First-order highpass digital filter network, simplified.

Example 7.4

We wish to design a first-order digital highpass filter having half-power frequency $\omega_0 = 2.5$ rad/sec. From Eq. (7.59), $C_2 \cong 1/4$ and $\alpha \cong 1/2$. Using these parameters in Eq. (7.58) gives the desired transfer function:

$$H_{HPF}(e^{j\omega}) = \frac{\frac{1}{4} - \frac{1}{4}e^{-j\omega}}{1 + \frac{1}{2}e^{-j\omega}}. \tag{7.62}$$

The graphs in Fig. 7.24 show $|H_{HPF}(e^{j\omega})|^2$ in dB, and $\angle H_{HPF}(e^{j\omega})$ in radians, over frequency range $0 \leq \omega \leq \pi$ rad/sec.

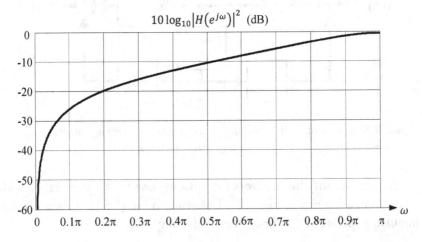

Figure 7.24 (a). Magnitude-squared value of $H(e^{j\omega})$ in Example 7.4 (first-order digital highpass filter).

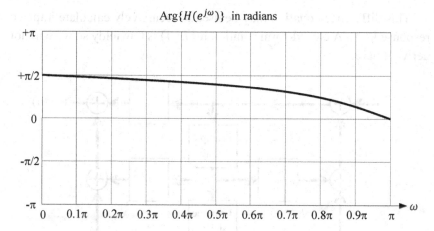

Arg$\{H(e^{j\omega})\}$ in radians

Figure 7.24 (b). Phase of $H(e^{j\omega})$ in Example 7.4 (first-order digital highpass filter).

7.2.2 Second-order digital filters

Just as many useful analog filter circuits may be designed using only one capacitor, one inductor and one resistor, which have 2nd order differential equations describing their time-domain operation, 2nd order discrete-time networks are useful for the same reason: the 2nd order difference equation can implement simple bandpass, notch, and allpass frequency responses. Substituting $M = 2$ into Eq. (7.44), we see that the frequency-domain transfer function of the 2nd order discrete-time network is:

$$H(e^{j\omega}) = \frac{b_0 + b_1 e^{-j\omega} + b_2 e^{-j2\omega}}{1 + a_1 e^{-j\omega} + a_2 e^{-j2\omega}}. \qquad (7.63)$$

The general discrete-time network for $M = 2$ (composed of two sample delay units, 5 multipliers and 4 adders) is shown in Fig. 7.25. The difference equation being implemented by this network is:

$$y(n) = b_0 x(n) + b_1 x(n-1) + b_2 x(n-2)$$

$$-a_1 y(n-1) - a_2 y(n-2). \qquad (7.64)$$

This difference equation may be used to recursively calculate impulse response $h(n)$. A closed-form solution for $h(n)$ is unwieldy so we will not derive it here.

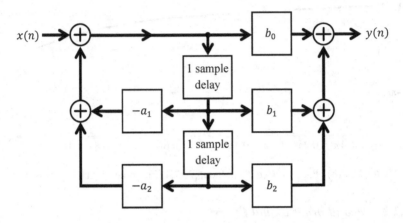

Figure 7.25. General 2nd order discrete-time network.

7.2.2.1 *Bandpass filter*

The following transfer function realizes a practical digital bandpass filter:

$$H_{BPF}(e^{j\omega}) = Y(e^{j\omega})/X(e^{j\omega}) = \frac{S.F.\times(1-e^{-j2\omega})}{1+2\alpha\cos(\omega_0)e^{-j\omega}+\alpha^2e^{-j2\omega}}. \qquad (7.65)$$

Note this matches the form of a second-order digital filter in Eq. (7.63) when $b_0 = 1$, $b_1 = 0$, $b_2 = -1$, $a_1 = -2\alpha\cos(\omega_0)$ and $a_2 = \alpha^2$. The peak frequency will be approximately ω_0 ($0 < \omega_0 < \pi$), and bandwidth is determined by α ($0 < \alpha < 1$): higher α results in narrower bandwidth. Once ω_0 and α are chosen, Eq. (7.65) may be evaluated to determine the value of scale factor $S.F.$ that gives unity gain at the peak response frequency.[v] Greater insight into this 2nd order digital bandpass filter is given in Ch. 11 with the aid of a pole-zero diagram in the z-plane.

---

[v] The exact formulas relating $\{\alpha, \omega_0\}$ to {peak gain, center frequency, bandwidth} for this digital bandpass filter are not shown here.

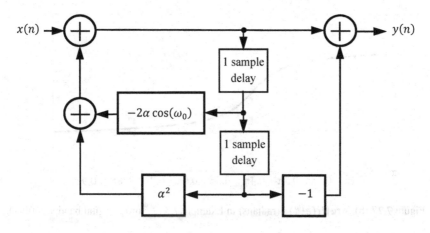

Figure 7.26. Discrete-time network to implement a simple bandpass filter having peak gain at frequency $\cong \omega_0$ $(0 < \omega_0 < \pi)$.

Example 7.5

Consider the case that $\omega_0 = 0.4\pi$, $\alpha = 0.8$ and $S.F. = 0.18$ (which was chosen to normalize the peak magnitude gain to 1):

$$H_{BPF}(e^{j\omega}) = \frac{0.18(1-e^{-j2\omega})}{1-0.4944e^{-j\omega}+0.64e^{-j2\omega}}. \tag{7.66}$$

The graphs in Fig. 7.27 show $|H_{BPF}(e^{j\omega})|^2$ in dB and $\angle H_{BPF}(e^{j\omega})$ in radians, over frequency range $0 \le \omega \le \pi$ rad/sec:

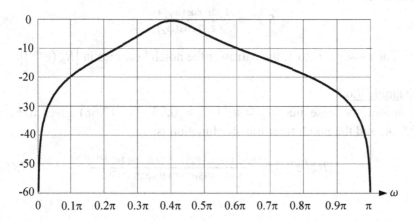

Figure 7.27 (a). $10\log_{10}|H(e^{j\omega})|^2$ (dB) in Example 7.5 (2nd order digital bandpass filter).

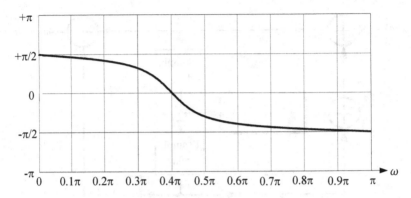

Figure 7.27 (b). Arg$\{H(e^{j\omega})\}$ (radians) in Example 7.5 (2nd order digital bandpass filter).

7.2.2.2 Notch filter

The following 2nd order transfer function realizes a digital notch filter:

$$H_N(e^{j\omega}) = S.F. \times \frac{1-2\cos(\omega_z)e^{-j\omega}+e^{-j2\omega}}{1-2\alpha\cos(\omega_z)e^{-j\omega}+\alpha^2e^{-j2\omega}}. \qquad (7.67)$$

The frequency response of a notch filter has a zero at the notch frequency ω_z $(0 < \omega_z < \pi)$. Parameter α determines the width of the notch (may be anywhere in the range $0 \leq \alpha < 1$, but usually chosen close to 1), and the scale factor may be selected to give unity gain at zero frequency:

$$S.F. = \frac{1-2\alpha\cos(\omega_z)+\alpha^2}{2-2\cos(\omega_z)}. \qquad (7.68)$$

The closer α is to 1, the narrower the notch becomes in $\left|H_N(e^{j\omega})\right|$.

Example 7.6

Consider the case that $\omega_z = \pi/4$, $\alpha = 0.95$. Eq. (7.68) gives $S.F. = 0.954$, and the notch filter transfer function is:

$$H_N(e^{j\omega}) = \frac{0.9543-1.3495e^{-j\omega}+0.9543e^{-j2\omega}}{1-1.3435e^{-j\omega}+0.9025e^{-j2\omega}}. \qquad (7.69)$$

Figure 7.28 shows $|H_N(e^{j\omega})|^2$, and Fig. 7.29 shows $\angle H_N(e^{j\omega})$ over frequencies $0 \leq \omega \leq \pi$ rad/sec.

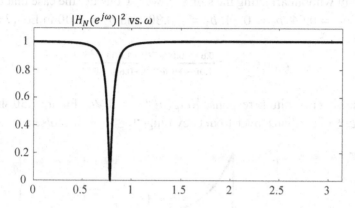

Figure 7.28. $|H_N(e^{j\omega})|^2$ of the notch filter in Example 7.6.

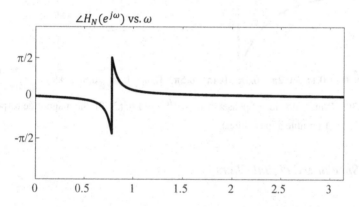

Figure 7.29. $\angle H_N(e^{j\omega})$ of the notch filter in Example 7.6.

7.2.2.3 *Allpass filter*

In the discrete time domain, it is possible to easily realize a filtering characteristic that is practically impossible[w] to achieve using analog circuitry in the continuous-time domain: the *allpass* response. The digital allpass filter gets its name from the fact that $\left|H_{AP}(e^{j\omega})\right| = 1$ at all

---

[w] Other than the trivial $H_{AP}(e^{j\omega}) = 1$.

frequencies. The phase response of this filter is not zero (or even linear), however. Passing a signal through this type of filter will change its waveform without affecting the signal power. Consider the case that $a_1 = -0.80$, $a_2 = 0.64$, $b_0 = 0.64$, $b_1 = -0.80$, and $b_2 = 1.00$ in Eq. (7.63):

$$H_{AP}(e^{j\omega}) = \frac{0.64 - 0.80e^{-j\omega} + 1.00e^{-j2\omega}}{1.00 - 0.80e^{-j\omega} + 0.64e^{-j2\omega}}. \qquad (7.70)^x$$

The resulting magnitude response $|H_{AP}(e^{j\omega})| = 1 \ \forall \omega$. Figure 7.30 shows $\angle H_{AP}(e^{j\omega})$ in radians, over frequency range $0 \leq \omega \leq \pi$ rad/sec.

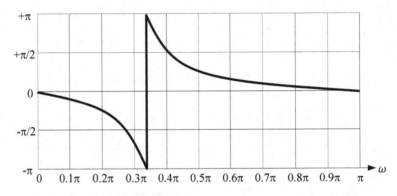

Figure 7.30. Principal value of phase[y] of $H_{AP}(e^{j\omega})$ in Eq. (7.70); the magnitude response $|H_{AP}(e^{j\omega})| = 1$ (simple allpass filter).

7.2.3 *Specialized digital filters*

7.2.3.1 *Comb filter*

A simple digital comb filter may easily be identified by observing its impulse response:

$$h_{CF}(n) = 1 \pm \delta(n - k). \qquad (7.71)$$

---

[x] This type of response has the characteristic that numerator and denominator polynomial coefficients are reversed copies of one another.

[y] The discontinuity in this curve is an artifact of displaying principal value of phase.

When some sequence $x(n)$ passes through the comb filter the output signal will be $y(n) = x(n) * h_{CF}(n) = x(n) \pm x(n - k)$. That is, the output is the input signal $\pm$ a delayed version of itself. Part of what makes the comb filter so popular is its ease of implementation: required are only one k-sample delay, and one adder/subtractor. Observe the magnitude frequency response of a comb filter having impulse response $h_{CF}(n) = 1 - \delta(n - 5)$:

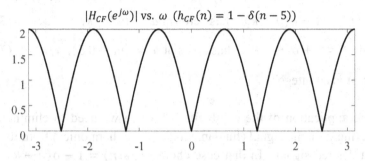

Figure 7.31. Magnitude response of comb filter having impulse response $h_{CF}(n) = 1 - \delta(n - 5)$. The frequency range is $-\pi \le \omega \le \pi$.

First, you will notice how the comb filter gets its name: Fig. 7.31 clearly shows the comb-like nature of its magnitude frequency response. Also, note that the filter has five peaks over one period of its frequency response, and that the gain at $\omega = m2\pi/5, m \in \mathbb{Z}$, is zero. Peaks of magnitude 2 occur at frequencies $\omega = (m + 0.5)2\pi/5, m \in \mathbb{Z}$. Next observe the magnitude frequency response of a comb filter having impulse response $h_{CF}(n) = 1 + \delta(n - 4)$:

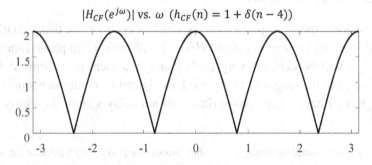

Figure 7.32. Magnitude response of comb filter having impulse response $h_{CF}(n) = 1 + \delta(n - 4)$. The frequency range is $-\pi \le \omega \le \pi$.

We again see a comb-like structure in magnitude frequency response, having four peaks over one period of its frequency response, and this time the gain at $\omega = m2\pi/4, m \in \mathbb{Z}$, is 2. Zeros occur at frequencies $\omega = (m + 0.5)2\pi/4, m \in \mathbb{Z}$.

One may generalize these results to any number of peaks in one period of frequency magnitude response function $\left|H_{CF}\left(e^{j\omega}\right)\right|$:

$$h_{CF}(n) = 1 - \delta(n - k): \quad \text{has zeros at } \omega = m2\pi/k, \tag{7.72}$$

$$h_{CF}(n) = 1 + \delta(n - k): \quad \text{has zeros at } \omega = (m + 0.5)2\pi/k, \tag{7.73}$$

where m is an integer.

One application of the comb filter is when we need to eliminate an interfering periodic signal (fundamental ω_0 + its harmonics) from its sum with a desired signal. In that case choose $h_{CF}(n) = 1 - \delta(n - k)$, and select value k so that $\omega_0 = 2\pi/k$.[z]

Another application of the comb filter is to eliminate only the even or only the odd harmonics of a periodic signal. To eliminate only the even harmonics of ω_0 choose $h_{CF}(n) = 1 + \delta(n - k)$, and select integer k so that $\omega_0 = \pi/k$. To eliminate only the odd harmonics (including fundamental frequency ω_0) choose $h_{CF}(n) = 1 + \delta(n + k)$, and select integer k so that $\omega_0 = \pi/k$.

7.2.3.2 *Linear-phase filter*

Assume that impulse response $h(n)$ is real. If it is also even, that is $h(n) = h(-n)$, then its Fourier transform $H(e^{j\omega})$ will be real and phase function $\angle H(e^{j\omega})$ will be zero at every ω.[aa] Phase is a measure of time delay, normalized to the length of one period, of sinusoidal components passing through the filter. A zero-phase filter will not delay a sinusoid – only the

---

[z] For an arbitrary sampling frequency ω_s, this zero will lie at $\omega = \omega_s/k$ (replace 2π with ω_s). This gives us the freedom to adjust ω_s slightly so that ω_0 exactly equals ω_s/k.

[aa] See p. 4-12.

amplitude may change. *Zero-phase* digital filters are desirable when waveform shape must be preserved as much as possible. Only trivial cases[bb] of zero-phase filters exist in the real world due to the requirement that they be causal ($h(n) = 0$ for $n < 0$). However, if $h(n)$ is even and finite-length then it may be delayed to make it causal while affecting phase in a predictable way. Consider the even sequence $h(n)$ having length $2N + 1$ samples:

$$h(n) = h(-n); \quad h(n) = \text{rect}_N(n)h(n). \tag{7.74}$$

Delay $h(n)$ by N samples to form a new sequence:

$$h_1(n) = h(n - N). \tag{7.75}$$

We know that $h(n)$ is the impulse response of a zero-phase filter, but what is the phase characteristic of a filter with impulse response $h_1(n)$? To answer this question, note that $h_1(n) = h(n) * \delta(n - N)$. Then, the Fourier transform gives us:

$$H_1\left(e^{j\omega}\right) = H\left(e^{j\omega}\right) e^{-jN\omega}. \tag{7.76}$$

From this we see that

$$\left|H_1\left(e^{j\omega}\right)\right| = \left|H\left(e^{j\omega}\right)\right| \left|e^{-jN\omega}\right| = \left|H\left(e^{j\omega}\right)\right|, \text{ and}$$

$$\angle H_1\left(e^{j\omega}\right) = \angle H\left(e^{j\omega}\right) + \angle e^{-jN\omega} = 0 - \omega N = -\omega N. \tag{7.77}$$

Because of the delay of $h(n)$ by N samples to give $h_1(n)$, the phase has gone from zero to a linear function of ω: a line intersecting $\omega = 0$ having slope $-N$. This phase function is called *linear phase* and it delays all frequencies equally (in this example by N samples).[cc] Therefore, as with a zero-phase characteristic, there is no frequency-dependent delay and any waveform distortion (other than an overall delay) is due solely to non-uniform magnitude response.

[bb] The only causal zero-phase filter is gain stage having impulse response $h(n) = a\delta(n)$.
[cc] *Group delay* is defined as $-\, d\angle H\left(e^{j\omega}\right)/d\omega$, which here is N samples independent of ω.

To summarize: one can obtain the impulse response for a realizable *linear-phase filter* by starting with $h(n)$ – the finite-length impulse response of a zero-phase filter – and delaying it until it is causal. More commonly: we modify an even and infinite-length $h(n)$ to make it even and finite-length, and then delay it so the result is causal. That is the topic of the following section.

Here is an interesting final note: the convolution product of $h(n)$ and $h(-n)$ will always be an even function of time. This has a zero-phase characteristic when $h(n)$ is real. Proof:

$$h(n) \Leftrightarrow H\left(e^{j\omega}\right)$$

$$h(-n) = h^*(-n) \Leftrightarrow H^*\left(e^{j\omega}\right)$$

$$h(n) * h(-n) \Leftrightarrow H\left(e^{j\omega}\right)H^*\left(e^{j\omega}\right) = \left|H\left(e^{j\omega}\right)\right|^2, \qquad (7.78)$$

which is purely real and thus has zero phase.

Let $g(n) = h(n) * h(-n)$. Then $y(n)$, the result of filtering sequence $x(n)$ by passing it through a digital filter having impulse response $g(n)$, may be obtained as follows:

a) Pass $x(n)$ through filter having impulse response $h(n)$
b) Time-reverse the result from (a)
c) Pass result from (b) through filter having impulse response $h(n)$
d) Time-reverse the result from (c) to yield $y(n)$

Time-reversing a finite-length signal is a trivial task in the discrete-time domain: it is simply reversing the order of values in a list. Therefore, this method may be used to obtain zero-phase[dd] filtering with magnitude response being squared.

[dd] We need to wait until sequences are produced in steps (a) and (d) before reversing them in time, so a delay is introduced. Thus, our "zero-phase" filtering is actually linear-phase filtering.

7.2.4 *Interpolation and Decimation*

In this section, we present digital signal processing operations that increase or decrease the sampling frequency of a discrete-time signal. We begin with a technique to raise effective sampling rate known as *interpolation*.

7.2.4.1 *Interpolation by factor a*

A Fourier Transform property previously discussed in Chapter 4 is *up-sampling a sequence by factor a*. From Eq. (4.46): when $(a-1)$ zeros are inserted between each pair of consecutive samples of $f(n)$ to give a new sequence $g(m)$,[ee] then $G(e^{j\omega}) = F(e^{ja\omega})$. This tells us that the spectrum $F(e^{j\omega})$ is compressed towards $\omega = 0$ by scale factor a:

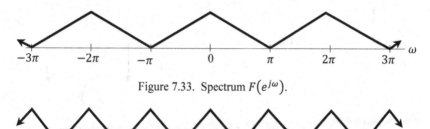

Figure 7.33. Spectrum $F(e^{j\omega})$.

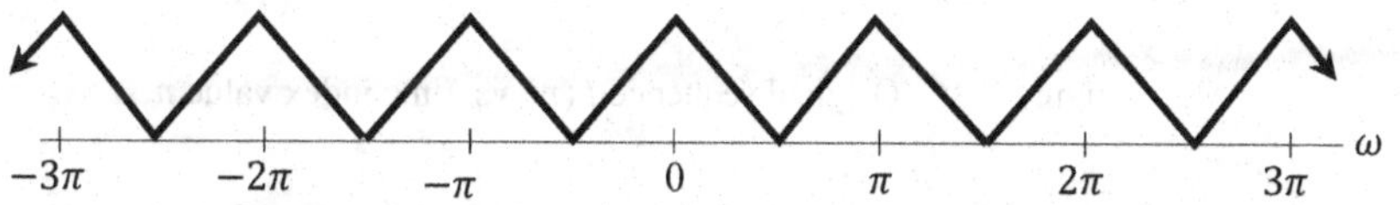

Figure 7.34. Spectrum $G(e^{j\omega}) = F(e^{j2\omega})$ $(a = 2)$.

Up-sampling is the first step in a process called *interpolation*. The second step is to digitally filter $g(m)$, using a lowpass characteristic having bandwidth $\omega_s/2a = 2\pi/2a = \pi/a$ rad/sec, to pass only the baseband spectrum as shown in Fig. 7.35.

Thus, from a frequency domain perspective, interpolation increases the sampling rate of a signal by factor a.[ff] The name *interpolation* is more

---

[ee] n will be used to represent time index values before up-sampling, and m will be used to represent time index values after up-sampling.

[ff] It is as if analog signal $f(t)$ is sampled at sampling frequency f_s to give sequence $f(n)$, and sampled at sampling frequency af_s to give sequence $g(m)$.

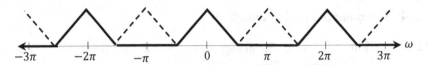

Figure 7.35. Spectrum of $Y(e^{j\omega}) = \text{LPF}\{G(e^{j\omega})\}$ (dotted lines indicate the spectral copies eliminated by the lowpass filter when $a = 2$).

meaningful in the time domain, however, where we see that output sequence $y(m)$ is formed by computing $(a - 1)$ new data samples and inserting them between every pair of sequential samples in $f(n)$; the amplitudes of these new data samples are interpolated from the original information in $f(n)$.

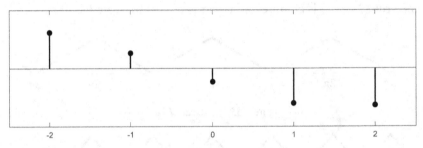

Figure 7.36. Original sequence $f(n)$ vs. time index value n.

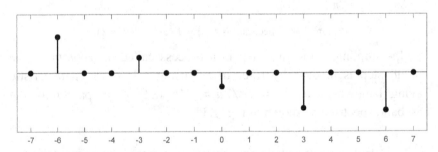

Figure 7.37. Sequence $g(m)$ ($f(n)$ after up-sampling by factor $a = 3$) vs. time index value m.

The lowpass filter that is used in the second step of the process is known as the *interpolation filter*. Since every $(a - 1)$ out of a samples at

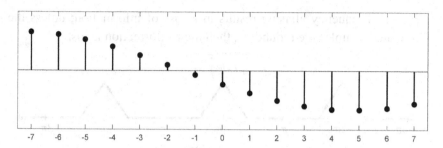

Figure 7.38. Sequence $y(m)$ ($f(n)$ after up-sampling by factor $a = 3$ and lowpass filtering with bandwidth $\omega_s/2a = \pi/3$) vs. time index value m.

the input to the interpolation filter are equal to zero, this cuts the number of computations required for FIR lowpass filtering by factor a.

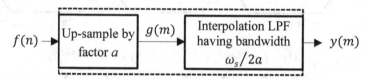

Figure 7.39. Interpolator that operates on $f(n)$ to produce $y(m)$.

7.2.4.2 Decimation by factor b

Chapters 2 and 4 describe an operation called "down-sampling a sequence by an integer factor," which is used as part of the sampling-rate reduction method called *decimation*. From Eq. (4.45): when 1 out of b samples in $f(n)$ are retained to form a new sequence $g(n)$, then $G(e^{j\omega})$ is related to $F(e^{j\omega})$ as follows:

$$g(n) = f(bn) \Leftrightarrow \frac{1}{b}X\left(e^{j\omega/b}\right), \quad b \in \mathbb{Z}^+,$$

$$\text{where } X\left(e^{j\omega}\right) = F\left(e^{j\omega}\right) * \sum_{k=0}^{b-1}\delta\left(\omega - \frac{k}{b}2\pi\right). \quad (7.79)$$

To put it simply, down-sampling $f(n)$ by integer factor b expands the spectrum away from $\omega = 0$ by factor b (with the possibility that aliasing occurs). Frequency aliasing results in a loss of information; unless the decimated samples are redundant, then their information is lost.

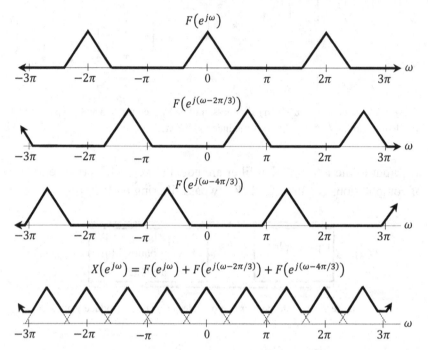

Figure 7.40. Graphical depiction of Eq. (7.79) when up-sampling factor $b = 3$.

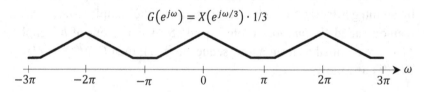

Figure 7.41. Aliased result after decimation described in Fig. 7.39 ($b = 3$).

As with the analog-to-digital conversion described in Ch. 6, anti-alias pre-filtering can be used to eliminate spectral component before aliasing can occur in signal decimation. Thus, a decimator typically includes a lowpass filter (having bandwidth $\omega_s/2b = 2\pi/2b = \pi/b$) to pre-filter the data before sending it to a down-sampling unit:

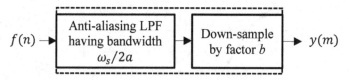

Figure 7.42. Decimator that operates on $f(n)$ to produce $y(m)$.

Note that one sample of $y(m)$ is produced for every b samples of $f(n)$; this means that if an FIR anti-aliasing filter is used, its computational load is reduced by factor b from what would normally be the case.

Figure 7.43 shows the decimator output spectrum for the same case as in Fig. 7.41, except without aliasing due to pre-filtering:

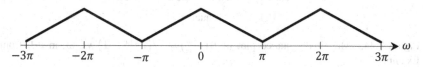

Figure 7.43. Resulting spectrum after the decimation described in Fig. 7.41, this time with proper anti-aliasing lowpass filtering prior to down-sampling by factor ($b = 3$).

7.2.5 *Nyquist frequency response plot*

An interesting alternative to the magnitude-phase description of frequency response function $H(e^{j\omega})$ is the Nyquist plot, where the imaginary part of $H(e^{j\omega})$ is plotted versus the real part of $H(e^{j\omega})$. Although this type of plot obscures the dependence of $H(e^{j\omega})$ on ω, with proper labeling it still contains all the information of $H(e^{j\omega})$. The Nyquist plot is useful for analyzing the stability of systems where feedback is present (control systems).

Example 7.7
Consider the case that $H(e^{j\omega}) = 1/(2e^{j\omega} - 1)$. This transfer function has a non-ideal lowpass filter characteristic, as seen in the plot of

$\left|H(e^{j\omega})\right|^{2}$ (in dB) vs. ω in Fig. 7.44.   The dots represent gain values measured every 0.1π increment along the frequency axis.

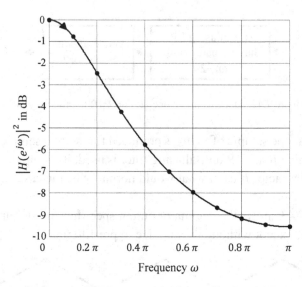

Figure 7.44.   Magnitude-squared plot of $H(e^{j\omega}) = 1/(2e^{j\omega} - 1)$ vs. ω, in dB, from Example 7.7.

The corresponding Nyquist plot, which recall is a plot of $Im\{H(e^{j\omega})\}$ along the vertical direction vs. $Re\{H(e^{j\omega})\}$ along the horizontal direction, is shown in Fig. 7.45. In this plot a star marks the origin of the complex plane. Thus, $|H(e^{j\omega})|$ at some frequency is the distance between the star and the corresponding point on the curve, and $\angle H(e^{j\omega})$ is the angle of a vector drawn from the star to the same point on the curve. (The dots on the curve mark every $0.1\,\pi$ rad/sec increment in ω, the same as in Fig. 7.44.)

Interestingly, the curve is a semicircle! As ω goes from 0 to π, magnitude decreases from 1.0 to approximately 0.3, and phase decreases from 0 to $-90°$. Both magnitude and phase are represented in a single curve, with the disadvantage that frequency values must be marked along the curve.

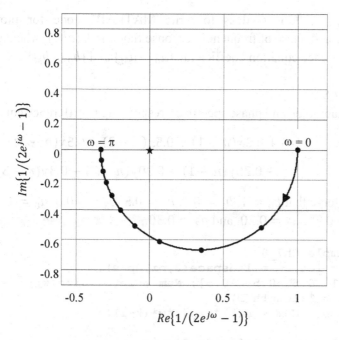

Figure 7.45. Nyquist plot of $H(e^{j\omega}) = 1/(2e^{j\omega} - 1)$ vs. ω, from Example 7.7.

7.3 Useful MATLAB® Code

7.3.1 *Plotting frequency response of filter described by a difference equation*

As presented in Fig. 7.8, a class of linear shift-invariant systems have their behavior described by an Mth-order difference equation:

$$y(n) = \sum_{k=0}^{M} b_k x(n - k) - \sum_{k=1}^{M} a_k y(n - k). \qquad (7.80)$$

The frequency response of such system is found by taking the Fourier transform of both sides, to give

$$Y(e^{j\omega}) = \sum_{k=0}^{M} b_k X(e^{j\omega})e^{-jk\omega} - \sum_{k=1}^{M} a_k Y(e^{j\omega})e^{-jk\omega}, \quad \text{or}$$

$$H(e^{j\omega}) = \frac{Y(e^{j\omega})}{X(e^{j\omega})} = \frac{\sum_{k=0}^{M} b_k e^{-jk\omega}}{1 + \sum_{k=1}^{M} a_k e^{-jk\omega}} = \frac{\sum_{k=0}^{M} b_k e^{-jk\omega}}{\sum_{k=0}^{M} a_k e^{-jk\omega}}, \qquad (7.81)$$

where $a_0 = 1$. Our goal is to write MATLAB® code for plotting magnitude and phase of frequency response function $H(e^{j\omega})$, when given linear difference equation coefficients $\{a_0 \cdots a_M\}$ and $\{b_0 \cdots b_M\}$.

Example 7.8

Plot the magnitude and phase spectra corresponding to difference equation

$$y(n) = x(n) + 0.5x(n-1) + 0.5x(n-2) - 0.5x(n-3)$$

$$+ 0.25y(n-1) - 0.20y(n-2) - 0.3y(n-3).$$

We observe that $b_0 = 1$, $b_1 = 0.5$, $b_2 = 0.5$, $b_3 = -0.5$, $a_0 = 1$, $a_1 = -0.25$, $a_2 = 0.20$, and $a_3 = 0.30$.

```
% Example Ch7_8
Npts = 1e4; w = linspace(0,pi,Npts);
B = [1, 0.5, 0.5, -0.5]; Num = zeros(size(w));
for k = 1:length(B)
    Num = Num + B(k)*exp(j*w*(k-1));
end
A = [1, -0.25, 0.20, 0.30];
Den = zeros(size(w));
for k = 1:length(A)
    Den = Den + A(k)*exp(j*w*(k-1));
end
H = Num./Den;
plot(w,abs(H))
xlabel('frequency in radians/sec')
ylabel('magnitude response')
```

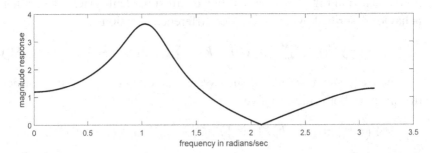

Figure 7.46. Magnitude spectrum corresponding to difference equation in Example 7.7.

```
figure; plot(w,angle(H))
xlabel('frequency in radians/sec')
ylabel('phase angle in radians')
```

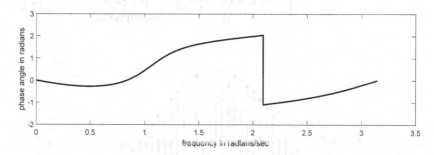

Figure 7.47. Phase spectrum corresponding to difference equation in Example 7.7.

7.3.2 *FIR filter design by windowing the ideal filter's impulse response*

As we saw in Section 7.1.3, expressions for the impulse responses of ideal filters are found as the inverse Fourier transforms of idealized frequency-domain responses. The ideal filters cannot be realized in practice because (a) they are not causal, and (b) the filter order is infinite. Both limitations can be eliminated by multiplying the ideal filter impulse response by a time-domain window (making it finite-duration, and hence of finite order), and then delaying the result so that no nonzero samples lie to the left of $n = 0$ (making $h(n)$ causal). In the following examples, we will use a windowing function $w(n)$ that gradually tapers to zero at both ends called a *half-raised cosine* or *Hanning* window. We define $w(n)$ to be an odd number of samples in width ($M = 2N + 1$) and centered at time index $n = 0$ so that it is an even function of n.[gg] Since the ideal filter's impulse response is also an even function of n, $h(n)w(n)$ will be even and real, giving the FIR filters designed this way a phase function equal to zero.[hh]

[gg] Because $w(n)$ is both real and even, its spectrum has zero phase.

[hh] When implemented in practice they are delayed to become causal, so phase is linear.

<u>Half-raised cosine (Hanning) window:</u>

$$w(n) = \begin{cases} 0.5\left(1 + \cos\left(\frac{\pi n}{N+1}\right)\right), & |n| \leq N \\ 0, & \text{otherwise.} \end{cases} \qquad (7.82)$$

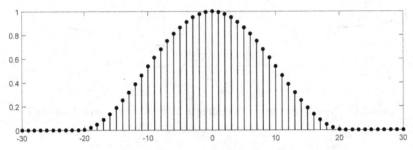

Figure 7.48. A plot of the half-raised (Hanning) window for $M = 41$.

Consider the ideal lowpass filter impulse response (Eq. (7.13)):

$$h_{ILP}(n) = \frac{\omega_0}{\pi} \text{sinc}(\omega_0 n).$$

Figure 7.49 shows a plot of $h_{ILP}(n)$ when $\omega_0 = \pi/4$:

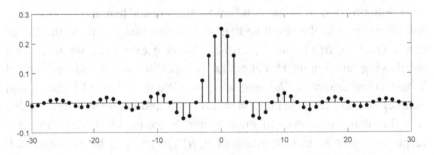

Figure 7.49. A plot of an ideal lowpass filter's impulse response $h_{ILP}(n)$ vs. n, over $-30 \leq n \leq 30$, when cutoff frequency $\omega_0 = \pi/4$.

Figure 7.50 shows the same sinc function after time windowing:

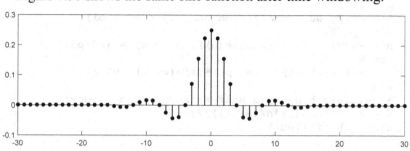

Figure 7.50. A plot of $h_{ILP}(n)w(n)$ vs. n, over $-30 \le n \le 30$, where $w(n)$ is a 41-pt. Hanning window. Figure 7.49 shows $h_{ILP}(n)$ without windowing.

Because of the windowing, the frequency response is no longer ideal. Figure 7.49 shows a plot of the resulting FIR lowpass filter's frequency response together with the original ideal filter's frequency response:

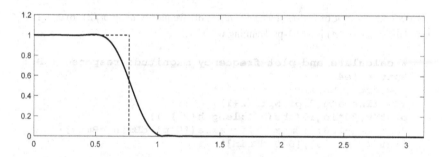

Figure 7.51. Frequency response $H_{LP}(e^{j\omega}) = \mathcal{F}\{h_{ILP}(n)w(n)\}$ of FIR filter having cutoff frequency $\omega_0 = \pi/4$, which was obtained by multiplying an ideal LPF's impulse response with a 41-pt. Hanning window. The ideal LPF's magnitude response is the dotted line.

Here is MATLAB® code to generate and plot a Hanning-windowed impulse response of ideal lowpass filter, an even function having length $M = 2N + 1$ samples (M is odd), and its frequency response:

```
% Generate M samples of ideal LPF impulse response:
%    hILP(n) = (w0/pi)*sinc(w0*n)
```

Practical Signal Processing and its Applications

```
% specify N (M = 2*N+1, an odd integer):
N = 100;
% specify w0, cutoff frequency (0 < w0 < pi):
w0 = pi/2;
n = -N:N; h = zeros(size(n)); h(n==0) = (w0/pi);
for k = 1:N
    h(abs(n)==k) = (w0/pi)*(sin(w0*k)/(w0*k));
end
% multiply h(n) by a half-raised cosine window:
h = h.*(1+cos(pi*n/(N+1)))/2;
stem(n,h,'filled')
```

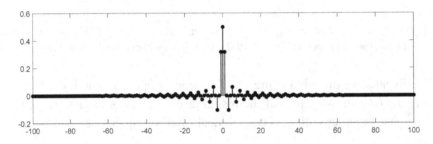

Figure 7.52. A plot of $h_{ILP}(n)w(n)$ vs. n (cutoff frequency $\omega_0 = \pi/2$), over $-100 \leq n \leq 100$, where $w(n)$ is a 201-pt. Hanning window.

```
% calculate and plot frequency magnitude response (dB):
Npts = 1e4;
H = fft(h,Npts);
w = linspace(0,pi,Npts/2+1);
plot(w,20*log10(abs(H(1:length(w)))))
dBmin = -100; dBmax = 20; axis([0 pi dBmin dBmax])
line([0 w0 w0],[0 0 dBmin])
```

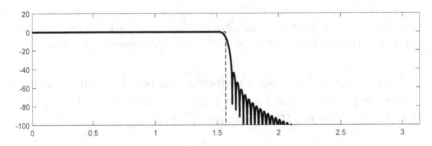

Figure 7.53. $\left|H_{LP}(e^{j\omega})\right|$ vs. ω in dB, where $w(n)$ is a 201-pt. Hanning window, corresponding to the impulse response function shown in Fig. 7.52.

When M is increased to 1001, the frequency response better resembles that of the ideal lowpass filter:

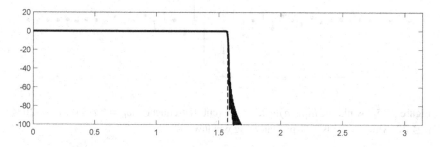

Figure 7.54. $\left|H_{LP}(e^{j\omega})\right|$ vs. ω in dB (cutoff frequency $\omega_0 = \pi/2$), where $w(n)$ is a 1001-pt. Hanning window.

Next is MATLAB® code to generate and plot a windowed ideal *high-pass* filter impulse response, an even function of length $M = 2N + 1$ of samples (M is odd). In this case, we use the *Hamming* window. It is like the half-raised cosine window, except that some constants are optimized to reduce the sidelobe levels:

<u>Hamming window:</u>

$$w(n) = \begin{cases} 0.54 + 0.46\cos\left(\frac{\pi n}{N}\right), & |n| \leq N \\ 0, & \text{otherwise.} \end{cases} \qquad (7.83)$$

```
% Generate M samples of ideal HPF impulse response:
%   hIHP(n) = d(n)-(w0/pi)*sinc(w0*n)
%
% specify N (M = 2*N+1, an odd integer):
N = 20;
% specify w0, cutoff frequency (0 < w0 < pi):
w0 = 3*pi/4;
n = -N:N; h = zeros(size(n));
h(n==0) = 1-(w0/pi);
for k = 1:N
    h(abs(n)==k) = -(w0/pi)*(sin(w0*k)/(w0*k));
end
% multiply h(n) by a Hamming window:
h = h.*(0.54+0.46*cos(pi*n/N));
stem(n,h,'filled')
```

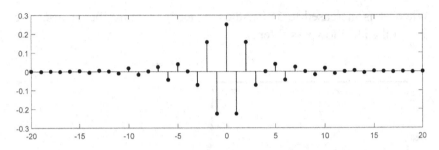

Figure 7.55. A plot of $h_{IHP}(n)w(n)$ vs. n (cutoff frequency $\omega_0 = 3\pi/4$), over $-20 \leq n \leq 20$, where $w(n)$ is a 41-pt. Hamming window.

```
% calculate and plot frequency magnitude response (dB):
Npts = 1e4;
H = fft(h,Npts);
w = linspace(0,pi,Npts/2+1);
plot(w,20*log10(abs(H(1:length(w)))))
dBmin = -100; dBmax = 20;
axis([0 pi dBmin dBmax])
line([w0 w0 pi],[dBmin 0 0])
```

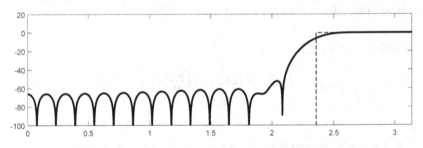

Figure 7.56. $|H_{HP}(e^{j\omega})|$ vs. ω in dB, where $w(n)$ is a 41-pt. Hamming window, corresponding to the impulse response function shown in Figure 7.55.

Here is MATLAB® code to generate and plot a windowed ideal band-pass filter impulse response, an even function of length $M = 2N + 1$ of samples (M is odd). We again use the Hamming window:

```
% Generate M samples of ideal BPF impulse response:
%    hIBP(n) = (w2/pi)*sinc(w2*n)-(w1/pi)*sinc(w1*n)
```

```
% specify N (M = 2*N+1, an odd integer):
N = 100;
% specify passband edges w1, w2 (0 < w1 < w2 < pi)
w1 = pi/4; w2 = pi/2;
n = -N:N; h = zeros(size(n)); h(n==0) = (w2/pi)-(w1/pi);
for k = 1:N
    h(abs(n)==k) = (w2/pi)*(sin(w2*k)/(w2*k)) ...
                 - (w1/pi)*(sin(w1*k)/(w1*k));
end
% multiply h(n) by a Hamming window:
h = h.*(0.54+0.46*cos(pi*n/N)); stem(n,h,'filled')
```

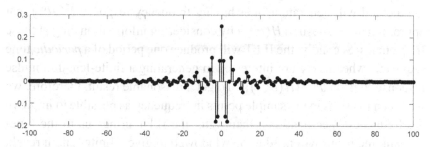

Figure 7.57.   A plot of $h_{IBP}(n)w(n)$ vs. n (passband: $\pi/4 < \omega < \pi/2$), over $-100 \leq n \leq 100$, where $w(n)$ is a 201-pt. Hamming window.

```
% calculate and plot frequency magnitude response (dB):
Npts = 1e4; H = fft(h,Npts);
w = linspace(0,pi,Npts/2+1);
plot(w,20*log10(abs(H(1:length(w)))))
dBmin = -100; dBmax = 20; axis([0 pi dBmin dBmax])
line([w1 w1 w2 w2],[dBmin 0 0 dBmin])
```

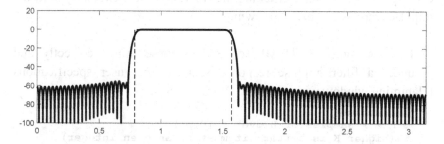

Figure 7.58. $|H_{BP}(e^{j\omega})|$ vs. ω in dB, where $w(n)$ is a 201-pt. Hamming window, corresponding to the impulse response function shown in Fig. 7.57.

7.3.3 *FIR filter design by frequency sampling*

Since the FFT makes it practical to transform data back and forth between the time and frequency domains, one may ask: may we design a digital filter by specifying a desired $H(e^{j\omega})$, then calculating an inverse FFT to obtain $h(n)$? The answer is *yes*, but with some strings attached. Firstly, the IFFT assumes that it is operating on one period of a *sampled* spectrum $H_p(k)$ (not continuous spectrum $H(e^{j\omega})$); we will therefore be able to specify the spectrum only at the sampled frequencies, which means that the method will generate a filter having frequency response $\tilde{H}(e^{j\omega})$ that approximates the desired $H(e^{j\omega})$ by considering information only at those frequencies. Secondly, the IFFT will produce one period of a *periodic* time sequence, whereas we are interested in generating a finite-length impulse response $\tilde{h}(n) = \mathcal{F}^{-1}\{\tilde{H}(e^{j\omega})\}$. To get a reasonable result, therefore, we will need to use as many sample points in frequency as possible to improve the frequency specifications; then, after the IFFT brings us to the time-domain, the result will need to be windowed to give a finite-length result. This method has the advantage of giving the filter designer full freedom in specifying frequency magnitude response (the phase is assumed zero). As with any FIR filter design method, best results are obtained when the filter order is high.

A tricky part of this *frequency sampling* filter design approach is to properly specify the desired $|H(e^{j\omega})|$: for $\tilde{h}(n) \cong \mathcal{F}^{-1}\{H(e^{j\omega})\}$ to be real, the input data to the IFFT algorithm must represent one period of an even spectrum. Also, after taking the IFFT, one must circularly shift the impulse response before windowing.

Here is some MATLAB® code does these things correctly and produces a filter impulse response sequence per user specifications (Hamming window width $M = 2 \cdot 25 + 1 = 51$):

```
% K = # samples of H over w=[0,pi]
% (higher K is better; it must be an even integer)
K = 15000;
% Define the desired magnitude spectrum over w=[0,pi]:
w = linspace(0,pi,K+1);
w1 = 1; w2 = 2; w3 = 2.5;
```

```
H = zeros(size(w));
range1 = find([w>=w1]&[w<=w2]);
range2 = find([w>=w2]&[w<=w3]);
H(range1) = 1.0;
H(range2) = 2.0 - (range2-range2(1))/(length(range2)-1);
% Define the remainder of the spectrum to make it even:
H = [H, fliplr(H(2:(K-1)))];
h = fftshift(ifft(H));
% time window will be Hamming of width M = 2N + 1:
N = 25; n = -N:N;
h = h(n+K).*(0.54+0.46*cos(pi*n/N));
stem(n,h,'filled')
```

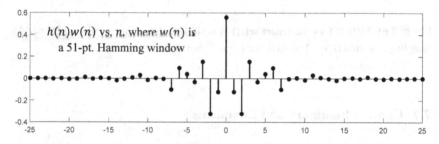

Figure 7.59. A plot of $h(n)w(n)$ vs. n, over $-25 \leq n \leq 25$, where $w(n)$ is a 51-pt. Hamming window. The filter was designed by sampling a desired $H(e^{j\omega})$.

```
% calculate and plot frequency magnitude response:
Npts = 1e4; H = fft(h,Npts);
w = linspace(0,pi,Npts/2+1);
plot(w, abs(H(1:length(w))))
line([1 1 2 2 2.5 2.5],[0 1 1 2 1 0]);axis([0 pi 0 2.5])
```

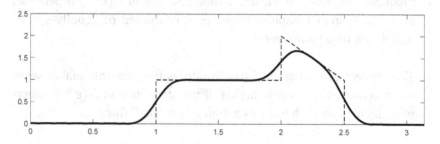

Figure 7.60. $H(e^{j\omega})$ vs. ω, where $w(n)$ is a 51-pt. Hamming window, corresponding to the impulse response function shown in Fig. 7.59.

Here is the resulting frequency response of the same filter, only this time for impulse response width = 1001 pts:

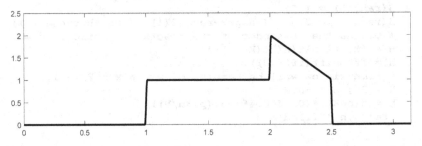

Figure 7.61. $H(e^{j\omega})$ vs. ω, where $w(n)$ is a 1001-pt. Hamming window, obtained by sampling a desired $H(e^{j\omega})$ (dotted line in Fig. 7.60) in frequency.

7.4 Chapter Summary and Comments

- Just as with continuous-time systems, discrete-time systems have input and output sequences related by convolution when these systems are linear and shift-invariant: $y(n) = x(n) * h(n)$.

- Sequence $e^{j\omega_0 n}$ is an *eigenfunction* of an LSI system. When $e^{j\omega_0 n}$ is applied as input to the system, $ce^{j\omega_0 n}$ is the output. The constant $= H(e^{j\omega})\big|_{\omega=\omega_0}$.

- Frequency response of a discrete-time LSI system $H(e^{j\omega})$ is periodic, since the system's impulse response is composed of equally-spaced samples in time (is *discrete*).

- The concept of lowpass/highpass/bandpass filtering only makes sense when considering a single period of transfer function $H(e^{j\omega})$; otherwise, there is no such thing as a digital "lowpass" filter.

- An ideal digital lowpass filter has frequency response $H_{ILP}(e^{j\omega}) = \text{rect}(\omega/2\omega_0) * \delta_{2\pi}(\omega)$, where $\omega = \omega_0$ marks the upper edge of the passband.

- An ideal digital lowpass filter has impulse response

$$h_{ILP}(n) = \frac{\omega_0}{\pi} \operatorname{sinc}(\omega_0 n).$$

- An ideal digital highpass filter has frequency response $H_{IHP}(e^{j\omega}) = 1 - \operatorname{rect}(\omega/2\omega_0) * \delta_{2\pi}(\omega)$, where $\omega = \omega_0$ marks the lower edge of the passband.

- An ideal digital highpass filter has impulse response

$$h_{IHP}(n) = \delta_{2\pi}(n) - \frac{\omega_0}{\pi} \operatorname{sinc}(\omega_0 n).$$

- An ideal digital bandpass filter has frequency response $H_{IBP}(e^{j\omega}) = \{\operatorname{rect}(\omega/2\omega_2) - \operatorname{rect}(\omega/2\omega_1)\} * \delta_{2\pi}(\omega)$, where frequencies ω_1 and ω_2 mark the passband edges ($\omega_1 < \omega_2$).

- An ideal digital bandpass filter has impulse response

$$h_{IBP}(n) = \frac{\omega_2}{\pi} \operatorname{sinc}(\omega_2 n) - \frac{\omega_1}{\pi} \operatorname{sinc}(\omega_1 n).$$

- One class of LSI systems may be described using difference equation:

$$y(n) = b_0 x(n) + b_1 x(n-1) + b_2 x(n-2) + b_3 x(n-3) + \cdots$$

$$-\{a_1 y(n-1) + a_2 y(n-2) + a_3 y(n-3) + \cdots\}.$$

- When taking the discrete-time Fourier Transform of an Mth-order difference equation we obtain this frequency response:

$$H(e^{j\omega}) = \left(\sum_{k=0}^{M} b_k e^{-j\omega k}\right) / \left(1 + \sum_{k=1}^{M} a_k e^{-j\omega k}\right).$$

- Any causal difference equation may be implemented using the network structure shown in Fig. 7.16. That network realizes a causal LSI system because only delay units are present (no "advance units", which look into the future).

- Digital filters based on the network diagram in Fig. 7.16 may be realized using specialized hardware (fast, not easily reconfigurable) or with software running on a general-purpose computer (slower, easily reconfigurable). These methods calculate output sequence $y(n)$ one sample at a time in the time domain.

- Another approach is to do digital filtering in the frequency domain. The input sequence is chopped up into blocks of sample points, these are converted to the frequency domain using the FFT, operations are done in that domain, and the results are sent back to the time domain using the IFFT. Results are calculated N points at a time. This too can be done either using special-purpose hardware or on a general-purpose computer.

7.5 Homework Problems

P7.1 Eigenfunction e^{jn} is applied as input to a system having frequency response $H(e^{j\omega}) = \text{rect}(\omega/3) * \delta_{2\pi}(\omega)$. What will be the output sequence in response to this input?

P7.2 Sequence $5e^{jn}$ is applied as input to an LSI system whose impulse response $h(n) = (0.1)^n u(n)$. What will be the output sequence in response to this input?

P7.3 Sequence $\cos(2n)$ is applied as input to an LSI system whose impulse response $h(n) = 3\text{sinc}(3n) - \text{sinc}(n)$. What will be the output sequence in response to this input?

P7.4 Write an expression for the impulse response of an ideal digital lowpass filter having bandwidth $= \pi/2$ rad/sec.

P7.5 Write an expression for the frequency response of an ideal digital lowpass filter having bandwidth $= \pi/3$ rad/sec. Sketch $H(e^{j\omega})$ over frequency range $-2\pi \le \omega \le 2\pi$ rad/sec.

P7.6 Write an expression for the impulse response of an ideal digital highpass filter having cutoff frequency $= \pi/4$ rad/sec.

P7.7 Write an expression for the frequency response of an ideal digital highpass filter having cutoff frequency $= \pi/2$ rad/sec. Sketch $H(e^{j\omega})$ over frequency range $0 \leq \omega \leq 2\pi$ rad/sec.

P7.8 Write an expression for the impulse response of an ideal digital bandpass filter having passband edge frequencies $= \{\pi/8, \pi/4\}$ rad/sec.

P7.9 Write an expression for the frequency response of an ideal digital bandpass filter having passband edge frequencies $= \{\pi/4, 3\pi/4\}$ rad/sec. Sketch $H(e^{j\omega})$ over frequency range $0 \leq \omega \leq \pi$ rad/sec.

P7.10 Draw a discrete-time network that implements difference equation $y(n) = x(n) + 2x(n-1) + 3x(n-2) + y(n-1) - 0.9y(n-2)$.

P7.11 Draw a discrete-time network that implements difference equation $y(n) = 0.9x(n) - x(n-1) + x(n-2) + y(n-1) - 0.9y(n-2)$. Plot the corresponding frequency phase response $\angle H(e^{j\omega})$ vs. ω, for $0 \leq \omega \leq \pi$, using MATLAB®.

P7.12 Write the difference equation for a causal lowpass filter having bandwidth $= \pi/3$ rad/sec. Sketch a discrete-time network that implements this difference equation.

P7.13 Write the difference equation for a causal highpass filter having cutoff frequency $= \pi/7$ rad/sec. Sketch a discrete-time network that implements this difference equation.

P7.14 Write the difference equation for a causal bandpass filter having peak frequency $= \pi/2$ rad/sec (any bandwidth and passband gain are acceptable). Plot the corresponding frequency magnitude response $|H(e^{j\omega})|$ vs. ω, for $0 \leq \omega \leq \pi$, using MATLAB®.

P7.15 A comb filter has impulse response $h_{CF}(n) = 1 + \delta(n-10)$. Plot this filter's frequency magnitude response $|H(e^{j\omega})|$ vs. ω, over the frequency range $0 \leq \omega \leq \pi$, using MATLAB®.

P7.16 Design an FIR bandpass filter, having impulse response of length 201 samples, by multiplying an ideal bandpass filter's impulse response function $h_{IBP}(n)$ by a half-raised cosine window. The passband edge frequencies of the ideal filter are $\omega = \{1, 2\}$ rad/sec. Plot your filter's frequency magnitude response $|H(e^{j\omega})|$ (-80 dB to 10 dB) vs. ω (0 to π rad/sec), using MATLAB®.

P7.17   Design an FIR filter, using the frequency sampling method, to meet this specification: $|H(e^{j\omega})| \cong |\pi - 2\omega|/\pi$ for $0 \le \omega \le \pi$. Use a Hamming window to produce an impulse response function that is 101 samples long. Plot your filter's frequency magnitude response $|H(e^{j\omega})|$ vs. ω, over the frequency range $0 \le \omega \le \pi$, using MATLAB®.

Chapter 8

Frequency Analysis of Continuous-Time Systems

8.1 Theory

8.1.1 *Introduction*

In previous chapters, we have introduced a number of useful time-domain signals, and considered how they may be represented as a weighted sum (that is, a linear combination) of sinusoids at various frequencies – this being a signal's spectrum. In the engineering field, practical systems and devices are often modeled using linear components. Such mathematical models are described via ordinary differential equations. The coefficients of the differential equations are mostly constant. In some instances, the coefficients may be time varying or the differential equations may be nonlinear. In those situations, it is possible to develop approximate linear models. In this text, we will concern ourselves only with systems that may be modeled by constant coefficient linear differential equations.

This chapter considers how a time varying signal is modified as it passes through a linear system. Specifically, our interest will be to look at how the response of some system will change the frequency content of the input signal. First we will consider the characterization of a general linear system (as a black box), and then discuss some examples from electrical circuit analysis that you may be familiar with.

8.1.2 *Linear Time-Invariant Continuous System*

Let us represent a LTIC system as a black box, with one input and one output. Such a system is completely described by its response (output) to an impulse input, $h(t)$. If this system's behavior may be described with a constant-coefficient linear differential equation, then $h(t)$ may be found by solving that differential equation. We will not be concerned with that aspect here. For the time being we will assume that we have knowledge of the impulse response. A block diagram of such a system with impulse response $h(t)$, input signal $f(t)$, and the resulting output signal $g(t)$ are shown in Fig. 8.1:

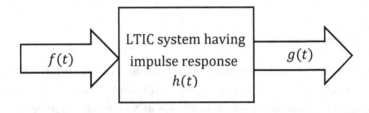

Figure 8.1.   A linear and time-invariant continuous-time system.

Two questions naturally arise in our mind: "Why does the impulse response completely characterize a linear system? How can we relate output signal $g(t)$ to input signal $f(t)$ from that knowledge?" The answer to the first question lies in the frequency components of an impulse. We recall that the Fourier transform of $\delta(t)$ is constant 1. This signifies that it contains sinusoids at all frequencies with equal weighting. Hence the output to an impulse input will have information about the behavior of the system to inputs all frequencies. That is to say, $\mathcal{F}\{h(t)\} = H(\omega)$ is the output signal spectrum (frequency response) when the input signal spectrum contains all frequency components with unity weights.

In a previous chapter a definition for linearity was discussed. That characteristic holds true for linear systems as well, and may be simply stated as follows:

> *If, for a given linear system, input $f_1(t)$ results in output $g_1(t)$ and input $f_2(t)$ results in output $g_2(t)$, then in response to input $af_1(t) + bf_2(t)$ this linear system will produce output $ag_1(t) + bg_2(t)$ (a, b are constants).*

We now also introduce the concept of time-invariance for a system:

> *If, for a given time-invariant system, input $f(t)$ results in output $g(t)$, then in response to input $f_1(t - \tau)$ this system will produce output $g_1(t - \tau)$ (τ is a constant). That is, when the input is delayed by τ sec then the only change in the output will be the same delay.*

We can now readily determine the relationship between the input $f(t)$ and output $g(t)$ of the system from the knowledge of $h(t)$ and/or $H(\omega)$. Since $F(\omega)$ represents the coefficients in a linear combination of all frequencies, the output frequency coefficients $G(\omega)$ should represent the coefficients in the linear combination of all frequencies in the product $F(\omega)H(\omega)$. Therefore, we may conclude[a] that

$$g(t) \Longleftrightarrow G(\omega) = F(\omega) \times H(\omega). \qquad (8.1)$$

And, from the convolution property of Fourier transforms we have

$$g(t) = \mathcal{F}^{-1}\{G(\omega)\} = f(t) * h(t). \qquad (8.2)$$

Figure 8.2 symbolically represents this process of determining the output of a system having impulse response $h(t)$ to any input $f(t)$. With knowledge of the linear, time-invariant system's impulse response one could determine its output via convolution in the time domain. Or, if it is advantageous, one may Fourier transform $f(t)$ and $h(t)$ to the frequency domain, multiply together the resulting spectra, and then convert the product back to the time domain using an Inverse Fourier transform.

---

[a] For this to be true the system must not only be linear but also time-invariant. Otherwise, both system impulse response $h(t)$ and its Fourier transform $H(\omega)$ may change depending on when they are being measured.

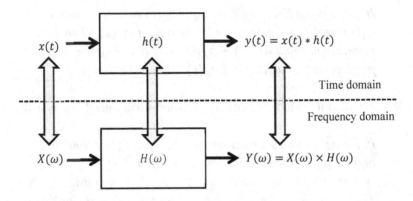

Figure 8.2. The input/output relationships of an LTIC system, shown in both time and frequency domains.

It is interesting to note that $H(\omega)$ is called *transfer function* of the system. Frequently the system behavior may be specified more conveniently through its transfer function than through its impulse response, as we will observe in the case of filters. In the following subsection we formally derive Eq. (8.2).

8.1.2.1 *Input/output relation*

Recall the sifting property of the impulse function, and the linear super-position property. From the knowledge of its impulse response, we derive the output of a LTIC system as follows:

input $\rightarrow$ output

$\delta(t) \rightarrow h(t)$ (system impulse response)

$\delta(t - \tau) \rightarrow h(t - \tau)$ (time-invariance)

$x(\tau)\delta(t - \tau) \rightarrow x(\tau)h(t - \tau)$ (linearity)

$x(t)\delta(t - \tau) \rightarrow x(\tau)h(t - \tau)$ (sifting property)

$\int_{\tau=-\infty}^{\infty} x(t)\delta(t - \tau)d\tau \rightarrow \int_{\tau=-\infty}^{\infty} x(\tau)h(t - \tau)d\tau$ (linearity)

$x(t) \int_{\tau=-\infty}^{\infty} \delta(t - \tau)d\tau \rightarrow \int_{\tau=-\infty}^{\infty} x(\tau)h(t - \tau)d\tau$ (simplified)

$$x(t) \longrightarrow \int_{\tau=-\infty}^{\infty} x(\tau)h(t-\tau)d\tau \qquad \text{(simplified)}$$

$$x(t) \longrightarrow x(t) * h(t) = y(t) \qquad \text{(definition of convolution)}$$

$$X(\omega) \longrightarrow X(\omega) \times H(\omega) = Y(\omega) \qquad \text{(Fourier Transform)}$$

Thus, the output signal $y(t)$ may be found in terms of the input signal $x(t)$ via the convolution operation in the time domain:

$$y(t) = x(t) * h(t). \tag{8.3}$$

This derivation relies on the fact that the system is time invariant. If it is not, then the result is a slight variation on the above: a time-delayed impulse at the system's input, $\delta(t-\tau)$, produces a time-dependent impulse response $h(\tau, t)$ at the output. As a result we can state that, for a time varying system,

$$x(t) \Longrightarrow y(t) = \int_{-\infty}^{\infty} x(\tau)h(t,\tau)\, d\tau. \tag{8.4}$$

8.1.2.2 *Response to $e^{j\omega_0 t}$*

Let us examine the output of a LTIC system when we input a complex exponential signal, which has energy only at frequency ω_0: $f(t) = e^{j\omega_0 t}$. Assume that the LTIC system has an impulse response $h(t)$. From the previous discussion and Eq. (8.3), we have output

$$y(t) = h(t) * e^{j\omega_0 t} = \mathcal{F}^{-1}\{2\pi H(\omega)\delta(\omega - \omega_0)\}$$

$$= H(\omega_0)\,\mathcal{F}^{-1}\{2\pi\delta(\omega - \omega_0)\} = H(\omega_0)e^{j\omega_0 t} \tag{8.5}$$

We reach a very important conclusion from the analysis in Eq. (8.5): A single-frequency input to an LTIC system results in the output of the same single-frequency signal scaled by the system's transfer function evaluated at that frequency. Thus, the system may be analyzed one frequency component at a time, as if only that signal frequency is present. This is the basis of the sinusoidal analysis that we are familiar with in electrical circuits and in other LTIC systems.

8.1.3 *Ideal filters*

Filters are found in many electrical systems, especially in communication systems. *Filtering*, the function of filters, is a frequency-domain concept that is best described by a linear system's transfer function $H(\omega)$ as opposed to its impulse response $h(t)$. Here we discuss filters having idealized frequency response functions. An ideal filter passes some frequencies while stopping others from passing. Its transfer function, $H(\omega)$, multiplies the input signal's spectrum by either 1 or 0. The range of frequencies where $H(\omega) = 1$ is called the *passband* and the range of frequencies where $H(\omega) = 0$ is called the *stopband*. Figure 8.3 shows the four different types of ideal analog filters: *lowpass, highpass, bandpass,*

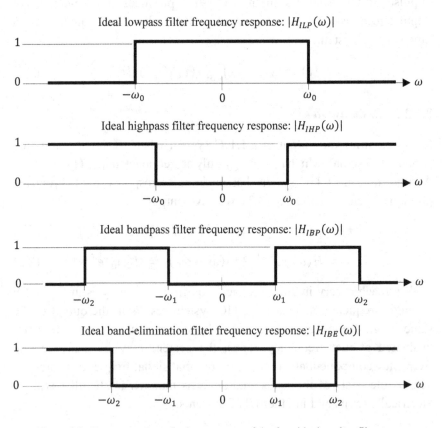

Figure 8.3. Frequency magnitude responses of the four ideal analog filter types.

and *band-elimination* filters.[b] Note that $|H(\omega)|$ is shown, but not $\angle H(\omega)$. This is because we assume that either phase $\angle H(\omega) = 0$, or is a linear function of ω (zero phase after some delay).

8.1.3.1 *Ideal lowpass filter*

The names of these filters are quite representative of their frequency response characteristics. A lowpass filter passes all frequency signals below a certain specified *cutoff frequency* ω_0. If the input to a lowpass filter has energy above ω_0, then those components are eliminated from the output of the filter. The ideal lowpass filter's frequency response $H_{ILP}(\omega)$ is defined as a rectangular pulse function that is centered at $\omega = 0$:

$$H_{ILP}(\omega) = \mathrm{rect}\left(\frac{\omega}{2\omega_0}\right). \tag{8.6}$$

From Table 5.1, the inverse Fourier Transform of $H_{ILP}(\omega)$ is the impulse response of an ideal lowpass filter:

$$h_{ILP}(t) = \frac{\omega_0}{\pi}\,\mathrm{sinc}(\omega_0 t). \tag{8.7}$$

8.1.3.2 *Ideal highpass filter*

The ideal highpass filter transfer function is *complementary* to that of the ideal lowpass filter, which is to say $H_{IHP}(\omega) = 1 - H_{ILP}(\omega)$:

$$H_{IHP}(\omega) = 1 - \mathrm{rect}\left(\frac{\omega}{2\omega_0}\right). \tag{8.8}$$

Multiplication by transfer function $H_{IHP}(\omega)$ passes all frequencies above a certain specified *cutoff frequency* ω_0. Just as for the lowpass filter, the impulse response of an ideal highpass filter is noncausal:

---

[b] Given that an ideal filter's transfer function $H(\omega)$ is either zero or one at any frequency, two other trivial filter types may be defined: the *all-stop* filter where $H(\omega) = 0$, and the *all-pass* filter where $H(\omega) = 1$. We should note that another, less restrictive filter commonly bears the name *allpass* filter; it is characterized by magnitude $|H(\omega)| = 1$ and arbitrary phase.

$$h_{IHP}(t) = \delta(t) - \frac{\omega_0}{\pi}\operatorname{sinc}(\omega_0 t). \qquad (8.9)$$

8.1.3.3 *Ideal bandpass filter*

The ideal bandpass filter passes all frequency components in the range $\omega_1 < |\omega| < \omega_2$. An ideal bandpass filter can be realized by placing ideal lowpass and highpass filters in series (where ω_2 is the cutoff frequency of the lowpass filter, and ω_1 is the cutoff frequency of the highpass filter):

$$H_{IBP}(\omega) = H_{ILP}(\omega)H_{IHP}(\omega), \qquad (8.10)$$

$$= \operatorname{rect}\left(\frac{\omega}{2\omega_2}\right)\left\{1 - \operatorname{rect}\left(\frac{\omega}{2\omega_1}\right)\right\}. \qquad (8.11)$$

$$(\omega_2 > \omega_1)$$

Equation (8.11) may be manipulated to give an expression for ideal bandpass frequency response that is a difference of rectangular pulses:

$$H_{IBP}(\omega) = \operatorname{rect}\left(\frac{\omega}{2\omega_2}\right) - \operatorname{rect}\left(\frac{\omega}{2\omega_2}\right)\operatorname{rect}\left(\frac{\omega}{2\omega_1}\right) \qquad (8.12)$$

$$H_{IBP}(\omega) = \operatorname{rect}\left(\frac{\omega}{2\omega_2}\right) - \operatorname{rect}\left(\frac{\omega}{2\omega_1}\right),$$

because $\operatorname{rect}(\omega/2\omega_2)\operatorname{rect}(\omega/2\omega_1) = \operatorname{rect}(\omega/2\omega_1)$ when $\omega_2 > \omega_1$. The corresponding impulse response is the inverse Fourier transform of Eq. (8.12):

$$h_{IBP}(t) = \frac{\omega_2}{\pi}\operatorname{sinc}(\omega_2 t) - \frac{\omega_1}{\pi}\operatorname{sinc}(\omega_1 t). \qquad (8.13)$$

An equivalent formulation for $H_{IBP}(\omega)$ is the sum of rectangular pulses having width W, that are shifted $\pm\omega_c$ from the origin:

$$H_{IBP}(\omega) = \operatorname{rect}\left(\frac{\omega+\omega_c}{W}\right) + \operatorname{rect}\left(\frac{\omega-\omega_c}{W}\right). \qquad (8.14)$$

$$\left(W = \omega_2 - \omega_1, \ \omega_c = \frac{\omega_1+\omega_2}{2}\right)$$

The corresponding impulse response is the inverse Fourier transform of Eq. (8.14):

$$h_{IBP}(t) = \frac{W}{\pi} \operatorname{sinc}\left(\frac{Wt}{2}\right) \cos(\omega_c t). \qquad (8.15)$$

$$\left(W = \omega_2 - \omega_1, \ \omega_c = \frac{\omega_1 + \omega_2}{2}\right)$$

8.1.3.4 *Ideal band-elimination filter*

The band-elimination filter is the complement of the bandpass filter, making the stopband range $\omega_1 < |\omega| < \omega_2$. Referring to the ideal bandpass filter expressions above, the corresponding results for band-elimination filter are therefore:

$$H_{IBE}(\omega) = 1 - \left\{ \operatorname{rect}\left(\frac{\omega}{2\omega_2}\right) - \operatorname{rect}\left(\frac{\omega}{2\omega_1}\right) \right\}. \qquad (8.16)$$

Alternatively:

$$H_{IBE}(\omega) = 1 - \left\{ \operatorname{rect}\left(\frac{\omega + \omega_c}{W}\right) + \operatorname{rect}\left(\frac{\omega - \omega_c}{W}\right) \right\}. \qquad (8.17)$$

$$\left(W = \omega_2 - \omega_1, \ \omega_c = \frac{\omega_1 + \omega_2}{2}\right)$$

The corresponding time-domain impulse response expressions are

$$h_{IBP}(t) = \delta(t) - \left\{ \frac{\omega_2}{\pi} \operatorname{sinc}(\omega_2 t) - \frac{\omega_1}{\pi} \operatorname{sinc}(\omega_1 t) \right\} \qquad (8.18)$$

and
$$h_{IBE}(t) = \delta(t) - \frac{W}{\pi} \operatorname{sinc}\left(\frac{Wt}{2}\right) \cos(\omega_c t). \qquad (8.19)$$

$$\left(W = \omega_2 - \omega_1, \ \omega_c = \frac{\omega_1 + \omega_2}{2}\right)$$

All the ideal filters discussed above are not causal and thus cannot be implemented in practice. Another reason they cannot be realized is that infinite circuitry is required to achieve the sudden transitions between passband and stopband in frequency response. Later in the chapter we will consider some simple analog filters that are practical and inexpensive to build, but whose frequency transfer functions only approximate the ideal case.[c]

8.2 Practical Applications

8.2.1 *RLC circuit impedance analysis*

Let us now apply transform techniques to analyzing electrical circuits. We will first consider behavior of the three basic element types (resistor, inductor, and capacitor) from the Fourier transform point of view, and then will look at some circuits where these elements are interconnected (RLC networks).

In the following, for convenience, we will assume that voltage $v(t)$ is the stimulus (input) and that current $i(t)$ is the response (output). Our LTIC system is defined by a component's voltage-current characteristic. For a resistor of R Ω the voltage-current characteristic is:

$$v(t) = Ri(t) \qquad \text{(time domain)}, \qquad (8.20)$$

$$V(\omega) = RI(\omega) \qquad \text{(frequency domain)}, \qquad (8.21)$$

$$\therefore H(\omega) = V(\omega)/I(\omega) = R \ \Omega. \qquad (8.22)$$

The transfer function $H(\omega)$ has impedance units (Ω), and in sinusoidal steady state circuit analysis it is termed impedance $Z(j\omega)$.[d] If one assumes that the current $i(t)$ is the input and voltage $v(t)$ is the output, the transfer function will have units of Siemens and it is admittance (the inverse of impedance), denoted by $Y(j\omega) = 1/Z(j\omega)$.

---

[c] The presence of an instantaneous jump in the magnitude frequency response characteristic (at passband-to-stopband transition) requires an infinite number of circuit components to be exactly realized as an analog filter.

[d] May also be called "$Z(\omega)$". In a later chapter, when relating the Laplace transform to the Fourier transform, we will see why notation $j\omega$ is often used in place of ω.

Consider next the energy storage element: an inductor of L Henrys. Following an analogous derivation, we obtain the voltage-current relationship:

$$v(t) = L\frac{di(t)}{dt} \quad \text{(time domain)}, \quad (8.23)$$

$$V(\omega) = j\omega L\, I(\omega) \quad \text{(frequency domain)}, \quad (8.24)$$

$$\therefore H(\omega) = V(\omega)/I(\omega) = j\omega L \ \Omega. \quad (8.25)$$

Thus, the impedance of an inductor is the familiar value $Z(j\omega) = j\omega L$ Ω. Note that we have used the Fourier transform property of differentiation (Table 5.2) to arrive at this frequency domain relationship. Similarly, another energy storage element – a capacitor of C Farads – has impedance $Z(j\omega) = -j/\omega C$ Ω. This may be derived as follows:

$$i(t) = C\frac{dv(t)}{dt} \quad \text{(time domain)}, \quad (8.26)$$

$$I(\omega) = j\omega C\, V(\omega) \quad \text{(frequency domain)}, \quad (8.27)$$

$$\therefore H(\omega) = V(\omega)/I(\omega) = 1/j\omega C \ \Omega = -j/\omega C \ \Omega. \quad (8.28)$$

Table 8.1. Voltage-current characteristics of R,L,C components in time and frequency domains.

Element	$v - i$ Characteristic (time domain)	$V - I$ Characteristic (frequency domain)	Impedance $Z(\omega)$ Ω
R, $i(t)$, $v(t)$, $+$ $-$	$v(t) = Ri(t)$	$V(\omega) = R\, I(\omega)$	R
L, $i(t)$, $v(t)$, $+$ $-$	$v(t) = L\dfrac{di(t)}{dt}$	$V(\omega) = j\omega L\, I(\omega)$	$j\omega L$
C, $i(t)$, $v(t)$, $+$ $-$	$i(t) = C\dfrac{dv(t)}{dt}$	$V(\omega) = \left(\dfrac{1}{j\omega C}\right) I(\omega)$	$1/j\omega C$

The impedance functions of R, L and C components are summarized in Table 8.1. We will use them when working with electrical circuits in examples that follow. Fortunately, rules of node and mesh analysis from circuit theory apply and may be used to carry out the analysis in the frequency domain.

8.2.2 *First order passive filter circuits*

8.2.2.1 *Highpass filter*

Example 8.1
Determine the voltage transfer function (output voltage in response to input voltage) across the resistor for the RC circuit shown in Fig. 8.4 (a).

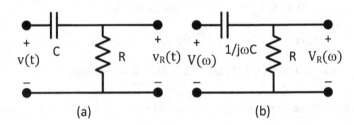

(a)   (b)

Figure 8.4. (a) Basic RC highpass filter network; (b) network transformed to impedance values.

There are different approaches to finding the transfer function. We will use the direct method of sinusoidal steady state analysis. For that, we will first convert the element values to their respective impedance values (creating a transformed network) as shown in Fig. 8.4 (b). Then we use the principle of voltage division across series elements to obtain the voltage across the resistor, just as it would be done for purely resistive circuits:

$$Z_C(\omega) = \frac{1}{j\omega C} \ \Omega, \quad Z_R(\omega) = R \ \Omega$$

$$V_R(\omega) = \frac{Z_R(\omega)}{Z_C(\omega) + Z_R(\omega)} V(\omega) = \frac{R}{1/j\omega C + R} V(\omega)$$

$$\therefore H_{HP}(\omega) = \frac{V_R(\omega)}{V(\omega)} = \frac{j\omega}{j\omega + 1/RC}. \tag{8.29}$$

This is an example of a first-order highpass filter. Because there is no sharp transition between the passband and stopband in this filter's frequency response, it is customary to assign the label ω_H to the frequency where $|H_{HP}(\omega)|^2/|H_{HP}(\omega)|^2_{max} = 1/2$ (-3.01 dB). This cutoff frequency corresponds to $\omega_H = 1/RC$ rad/sec. $\tau = RC = 1/\omega_H$ is called the *time constant* of the circuit. A plot of voltage transfer function magnitude $|H_{HP}(\omega)|$ of Eq. (8.17) for $\tau = 0.1$ ($\omega_H = 10$) is shown in Fig. 8.5 and the corresponding phase plot is shown in Fig. 8.6. Note that at the cutoff frequency $\omega_H = 10$ rad/sec, the magnitude of the transfer function is $\sqrt{1/2} = 0.707$, defining the "half power frequency." At that frequency, the phase shift is 45 degrees.

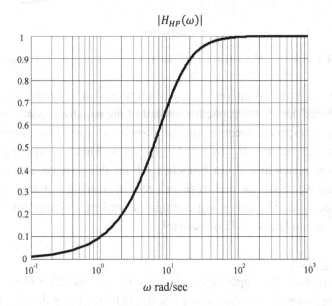

Figure 8.5. $|H_{HP}(\omega)| = |j\omega/(j\omega + \omega_H)|$ vs. ω, for $\omega_H = 10$ rad/sec.

Many textbooks and filter design handbooks dealing with the design of higher-order filters are available; the interested reader should refer to such books to gain more understanding with filter design.

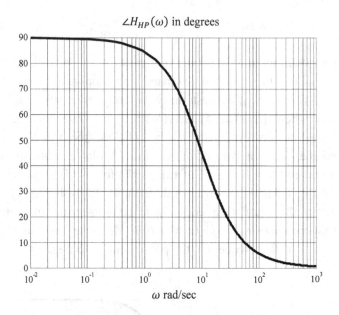

Figure 8.6. $\angle H_{HP}(\omega) = \angle(j\omega/(j\omega + \omega_H))$ vs. ω, for $\omega_H = 10$ rad/sec.

Another point of interest in this example is that when the output signal is measured across the capacitor instead of the resistor ($V_C = V - V_R$), the frequency response becomes that of a lowpass filter – the complement of what it was before: $H_{LP}(\omega) = 1 - H_{HP}(\omega)$. This can also be done by swapping series and shunt branches within the passive ladder network (exchanging capacitor with resistor), and measuring output signal at the usual location. The following example describes a simple lowpass filter having one inductor and one resistor.

<u>Example 8.2</u>
Let us replace the capacitive element with an inductive element in the previous example, as shown in Figs. 8.7 (a) and (b), and determine once again the voltage transfer function for this simple LR network.

We proceed with the analysis as in the previous example.

$$Z_L(\omega) = j\omega L \ \Omega, \quad Z_R(\omega) = R \ \Omega$$

$$V_R(\omega) = \frac{Z_R(\omega)}{Z_L(\omega)+Z_R(\omega)}\, V(\omega) = \frac{R}{j\omega L+R}\, V(\omega)$$

$$\therefore H_{LP}(\omega) = \frac{V_R(\omega)}{V(\omega)} = \frac{R/L}{j\omega+R/L}. \tag{8.30}$$

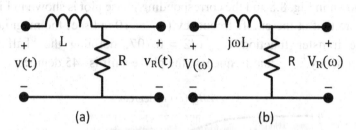

Figure 8.7. (a) Basic LR lowpass filter network; (b) network transformed to impedance values.

This is an example of a first-order lowpass filter. Because there is no sharp transition between the passband and stopband in this filter's frequency response, it is customary to assign the label ω_L to the

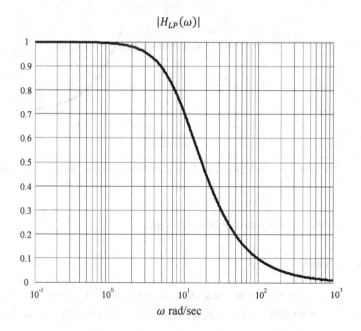

Figure 8.8. $|H_{LP}(\omega)| = |\omega_L/(j\omega + \omega_L)|$ vs. ω, for $\omega_L = 10$.

frequency where $|H_{LP}(\omega)|^2/|H_{LP}(\omega)|^2_{\max} = 1/2$ (−3.01 dB). This cutoff frequency corresponds to $\omega_L = R/L$ rad/sec. $\tau = L/R = 1/\omega_L$ is called the *time constant* of the circuit. A plot of voltage transfer function magnitude $|H_{LP}(\omega)|$ of Eq. (8.12) for $\tau = 0.1$ ($\omega_L = 10$) is shown in Fig. 8.8 and the corresponding phase plot is shown in Fig. 8.9. Note that at the cutoff frequency ($\omega_L = 10$ rad/sec) the magnitude of the transfer function is $\sqrt{1/2} = 0.707$, defining the "half power frequency." At that frequency, the phase shift is −45 degrees.

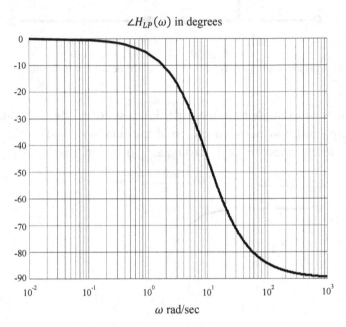

Figure 8.9. $\angle H_{LP}(\omega) = \angle(\omega_L/(j\omega + \omega_L))$ in degrees, vs. ω, for $\omega_L = 10$ rad/sec.

8.2.3 *Second order passive filter circuits*

8.2.3.1 *Bandpass filter*

A bandpass filter frequency response characteristic requires that the filter order is 2 or higher. We next consider the simplest-possible bandpass filter, a 2nd-order circuit composed of one resistor and two energy storage elements (L and C).

Example 8.3

Find transfer function $H_{BP}(\omega) = V_R(\omega)/V(\omega)$ of the bandpass RLC filter in Fig. 8.10 (a).

From the transformed version of the circuit shown in Fig. 8.10 (b):

$$Z_C(\omega) = \frac{1}{j\omega C} \ \Omega, \quad Z_L(\omega) = j\omega L \ \Omega, \quad Z_R(\omega) = R \ \Omega$$

$$V_R(\omega) = \frac{Z_R(\omega)}{Z_C(\omega) + Z_L(\omega) + Z_R(\omega)} V(\omega)$$

$$= \frac{R}{\frac{1}{j\omega C} + j\omega L + R} V(\omega) = \frac{j\omega RC}{1 - \omega^2 LC + j\omega RC} V(\omega)$$

$$\therefore H_{BP}(\omega) = \frac{V_R(\omega)}{V(\omega)} = \frac{j\omega RC}{(1 - \omega^2 LC) + j\omega RC} = \frac{j\omega(R/L)}{-\omega^2 + j\omega(R/L) + \frac{1}{LC}} \quad (8.31)$$

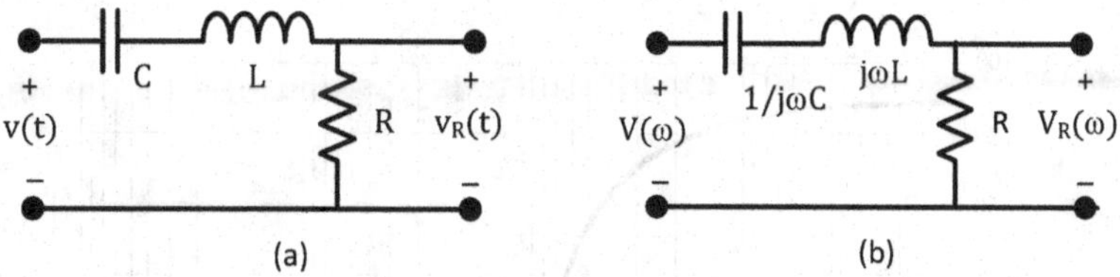

Figure 8.10. (a) Basic RLC bandpass filter network; (b) network transformed to impedance values.

This circuit's transfer function magnitude and phase plots are shown in Figs. 8.11 and 8.12. The component values were selected[e] to provide the following -3 dB cutoff frequencies: $\omega_L = 5$ rad/sec, $\omega_H = 15$ rad/sec, thus giving a bandpass filter bandwidth of $15 - 5 = 10$ rad/s. We note that at lower and higher cutoff frequencies, the transfer function magnitudes are $1/\sqrt{2} = 0.707$, whereas at the peak frequency of 10 rad/s the transfer function magnitude is 1.

---

[e] The filter design method is not shown, but one can derive the following relationships: 3dB bandwidth $\omega_H - \omega_L = R/L$, peak frequency $\omega_0 = \sqrt{1/LC} = \sqrt{\omega_H \omega_L}$.

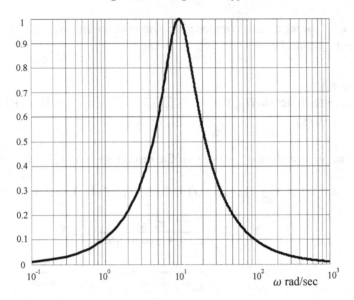

Figure 8.11. $|H_{BP}(\omega)| = |j\omega RC/((1 - \omega^2 LC) + j\omega RC)|$ vs. ω, for peak frequency $\omega_0 = 10$ rad/sec. $(R = 10\ \Omega, L = 1$ H$, C = 4/375$ F$)$.

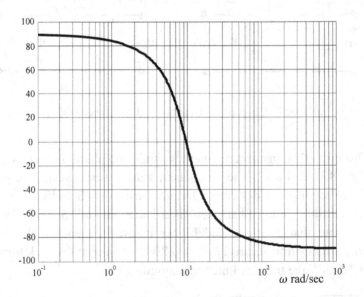

Figure 8.12. $\angle H_{BP}(\omega) = \angle(j\omega RC/((1 - \omega^2 LC) + j\omega RC))$ vs. ω, for peak frequency $\omega_0 = 10$ rad/sec. $(R = 10\ \Omega, L = 1$ H$, C = 4/375$ F$)$.

8.2.3.2 *Band-elimination filter*

If the series and shunt branches are swapped in Fig. 8.10, then the resulting voltage transfer function will have a band-elimination (also called *notch*[f]) filter characteristic:

$$H_{BE}(\omega) = 1 - H_{BP}(\omega) = \frac{(1-\omega^2 LC)}{(1-\omega^2 LC)+j\omega RC} = \frac{-\omega^2+\frac{1}{LC}}{-\omega^2+j\omega(R/L)+\frac{1}{LC}}. \qquad (8.32)$$

8.2.4 *Active filter circuits*

In the previous section, we considered several basic passive RLC filter circuits. However, modern analog filter designs frequently use active circuits for better filter performance, reduced size and lower cost using integrated circuit fabrication technology. A basic building block of active analog circuits is the *operational amplifier* (or "*op-amp*" for short), which is designed to contain critical components of a linear feedback system. In this section, we introduce the basic linear feedback network and its implementation using an operational amplifier. For simplicity in analysis, we will use ideal op-amps.[g] We introduce basic op-amp circuit topologies, and then conclude with op-amp circuits that may be used as building-blocks for active analog filters.

8.2.4.1 *Basic feedback network*

A feedback system operates by *feeding back* its output signal so that it, together with input signal, determines future system outputs. Consider the following network, which is a model for the simplest type of linear feedback system:

---

[f] When plotted using a log magnitude scale, $H_{BE}(\omega)$ appears to have a sharp notch at $\omega = \omega_0$.

[g] For more accurate data comparison with practical circuits, and especially at higher frequencies, it is important to simulate/analyze the circuits using real-world op-amp models.

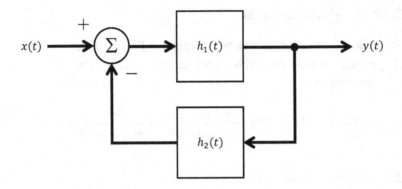

Figure 8.13. Time domain description of a basic continuous-time feedback network.

The two square blocks depict linear, time-invariant continuous-time systems having impulse responses $h_1(t)$ and $h_2(t)$, and the circular unit is a subtractor. We now write an expression for output signal $y(t)$ in terms of itself and the input signal $x(t)$:

$$y(t) = h_1(t) * \big(x(t) - y(t) * h_2(t)\big)$$
$$= h_1(t) * x(t) - h_1(t) * y(t) * h_2(t). \qquad (8.33)$$

This expression cannot be further simplified in the time domain. However, taking the Fourier transform of both sides gives us a frequency-domain expression that is easier to manipulate because in that domain convolutions are replaced by multiplications:

$$Y(\omega) = H_1(\omega)X(\omega) - H_1(\omega)Y(\omega)H_2(\omega). \qquad (8.34)$$

This simplifies to:

$$\frac{Y(\omega)}{X(\omega)} = \frac{H_1(\omega)}{1+H_1(\omega)H_2(\omega)}. \qquad (8.35)$$

Shown below is the same feedback system, only labeled in terms of its frequency-domain[h] parameters:

---

[h] Due to linearity of the transform, addition in time results in addition in frequency; that is why the adder block remains unchanged.

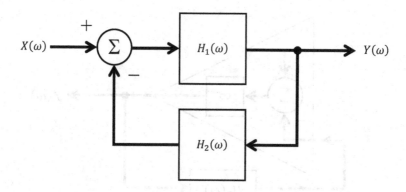

Figure 8.14. Frequency domain description of the basic feedback network in Fig. 8.13, having transfer function $Y(\omega)/X(\omega) = H_1(\omega)/(1 + H_1(\omega)H_2(\omega))$.

Rewrite Eq. (8.35) as follows:

$$\frac{Y(\omega)}{X(\omega)} = \frac{H_1(\omega)}{1 + H_1(\omega)H_2(\omega)} = \frac{1}{1/H_1(\omega) + H_2(\omega)}. \tag{8.36}$$

A very important class of feedback systems results when the system having frequency response $H_1(\omega)$ has infinite gain at every ω:

$$\frac{Y(\omega)}{X(\omega)} = \frac{1}{1/\infty + H_2(\omega)} = 1/H_2(\omega). \tag{8.37}$$

Note that despite $H_1(\omega) = \infty$, which is clearly a nonlinear behavior, the overall feedback system is linear! The following example demonstrates how a linear amplifier may be constructed using this feedback system approach.

Example 8.4

Referring to Fig. 8.14: Select $H_1(\omega) = \infty$ and $H_2(\omega) = a$, where $0.1 \leq a \leq 1$. What is the transfer function of this feedback system?

$$\frac{Y(\omega)}{X(\omega)} = \frac{1}{1/\infty + a} = \frac{1}{a} = G, \text{ where } 1 \leq G \leq 10.$$

8.2.4.2 *Operational amplifier*

When $H_1(\omega) = \infty$, we may redraw Fig. 8.14 as shown in Fig. 8.15.

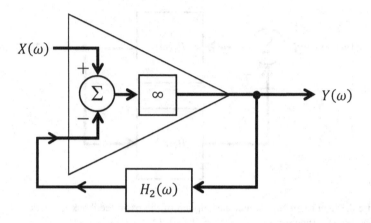

Figure 8.15. Frequency domain description of the basic feedback network in Fig. 8.14 having transfer function $Y(\omega)/X(\omega) = 1/H_2(\omega)$. The blocks inside of the triangle define an operational amplifier.

In Fig. 8.15 above, the subtractor and infinite gain blocks appearing within the triangle define an ideal operational amplifier.[i] The op-amp is not a stable linear circuit when used without any feedback path between its output terminal and the subtracting input (called *negative feedback*) because of its *open-loop gain* $= \infty$. In circuits having sufficient negative feedback, however, $Y(\omega)$ (and hence $y(t)$) will be finite despite being outputted by an amplifier having infinite gain. This is only possible if the difference between the two signals entering the subtractor is zero. Think of it this way:

From Fig. 8.14, $Y(\omega)/H_1(\omega) = X(\omega) - Y(\omega)H_2(\omega).$ (8.38)

As $H_1(\omega) \to \infty$ the left side of this expression goes to zero (assuming finite $Y(\omega)$). Therefore, the right side, which is the frequency domain description of the difference between the two signals entering the subtractor, must also go to zero.

---

[i] In a practical op-amp, the infinite gain stage is replaced by an amplifier having gain approaching 10^6.

Figure 8.16 shows an ideal operational amplifier circuit described by its instantaneous voltages and currents in the time domain. A practical op-amp circuit also has connections to an external power supply.

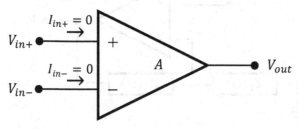

$$V_{out}(t) = A\big(V_{in+}(t) - V_{in-}(t)\big); \ A \to \infty \ ; \ \text{therefore, when } V_{out} \text{ is finite then}$$
$$V_{in+} = V_{in-} \ .$$

Figure 8.16. Time domain description of the ideal operational amplifier. (All voltages are measured relative to a ground node that is not shown.)

The op-amp is a useful building block for both linear and nonlinear analog systems. Of special interest to us, for signal processing purposes, are op-amps circuits that may be used for frequency-dependent amplification (filtering).

8.2.4.3 *Noninverting topology*

As seen in Fig. 8.17, the non-inverting topology of the op-amp has input signal V_{in} sent to the non-inverting input terminal. Because $I_{in-} = 0$, the voltage at the inverting input terminal is simply found via voltage division: $V_{in-} = V_{out} Z_1/(Z_1 + Z_2)$. Comparing this to the feedback system shown in Fig. 8.15, we see that $H_2(\omega) = Z_1/(Z_1 + Z_2)$. Thus, given that the op-amp is ideal with open-loop gain equal to ∞, we find that the voltage transfer function of this circuit is:

$$\frac{V_{out}}{V_{in}} = \frac{1}{H_2(\omega)} = \frac{Z_1(\omega) + Z_2(\omega)}{Z_1(\omega)} = 1 + \frac{Z_2(\omega)}{Z_1(\omega)}. \tag{8.39}$$

Another way to obtain this result is to assume that the system is operating in a stable feedback mode such that $V_{in+} = V_{in-}$, as mentioned in

Fig. 8.16. Then $V_{in-} = V_{out} Z_1/(Z_1 + Z_2) = V_{in+} = V_{in}$, so that $V_{out}/V_{in} = (Z_1(\omega) + Z_2(\omega))/Z_1(\omega) = 1 + Z_2(\omega)/Z_1(\omega)$.

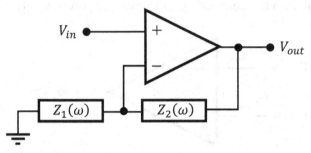

Figure 8.17.   Non-inverting topology of the operational amplifier: $\frac{V_{out}}{V_{in}} = 1 + \frac{Z_2(\omega)}{Z_1(\omega)}$. (Voltages V_{in} and V_{out} are measured relative to ground.)

A commonly-used circuit having non-inverting topology is the *voltage follower*. As implied by its name, $V_{out} = V_{in}$. This is achieved by selecting $Z_2(\omega) = 0$, or $Z_1(\omega) = \infty$, or both (usually the case). This circuit sends current to a load without drawing any current from the source, while maintaining the same voltage level:

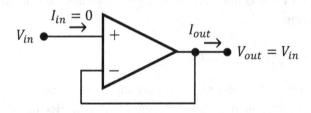

Figure 8.18.   Voltage follower circuit: $V_{out} = V_{in}$. (Voltages V_{in} and V_{out} are measured relative to ground.)

8.2.4.4   *Inverting topology*

As seen in Fig. 8.19, the inverting topology of the op-amp has input signal V_{in} sent to its inverting input terminal. Although this circuit is more difficult to analyze from the perspective of a feedback system, we may easily find its voltage transfer function by assuming that negative feedback establishes the condition $V_{in+} = V_{in-}$:

$$V_{out} = 0 - I_{in}Z_2 = -\left(\frac{V_{in}}{Z_1}\right)Z_2, \text{ or} \qquad (8.40)$$

$$\frac{V_{out}}{V_{in}} = -\frac{Z_2(\omega)}{Z_1(\omega)}.$$

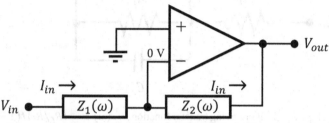

Figure 8.19. Inverting topology of the operational amplifier: $\frac{V_{out}}{V_{in}} = -\frac{Z_2(\omega)}{Z_1(\omega)}$. (Voltages V_{in} and V_{out} are measured relative to ground.)

The circuit shown in Fig. 8.20 below has $Z_1 = Z_2 = R$, and is called the op-amp *inverter* ($V_{out}/V_{in} = -R/R = -1$):

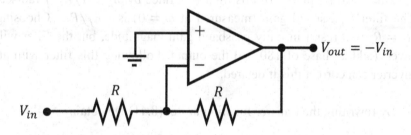

Figure 8.20. Op-amp inverter circuit: $\frac{V_{out}}{V_{in}} = -1$. (Voltages V_{in} and V_{out} are measured relative to ground.)

8.2.4.5 *First-order active filter*

A filter whose behavior may be described using a first-order differential equation in time is called a *first-order* filter. Figure 8.21 shows a first-order active lowpass filter circuit. Comparing Fig. 8.21 with

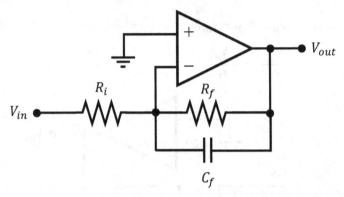

Figure 8.21. Lowpass filter having transfer function $H_{LP}(\omega) = -(R_f/R_i)/(1 + j\omega R_f C_f)$.

Fig. 8.19, we note that in this case $Z_1 = R_i \ \Omega$ and $Z_2 = R_f \ || \ (1/j\omega C_f) \ \Omega$. Thus, we obtain (using Eq. (8.40)) the transfer function:

$$H_{LP}(\omega) = \frac{V_{out}}{V_{in}} = -\frac{Z_2(\omega)}{Z_1(\omega)} = -\left(\frac{R_f}{R_i}\right)\frac{1}{1 + j\omega R_f C_f}. \qquad (8.41)$$

The cutoff frequency of this filter is defined by $\omega_L = 1/R_f C_f$ rad/sec. The filter's passband gain (measured at $\omega = 0$) is $-R_f/R_i$. Choosing $R_f = R_i$ will result in unity passband gain magnitude, but the filter will invert (add a phase of $180°$ to) the output. Following this filter with an inverter can correct this if desired.

By rewriting the transfer function in Eq. (8.41) we obtain

$$H_{LP}(\omega) = \left(\frac{-1}{R_i C_f}\right)\frac{1}{\frac{1}{R_f C_f} + j\omega} = c\ \frac{1}{b + j\omega}, \qquad (8.42)$$

where constants $c = -1/R_i C_f$, $b = 1/R_f C_f$. But for the scaling factor c, this matches the Fourier transform pair shown as Entry #2 in Table 5.1:

$$ce^{-bt}u(t) \Leftrightarrow c\ \frac{1}{b + j\omega}. \qquad (8.43)$$

Hence, the impulse response of the active lowpass filter shown in Fig. 8.21 is

$$h_{LP}(t) = \left(\frac{-1}{R_i C_f}\right) e^{-(1/R_f C_f)t} u(t). \tag{8.44}$$

8.2.4.6 *Second-order active filter*

A topology often used to realize 2nd-order active filters is the Sallen-Key circuit [Chen]. Figure 8.22 shows a Sallen-Key circuit that uses a unity-gain amplifier (voltage follower op-amp):

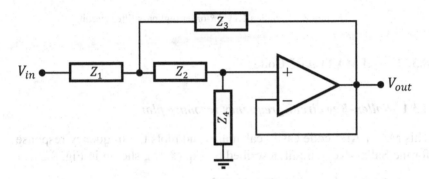

Figure 8.22. Sallen-Key circuit using a voltage-follower op-amp.

Circuit analysis yields the following voltage transfer function for the Sallen-Key circuit:

$$H(\omega) = \frac{V_{out}(\omega)}{V_{in}(\omega)} = \frac{Z_3 Z_4}{Z_1 Z_2 + Z_3(Z_1+Z_2) + Z_3 Z_4}. \tag{8.45}$$

For example, a second-order highpass filter may be constructed by choosing elements having impedances Z_1 and Z_2 to be capacitors and elements having impedances Z_3 and Z_4 to be resistors. Let $C_1 = C_2 = C$ and $R_3 = R_4 = R$, as shown in Fig. 8.23. From Eq. (8.45):

$$H_{HP}(\omega) = \frac{V_{out}(\omega)}{V_{in}(\omega)} = \frac{R^2}{\left(\frac{1}{j\omega C}\right)^2 + R\left(\frac{2}{j\omega C}\right) + R^2} = \frac{\omega^2}{\omega^2 - j\omega\left(\frac{2}{RC}\right) - \frac{1}{R^2 C^2}}. \tag{8.46}$$

Note that $H_{HP}(\omega)$ is a ratio of 2nd-order polynomials in ω, making it the transfer function of a 2nd-order filter.

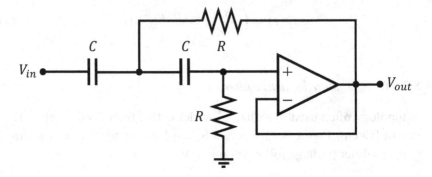

Figure 8.23. Sallen-Key 2nd-order highpass filter circuit.

8.3   Useful MATLAB® Code

8.3.1   *Sallen-Key circuit frequency response plot*

This MATLAB® code easily calculates and plots the frequency response for the Sallen-Key circuit described by Eq. (8.45), shown in Fig. 8.23:

```
% Choose R=1 Ohm, C=1 Farad
R = 1;
C = 1;
w = logspace(-2,2,1e4);
Z1 = R*ones(size(w));
Z2 = R*ones(size(w));
Z3 = 1./(j*w*C);
Z4 = 1./(j*w*C);
H = Z1.*Z2./(Z1.*Z2 + Z3.*(Z1+Z2) + Z3.*Z4);
semilogx(w,abs(H).^2)
grid on
```

The resulting graph is shown in Fig. 8.24.

8.3.2   *Calculating and plotting impedance of a one-port network*

The term *one-port network* describes any circuit that may be put into the form shown in Fig. 8.25. The relationship between externally applied voltage V_{in} and the resulting current I_{in} is described in the frequency domain as impedance $Z_{in}(\omega) = V_{in}(\omega)/I_{in}(\omega)$.

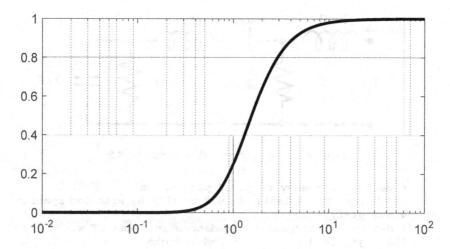

Figure 8.24. $|H(\omega)|^2$ vs. ω for the highpass Sallen-Key circuit shown in Fig. 8.23 ($C = 1$ Farad, $R = 1$ Ohm).

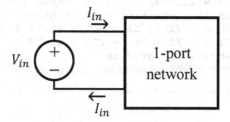

Figure 8.25. One-port network.

Inside the box shown as the network, there may be R, L, C components in various series and parallel combinations. MATLAB® makes it easy to calculate the input impedance of that type of network as a function of frequency. We demonstrate that method with the following example.

Example 8.5
Calculate and plot $|Z_{in}(f)|$ vs. f for the one-port network shown in Fig. 8.26.

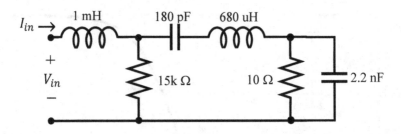

Figure 8.26.   One-port network from Example 8.5.

```
% define frequency range of plot in Hz
f = logspace(2,8,1e4);  % 10^2 to 10^8 Hz with log spacing
w = 2*pi*f;
% define all component impedances in terms of w:
ZL1 = j*w*1e-3;         % 1mH inductor
ZR1 = 15e3;             % 15k Ohm resistor
ZC1 = 1./(j*w*180e-12); % 180pF capacitor
ZL2 = j*w*680e-6;       % 680uH inductor
ZR2 = 10;               % 10 Ohm resistor
ZC2 = 1./(j*w*2.2e-9);  % 2.2nF capacitor
% Construct Zin, starting from right-most component:
Y1 = 1./ZR2 + 1./ZC2; Z2 = ZC1 + ZL2 + 1./Y1;
Y3 = 1./ZR1 + 1./Z2; Zin = ZL1 + 1./Y3;
loglog(f,abs(Zin)); grid on
```

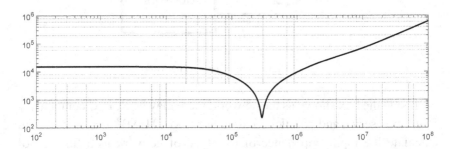

Figure 8.27. MATLAB® calculated plot of $|Z_{in}(f)|$ vs. f for the one-port network shown in Fig. 8.26.

From this plot of $|Z_{in}(f)|$ vs. f we see the following characteristic features: (a) $|Z_{in}| \cong 15\,k\Omega$ when $f < 100$ kHz; (b) $|Z_{in}|$ dips to $\cong 200\,\Omega$ at $f = 300$ kHz, and (c) $|Z_{in}|$ rises above $10\,k\Omega$ (and keeps rising) for $f > 1$ MHz.

8.4   Chapter Summary and Comments

- If a system is linear and time-invariant, then its output signal $g(t)$ is a convolution product between the system impulse response $h(t)$ and the input signal $f(t)$: $g(t) = h(t) * f(t)$.

- Stated in the frequency domain, $G(\omega) = H(\omega)F(\omega)$.

- Using the characteristics of system frequency response $H(\omega)$ to pass certain frequency components of $F(\omega)$ while blocking other components from passing is called *filtering*.

- An ideal lowpass filter has frequency response
$$H_{ILP}(\omega) = \text{rect}\left(\frac{\omega}{2\omega_L}\right).$$
(ω_L is the cutoff frequency = right edge of the passband)

- An ideal lowpass filter has impulse response:
$$h_{ILP}(t) = \mathcal{F}^{-1}\left\{\text{rect}\left(\frac{\omega}{2\omega_L}\right)\right\} = \frac{\omega_L}{\pi}\text{sinc}(\omega_L t).$$

- An ideal highpass filter has frequency response:
$$H_{IHP}(\omega) = 1 - \text{rect}\left(\frac{\omega}{2\omega_H}\right).$$
(ω_H is the cutoff frequency = left edge of the passband)

- An ideal highpass filter has impulse response:
$$h_{IHP}(t) = \mathcal{F}^{-1}\left\{1 - \text{rect}\left(\frac{\omega}{2\omega_H}\right)\right\} = \delta(t) - \frac{\omega_H}{\pi}\text{sinc}(\omega_H t).$$

- An ideal bandpass filter may be designed by cascading an ideal lowpass filter with an ideal highpass filter. The resulting transfer function is then:
$$H_{IBP}(\omega) = \text{rect}\left(\frac{\omega}{2\omega_2}\right)\left\{1 - \text{rect}\left(\frac{\omega}{2\omega_1}\right)\right\} = \text{rect}\left(\frac{\omega}{2\omega_2}\right) - \text{rect}\left(\frac{\omega}{2\omega_1}\right).$$
$$(\omega_2 > \omega_1)$$

- The impulse response of an ideal bandpass filter is:

$$h_{IBP}(t) = \mathcal{F}^{-1}\left\{\text{rect}\left(\frac{\omega}{2\omega_2}\right) - \text{rect}\left(\frac{\omega}{2\omega_1}\right)\right\}$$

$$= \frac{\omega_2}{\pi}\text{sinc}(\omega_2 t) - \frac{\omega_1}{\pi}\text{sinc}(\omega_1 t) \qquad (\omega_2 > \omega_1).$$

- When an LTIC system has input signal $e^{j\omega_0 t}$, the output signal will be the same except for a scaling factor: $e^{j\omega_0 t}H(\omega_0)$. Therefore $e^{j\omega_0 t}$ is called an eigenfunction of the system.

- For any LTIC system with real $h(t)$, input signal $x(t) = \cos(\omega_0 t)$ gives output signal $|H(\omega_0)|\cos(\omega_0 t + \angle H(\omega_0))$: a sinusoid at the same frequency but possibly at different amplitude and/or phase. A LTIC system does not produce any new frequencies at its output that were not already present at its input.

- First-order differential equations relate voltage and current in capacitors and inductors.

- Impedance $Z(\omega)$ is defined as $V(\omega)/I(\omega)$ in a circuit component.

- $Z(\omega) = R(\omega) + jX(\omega)$: Impedance = Resistance + j Reactance.

- Admittance $Y(\omega)$ is defined as $I(\omega)/V(\omega)$ in a circuit component ($Y(\omega) = 1/Z(\omega)$).

- $Y(\omega) = G(\omega) + jB(\omega)$: Admittance = Conductance + j Susceptance.

- The impedance of an L-Henry inductor is: $Z_L(\omega) = j\omega L$ Ω,
  The impedance of a C-Farad capacitor is: $Z_C(\omega) = 1/(j\omega C)$ Ω,
  The impedance of an R-Ohm resistor is: $Z_R(\omega) = R$ Ω.

- Circuit analysis in the frequency domain is done by converting all components to their impedance values and proceeding as if dealing with resistances (using KVL, KCL).

8.5 Homework Problems

P8.1 Eigenfunction e^{j2t} is applied as input to a system having frequency response $H(\omega) = \text{sinc}(\omega)$. What will be the output signal $y(t)$ in response to this input?

P8.2 Signal $5e^{jt}$ is applied as input to an LTIC system whose impulse response $h(t) = e^{-2t}u(t)$. What will be the output signal $y(t)$ in response to this input?

P8.3 Signal $\cos(2t)$ is applied as input to an LTIC system whose impulse response $h(t) = 3\text{sinc}(1.5t) - 2\text{sinc}(t)$. What will be the output signal $y(t)$ in response to this input?

P8.4 Write an expression for the impulse response of an ideal analog lowpass filter having bandwidth $= 20\pi$ rad/sec.

P8.5 Write an expression for the frequency response of an ideal analog lowpass filter having bandwidth $= 5$ rad/sec. Sketch $H(\omega)$ over frequency range $-10 \leq \omega \leq 10$ rad/sec.

P8.6 Write an expression for the impulse response of an ideal analog highpass filter having cutoff frequency $= 4\pi$ rad/sec.

P8.7 Write an expression for the frequency response of an ideal analog highpass filter having cutoff frequency $= 10$ rad/sec. Sketch $H(\omega)$ over frequency range $0 \leq \omega \leq 40$ rad/sec.

P8.8 Write an expression for the impulse response of an ideal analog bandpass filter having passband edge freqs. $= \{20, 30\}$ rad/sec.

P8.9 Write an expression for the frequency response of an ideal analog bandpass filter having passband edge frequencies $= \{5, 10\}$ rad/sec. Sketch $H(\omega)$ over $-15 \leq \omega \leq 15$ rad/sec.

P8.10 Design an LR lowpass filter having resistor value $1\ \Omega$ and cutoff frequency 100 rad/sec. Draw the circuit and plot its frequency magnitude response $|H(\omega)|$ vs. ω, over $10^1 \leq \omega \leq 10^3$ rad/sec (log frequency scale).

P8.11 Design an RC highpass filter having capacitor value 0.22 uF and cutoff frequency 100 kHz. Draw the circuit and plot its frequency phase response $\angle H(f)$ vs. f, over frequency range 0-500 kHz (linear frequency scale).

P8.12 Design an RLC bandpass filter having R = 10 Ω, L = 0.2 Henry, and C = 5 Farad. Draw the circuit and plot its frequency magnitude response $|H(\omega)|$ (dB) vs. ω, over $10^{-2} \le \omega \le 10^2$ rad/sec (log frequency scale).

P8.13 Draw an op-amp circuit having frequency response function $H(\omega) = 1 + R_1/(1/sC + R_2)$.

P8.14 Draw an op-amp circuit having frequency response function $H(\omega) = -(1/R_2 + sC_2)^{-1}/(R_1 + 1/sC_1)$.

P8.15 Draw a Sallen–Key circuit having $Z_1 = R_1 = 10 \,\Omega$, $Z_2 = R_2 = 0.1 \,\Omega$, and $Z_3 = Z_4 = \dfrac{1}{j\omega C} = \dfrac{1}{j\omega(1 \text{ Farad})}$. Plot its frequency magnitude response $|H(\omega)|$ vs. ω, over frequency range $10^{-1} \le \omega \le 10^1$ rad/sec (log frequency scale).

Chapter 9

Z-Domain Signal Processing

9.1 Theory

9.1.1 *Introduction*

The Z transform – a generalization of the Discrete-Time Fourier transform – is useful for many reasons. One reason is to make possible the analysis of signals whose Fourier transform does not exist due to lack of convergence, such as the ramp sequence. Another is to gain insight regarding discrete-time linear system stability. Very importantly, the Z transform greatly simplifies the design and analysis of linear difference equations used for digital filtering.

One may generalize the Fourier transform by extending real variable ω to the complex form $\omega \to (\omega_1 + j\omega_2)$. This is done in mathematical, scientific and some engineering literature. With proper choice of the value of ω_2 the Discrete-Time Fourier summation may be made to converge absolutely, and the Fourier transform of many growing functions would exist. A similar approach is to apply the Z transform, which can be traced back to the work of Abraham de Moivre[a] (1667-1754) that was later formalized by electrical engineers John R. Ragazzini and Lotfi A. Zadeh[b] (1952). The Z transform extends the Discrete-Time Fourier transform by

---

[a] https://www.encyclopediaofmath.org/index.php/Z-transform.

[b] J. R. Ragazzini and L. A. Zadeh, "The analysis of sampled-data systems". *Trans. Am. Inst. Elec. Eng.* 71(II): 225–234 (1952).

replacing unit-magnitude vector $e^{j\omega}$ with magnitude-unrestricted complex value $z = re^{j\omega}$, which is essentially the same as the above-mentioned generalization[c] of frequency variable ω.

The Z transform provides a formal mathematical framework for solving constant-coefficient linear difference equations. The technique is a powerful tool for designing and analyzing linear discrete systems that are modeled using difference equations. In the Z transform, complex variable $z = re^{j\omega}$ defines a complex z-plane that is circularly warped about the origin as compared to that defined by considering complex ω of the previous paragraph. The *bilateral Z transform* summation expression results when substituting z for $e^{j\omega}$ in the discrete-time Fourier transform. When a lower limit of summation 0 is chosen in this definition, which is appropriate for causal sequences, the transform is termed the *unilateral Z transform*. When only causal signals are considered the inverse Z-transformation is greatly simplified.

9.1.2 *The Z transform*

Z transform analysis, as it is now known, was first used in 1730 by French mathematician Abraham de Moivre as applied to probability theory. The Z transform may be viewed as a mapping of sequence $f(n)$ in the time domain to its uniquely associated complex function $F(z)$ in the complex frequency domain. In its general (bilateral) form, the Z transform is the following mathematical relationship between functions $f(n)$ and $F(z)$:

Bilateral Z transform	$F(z) = \sum_{n=-\infty}^{\infty} f(n)z^{-n},$	(9.1)
Inverse Z transform	$f(n) = \frac{1}{2\pi j} \oint_{\Gamma} F(z)z^{n-1}dz.$	(9.2)

Or, as written in shorthand notation:

---

[c] $e^{j\omega} \rightarrow e^{j(\omega_1 + j\omega_2)} = e^{-\omega_2 + j\omega_1} = re^{j\omega_1}$, where $r = e^{-\omega_2}$.

$$F(z) = \mathcal{Z}\{f(n)\}, \quad f(n) = \mathcal{Z}^{-1}\{F(z)\}. \tag{9.3}$$

When evaluating the inverse Z transform, contour of integration Γ in the complex z-plane is counterclockwise direction about the origin; it must lie entirely within the region of the z-plane where $F(z)$ converges. If the integral defining inverse Z transform converges, it may do so only for a specific range of radii of circular path Γ. Direct evaluation of Eq. (9.2) generally requires knowledge of complex variable theory. However, we can easily invert functions $F(z)$ that are ratios of polynomials in z (these are called *rational $F(z)$*), which are often encountered in signal and system analysis, using partial fraction expansion based on the Cauchy Residue Theorem.

The Z transform of causal signals is called the *unilateral* Z transform. Since causal $f(n) = 0$ for $n < 0$, its bilateral Z transform becomes:

Unilateral Z transform

$$F(z) = \sum_{n=0}^{\infty} f(n)z^{-n}, \tag{9.4}$$

Inverse Z transform for causal $f(n)$

$$f(n)u(n) = \frac{1}{2\pi j}\oint_{\Gamma} F(z)z^{n-1}dz. \tag{9.5}$$

To avoid any confusion, unless otherwise stated, we will assume to be dealing with the unilateral Z transform in this text. When needed, we will indicate this by using the shorthand notation:

$$F(z) = \mathcal{Z}\{f(n)u(n)\}, \quad f(n)u(n) = \mathcal{Z}^{-1}\{F(z)\}. \tag{9.6}$$

We are interested in signals whose Z transforms are *invertible*, meaning that both the Z transform and its inverse exist to give a unique mapping between $f(n)u(n)$ and $F(z)$. When that is true, the following notation is used:

$$f(n) \Leftrightarrow F(z). \tag{9.7}$$

Practically speaking, signals that may be measured and recorded are those having invertible Z transforms. Engineers often call these *real-world* signals in the sense that they are of finite length, have finite amplitudes, and have finite energy. The Z transforms of such signals are invertible.

We should point out that both the Discrete Fourier transform and the Z transform are among a class of transforms that form a powerful technique for solving many problems in engineering and science. In some cases, the two independent variables n and z have different physical interpretations. In some applications, transform analysis is applied to multidimensional functions in both domains.

9.1.3 *Region of convergence*

From a mathematical point of view, the Z transform of a discrete sequence may be thought of as a Laurent Series in the complex variable z. Thus, all properties and theorems of complex variables that apply to the Laurent Series also apply to the Z transform. In the definitions of both bilateral and unilateral Z transforms, specific values of $|z| = r$ may be necessary for the integrals to converge. The range of r values leading to convergence defines a region in the complex z-plane, which is called the *region of convergence*, or ROC. For the bilateral Z transform, the ROC is an annular ring of the form $r_1 < |z| < r_2$. Boundary circles $|z| = r_1$ and $|z| = r_2$ are not part of the ROC except in certain cases.[d] It should be stressed that a specification of the ROC is necessary to completely describe $Z\{f(n)\}$, because different $f(n)$ may have the same expression for $F(z)$ but have different regions of convergence. For example:

Causal: $a^n u(n) \Leftrightarrow \dfrac{z}{z-a}$,   ROC: $|z| > |a|$,   (9.8)

Anticausal: $a^n(u(n) - 1) \Leftrightarrow \dfrac{z}{z-a}$,   ROC: $|z| < |a|$.   (9.9)

The region of convergence for $F(z)$ when $f(n)$ is causal is illustrated in Fig. 9.1.

---

[d] When $f(n)$ is anticausal, $r_1 = 0$ and the ROC is $0 \leq |z| < r_2$;
when $f(n)$ is causal, $r_2 = \infty$ and the ROC is $r_1 < |z| \leq \infty$.

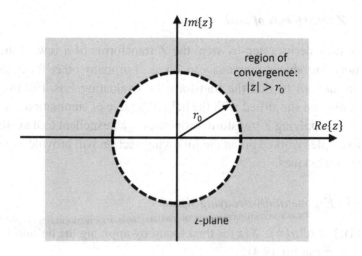

Figure 9.1. Z transform region of convergence for causal $f(n)$.

Table 9.1. Regions of convergence for the Z transforms of various types of sequences.

$f(n)$:	ROC of $F(z)$, if it exists:
Left-sided: continues indefinitely to the left of $n = n_1$	$0 < |z| < r_2$ ROC includes $z = 0$ when anticausal
Right-sided: continues indefinitely to the right of $n = n_2$	$r_1 < |z| < \infty$ ROC includes $z = \infty$ when causal
Two-sided: a sum of left-sided and right-sided sequences	$r_1 < |z| < r_2$
Finite-length: a product of left-sided and right-sided sequences	$0 < |z| < \infty$ ROC includes $z = 0$ when anticausal, $z = \infty$ when causal

9.1.4 *Z transforms of basic signals*

Let us now derive, step-by-step, the Z transforms of a few of the basic functions introduced previously in Ch. 2. For many other functions, it is straightforward to find the transform by evaluating Eqs. (9.1) or (9.4), which may be simplified with the help of a table of summations. In other instances, applying Z transform properties is an excellent tool available to us. Examples worked out in the following section will provide experience in such techniques.

9.1.4.1 *Exponentially decaying signal*

Let $x(n) = a^n u(n)$. $X(z)$ is then found by applying the definition of the Z transform in Eq. (9.4):[e]

$$X(z) = \mathcal{Z}\{a^n u(n)\} = \sum_{n=0}^{\infty} a^n z^{-n} = \sum_{n=0}^{\infty} (az^{-1})^n$$

$$= \frac{1}{1-az^{-1}} = \frac{z}{z-a}, \quad |z| > |a|. \tag{9.10}$$

It may be shown that $\mathcal{Z}^{-1}\{z/(z-a), |z| > |a|\} = a^n u(n)$, and so these comprise a "Z transform pair":

$$\boxed{\;a^n u(n) \Leftrightarrow \frac{z}{z-a} \quad \text{ROC: } |z| > |a|\;} \tag{9.11}$$

The region of convergence boundary is $|z| = a$, and the region of convergence is $|z| > a$. The transform function $F(z) = z/(z-a)$ is singular (goes to infinity) at $z = a$. This point in the z-plane is called a *pole* of the complex function $X(z)$, and it obviously cannot lie within the region of convergence. Also, because $X(z) = 0$ at $z = 0$, that point in the z-plane is called a *zero* of $X(z)$.

---

[e] The Geometric Series $\sum_{n=0}^{\infty} r^n = \frac{1}{1-r}$ when $|r| < 1$. When $r = az^{-1}$, $|r| = |a|/|z|$.
∴ $|r| < 1 \rightarrow |z| > |a|$.

9.1.4.2 Impulse sequence

The Z transform of a discrete impulse sequence is straightforward using the sifting property:

$$F(z) = \sum_{n=0}^{\infty} \delta(n)z^{-n} = \sum_{n=0}^{\infty} \delta(n)z^{-0} = \sum_{n=0}^{\infty} \delta(n) = 1. \quad (9.12)$$

It may be shown, using complex variable theory, that the inverse Z transform of 1 returns $\delta(n)$. Therefore, we add this entry to our list of Z-transform pairs:

$$\boxed{\delta(n) \Leftrightarrow 1 \quad \text{ROC: } \forall z} \quad (9.13)$$

Note that $Z\{\delta(n)\} = 1$ converges in the entire z-plane; it has no poles.

9.1.4.3 Delayed impulse sequence

Similarly, and just as easily, we may find the Z transform of $\delta(n - n_0)$:

$$Z\{\delta(n - n_0)\} = \sum_{n=0}^{\infty} \delta(n - n_0)z^{-n} = \sum_{n=0}^{\infty} \delta(n - n_0)z^{-n_0}$$

$$= z^{-n_0} \sum_{n=0}^{\infty} \delta(n - n_0) = z^{-n_0}(1) = z^{-n_0}. \quad (9.14)$$

This gives us the Z transform pair:

$$\boxed{\delta(n - n_0) \Leftrightarrow z^{-n_0} \qquad \text{ROC: } |z| > 0} \quad (9.15)$$

9.1.4.4 Unit step sequence

The Z transform of $u(n)$ can be readily found by considering it as limiting case of the exponentially decaying signal $a^n u(n)$:

$$Z\{u(n)\} = \lim_{a \to 1} Z\{a^n u(n)\}. \quad (9.16)$$

The Z transform pair is then

$$\boxed{u(n) \Leftrightarrow \frac{z}{z-1} \qquad \text{ROC: } |z| > 1} \quad (9.17)$$

This Z transform of the unit step sequence has a pole at $z = 1$, and thus the region of convergence is outside of the unit circle in the z plane.

9.1.4.5 *Causal complex exponential sequence*

The Z transform of sequence $e^{j\omega_0 n}u(n)$, which for $n \geq 0$ is essentially a phasor having frequency ω_0, can be readily obtained by substituting $a = e^{j\omega_0}$ in the Z transform pair for $a^n u(n)$ that was derived previously. Since $e^{j\omega_0 n}$ has magnitude 1, its Z transform $\mathcal{Z}\{e^{j\omega_0 n}u(n)\} = \sum_{n=0}^{\infty} e^{j\omega_0 n}z^{-n}$ will converge only when $|z| > 1$.

$$e^{j\omega_0 n}u(n) \Leftrightarrow \frac{z}{z - e^{j\omega_0}} \qquad \text{ROC: } |z| > 1 \qquad (9.18)$$

Clearly, the transformed function has a pole at $z = e^{j\omega_0}$.

9.1.4.6 *Causal sinusoidal sequence*

Because, as does the Discrete Fourier transform, the Z transform has the property of linear superposition, we may easily find the transform of $\cos(\omega_0 n)u(n)$ by expressing it in terms of complex exponential functions using Euler's identity: $\cos(\omega_0 n)u(n) = \frac{1}{2}(e^{j\omega_0 n} + e^{-j\omega_0 n})u(n)$, which has Z transform

$$= \frac{1}{2}\mathcal{Z}\{e^{j\omega_0 n}u(n)\} + \frac{1}{2}\mathcal{Z}\{e^{-j\omega_0 n}u(n)\} = \frac{1}{2}\left(\frac{z}{z - e^{j\omega_0}}\right) + \frac{1}{2}\left(\frac{z}{z - e^{-j\omega_0}}\right)$$

$$= \frac{\frac{z}{2}(z - e^{-j\omega_0}) + \frac{z}{2}(z - e^{j\omega_0})}{(z - e^{j\omega_0})(z - e^{-j\omega_0})} = \frac{z\left(z - \frac{1}{2}(e^{j\omega_0} + e^{-j\omega_0})\right)}{(z - e^{j\omega_0})(z - e^{-j\omega_0})}$$

$$= \frac{z(z - \cos(\omega_0))}{(z - e^{j\omega_0})(z - e^{-j\omega_0})}. \qquad (9.19)$$

Note that this z-domain function has two poles at $z = \{e^{j\omega_0}, e^{-j\omega_0}\}$, and the pole magnitudes are both equal to 1. The function also has a zero at $z = 0$ and another zero at $z = \cos(\omega_0)$. The region of convergence is solely determined by the location of the poles; the locations of the zeros have no effect. We may simplify further to obtain the Z transform pair:

$$\cos(\omega_0 n)u(n) \Leftrightarrow \frac{z^2-\cos(\omega_0)z}{z^2-2\cos(\omega_0)z+1} \qquad \text{ROC: } |z| > 1 \qquad (9.20)$$

Following the derivation for the Z transform of $\cos(\omega_0 n)u(n)$, we can readily determine the Z transform of $\sin(\omega_0 n)u(n)$ as follows:

$$
\begin{aligned}
Z\{\sin(\omega_0 n)u(n)\} &= Z\left\{\tfrac{1}{2j}\left(e^{j\omega_0 n} - e^{-j\omega_0 n}\right)u(n)\right\} \\
&= \tfrac{1}{2j} Z\{e^{j\omega_0 n}u(n)\} - \tfrac{1}{2j} Z\{e^{-j\omega_0 n}u(n)\} \\
&= \tfrac{1}{2j}\left(\frac{z}{z-e^{j\omega_0}}\right) - \tfrac{1}{2j}\left(\frac{z}{z-e^{-j\omega_0}}\right) = \frac{\frac{z}{2j}(z-e^{-j\omega_0})-\frac{z}{2j}(z-e^{j\omega_0})}{(z-e^{j\omega_0})(z-e^{-j\omega_0})} \\
&= \frac{z\tfrac{1}{2j}(e^{j\omega_0}-e^{-j\omega_0})}{(z-e^{j\omega_0})(z-e^{-j\omega_0})} = \frac{\sin(\omega_0)z}{(z-e^{j\omega_0})(z-e^{-j\omega_0})} \\
&= \frac{\sin(\omega_0)z}{z^2-2\cos(\omega_0)z+1}. \qquad (9.21)
\end{aligned}
$$

This results in the transform pair:

$$\sin(\omega_0 n)u(n) \Leftrightarrow \frac{\sin(\omega_0)z}{z^2-2\cos(\omega_0)z+1} \qquad \text{ROC: } |z| > 1 \qquad (9.22)$$

This z-domain function has two poles at $z = \{e^{j\omega_0}, e^{-j\omega_0}\}$, and the pole magnitudes are both equal to 1. There is a zero at $z = 0$.[f]

9.1.4.7 *Discrete ramp sequence*

The Z transform of a ramp sequence $r(n) = n \cdot u(n)$ is derived by direct integration, applying the definition given by Eq. (9.4):

$$F(z) = Z\{n \cdot u(n)\} = \sum_{n=0}^{\infty} n z^{-n} = \frac{1}{z} + \frac{2}{z^2} + \frac{3}{z^3} + \cdots \qquad (9.23)$$

Dividing both sides by z we obtain:[g]

---

[f] The number of zeros is always equal to the number of poles in a rational function. Although we do not mention it, here the second zero is at $z = \infty$.

[g] This derivation was found in: http://www.uv.es/~soriae/Section_2.pdf

$$\frac{F(z)}{z} = \frac{1}{z^2} + \frac{2}{z^3} + \frac{3}{z^4} + \cdots. \tag{9.24}$$

Then we multiply by dz and integrate:

$$\int \frac{F(z)}{z} dz = -\frac{1}{z} - \frac{1}{z^2} - \frac{1}{z^3} - \cdots + K. \tag{9.25}$$

K is a constant of integration. The sum of terms in z converge as a geometric series when $|z| > 1$:

$$\int \frac{F(z)}{z} dz = -\left(\frac{1}{z} + \frac{1}{z^2} + \frac{1}{z^3} + \cdots\right) + K$$

$$= -\left(\frac{1}{z-1}\right) + K \quad (|z| > 1). \tag{9.26}$$

Finally, differentiate both sides and multiply through by z:

$$F(z) = z \frac{d}{dz}\left\{-\left(\frac{1}{z-1}\right)\right\} = \frac{z}{(z-1)^2} \quad (|z| > 1). \tag{9.27}$$

We now have the Z transform pair for the discrete ramp sequence:

$$\boxed{n \cdot u(n) = r(n) \Leftrightarrow \frac{z}{(z-1)^2} \qquad \text{ROC: } |z| > 1} \tag{9.28}$$

Observe that the transformed function has a multiple (second order) pole at $z = 1$.

Using the same approach, it may be shown that when $f(n) \Leftrightarrow F(z)$, $nf(n) \Leftrightarrow -z\, dF(z)/dz$. Therefore, since $u(n) \Leftrightarrow z/(z-1)$ ($|z| > 1$), repeatedly multiplying $u(n)$ by n gives:

$$\boxed{\begin{array}{l} n^k \cdot u(n) \Leftrightarrow \left(-z\frac{d}{dz}\right)^k \left\{\frac{z}{z-1}\right\} \qquad \text{ROC: } |z| > 1 \\ (k > 0) \end{array}} \tag{9.29}$$

The resulting function will have a kth-order pole at $z = 1$.

9.1.5　*Table of Z transforms*

Instead of solving the Z transforms for commonly used signals each time, you may refer to the Z transform pairs shown in Table 9.2 that follows.

We encourage you to work through, as an exercise, and verify the entries in the table. Remember that each $F(z)$ will have a corresponding region of convergence. As all are causal signals, the ROC of the Z transform is the area outside of the circle whose radius is the highest-magnitude pole of the transfer function.

Table 9.2. Table of Z transform pairs. (region of convergence is for a causal time signal).

Entry #	$f(n) = f(n)u(n)$	$F(z)$	Refer to Eq. #
1.	$\dfrac{1}{2\pi j}\oint_\Gamma F(z)z^{n-1}dz$	$\displaystyle\sum_{n=0}^{\infty} f(n)z^{-n}$	(9.5), (9.4)
2.	$r^n u(n)$	$\dfrac{z}{z-r}$	(9.11)
3.	$\delta(n)$	1	(9.13)
4.	$\delta(n-n_0)$	z^{-n_0}	(9.15)
5.	$u(n)$	$\dfrac{z}{z-1}$	(9.17)
6.	$e^{j\omega_0 n}u(n)$	$\dfrac{z}{z-e^{j\omega_0}}$	(9.18)
7.	$\cos(\omega_0 n)u(n)$	$\dfrac{z^2 - \cos(\omega_0)\,z}{z^2 - 2\cos(\omega_0)\,z + 1}$	(9.20)

(Continued)

Table 9.2. (*Continued*)

Entry #	$f(n) = f(n)u(n)$	$F(z)$	Refer to Eq. #
8.	$\sin(\omega_0 n)u(n)$	$\dfrac{\sin(\omega_0)\,z}{z^2 - 2\cos(\omega_0)\,z + 1}$	(9.22)
9.	$n \cdot u(n) = r(n)$	$\dfrac{z}{(z-1)^2}$	(9.28)
10.	$n^k \cdot u(n) \quad (k \geq 0)$	$\left(-z\dfrac{d}{dz}\right)^k \left\{\dfrac{z}{z-1}\right\}$	(9.29)
11.	$r^n \cos(\omega_0 n)\,u(n)$	$\dfrac{z^2 - r\cos(\omega_0)\,z}{z^2 - 2r\cos(\omega_0)\,z + r^2}$	Jackson
12.	$r^n \sin(\omega_0 n)\,u(n)$	$\dfrac{r\sin(\omega_0)\,z}{z^2 - 2r\cos(\omega_0)\,z + r^2}$	Jackson
13.	$\cos(\omega_0 n + \theta)u(n)$	$\dfrac{\cos(\theta)\,z^2 - \cos(\omega_0 - \theta)\,z}{z^2 - 2\cos(\omega_0)\,z + 1}$	Lathi
14.	$r^n \cos(\omega_0 n + \theta)u(n)$	$\dfrac{\cos(\theta)\,z^2 - r\cos(\omega_0 - \theta)\,z}{z^2 - 2r\cos(\omega_0)\,z + r^2}$	Lathi
15.	$\mathrm{rect}_k(n - k)u(n)$ pulse width = $2k + 1$ samples	$\dfrac{z^{2k+1} - 1}{z^{2k}(z - 1)}$	(9.36)

9.1.6 *Z transform properties*

In most applications, we will need to derive Z transforms of many different functions. Frequently it is convenient to use some of the mathematical properties of the unilateral Z transform summation given in Eq. (9.4). Therefore, in this section we will discuss some of the important properties and then show all other properties that may be useful (Table 9.3).

9.1.6.1 *Linearity*

Let us consider two causal functions of time, $f_1(n)$ and $f_2(n)$, whose Z transforms are known to be $F_1(z)$ and $F_2(z)$, respectively:

$$f_1(n)u(n) \Leftrightarrow F_1(z)$$
$$f_2(n)u(n) \Leftrightarrow F_2(z)$$

Then, from Eq. (9.1), for two arbitrary constants a and b, the linear superposition relationship[h] holds true:

$$af_1(n)u(n) + bf_2(n)u(n) \Leftrightarrow aF_1(z) + bF_2(z) \qquad (9.30)$$

With this property, we may find Z transforms of signals that are linear combinations of signals having known transforms. Example 9.1 illustrates this property, and shows the plot of the function in time domain and its transform's pole-zero diagram in the z-plane.

Example 9.1
Find the Z transform of $x(n) = \{2 + 3\sin(n)\}u(n)$. Plot the function in the time domain and its pole-zero diagram in the z-plane.

To find $X(z)$, first express $x(n)$ as the sum of two functions whose transforms are known or can be readily obtained. Rewrite $x(n)$:

---

[h] This is called the *linear superposition property* of the Z transform. It combines the scaling property $(Z\{af(n)u(n)\} = aZ\{f(n)u(n)\})$ and the additive property $(Z\{f_1(n)u(n) + f_2(n)u(n)\} = Z\{f_1(n)u(n)\} + L\{f_2(n)u(n)\})$ into a single expression.

$$x(n) = 2u(n) + 3\sin(n)\,u(n) \qquad (9.31)$$

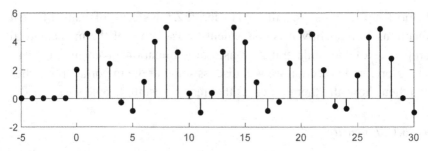

Figure 9.2. Plot of $x(n) = \{2 + 3\sin(n)\}u(n)$ vs. n, for $-5 \le n \le 30$, in Example 9.1.

Then let $a = 2, f_1(n) = u(n), b = 3$ and $f_2(n) = \sin(n)\,u(n)$. Now, from Table 9.1 of Laplace transforms we have:

$$u(n) \Leftrightarrow \frac{z}{z-1}, \qquad (9.32)$$

$$\sin(n)\,u(n) \Leftrightarrow \frac{\sin(1)z}{z^2 - 2\cos(1)z + 1}. \qquad (9.33)$$

And thus,

$$X(z) = \mathcal{Z}\{x(n)\} = 2\mathcal{Z}\{u(n)\} + 3\mathcal{Z}\{\sin(n)\,u(n)\}$$

$$= \frac{2z}{z-1} + \frac{3\sin(1)z}{z^2 - 2\cos(1)z + 1}$$

$$= \frac{2z^3 - (4\cos(1) + 3\sin(1))z^2 + (2 - 3\sin(1))z}{z^3 - (1 + 2\cos(1))z^2 + (1 + 2\cos(1))z - 1},$$

or $x(n) = \{2 + 3\sin(n)\}u(n) \Leftrightarrow \dfrac{2z^3 + 4.69z^2 - 0.524z}{z^3 - 2.08z^2 + 2.08z - 1} = X(z)$

$$= 2\frac{(z + 2.45)(z - 0.107)(z - 0)}{(z - (0.540 + j0.842))(z - (0.540 - j0.842))(z - 1)}. \qquad (9.34)$$

The numerator and denominator polynomials were factored using MATLAB's *roots* function. Figure 9.3 shows the location of poles and zeros of $X(z)$; a plot of this kind is called a *pole-zero diagram*. When $X(z)$ is rational, the pole-zero diagram describes it within a constant;

for completeness, this constant must also be specified as the *scale factor* of the diagram.

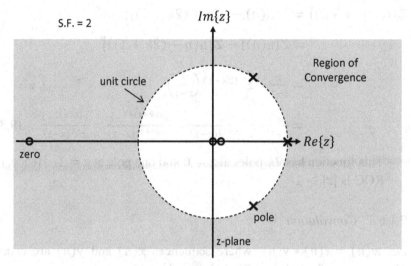

Figure 9.3. Pole-zero diagram for the $X(z)$ in Example 9.1.

9.1.6.2 *Time shifting*

When causal sequence $f(n)$ is shifted to the right by n_0 samples, the result is another causal sequence $f(n - n_0)$ whose Z transform is

$$\mathcal{Z}\{f(n - n_0)u(n)\} = \sum_{n=0}^{\infty} f(n - n_0)z^{-n}$$

$$= \sum_{m=-n_0}^{\infty} f(m)z^{-(m+n_0)} = \sum_{m=-n_0}^{\infty} f(m)u(m)z^{-m}z^{-n_0}$$

$$= z^{-n_0} \sum_{m=0}^{\infty} f(m)z^{-m} = z^{-n_0} F(z). \tag{9.35}$$

Thus, the effect of an n_0-sample delay is to scale the Z transform by z^{-n_0}. Advancing $f(n)$ n_0 samples to the left to give $f(n + n_0)$ will scale $F(z)$ by z^{n_0}, but to use the unilateral Z transform be sure that $f(n + n_0)$ is causal.

Example 9.2
Find the Z transform of $\text{rect}_k(n - k) = \text{rect}_k(n - k)u(n)$.

To solve this problem, note that the delayed rectangular pulse may be expressed as $u(n) - u(n - (2k + 1))$:

$$\mathcal{Z}\{rect_k(n - k)\} = \mathcal{Z}\{u(n) - u(n - (2k + 1))\}$$

$$= \mathcal{Z}\{u(n)\} - \mathcal{Z}\{u(n - (2k + 1))\}$$

$$= \frac{z}{z-1} - z^{-(2k+1)}\left(\frac{z}{z-1}\right)$$

$$= \frac{z}{z-1} - \frac{z}{z^{(2k+1)}(z-1)} = \frac{z^{(2k+1)+1} - z}{z^{(2k+1)}(z-1)} = \frac{z^{(2k+1)} - 1}{z^{2k}(z-1)}. \quad (9.36)$$

This function has $2k$ poles at $z = 0$ and one pole at $z = 1$. Hence, the ROC is $|z| > 1$.

9.1.6.3 *Convolution*

Let $w(n) = x(n) * y(n)$, where sequences $x(n)$ and $y(n)$ are causal. Then, solving for $W(z) = \mathcal{Z}\{x(n) * y(n)\}$:

$$W(z) = \sum_{n=0}^{\infty}\{x(n) * y(n)\}z^{-n} = \sum_{n=0}^{\infty}\{\sum_{m=-\infty}^{\infty} x(n - m)y(m)\}z^{-n}$$

$$= \sum_{m=-\infty}^{\infty} y(m)\{\sum_{n=0}^{\infty} x(n - m)z^{-n}\} = \sum_{m=-\infty}^{\infty} y(m)z^{-m}X(z)$$

$$= X(z)\sum_{m=-\infty}^{\infty} y(m)z^{-m} = X(z)Y(z). \quad (9.37)$$

As with Fourier transforms, we have the result that convolution in time results in multiplication in the z (complex frequency) domain.

9.1.6.4 *Time multiplication*

Let $w(n) = x(n)y(n)$, where sequences $x(n)$ and $y(n)$ are causal. Then, solving for $W(z) = \mathcal{Z}\{x(n)y(n)\}$:

$$W(z) = \sum_{n=0}^{\infty}\{x(n)y(n)\}z^{-n}$$

$$= \sum_{n=0}^{\infty}\left\{x(n)\left(\frac{1}{2\pi j}\oint_{\Gamma_y} Y(v)v^{n-1}dv\right)\right\}z^{-n}$$

$$= \frac{1}{2\pi j}\oint_{\Gamma_y} Y(v)\{\sum_{n=0}^{\infty} x(n)z^{-n} v^{n-1}\}dv$$

$$= \frac{1}{2\pi j} \oint_{\mathcal{T}_y} Y(v)\{\sum_{n=0}^{\infty} x(n) z^{-n} v^n \; v^{-1}\} dv$$

$$= \frac{1}{2\pi j} \oint_{\mathcal{T}_y} Y(v) \left\{\sum_{n=0}^{\infty} x(n) \left(\frac{z}{v}\right)^{-n}\right\} v^{-1} dv$$

$$= \frac{1}{2\pi j} \oint_{\mathcal{T}_y} Y(v) X\left(\frac{z}{v}\right) v^{-1} dv. \tag{9.38}$$

By interchanging $x(n)$ and $y(n)$, we would have obtained the equivalent result $W(z) = Z\{y(n)x(n)\} = \frac{1}{2\pi j} \oint_{\mathcal{T}_x} Y\left(\frac{z}{v}\right) X(v) v^{-1} dv$.

9.1.6.5 *Conjugation*

The Z transform of $f^*(n)$ is [i]

$$Z\{f^*(n)\} = \sum_{n=0}^{\infty} f^*(n) z^{-n} = \left(\sum_{n=0}^{\infty} f(n)(z^{-n})^*\right)^*$$

$$= \left(\sum_{n=0}^{\infty} f(n)(z^*)^{-n}\right)^* = F^*(z^*). \tag{9.39}$$

9.1.6.6 *Multiplication by n in the time domain*

The Z transform of $nf(n)$, when $f(n)$ is causal, is

$$Z\{nf(n)\} = \sum_{n=0}^{\infty} nf(n) z^{-n} = \sum_{n=0}^{\infty} nf(n) z^{-n-1} z$$

$$= -z \sum_{n=0}^{\infty} (-nf(n) z^{-n-1})$$

$$= -z \frac{d}{dz} \{\sum_{n=0}^{\infty} f(n) z^{-n}\} = -z \frac{dF(z)}{dz}. \tag{9.40}$$

This property may have been used to find the Z transform of the discrete ramp sequence $r(n)$:

$$Z\{r(n)\} = Z\{n \, u(n)\} = -z \frac{dZ\{u(n)\}}{dz} = -z \frac{d}{dz}\left(\frac{z}{z-1}\right)$$

$$= -z \left(\frac{-1}{(z-1)^2}\right) = \frac{z}{(z-1)^2}. \tag{9.41}$$

---

[i] This derivation relies on the property that $(z^m)^* = (z^*)^m$ when m is an integer.

9.1.6.7 *Multiplication by a^n in the time domain*

The Z transform of $a^n f(n)$, when $f(n)$ is causal, is

$$Z\{a^n f(n)\} = \sum_{n=0}^{\infty} a^n f(n) z^{-n}$$

$$= \sum_{n=0}^{\infty} f(n)(z/a)^{-n} = F(z/a). \tag{9.42}$$

9.1.6.8 *Backward difference*

Assume $f(n)$ is a causal sequence. Then, subtracting from it $f(n-1)$, which is another causal sequence, and transforming the result to the z domain, we obtain:

$$Z\{f(n) - f(n-1)\} = F(z) - z^{-1}F(z)$$

$$= F(z)(1 - z^{-1}) = F(z)\left(\frac{z-1}{z}\right). \tag{9.43}$$

For example:

$$u(n) \Leftrightarrow \frac{z}{z-1}$$

$$\therefore \; u(n) - u(n-1) \Leftrightarrow \frac{z}{z-1}\left(\frac{z-1}{z}\right) = 1$$

We could have also found $Z\{u(n) - u(n-1)\}$ this way:

$$u(n) \Leftrightarrow \frac{z}{z-1}$$

$$u(n-1) \Leftrightarrow z^{-1}\left(\frac{z}{z-1}\right) = \frac{1}{z-1}$$

$$u(n) - u(n-1) \Leftrightarrow \frac{z}{z-1} - \frac{1}{z-1} = 1$$

(Note that $u(n) - u(n-1) = \delta(n) \Leftrightarrow 1$.)

9.1.6.9 *Cumulative sum*

Assume $f(n)$ is a causal sequence whose cumulative sum is $\sum_{m=-\infty}^{n} f(m)$

$= \sum_{m=-\infty}^{n} f(m) \, u(m) = \sum_{m=0}^{n} f(m)$, which is equal to $f(n) * u(n)$. The Z transform of the cumulative sum of causal sequence $f(n)$ is therefore

$$\mathcal{Z}\{\sum_{m=0}^{n} f(m)\} = \mathcal{Z}\{f(n) * u(n)\}$$

$$= F(z) \left(\frac{z}{z-1}\right) = \frac{F(z)}{1-z^{-1}}. \qquad (9.44)$$

9.1.7 Table of Z transform properties

Table 9.3. Table of Z transform properties.

1. Linearity:	(Eq. (9.30))
$af_1(n) + bf_2(n) \Leftrightarrow aF_1(z) + bF_2(z)$	
2. Time shift:	(Eq. (9.35))
$f(n - n_0)u(n - n_0) \Leftrightarrow z^{-n_0} F(z), \quad n_0 > 0$	
3. Conjugation in time:	(Eq. (9.39))
$f^*(n) \Leftrightarrow F^*(z^*)$	
4. Multiplication by n:	(Eq. (9.40))
$nf(n) \, u(n) \Leftrightarrow -z \dfrac{dF(z)}{dz}$	
5. Multiplication by a^n:	(Eq. (9.42))
$a^n f(n) \, u(n) \Leftrightarrow F\left(\dfrac{z}{a}\right)$	

(*Continued*)

Table 9.3. (*Continued*)

6. Time multiplication:	(Eq. (9.38))

$$f_1(n)u(n) \times f_2(n)u(n) \Leftrightarrow \frac{1}{2\pi j}\oint_{\Gamma_1} F_1(v)F_2\left(\frac{z}{v}\right)v^{-1}dv$$

$$= \frac{1}{2\pi j}\oint_{\Gamma_2} F_2(v)F_1\left(\frac{z}{v}\right)v^{-1}dv$$

7. Time convolution:	(Eq. (9.37))

$$f_1(n) * f_2(n) \Leftrightarrow F_1(z) \times F_2(z)$$

8. Backward difference:	(Eq. (9.43))

$$f(n) - f(n-1) \Leftrightarrow F(z) \cdot (1 - z^{-1}) = F(z)\left(\frac{z-1}{z}\right)$$

9. Cumulative sum:	(Eq. (9.44))

$$\sum_{k=0}^{n} f(k)u(k) = f(n) * u(n) \Leftrightarrow F(z)\left(\frac{z}{z-1}\right) = \frac{F(z)}{(1-z^{-1})}$$

9.1.8 *Z transform of linear difference equations*

Recall, from Fig. 7.8 in Ch. 7, that some LSI discrete-time systems may be described in the time domain using linear difference equations:

$$y(n) = \sum_{k=0}^{M} b_k x(n-k) - \sum_{k=1}^{M} a_k y(n-k). \tag{9.45}$$

Taking the Z transform of both sides we obtain:

$$Y(z) = \sum_{k=0}^{M} b_k X(z)z^{-k} - \sum_{k=1}^{M} a_k Y(z)z^{-k}. \tag{9.46}$$

Then, solving for $H(z) = Y(z)/X(z)$:

$$H(z) = \left(\sum_{k=0}^{M} b_k z^{-k}\right) / \left(1 + \sum_{k=1}^{M} a_k z^{-k}\right). \tag{9.47}$$

This is the transfer function of the discrete-time system as expressed in the z-domain. Since it is the ratio of polynomials in z, it is called a *rational function*. Most practical applications of digital signal processing yield rational transfer functions, and the roots of numerator and especially the denominator in $H(z)$ offer great insight into system behavior. In Ch. 11 we will relate poles and zeros of $H(z)$ to the filtering characteristics and stability of the system that it describes.

9.1.9 *Inverse Z transform of rational functions*

Z transform pairs of causal sequences are tabulated in Table 9.2, and together with the Z transform properties given in Table 9.3, can be applied to find the inverse Z transforms of many z-domain expressions. In the special case that $F(z)$ is rational, we can systematically determine $Z^{-1}\{F(z)\}$ using other methods. These methods include long division and partial fraction expansion, which we overview in this section.

In our study of the unilateral Z transform, the function $f(n)$ is assumed to be a causal function. Thus, when finding the inverse transform of $F(z)$, we expect to recover sequence $f(n)$ that is causal. We emphasize this in Table 9.2 by writing: $f(n) = f(n)u(n)$.

If $Z^{-1}\{F(z)\}$ cannot be obtained using the methods discussed above, then one must evaluate the inverse Z transform contour integral (Eq. (9.5)) using techniques from the theory of complex variables. We shall not consider that here. Fortunately, applications in linear systems and signal analysis usually lead to z-domain functions that are rational.

9.1.9.1 *Inverse Z transform yielding finite-length sequences*

Whenever sequence $h(n)$ has finite length ($h(n) = 0$ outside of the range $n_1 \leq n \leq n_2$), then it may be written as a weighted sum of a finite number of time-shifted impulse functions:

$$h(n) = \sum_{k=n_1}^{n_2} h(k)\delta(n - k). \tag{9.48}$$

Since $Z\{c\delta(n - k)\} = cz^{-k}$, the Z transform of $h(n)$ is:

$$H(z) = \sum_{k=n_1}^{n_2} h(k)z^{-k}. \tag{9.49}$$

Thus, when $H(z)$ has this form, $h(n) = Z^{-1}\{H(z)\}$ is easily found by inspection. For example, recall from Ch. 7 (Eq. (7.42)) the difference equation describing a feed-forward (FIR) causal network in the time-domain:

$$y(n) = \sum_{k=0}^{M} b_k x(n-k).$$

Taking the Z transform and solving for $H(z) = Y(z)/X(z)$ we obtain:

$$H(z) = \sum_{k=0}^{M} b_k z^{-k}. \tag{9.50}$$

By inspection, we find the inverse Z transform of $H(z)$ to be

$$h(n) = \sum_{k=0}^{M} b_k \delta(n-k), \tag{9.51}$$

which is the impulse response of a causal M^{th}-order FIR digital filter.

9.1.9.2 *Long division method*

When $H(z) = N(z)/D(z)$ represents a system that is both causal and stable (Ch. 11), then it must be a *proper* transfer function: degree of $N(z)$ is less than or equal to the degree $D(z)$. In this case, we may order all polynomial coefficients in decreasing powers of z and perform long division to obtain a frequency-domain description that is easily convertible to the time domain.

We demonstrate the polynomial long division method with the following example.

<u>Example 9.3</u>
Let $H(z) =$

$$\frac{N(z)}{D(z)} = \frac{2z}{(8z^2 - 6z + 1)},$$

for which the condition that $H(z)$ be a proper transfer function is satisfied. Find the values of causal $h(n)$ for $0 \le n \le 4$ using the long division method.

Dividing numerator polynomial by denominator polynomial:

$$\frac{1}{4}z^{-1} + \frac{3}{16}z^{-2} + \frac{7}{64}z^{-3} + \frac{15}{256}z^{-4} + \cdots$$

$$8z^2 - 6z + 1 \overline{)\, 2z}$$

$$-\left(2z - \frac{3}{2} + \frac{1}{4}z^{-1}\right)$$

$$0 + \frac{3}{2} - \frac{1}{4}z^{-1}$$

$$-\left(\frac{3}{2} - \frac{9}{8}z^{-1} + \frac{3}{16}z^{-2}\right)$$

$$0 + \frac{7}{8}z^{-1} - \frac{3}{16}z^{-2}$$

$$-\left(\frac{7}{8}z^{-1} - \frac{21}{32}z^{-2} + \frac{7}{64}z^{-3}\right)$$

$$0 \quad + \frac{15}{32}z^{-2} - \frac{7}{64}z^{-3}$$

$$-\left(\frac{15}{32}z^{-2} - \frac{45}{128}z^{-3} + \frac{15}{256}z^{-4}\right)$$

$$0 \quad + \frac{31}{128}z^{-3} - \frac{15}{256}z^{-4}$$

Thus, $H(z)$ may be expressed as:

$$H(z) = \frac{2z}{8z^2 - 6z + 1} = \frac{1}{4}z^{-1} + \frac{3}{16}z^{-2} + \frac{7}{64}z^{-3} + \frac{15}{256}z^{-4} + \cdots$$

Taking the inverse Z transform yields:

$$h(n) = \frac{1}{4}\delta(n-1) + \frac{3}{16}\delta(n-2) + \frac{7}{64}\delta(n-3) + \frac{15}{256}\delta(n-4) + \cdots$$

The values of $h(n)$ for $0 \le n \le 4$ are:

$$h(0) = 0; \quad h(1) = \frac{1}{4}; \quad h(2) = \frac{3}{16}; \quad h(3) = \frac{7}{64}; \quad h(4) = \frac{15}{256}. \qquad (9.52)$$

A disadvantage of the long division method is that it does not produce a closed-form solution for $h(n)$, but instead calculates one coefficient at a time. Due to the recursive solution, computer methods used to perform long division will have an accumulating numerical round-off error.

9.1.9.3 *Partial fraction expansion method*

When the denominator of rational $H(z)$ can be accurately factored, *partial fraction expansion* may be used to obtain a closed-form solution for $h(n)$. The method is described in greater detail in Ch. 10, but there are some differences in the way it is implemented in the z-domain vs. the s-domain.

The method is basically this: as in the method of long division, $H(z) = N(z)/D(z)$ must be *proper* to describe a system that is both causal and stable. We assume $D(z)$ has degree N, and that its highest power term has scaling factor $a_N = 1$:

$$D(z) = z^N + a_{N-1}z^{N-1} + a_{N-2}z^{N-2} + \cdots + a_2z^2 + a_1z + a_0.$$

$D(z)$ is factored into a product of terms $(z - p_k)$ where p_k is one of its n roots:

$$H(z) = \frac{N(z)}{D(z)} = \frac{N(z)}{(z-p_1)(z-p_2)\cdots(z-p_N)}. \tag{9.53}$$

Next, divide both sides by z:

$$\frac{H(z)}{z} = \frac{N(z)/z}{(z-p_1)(z-p_2)\cdots(z-p_N)}. \tag{9.54}$$

Perform partial fraction expansion on $H(z)/z$:

$$\frac{H(z)}{z} = \frac{C_1}{(z-p_1)} + \frac{C_2}{(z-p_2)} + \cdots + \frac{C_N}{(z-p_N)}, \tag{9.55}$$

$$\text{where } C_k = (z - p_k)\frac{H(z)}{z}\bigg|_{z=p_k} \quad \text{for } k = 1, 2, \cdots, N.$$

(We have assumed that all roots $\{p_1, p_2, ..., p_N\}$ are unique. See Example 10.6 in Ch. 10 for how to handle the case of repeated roots.)

Multiply both sides by z to obtain an expression for $H(z)$:

$$H(z) = \frac{C_1 z}{(z-p_1)} + \frac{C_2 z}{(z-p_2)} + \cdots + \frac{C_N z}{(z-p_N)}. \tag{9.56}$$

Using Z transform pair $p^n u(n) \Leftrightarrow z/(z - p)$, $|z| > |p|$, in Table 9.2, together with the linearity property of the Z transform, take the inverse Z transform of both sides to obtain a closed-form solution for $h(n)$:

$$h(n) = C_1 p_1^n u(n) + C_2 p_2^n u(n) + \cdots + C_N p_N^n u(n). \tag{9.57}$$

We demonstrate the partial fraction expansion method with an example. Let $H(z) = 2z/(8z^2 - 6z + 1)$, for which the condition that $H(z)$ be proper is satisfied. Divide numerator and denominator by 8 to normalize the denominator's highest-power term scaling factor to 1:

$$H(z) = \frac{2z}{8z^2 - 6z + 1} = \frac{\frac{1}{4}z}{z^2 - \frac{3}{4}z + \frac{1}{8}} = \frac{N(z)}{D(z)}.$$

Divide both sides by z, factor $D(z)$, and expand using partial fractions:

$$\frac{H(z)}{z} = \frac{\frac{1}{4}}{z^2 - \frac{3}{4}z + \frac{1}{8}} = \frac{\frac{1}{4}}{\left(z - \frac{1}{2}\right)\left(z - \frac{1}{4}\right)} = \frac{C_1}{\left(z - \frac{1}{2}\right)} + \frac{C_2}{\left(z - \frac{1}{4}\right)}.$$

Solve for residues C_1 and C_2:

$$C_1 = \left(z - \frac{1}{2}\right) \frac{H(z)}{z} \bigg|_{z=\frac{1}{2}} = \frac{\frac{1}{4}}{\left(z - \frac{1}{4}\right)} \bigg|_{z=\frac{1}{2}} = 1,$$

$$C_2 = \left(z - \frac{1}{4}\right) \frac{H(z)}{z} \bigg|_{z=\frac{1}{4}} = \frac{\frac{1}{4}}{\left(z - \frac{1}{2}\right)} \bigg|_{z=\frac{1}{4}} = -1,$$

therefore $\dfrac{H(z)}{z} = \dfrac{1}{\left(z - \frac{1}{2}\right)} + \dfrac{-1}{\left(z - \frac{1}{4}\right)}$, and $H(z) = \dfrac{z}{\left(z - \frac{1}{2}\right)} - \dfrac{z}{\left(z - \frac{1}{4}\right)}.$

$\mathcal{Z}^{-1}\{H(z)\}$ is found by applying a Z transform pair from Table 9.2:

$$h(n) = \left(\frac{1}{2}\right)^n u(n) - \left(\frac{1}{4}\right)^n u(n). \qquad (9.58)$$

This was the same $H(z)$ as used in the polynomial division example; note that $h(n)$ found here matches the values of $h(n)$ for $0 \le n \le 4$ that were calculated there (Eq. (9.52)):

$$n = 0: \quad \left(\frac{1}{2}\right)^0 - \left(\frac{1}{4}\right)^0 = 0,$$

$$n = 1: \quad \left(\frac{1}{2}\right)^1 - \left(\frac{1}{4}\right)^1 = \frac{1}{4},$$

$$n = 2: \quad \left(\frac{1}{2}\right)^2 - \left(\frac{1}{4}\right)^2 = \frac{3}{16},$$

$$n = 3: \quad \left(\frac{1}{2}\right)^3 - \left(\frac{1}{4}\right)^3 = \frac{7}{64},$$

$$n = 4: \qquad \left(\tfrac{1}{2}\right)^4 - \left(\tfrac{1}{4}\right)^4 = \frac{15}{256}.$$

To conclude, we mention the case of non-repeating complex-conjugate roots:

$$H(z) = \frac{C_1 z}{(z-p)} + \frac{C_2 z}{(z-p^*)}. \qquad (9.59)$$

It can be shown that C_1 and C_2 will be complex conjugates of one another when the coefficients of $N(z)$ and $D(z)$ are real (as is the case in practical applications). Therefore, $C_1 = C_r + jC_i$ and $C_2 = C_r - jC_i$. Also, let $p = p_r + jp_i = re^{j\omega_0}$. Rewriting the expression:

$$H(z) = \frac{C_1 z}{(z-p)} + \frac{C_2 z}{(z-p^*)} = \frac{C_1 z(z-p^*) + C_2 z(z-p)}{(z-p)(z-p^*)} = \frac{C_1 z^2 - C_1 p^* z + C_2 z^2 - C_2 p z}{z^2 - z(p+p^*) + pp^*}$$

$$= \frac{z^2(C_1+C_2) - z(C_1 p^* + C_2 p)}{z^2 - z 2 p_r + r^2} = \frac{z^2 2 C_r - 2z(C_r p_r + C_i p_i)}{z^2 - 2 z p_r + r^2}$$

$$= \frac{2 C_r z^2 - 2z(C_r r \cos(\omega_0) + C_i r \sin(\omega_0))}{z^2 - 2 z r \cos(\omega_0) + r^2},$$

$$H(z) = 2 C_r \frac{z^2 - r \cos(\omega_0)\, z}{z^2 - 2 z r \cos(\omega_0) + r^2} - 2 C_i \frac{r \sin(\omega_0)\, z}{z^2 - 2 z r \cos(\omega_0) + r^2}. \qquad (9.60)$$

Taking the inverse Z transform via Table 9.2, entries 10 and 11 (p. 380), we obtain:

$$h(n) = 2 C_r r^n \cos(\omega_0 n) u(n) - 2 C_i r^n \sin(\omega_0 n)\, u(n)$$

$$= 2 r^n (C_r \cos(\omega_0 n) - C_i \sin(\omega_0 n))\, u(n) \qquad (9.61)$$

$$= 2 r^n |C_{1,2}| \cos(\omega_0 n + \tan^{-1}(C_i/C_r))\, u(n). \qquad (9.62)$$

As an example, a causal system has $H(z) = \dfrac{2z^2 + 4.53z}{z^2 - 1.13z + 0.64}$. We find $h(n)$ using the inverse Z transform method of partial fraction expansion:

$$\frac{H(z)}{z} = \frac{2z + 4.53}{z^2 - 1.13z + 0.64} = \frac{2z + 4.53}{\big(z - (0.566 + j0.566)\big)\big(z - (0.566 - j0.566)\big)},$$

$$\frac{H(z)}{z} = \frac{C_1}{z - (0.566 + j0.566)} + \frac{C_2}{z - (0.566 - j0.566)},$$

$$C_1 = \frac{2z+4.53}{z-(0.566-j0.566)}\bigg|_{z=0.566+j0.566} = 1 - 5j,$$

$$\left\{ C_2 = \frac{2z+4.53}{z-(0.566+j0.566)}\bigg|_{z=0.566-j0.566} = 1 + 5j \right\},$$

$$\therefore H(z) = \frac{(1-5j)z}{z-(0.566+j0.566)} + \frac{(1+5j)z}{z-(0.566-j0.566)}.$$

Note that $C_r = 1$, $C_i = -5$, $r = |0.566 + j0.566| = 0.8$, $\omega_0 = \angle p = \frac{\pi}{4}$.

Therefore, from Eq. (9.61),

$$h(n) - 2(0.8)^n \left(\cos\left(\frac{\pi}{4}n\right) + 5 \sin\left(\frac{\pi}{4}n\right) \right) u(n).$$

A plot of impulse response $h(n)$, over time index values $-5 \leq n \leq 30$, is shown in Fig. 9.4.

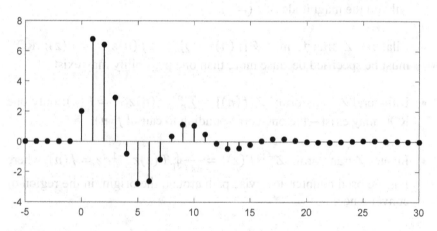

Figure 9.4. Plot of impulse response sequence calculated in the previous example: $h(n) = 2(0.8)^n \left(\cos(\pi n/4) + 5 \sin(\pi n/4) \right) u(n)$.

9.2 Chapter Summary and Comments

- The Z transform $F(z) = \mathcal{Z}\{f(n)\}$ is a generalization of the Discrete-Time Fourier transform $F(e^{j\omega}) = \mathcal{F}\{f(n)\}$.

- The z-domain is still the "frequency" domain, although $e^{j\omega}$ is generalized to be $re^{j\omega} = z$ ($r \in \mathcal{R}^+$). That results in the complex exponential signal $(e^{j\omega})^n = e^{j\omega n}$ (magnitude 1, angle ωn) becoming $(re^{j\omega})^n = r^n e^{j\omega n}$ (magnitude r^n, angle ωn). Instead of decomposing real-valued sequences into a sum of constant-amplitude sinusoidal sequences (as does the DTFT), the Z transform decomposes time signals into a sum of exponentially-decaying or growing sinusoidal sequences.

- Region of Convergence: the ROC is the range of $r = |z|$ where the summation defining $F(z) = \mathcal{Z}\{f(n)\}$ converges. The ROC depends only on the magnitude of z ($= r$).

- Bilateral Z transform: $\mathcal{Z}\{f(n)\} = \sum_{n=-\infty}^{\infty} f(n)z^{-n} = F(z)$; ROC must be specified because more than one possibility may exist.

- Unilateral Z transform: $\mathcal{Z}\{f(n)\} = \sum_{n=0}^{\infty} f(n)z^{-n} = F(z)$; only one ROC may exist – the one corresponding to causal $f(n)$.

- Inverse Z transform: $\mathcal{Z}^{-1}\{F(z)\} = \frac{1}{2\pi j} \oint_\Gamma F(z)z^{n-1}dz = f(n)$, where Γ is a closed counterclockwise path around the origin, in the region of convergence.

- Z transform pairs: we focus on causal $f(n)$, since that is the case for impulse response of a causal system.

- $H(z)$ is said to be *rational* when it is the ratio of polynomials in z: $H(z) = N(z)/D(z)$.

- Z transform pairs for the case of simple poles of rational $H(z)$:
  Causal: $a^n u(n)$ $\Leftrightarrow z/(z-a)$, ROC: $r = |z| > |a|$,
  Anticausal: $a^n(u(n) - 1) \Leftrightarrow z/(z-a)$, ROC: $r = |z| < |a|$.

- The ROC of $H(z) = \mathcal{Z}\{f(n)\}$ for causal $f(n)$ is $|z| > r_0$, where r_0 is the highest-magnitude pole in $H(z)$.

- The ROC of $H(z) = \mathcal{Z}\{f(n)\}$ for anticausal $f(n)$ is $|z| < r_0$, where r_0 is the lowest-magnitude pole in $H(z)$.

- Z transform properties are very much like the properties of the discrete-time Fourier transform (linearity, convolution in time is multiplication in z-domain, etc.).

- The inverse Z transform of rational $H(z)$ can be done via table look-up, long division, partial fraction expansion, or direct evaluation of the contour integral expression.

- Digital networks (formed of delay units, adders and multiplication by constants) gives finite-order constant-coefficient difference equations, which when transformed to the z-domain give rational transfer functions $H(z)$.

- Here are some advantages of the Z transform vs. discrete-time Fourier transform:

  a) $\mathcal{Z}\{f(n)\}$ may converge when $\mathcal{Z}\{f(n)\}$ does not, making it possible to analyze signals such as the ramp sequence,
  b) The Z transform makes it possible to analyze *unstable* systems when $\mathcal{Z}\{h(n)\}$ exists but $\mathcal{F}\{h(n)\}$ does not,
  c) When $H(z)$ is rational, its poles give us information about system stability,
  d) When $H(z)$ is rational, its pole/zero locations give us information that is useful for designing and analyzing digital filters.

- A one-sample delay unit has transfer function $H(z) = z^{-1}$.

9.3 Homework Problems

P9.1 When $(z) = \frac{z}{z+0.5}$, what ROC corresponds to causal $f(n)$?

P9.2 When $(z) = \frac{z}{z-2}$, what ROC corresponds to anticausal $f(n)$?

P9.3 Find $f(n)$ when $F(z) = \frac{z}{z-0.5}$ (ROC for causal $f(n)$).

P9.4 Find $f(n)$ when $F(z) = \frac{z}{z+2}$ (ROC for anticausal $f(n)$).

P9.5 Given: $F(z) = \frac{z}{z+0.5} + \frac{z}{z-2}$. What ROC's are possible?

P9.6 Given: $F(z) = \frac{2z}{2z+0.5} - \frac{3z}{3z-2}$. What ROC's are possible?

P9.7 Given: $F(z) = \frac{12z^2+2z}{4z^2-1}$. Find $f(n)$ when it is causal.

P9.8 Given: $(z) = \frac{z^2+z}{10z^2-10z+6}$. Draw the pole-zero diagram, with S.F.

P9.9 Find the Z transform of $f(n) = (0.5)^{n-1}u(n-1)$.

P9.10 Find the Z transform of $f(n) = (0.9)^n \cos(\pi n/4)\, u(n)$.

P9.11 Find the Z transform of $y(n) = x(n) - \sum_{k=1}^{M} a_k y(n-k)$.

P9.12 Find the Z transform of $f(n) = (0.5)^n \text{rect}_5(n-5)$.

P9.13 Find causal impulse response $h(n)$, for $0 \le n \le 4$, using the long division method, when $H(z) = \frac{z^2+6z}{6z^2+z-1}$.

P9.14 Find causal impulse response $h(n)$, for all n, using the partial fraction expansion method, when $H(z) = \frac{z^2+6z}{6z^2+z-1}$.

P9.15 Find causal impulse response $h(n)$, for all n, using the partial fraction expansion method, when $H(z) = \frac{z^2}{z^2-z+0.5}$.

Chapter 10

S-Domain Signal Processing

10.1 Theory

10.1.1 *Introduction*

There are several reasons, for system analysis purposes, to consider a generalization of the Fourier transform to include several important functions, such as the unit step, growing exponential and ramp. As is sometimes done in mathematics, science and engineering literature, this generalization can be achieved by extending real variable ω in the Fourier transform to $\omega = (\omega_1 + j\omega_2)$, defining a complex ω plane. With proper choice of imaginary part ω_2, the Fourier integral may be made to converge absolutely, and the Fourier transform of many growing functions can be defined. Instead, in system and signal analysis, one prefers to use Laplace transforms based on the pioneering work of Laplace (1749-1827) in his research in many branches of mathematical physics.

Another reason is that it provides formal mathematical justification to the s-domain technique for solution of constant coefficient linear integro-differential equations. The technique, attributed to the independent work of Heaviside (1850-1925), is a powerful method to solve many linear system problems modeled as differential equations. To provide mathematical rigor to successful s-domain technique of Heaviside to solve many engineering problems, the mathematicians of that era discovered the integral transform work previously carried out by Laplace. The variable $s = \sigma + j\omega$ defines a complex s-plane that is rotated by 90 degrees from that defined by considering complex ω of the previous paragraph. The transform integral defined substituting s for $j\omega$ is called the *bilateral* Laplace transform integral. When a lower limit of 0 is chosen in this definition, appropriate for causal functions, the transform is termed a *unilateral* Laplace transform (or simply Laplace transform, as common in the literature). This approach of using complex frequency s is quite familiar to all electrical engineers, and we will use that method of extending the Fourier transform pair defined in Eqs. (5.1) and (5.2).

10.1.2 *Laplace transform*

Laplace transform analysis is based on the work of Pierre-Simon Laplace dealing with integral transforms. We will view the Laplace transform as a mapping of a real signal $f(t)$ in the time domain to its uniquely associated complex function $F(s)$ in the complex (frequency) domain. The mathematical relationship between functions $f(t)$ and $F(s)$ is as follows:

Bilateral Laplace transform

$$F(s) = \int_{-\infty}^{\infty} f(t)e^{-st}dt, \tag{10.1}$$

Inverse Laplace transform

$$f(t) = \frac{1}{2\pi j}\int_{\sigma_0 - j\infty}^{\sigma_0 + j\infty} F(s)e^{st}ds. \tag{10.2}$$

Or, as written in shorthand notation:

$$F(s) = \mathcal{L}\{f(t)\}, \quad f(t) = \mathcal{L}^{-1}\{F(s)\}. \tag{10.3}$$

The value σ_0 in the limits defining the contour of integration in the complex s-plane in the inversion formula cannot be automatically determined by the behavior of the complex function $F(s)$ in s-plane. Suffice it to say that it is related to the value of σ necessary for the convergence of the integral in Eq. (10.1). Direct evaluation of the inversion integral of Eq. (10.2) generally requires knowledge of the theory of complex variables. However, we can find the inverse Laplace transform of many functions $F(s)$ encountered in signal and system analysis using Heaviside's partial fraction expansion method. We shall discuss this in more detail in the next section on the unilateral Laplace transform, which is of primary interest to us.

We can define the unilateral Laplace transform for causal functions as follows. Since causal $f(t) = 0$ for $t < 0$, its bilateral Laplace transform becomes:

Unilateral Laplace transform	$F(s) = \int_{0^-}^{\infty} f(t)e^{-st}dt,$	(10.4)

Inverse Laplace transform for causal $f(t)$	$f(t)u(t) = \frac{1}{2\pi j}\int_{\sigma_0-j\infty}^{\sigma_0+j\infty} F(s)e^{st}ds.$	(10.5)

To avoid any confusion, unless otherwise stated, we will assume to be dealing with the unilateral Laplace transform in this text. When needed, we will indicate this by using the shorthand notation:

$$F(z) = \mathcal{L}\{f(t)u(t)\}, \quad f(t)u(t) = \mathcal{L}^{-1}\{F(s)\}. \qquad (10.6)$$

Notice that the lower limit in the definition of Laplace transform in Eq. (10.4) is 0^-. This is chosen to ensure that the basic signals such as $u(t)$, $\delta(t)$ and functions that are discontinuous at $t = 0$ will have the discontinuity included in the definition. We will also see an advantage of this in Ch. 12, when we apply the Laplace transforms to solving problems dealing with networks and systems.

We are interested in signals whose Laplace transforms are *invertible*, meaning that both the Laplace transform integral and its inverse exist to give a unique mapping between $f(t)u(t)$ and $F(s)$. When that is true, the following notation is used:

$$f(t) \Leftrightarrow F(s). \qquad (10.7)$$

Practically speaking, signals that may be measured and recorded are those having invertible Laplace transforms. Engineers often call these *real-world* signals in the sense that they are piecewise continuous, have finite amplitudes, and have finite energy. The Laplace transforms of such signals are invertible.[a]

We should point out that both the Fourier transform and the Laplace transform are among a class of integral transforms that form powerful techniques for solution of many problems in engineering and science. In some cases, the two independent variables t and s have different physical interpretations. In some applications, transform analysis is applied to multidimensional functions in both domains.

10.1.3 *Region of convergence*

In the definitions of both bilateral and unilateral Laplace transforms, specific values of $Re(s) = \sigma$ may be necessary for the integrals to converge. The range of σ values leading to convergence defines a region in the complex s-plane, which is called the *region of convergence*, or ROC. For the bilateral Laplace transform, the ROC is the strip $\sigma_1 < Re\{s\} < \sigma_2$. Boundary lines $Re(s) = \sigma_1$ and $Re(s) = \sigma_2$ are not part of the ROC except in certain cases.[b] It should be stressed that a specification of the ROC is necessary to completely describe $\mathcal{L}\{f(t)\}$, because different $f(t)$ may have the same expression for $F(s)$ but different regions of convergence. For example, for $a \in \mathbb{R}^+$:

Causal:
$$e^{-at}u(t) \Leftrightarrow \frac{1}{s+a}, \quad \text{ROC: } Re\{s\} > -a, \quad (10.8)$$

Anticausal:
$$e^{-at}(u(t) - 1) \Leftrightarrow \frac{1}{s+a}, \quad \text{ROC: } Re\{s\} < -a. \quad (10.9)$$

The unilateral Laplace transform $F(s)$ is defined only for values of $Re(s) > \sigma_0$. In this region, function $F(s)$ is analytic. For a definition of

---

[a] Identifying the class of signals whose Laplace transforms exist and are invertible is an advanced mathematical topic. The interested reader is referred to references LePage, Papoulis, Bracewel, and Bateman.

[b] When $f(t)$ is anticausal, $\sigma_1 = -\infty$ and the ROC is $-\infty \le Re\{s\} < \sigma_2$, when $f(t)$ is causal, $\sigma_2 = \infty$ and the ROC is $\sigma_1 < Re\{s\} \le \infty$.

analytic functions, reader should refer to a text on complex variables such as Churchill or Pennisi.

The line $Re(s) = \sigma_0$ is called *abscissa of convergence*, and the region in the s-plane to the right of this line is the ROC. The value σ_0 can be negative, zero or positive depending on the behavior of the function $f(t)$ at infinity. For example, the unilateral Laplace transform integral (Eq. (10.4)) will converge for exponentially-growing function $e^{-at}u(t)$ only when $Re\{s\} = \sigma > -a$, as given in Eq. (10.8) (assumes $a > 0$). Therefore, the abscissa of convergence for this function is $\sigma_0 = -a$, and the region of convergence is the semi-infinite s-plane defined by $Re\{s\} > -a$. This is illustrated in Fig. 10.1.

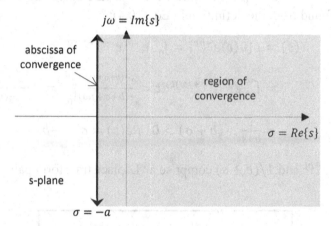

Figure 10.1. Region of convergence of the Laplace transform of $f(t) = e^{-at}u(t)$.

The Laplace transform of every causal function will have an associated region of convergence specified by σ_0. (If the ROC is an empty set, then the Laplace transform of that function is not defined.) We will see this when we evaluate the Laplace transforms of several important signals in the following section.

10.1.4 *Laplace transforms of basic signals*

Let us now derive, step-by-step, the Laplace transforms of a few of the basic functions previously introduced in Ch. 3. For many other functions, it is straightforward to find the transform by evaluating Eq. (10.4), which may be simplified with the help of a table of integrals.[c] In other instances, applying Laplace transform properties is an excellent tool available to us. Examples worked out in the following section will provide experience in such techniques.

10.1.4.1 *Exponentially decaying signal*

Let $x(t) = u(t)e^{-bt}$ (assuming real $b > 0$). Its Laplace transform $X(s)$ is then found from the definition in Eq. (10.4): [d]

$$X(s) = \mathcal{L}\{u(t)e^{-bt}\} = \int_{0^-}^{\infty} e^{-bt}e^{-(\sigma+j\omega)t}\,dt$$

$$= \int_{0^-}^{\infty} e^{-(b+\sigma+j\omega)t}\,dt = \frac{e^{-(b+\sigma+j\omega)t}}{-(b+\sigma+j\omega)}\bigg|_{0^-}^{\infty} = 0 - \frac{1}{-(b+\sigma+j\omega)}$$

$$= \frac{1}{b+s}, \quad (b+\sigma) > 0, \ Re(s) = \sigma > -b, \qquad (10.10)$$

or $u(t)e^{-bt}$ and $1/(b+s)$ comprise a "Laplace transform pair":

$$\boxed{u(t)e^{-bt} \Leftrightarrow \frac{1}{s+b} \qquad \text{ROC: } Re\{s\} > -b} \qquad (10.11)$$

The abscissa of convergence is $Re(s) = -b$, and the region of convergence is $Re(s) > -b$. The s-domain function $X(s) = 1/(s+b)$ is singular (goes to infinity) at $s = -b$. This location (a point) in the s-plane is called a *pole* of the complex function $X(s)$. Also, a point where $X(s) = 0$ is called a *zero* of $X(s)$.

---

[c] Abramowitz and Stegun, 1964; Dwight, 1961.
[d] In this derivation, note that $\lim_{t\to\infty} e^{-(\sigma+b+j\omega)t} = \lim_{t\to\infty} e^{-(\sigma+b)t}e^{-j\omega t} = 0 \cdot \lim_{t\to\infty} e^{-j\omega t} = 0$, since $|e^{-j\omega t}| = 1 < \infty \ \forall t$.

Note that if $b > 0$, the abscissa of convergence lies to the left of the imaginary ($j\omega$) axis. If $b < 0$, the axis lies to the right of $j\omega$-axis. For the case of $b = 0$, the abscissa of convergence coincides with the $j\omega$-axis and the region of convergence is the entire right half plane.

10.1.4.2 *Impulse function*

Finding the Laplace transform of an impulse function is straightforward. Substituting $f(t) = \delta(t)$ in Eq. (10.4), and evaluating the integral using the sifting property of an impulse function, we obtain

$$F(s) = \int_{0^-}^{\infty} \delta(t)\, e^{-st}\, dt = \int_{0^-}^{\infty} \delta(t)\, e^{-s \cdot 0}\, dt$$

$$= \int_{0^-}^{\infty} \delta(t)\, dt = 1 \tag{10.12}$$

Note that the choice of lower limit of 0^- in the definition of the Laplace transform includes the impulse function within the limits of integration. You may find some older textbooks using 0^+ as the lower limit leading to a different result for the transform of an impulse function. We will see the advantage of our choice when we apply the Laplace transform to linear system analysis. Thus,

$$\boxed{\delta(t) \Leftrightarrow 1 \quad \text{ROC: } \forall s} \tag{10.13}$$

The transformed function $F(s) = 1$ converges in the entire s-plane. It has no poles.

10.1.4.3 *Delayed impulse function*

Similarly, and just as easily, we may find the Laplace transform of delayed impulse $\delta(t - t_0)$:[e]

---

[e] Note that t_0 must be positive (shift to the right) for $\delta(t - t_0)$ to be causal. The unilateral Laplace transform is always used only for causal signals.

$$\mathcal{F}\{\delta(t - t_0)\} = \int_{0^-}^{\infty} \delta(t - t_0)e^{-st}\,dt = \int_{0^-}^{\infty} \delta(t - t_0)e^{-st_0}\,dt$$
$$= e^{-st_0}\int_{0^-}^{\infty}\delta(t - t_0)\,dt = e^{-st_0} \qquad (10.14)$$

This gives us the Laplace transform pair:

$$\boxed{\delta(t - t_0) \Leftrightarrow e^{-st_0} \qquad \text{ROC: } \forall s} \qquad (10.15)$$

10.1.4.4 *Unit step function*

As we did for the Fourier transform, the Laplace transform of $u(t)$ can be readily found by considering it as the limit of an exponentially decaying function with $b \to 0$ (p. 159). Alternatively, we can just evaluate the integral as follows:

$$F(s) = \int_{0^-}^{\infty} u(t)\,e^{-st}\,dt = \int_{0^-}^{\infty} 1\,e^{-st}\,dt = \left.\frac{e^{-st}}{-s}\right|_{t=0^-}^{t=\infty}$$

$$= 0 - \frac{1}{-s} = \frac{1}{s}, \quad \sigma > 0. \qquad (10.16)$$

The Laplace transform $1/s$ has a pole at $s = 0$; thus, the abscissa of convergence is $\sigma = 0$, and the region of convergence is the half-plane to the right of the $j\omega$-axis. The Laplace transform pair is:

$$\boxed{u(t) \Leftrightarrow \frac{1}{s} \qquad \text{ROC: } Re\{s\} > 0} \qquad (10.17)$$

10.1.4.5 *Complex exponential function*

The Laplace transform of complex exponential signal $e^{j\omega_0 t}u(t)$, which is essentially a phasor at frequency ω_0 for $t \geq 0$, can be readily obtained by substituting $b = -j\omega_0$ in Laplace transform pair given in Eq. 10.11. Since the phasor signal is bounded by magnitude 1, the value of σ must be greater than zero for the integral to converge.

$$\boxed{e^{j\omega_0 t}u(t) \Leftrightarrow \frac{1}{s - j\omega_0} \qquad \text{ROC: } Re\{s\} > 0} \qquad (10.18)$$

Clearly, the transformed function has an imaginary pole at $s = j\omega_0$.

10.1.4.6 *Sinusoid*

The Laplace transform of $\cos(\omega_0 t)\,u(t)$ is derived by expressing it in terms of complex exponential functions using Euler's identity:

$$\cos(\omega_0 t)u(t) = \left(\tfrac{1}{2}e^{j\omega_0 t} + \tfrac{1}{2}e^{-j\omega_0 t}\right)u(t). \qquad (10.19)$$

Then, taking advantage of Laplace transform pairs $e^{\pm j\omega_0 t} \Leftrightarrow \frac{1}{s\mp j\omega_0}$:

$$\mathcal{L}\{\cos(\omega_0 t)u(t)\} = \tfrac{1}{2}\mathcal{L}\{(e^{j(\omega_0 t)} + e^{-j(\omega_0 t)})u(t)\}$$

$$= \tfrac{1}{2}\mathcal{L}\{e^{j\omega_0 t}u(t)\} + \tfrac{1}{2}\mathcal{L}\{e^{-j\omega_0 t}u(t)\}$$

$$= \frac{1/2}{s-j\omega_0} + \frac{1/2}{s+j\omega_0} = \frac{s}{s^2+\omega_0^2}, \qquad (10.20)$$

or

$$\boxed{\cos(\omega_0 t)u(t) \Leftrightarrow \frac{s}{s^2+\omega_0^2} \qquad \text{ROC: } Re\{s\} > 0} \qquad (10.21)$$

In this case the transformed function has conjugate imaginary poles at $s = \pm j\omega_0$, and a zero at $s = 0$. The abscissa of convergence and region of convergence are determined by the location of the poles (they are on $j\omega$-axis), and the location of the zero in the s-plane has no effect on the region of convergence.[f] Following the above process for $\cos(\omega_0 t)\,u(t)$, we can readily determine the Laplace transform of $\sin(\omega_0 t)\,u(t)$:

$$\sin(\omega_0 t)\,u(t) = \left(\tfrac{1}{2j}e^{j\omega_0 t} - \tfrac{1}{2j}e^{-j\omega_0 t}\right)u(t)$$

$$\therefore \mathcal{L}\{\sin(\omega_0 t)\,u(t)\} = \frac{1/2j}{s-j\omega_0} - \frac{1/2j}{s+j\omega_0} \qquad (10.22)$$

This results in the transform pair:

$$\boxed{\sin(\omega_0 t)u(t) \Leftrightarrow \frac{\omega_0}{s^2+\omega_0^2} \qquad \text{ROC: } Re\{s\} > 0} \qquad (10.23)$$

The transformed function has poles at $s = \pm j\omega_o$ but does not have a zero at $s = 0$ as did $\mathcal{L}\{\cos(\omega_0 t)u(t)\}$.

---

[f] Only poles determine the region of convergence.

10.1.4.7 *Ramp function*

The Laplace transform of a ramp function $r(t) = tu(t)$ is derived by direct integration, applying the definition given by Eq. (10.4):

$$F(s) = \int_{0^-}^{\infty} t\, u(t)\, e^{-st} dt = \int_{0^-}^{\infty} t\, e^{-st} dt$$

$$= t\left.\frac{e^{-st}}{-s}\right|_{t=0^-}^{t=\infty} - \int_{0^-}^{\infty} \frac{e^{-st}}{-s} dt = 0 - 0 - \left.\frac{e^{-st}}{s^2}\right|_{t=0^-}^{t=\infty}$$

$$= 0 - 0 - 0 + \frac{1}{s^2} = \frac{1}{s^2}, \quad \sigma > 0 \qquad (10.24)^g$$

We now have the Laplace transform pair:

$$\boxed{t\, u(t) = r(t) \Leftrightarrow \frac{1}{s^2} \qquad \text{ROC: } Re\{s\} > 0} \qquad (10.25)$$

Observe that the transformed function has a multiple (second order) pole at $s = 0$. Repeating the integration by parts n times gives the Laplace transform pair:

$$\boxed{t^n u(t) \Leftrightarrow \frac{n!}{s^{n+1}} \qquad \text{ROC: } Re\{s\} > 0} \qquad (10.26)$$

The resulting function will have $n + 1$ poles at $s = 0$.

10.1.5 *Table of Laplace transforms*

Instead of solving the Laplace transforms for commonly used signals each time, you may refer to the table of Laplace transform pairs shown in Table 10.1. Each entry has a corresponding abscissa of convergence and region of convergence, both of which may easily be determined from the pole locations in the transfer function: the abscissa of convergence is a line parallel to the $j\omega$ axis passing through the right-most pole of the transfer function, and the region of convergence is the right-half plane to the right of the abscissa of convergence.

---

g Integration by parts.

Table 10.1. Table of Laplace transform pairs. (region of convergence is for a causal time signal).

Entry #	$f(t) = f(t)u(t)$	$F(s)$	Refer to Eq. #
1.	$= \dfrac{1}{2\pi j} \displaystyle\int_{\sigma_0 - j\infty}^{\sigma_0 + j\infty} F(s)e^{st}\,ds$	$= \displaystyle\int_{0^-}^{\infty} f(t)e^{-st}\,dt$	(10.5), (10.4)
2.	$e^{-at}u(t)$	$\dfrac{1}{s + a}$	(10.11)
3.	$\delta(t)$	1	(10.13)
4.	$\delta(t - t_0)$	e^{-st_0}	(10.15)
5.	$u(t)$	$\dfrac{1}{s}$	(10.17)
6.	$e^{j\omega_0 t}u(t)$	$\dfrac{1}{s - j\omega_0}$	(10.18)
7.	$\cos(\omega_0 t)\,u(t)$	$\dfrac{s}{s^2 + \omega_0^2}$	(10.21)
8.	$\sin(\omega_0 t)\,u(t)$	$\dfrac{\omega_0}{s^2 + \omega_0^2}$	(10.23)
9.	$t\,u(t) = r(t)$	$\dfrac{1}{s^2}$	(10.25)

(Continued)

Table 10.1. (*Continued*)

Entry #	$f(t) = f(t)u(t)$	$F(s)$	Refer to Eq. #
10.	$t^n u(t)$ $(n \geq 0)$	$\dfrac{n!}{s^{n+1}}$	(10.26)
11.	$e^{at} \cos(\omega_0 t)\, u(t)$	$\dfrac{s-a}{(s-a)^2 + \omega_0^2}$	LePage
12.	$e^{at} \sin(\omega_0 t) u(t)$	$\dfrac{\omega_0}{(s-a)^2 + \omega_0^2}$	LePage
13.	$\cos(\omega_0 t + \theta) u(t)$	$\dfrac{s \cos(\theta) - \omega_0 \sin(\theta)}{s^2 + \omega_0^2}$	Lathi
14.	$e^{at} \cos(\omega_0 t + \theta)\, u(t)$	$\dfrac{(s-a) \cos(\theta) - \omega_0 \sin(\theta)}{s^2 - 2as + (a^2 + \omega_0^2)}$	Lathi
15.	$\text{rect}\left(\dfrac{t-T/2}{T}\right) u(t)$ pulse width $= T$ sec	$\dfrac{1 - e^{-sT}}{s}$	(10.29)

10.1.6 *Laplace transform properties*

In most applications, we will need to derive Laplace transforms of many different functions. Frequently it is convenient to use some of the mathematical properties of the Laplace integral given in Eq. (10.4). Therefore, in this section we will discuss some of the important properties and then show all other properties that may be useful (Table 10.2).

10.1.6.1 *Linearity*

Let us consider two different functions of time, $f_1(t)$ and $f_2(t)$, whose Laplace transforms are known to be $F_1(s)$ and $F_2(s)$, respectively:

$$f_1(t)\,u(t) \Leftrightarrow F_1(s), \quad f_2(t)\,u(t) \Leftrightarrow F_2(s)$$

Then from Eq. (10.4), for two arbitrary constants a and b, the following relationship[h] holds true:

$$af_1(t)u(t) + bf_2(t)u(t) \Leftrightarrow aF_1(s) + bF_2(s). \qquad (10.27)$$

With this property, we may find Laplace transforms of signals that are linear combinations of signals having known transforms. The following example illustrates this and shows the plot of time-domain function and its transform's pole-zero diagram in the s-plane.

Example 10.1
Find the Laplace transform of $x(t) = \{2 + 3\sin(t)\}u(t)$. Plot the function in time and the poles/zeros of $X(s)$ in the s-plane.

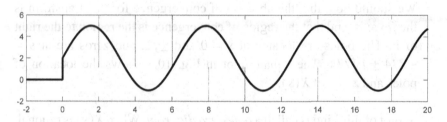

Figure 10.2. A plot of time-domain signal $x(t) = \{2 + 3\sin(t)\}u(t)$ in Example 10.1.

The function $x(t)$ is plotted in Fig. 10.2. To find $X(s)$, express $x(t)$ as the sum of two functions whose transforms are known or can be readily obtained. Rewrite $x(t)$:

---

[h] This is called the *linear superposition property* of the Laplace transform. It combines the scaling property ($\mathcal{L}\{af(t)u(t)\} = a\mathcal{L}\{f(t)u(t)\}$) and the additive property ($\mathcal{L}\{f_1(t)u(t) + f_2(t)u(t)\} = \mathcal{L}\{f_1(t)u(t)\} + \mathcal{L}\{f_2(t)u(t)\}$) into a single expression.

$$x(t) = 2u(t) + 3\sin(t)\,u(t)$$

Then let $a = 2, f_1(t) = u(t);\ b = 3$, and $f_2(t) = \sin(t)\,u(t)$ in Eq. 10.27. From Table 10.1 of Laplace transforms we have:

$$u(t) \Leftrightarrow \frac{1}{s} \qquad\qquad \text{(Table 10.1, Entry \#5)},$$

$$\sin(t)\,u(t) \Leftrightarrow \frac{1}{s^2+1} \qquad\qquad \text{(Table 10.1, Entry \#8)}.$$

And thus,

$$X(s) = \mathcal{L}\{x(t)\} = 2\mathcal{L}\{u(t)\} + 3\mathcal{L}\{\sin(t)\,u(t)\}$$

$$= \frac{2}{s} + \frac{3}{s^2+1} = \frac{2s^2 + 3s + 2}{s(s^2+1)}\,,$$

or $\qquad\qquad \{2 + 3\sin(t)\}u(t) \Leftrightarrow \dfrac{2s^2+3s+2}{s(s^2+1)}$

$$= 2\,\frac{\left(s-(-3/4+j\sqrt{7}/4)\right)\left(s-(-3/4-j\sqrt{7}/4)\right)}{(s-0)(s+j)(s-j)}.$$

We should note that the abscissa of convergence for this transform is the $j\omega$ axis, and that the region of convergence is the region to the right of it. The poles are located at $s = 0$ and $\pm j1$, and zeros are at $s = -3/4 \pm j\sqrt{7}/4$. The s-plane plot in Fig. 10.3 shows the location of poles and zeros of $X(s)$.

A plot of this kind is called a *pole-zero diagram*. When $X(s)$ is rational, the pole-zero diagram describes the function $X(s)$ within a constant; for completeness, this constant must also be specified as the *scale factor* of the diagram.[i]

Note that plotting complex-valued $X(s)$ vs. complex-valued s involves four dimensions, which would be difficult to accomplish on two-dimensional paper in any other way!

---

[i] In this plot the scale factor is 2, since $X(s) = \dfrac{2s^2+3s+2}{s(s^2+1)} = 2\left(\dfrac{s^2+1.5s+1}{s(s^2+1)}\right) = 2\dfrac{(s-z_1)(s-z_2)}{s(s-p_1)(s-p_2)}.$

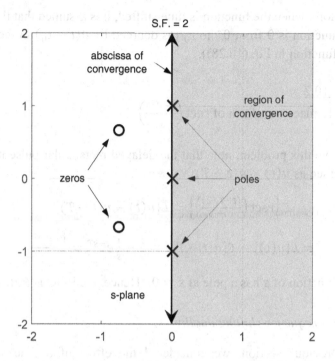

Figure 10.3. Pole-zero plot of $X(s) = (2s^2 + 3s + 2)/(s(s^2 + 1))$, the Laplace transform of time-domain signal $x(t) = \{2 + 3\sin(t)\}u(t)$ in Example 10.1.

10.1.6.2 *Time shifting*

In practice, quite often we encounter a signal whose Laplace transform is known but for the fact that it is delayed in time. Here we ask a question – is there a simple relationship between the two Laplace transforms? As with Fourier transforms, we can readily show that for time delay $t_0 > 0$,

$$f(t - t_0)\, u(t - t_0) \Leftrightarrow F(s)e^{-st_0}, \quad t_0 > 0. \qquad (10.28)$$

Caution must be exercised when time shifting prior to calculating the Laplace transform. It is necessary that the entire non-zero part of the function $f(t)$ is included in the integration from 0^- to ∞ in order the above result to be valid. This property is not valid for time advance, since in general a portion of the time advanced function may occur before $t = 0$ and may not be included in the integration limits of Laplace transform.

Furthermore, when the function is time shifted, it is assumed that the time shifted function is 0 from 0^- to t_0 (as denoted by $u(t - t_0)$ in the time-domain function in Eq. (10.28)).

Example 10.2

Find the Laplace transform of $\text{rect}\left(\frac{t-T/2}{T}\right)$.

To solve this problem, note that the delayed rectangular pulse may be expressed as $u(t) - u(t - T)$:

$$\mathcal{L}\left\{\text{rect}\left(\frac{(t-T/2)}{T}\right)\right\} = \mathcal{L}\{u(t) - u(t - T)\}$$

$$= \mathcal{L}\{u(t)\} - \mathcal{L}\{u(t - T)\} = \frac{1}{s} - \frac{1}{s}e^{-sT} = \frac{1-e^{-sT}}{s} \qquad (10.29)$$

This function of s has a pole at $s = 0$. Hence, the ROC is $Re\{s\} > 0$.

10.1.6.3 *Frequency shifting duality*

In the previous section, we considered the effect of a time shifting operation on the Laplace-transformed function. Now we consider a dual operation: an operation in the time domain that results in a complex frequency shift of transfer function in the s-domain. We consider the complex time domain function $f(t)e^{\pm s_0 t}$, where s_0 is a complex constant. The Laplace transform of such a function is determined by carrying out the integration using Eq. (10.4). The resulting transform pair is:

$$f(t)e^{\pm s_0 t} \Leftrightarrow F(s \mp s_0) \qquad (10.30)$$

10.1.6.4 *Time scaling*

Another interesting property of the Laplace transform is that of scaling in time. We consider a signal $f(t)$ and scale independent variable t by constant factor $a > 0$, resulting in signal $f(at)$. Depending on the numerical value of a, signal $f(t)$ is either expanded or compressed in time.[j] Unlike the case of Fourier transforms, negative values of a are not acceptable since that would result in time reversal of the function, making it non-causal. The unilateral Laplace transform requires the function to be nonzero only for t ≥ 0.

Finding the Laplace transform of $f(at)$:

$$\mathcal{L}\{f(at)\} = \int_{0^-}^{\infty} f(at)e^{-st}dt = \int_{\beta/a=0^-}^{\beta/a=\infty} f(\beta)e^{-s(\beta/a)}d(\beta/a)$$

$$= \int_{\beta=0^-}^{\beta=\infty} f(\beta)e^{-s(\beta/a)}\frac{1}{a}d\beta = \frac{1}{a}\int_{0^-}^{\infty} f(\beta)e^{-\beta(s/a)}d\beta = \frac{1}{a}F\left(\frac{s}{a}\right).$$

Or,
$$f(at) \Leftrightarrow \frac{1}{a}F\left(\frac{s}{a}\right). \quad (a > 0). \tag{10.31}$$

10.1.6.5 *Convolution*

Many signal analysis applications require either multiplying or convolving together two signals. As we shall now see, multiplication and convolution are closely related through the Laplace transform. What may be a difficult-to-simplify convolution of two signals in one domain is simply expressed as a multiplicative product in the other domain. Here we show the correspondence between multiplicative products and "convolution products." If $f_1(t) \Leftrightarrow F_1(s)$ and $f_2(t) \Leftrightarrow F_2(s)$, then:

$$f_1(t) * f_2(t) \Leftrightarrow F_1(s)F_2(s). \tag{10.32}[k]$$

10.1.6.6 *Time multiplication*

$$f_1(t)f_2(t) \Leftrightarrow \frac{1}{2\pi j} F_1(s) * F_2(s). \tag{10.33}$$

---

[j] $f(at)$ with $0 < a < 1$ is an expansion of $f(t)$, by factor a, away from $t = 0$; $f(at)$ with $a > 1$ is a compression of $f(t)$ in time, by factor a, toward $t = 0$.

[k] Proof: $\mathcal{L}\{f_1(t) * f_2(t)\} = \int_{0^-}^{\infty}(\int_{0^-}^{\infty} f_1(\tau)f_2(t-\tau)d\tau)e^{-st}dt =$
$\int_{0^-}^{\infty} f_1(\tau)(\int_{0^-}^{\infty} f_2(t-\tau)e^{-st}dt)d\tau = \int_{0^-}^{\infty} f_1(\tau)\mathcal{L}\{f_2(t-\tau)\}d\tau =$
$\int_{0^-}^{\infty} f_1(\tau)(e^{-s\tau}F_2(s))d\tau = \int_{0^-}^{\infty} f_1(\tau)e^{-s\tau}d\tau \times F_2(s) = F_1(s) \times F_2(s).$

(See footnote[1] for a proof.)

Example 10.3

Find the Laplace transform of:

$$x(t) = \text{rect}\left(\frac{t}{2} - \frac{1}{2}\right) u(t) * \sin(\pi t) u(t).$$

Solving: since $f_1(t) = \text{rect}\left(\frac{t-1}{2}\right) u(t) \Leftrightarrow \frac{1-e^{-2s}}{s}$, and $f_2(t) = \sin(\pi t) u(t) \Leftrightarrow \frac{\pi}{s^2+\pi^2}$, then

$$X(s) = \mathcal{L}\{f_1(t) * f_2(t)\} = F_1(s) \times F_2(s) = \frac{\pi(1-e^{-2s})}{s(s^2+\pi^2)}.$$

Taking the inverse Laplace transform, we have

$$\mathcal{L}^{-1}\left\{\frac{\pi(1-e^{-2s})}{s(s^2+\pi^2)}\right\} = \mathcal{L}^{-1}\left\{\frac{\pi}{s(s^2+\pi^2)}\right\} - \mathcal{L}^{-1}\left\{\frac{\pi e^{-2s}}{s(s^2+\pi^2)}\right\}$$

$$= f(t)u(t) - f(t-2)u(t-2),$$

where[m] $f(t)u(t) = \mathcal{L}^{-1}\left\{\frac{\pi}{s(s^2+\pi^2)}\right\} = \mathcal{L}^{-1}\left\{\frac{1}{\pi}\left(\frac{1}{s} - \frac{s}{s^2+\pi^2}\right)\right\}$

$$= \mathcal{L}^{-1}\left\{\frac{1}{\pi}\left(\frac{1}{s}\right)\right\} - \mathcal{L}^{-1}\left\{\frac{1}{\pi}\left(\frac{s}{s^2+\pi^2}\right)\right\} = \frac{1}{\pi}\{u(t) - \cos(\pi t) u(t)\}$$

$$= \frac{1}{\pi}u(t)\{1 - \cos(\pi t)\}.$$

Therefore, the convolution product is

$$x(t) = \frac{1}{\pi}u(t)\{1 - \cos(\pi t)\} - \frac{1}{\pi}u(t-2)\{1 - \cos(\pi(t-2))\}.$$

---

[1] Proof: $\mathcal{L}\{f_1(t)f_2(t)\} = \int_{0^-}^{\infty} f_1(t)f_2(t)e^{-st}dt = \int_{0^-}^{\infty} \mathcal{L}^{-1}\{F_1(s)\}f_2(t)e^{-st}dt = $
$\int_{0^-}^{\infty}\left(\frac{1}{2\pi j}\int_{\sigma_0-j\infty}^{\sigma_0+j\infty} F_1(\beta)e^{\beta t}d\beta\right) f_2(t)e^{-st}dt = $
$\frac{1}{2\pi j}\int_{\sigma_0-j\infty}^{\sigma_0+j\infty} F_1(\beta) \int_{0^-}^{\infty} e^{\beta t}f_2(t)e^{-st}dt\, d\beta = $
$\frac{1}{2\pi j}\int_{\sigma_0-j\infty}^{\sigma_0+j\infty} F_1(\beta) \int_{0^-}^{\infty} f_2(t)e^{-(s-\beta)t}dt\, d\beta = $
$\frac{1}{2\pi j}\int_{\sigma_0-j\infty}^{\sigma_0+j\infty} F_1(\beta)F_2(s-\beta)\, d\beta = \frac{1}{2\pi j}F_1(s) * F_2(s).$

[m] Using partial fraction expansion discussed in a later section.

Simplifying further: because $\cos(\pi t)$ has period $T_0 = 2$ sec, we see that $\cos(\pi(t-2)) = \cos(\pi t)$:

$$x(t) = \frac{1}{\pi}u(t)\{1 - \cos(\pi t)\} - \frac{1}{\pi}u(t-2)\{1 - \cos(\pi t)\}$$

$$= \frac{1}{\pi}\{1 - \cos(\pi t)\}\{u(t) - u(t-2)\}$$

$$= \frac{1}{\pi}\{1 - \cos(\pi t)\}\,\text{rect}\left(\frac{t-1}{2}\right).$$

10.1.6.7 *Time differentiation*

The relationship between the Laplace transform of a function and its time derivative is of considerable importance when analyzing linear systems that are described by differential equations. In the following we derive this relationship and then extend it to higher-order derivatives of a function $f(t)$.

Consider a Laplace transform pair $f(t) \Leftrightarrow F(s)$. Then, starting with Eq. (10.4) and integrating by parts,

$$\mathcal{L}\left\{\frac{df}{dt}\right\} = \int_{0^-}^{\infty} e^{-st}\left(\frac{df}{dt}\right) dt = [e^{-st}f(t)]_{0^-}^{\infty} + s\int_{0^-}^{\infty} f(t)e^{-st}\,dt$$

$$= -f(0^-) + sF(s),$$

we have the transform pair:

$$\frac{df}{dt} \Leftrightarrow sF(s) - f(0^-). \tag{10.34}$$

By repeated integration by parts, we may readily extend the result to higher order derivatives of function $f(t)$. We obtain the following transform pairs:

$$\frac{d^2f}{dt^2} \Leftrightarrow s^2F(s) - sf(0^-) - \left[\frac{df}{dt}\right]_{t=0^-}$$

$$\frac{d^2f}{dt^2} \Leftrightarrow s^2F(s) - sf(0^-) - f^{(1)}(0^-). \tag{10.35}$$

In general, the nth order derivative gives

$$\frac{d^n f}{dt^n} \Leftrightarrow s^n F(s) - s^{n-1} f(0^-) - \cdots - f^{(n-1)}(0^-). \qquad (10.36)$$

10.1.6.8 *Time integration*

Solution of integro-differential equations may require knowledge of the relationship between a transform of a function and the transform of the time integrated function. It is convenient to obtain this relationship, which we derive here.

Given a Laplace transform pair $f(t) \Leftrightarrow F(s)$. Let $g(t) = \int_{0^-}^{\infty} f(\tau) d\tau$. Therefore $dg/dt = f(t)$, $f(0^-) = 0$, and $\mathcal{L}\{dg/dt\} = \mathcal{L}\{f(t)\}$. From Eq. (10.34), we obtain $sG(s) = F(s)$ and the Laplace transform pair:

$$\int_{0^-}^{t} f(\beta) d\beta \Leftrightarrow \frac{1}{s} F(s). \qquad (10.37)$$

10.1.7 *Table of Laplace transform properties*

Table 10.2. Table of Laplace transform properties.

1. Linearity:	(Eq. (10.27))
$af_1(t) + bf_2(t) \Leftrightarrow aF_1(s) + bF_2(s)$	
2. Time scaling:	(Eq. (10.31))
$f(at) \Leftrightarrow \frac{1}{a} F\left(\frac{s}{a}\right), \qquad a > 0$	
3. Time shift:	(Eq. (10.28))
$f(t - t_0)u(t - t_0) \Leftrightarrow F(s)e^{-st_0}, t_0 > 0$	

(Continued)

Table 10.2. (*Continued*)

4. Modulation:	See footnote[n]
$$f(t)\cos(\omega_0 t) \Leftrightarrow \frac{1}{2}F(s - j\omega_0) + \frac{1}{2}F(s + j\omega_0)$$	
5. Modulation (arbitrary carrier phase) :	See footnote[o]
$$f(t)\cos(\omega_0 t + \phi) \Leftrightarrow \frac{e^{j\phi}}{2}F(s - j\omega_0) + \frac{e^{-j\phi}}{2}F(s + j\omega_0)$$	
6. Frequency shift:	(Eq. (10.30))
$$f(t)e^{\pm s_0 t} \Leftrightarrow F(s \mp s_0)$$	
7. Time multiplication:	(Eq. (10.33))
$$f_1(t) \times f_2(t) \Leftrightarrow \frac{1}{2\pi j}\{F_1(s) * F_2(s)\}$$	
8. Time convolution:	(Eq. (10.32))
$$f_1(t) * f_2(t) = F_1(s) \times F_2(s)$$	
9. Time differentiation:	(Eq. (10.34))
$$\frac{df(t)}{dt} \Leftrightarrow s\,F(s) - f(0^-)$$	
10. Time differentiation:	(Eq. (10.36))
$$\frac{d^n f}{dt^n} \Leftrightarrow s^n F(s) - s^{n-1}f(0^-) - s^{n-2}f^{(1)}(0^-) - \cdots - f^{(n-1)}(0^-)$$	
11. Time integration:	(Eq. (10.37))
$$\int_{0^-}^{t} f(\beta)d\beta \Leftrightarrow \frac{F(s)}{s}$$	

---

[n] These can be derived using the frequency shifting property.

10.1.8 *Inverse Laplace transform of rational functions*

The Laplace transform pairs tabulated in Table 10.1, together with the Laplace transform properties in Table 10.2, may be sufficient to find inverse Laplace transforms of simple functions via table look-up. But in problems of engineering interest, $F(s)$ will most likely be a *rational function* of s. We may systematically solve problems of this type by algebraically manipulating $F(s)$ using partial fraction expansion, followed by substituting entries from Table 10.1. We discuss this method in the following sections.

In our study of the unilateral Laplace transform $\mathcal{L}\{f(t)\}$, function $f(t)$ is assumed to be a causal function, which we explicitly denote by writing $f(t)u(t)$. Thus, when we determine the inverse transform of $F(s)$, it is expected that it will recover a causal function $f(t) = f(t)u(t)$. We have chosen to emphasize this in Table 10.1 by including $u(t)$ with $f(t)$ in the Laplace transform pairs.

If the inverse Laplace transform cannot be obtained as discussed above, then one must resort to the evaluation of inverse transform integral using integration techniques appropriately developed in the theory of complex variables. We shall not consider that here. Fortunately, applications in linear system and signal analysis usually require dealing with rational $F(s)$.[o]

10.1.8.1 *Partial fraction expansion method*

Rational functions are the ratio of two polynomials: $F(s) = P(s)/Q(s)$. If the degree of the polynomial $P(s)$ is less than that of $Q(s)$, then $F(s)$ is called a *strictly proper rational function*. If $P(s)$ and $Q(s)$ have the same

---

[o] It is interesting to note that one of the most basic Laplace transforms yields an irrational form: $\mathcal{L}\{\delta(t - t_0)\} = e^{-st_0}$. Because e^{-st_0} cannot be expressed as a ratio of finite-order polynomials in s, a pure delay cannot be realized using lumped-element circuits. Fortunately, every realizable linear circuit has transfer function that may be expressed as a rational function $F(s)$.

degree, then $F(s)$ is called a *proper rational function*. Otherwise, when degree $P(s) >$ degree $Q(s)$, $F(s)$ is called *improper*. Note that by a long division process, an improper rational function can be reduced to a sum of a strictly proper rational function and some terms forming a polynomial in powers of s. The polynomial in s may be inverse Laplace transformed using Table 10.2 entries #9 and #10. This section describes how to find the inverse Laplace transform of a strictly proper rational function in s.

Given strictly proper rational function $F(s) = P(s)/Q(s)$. Polynomial $Q(s)$ can be factored into products of terms $(s - s_k)$, where $s = s_k$ is a root of the polynomial. This root in general may be complex. From a fundamental theorem of algebra, the number of roots an nth-degree polynomial has is exactly n. If the coefficients of the polynomial are all real, then any complex roots must occur in complex conjugate pairs. It is also possible that some roots may be repeated. In the following section, we discuss three possible types of roots and the method of partial fraction expansion for each case: (1) simple real roots, (2) complex roots, and (3) repeated roots.

Simple real roots:

Let the nth-degree denominator polynomial $Q(s)$ have distinct roots $s_1, s_2, \cdots s_n$. Then we can write

$$Q(s) = (s - s_1)(s - s_2) \cdots (s - s_n) \qquad (10.38)$$

and
$$F(s) = \frac{P(s)}{Q(s)} = \frac{P(s)}{(s-s_1)(s-s_2)\cdots(s-s_n)}, \qquad (10.39)$$

which can be written as sum of partial fractions

$$F(s) = \frac{P(s)}{Q(s)} = \frac{C_1}{(s-s_1)} + \frac{C_2}{(s-s_2)} + \cdots + \frac{C_n}{(s-s_n)}, \qquad (10.40)$$

where $C_k = (s - s_k)F(s)|_{s=s_k}$ for $k = 1, 2, \cdots, n$.

Example 10.4

Express the following function in partial fraction form and then find causal $f(t)$: $F(s) = (s + 3)/(s^2 + 3s + 2)$

Solution: Roots of the denominator are $s = -1, -2$.

$$F(s) = \frac{s+3}{s^2+3s+2} = \frac{s+3}{(s+1)(s+2)} = \frac{C_1}{s+1} + \frac{C_2}{s+2},$$

$$C_1 = (s+1)F(s)|_{s=-1} = \left.\frac{s+3}{s+2}\right|_{s=-1} = \frac{-1+3}{-1+2} = 2,$$

$$C_2 = (s+2)F(s)|_{s=-2} = \left.\frac{s+3}{s+1}\right|_{s=-2} = \frac{-2+3}{-2+1} = -1,$$

$$\therefore F(s) = \frac{2}{s+1} + \frac{-1}{s+2}.$$

Using Table 10.1 and the linearity property, we find the inverse Laplace transform of $F(s)$ to be $f(t) = (2e^{-t} - e^{-2t})u(t)$.

Simple complex roots:

We shall study this case through a numerical example. The procedure is general and can be applied to any complex roots.

Example 10.5

Express the following rational function in partial fraction form, and then find the corresponding causal $f(t)$:

$$F(s) = \frac{(6s-14)}{(s^2+6s+25)}.$$

Solution: The roots of the denominator polynomial are complex and can be readily determined from the quadratic formula; they are $s = -3 \pm j4$.

$$F(s) = \frac{6s-14}{s^2+6s+25} = \frac{6s-14}{\left(s-(-3+j4)\right)\left(s-(-3-j4)\right)}$$

$$= \frac{C_1}{s+3-j4} + \frac{C_2}{s+3+j4},$$

$$C_1 = (s + 3 - j4)F(s)\big|_{s=-3+j4} = \frac{6s-14}{s+3+j4}\bigg|_{s=-3+j4}$$

$$= \frac{6(-3+j4)-14}{(-3+j4)+3+j4} = \frac{-32+j24}{j8} = 3 + j4.$$

If we are confident of above result, we can use the fact that $C_2 = C_1^*$. Or, we can solve for C_2 independently to obtain

$$C_2 = (s + 3 + j4)F(s)\big|_{s=-3-j4} = \frac{6s-14}{s+3-j4}\bigg|_{s=-3-j4}$$

$$= \frac{6(-3-j4)-14}{(-3-j4)+3-j4} = \frac{-32-j24}{-j8} = 3 - j4.$$

$$\therefore F(s) = \frac{3+j4}{s+3-j4} + \frac{3-j4}{s+3+j4}$$

From Table 10.1, the inverse Laplace transform is

$$f(t) = \{(3 + j4)e^{(-3+j4)t} + (3 - j4)e^{(-3-j4)t}\}u(t).$$

Using Euler's identity and some algebraic simplification we get:

$$f(t) = \{6e^{-3t}\cos(4t) - 8e^{-3t}\sin(4t)\}u(t) \qquad \text{(see footnote}^p)$$

$$\text{or } f(t) = 10e^{-3t}\cos(4t + 53.13°)\, u(t). \qquad \text{(see footnote}^q)$$

It may be shown that for simple complex conjugate poles $s_{1,2} = a \pm j\omega_0$ we obtain complex-conjugate residues $C_{1,2} = C_r \pm jC_i$, which leads us to a useful formula for obtaining a time-domain solution:

$$\mathcal{L}^{-1}\left\{\frac{C_r+jC_i}{(s-(a+j\omega_0))} + \frac{C_r-jC_i}{(s-(a-j\omega_0))}\right\} = \mathcal{L}^{-1}\left\{\frac{2C_r(s-a)}{(s-a)^2+\omega_0^2} + \frac{-2C_i\omega_0}{(s-a)^2+\omega_0^2}\right\}$$

$$= 2e^{at}(C_r\cos(\omega_0 t) - C_i\sin(\omega_0 t))\, u(t) \qquad (10.41)$$

$$= 2e^{at}|C_{1,2}|\cos(\omega_0 t + \tan^{-1}(C_i/C_r))\, u(t). \qquad (10.42)$$

---

p Rearranging $F(s)$ in the form suitable for using entries # 11 and 12 of Table 10.1 we could have arrived at this result.

q This form of $f(t)$ is obtained using entry #14 of Table 10.1.

Repeated roots:

We will consider the case of repeated real roots. The process can be readily extended to complex repeated roots. A general case of a k-times repeated root is discussed following the example.

Example 10.6

Find causal signal $f(t)$ whose Laplace transform is $(s) = \frac{2s^2+7s+4}{(s+1)(s+2)^2}$.

Solution: $F(s) = \frac{2s^2+7s+4}{(s+1)(s+2)^2} = \frac{C_1}{s+1} + \frac{C_{21}}{s+2} + \frac{C_{22}}{(s+2)^2}$,

where $C_1 = (s+1)F(s)|_{s=-1} = \frac{2(-1)^2+7(-1)+4}{(-1+2)^2} = -1$,

$$C_{22} = (s+2)^2 F(s)|_{s=-2} = \frac{2(-2)^2+7(-2)+4}{(-2+1)} = 2,$$

$$C_{21} = \frac{d}{ds}\{(s+2)^2 F(s)\}\Big|_{s=-2} = \frac{d}{ds}\left\{\frac{2s^2+7s+4}{(s+1)}\right\}\Big|_{s=-2}$$

$$= \frac{(4s+7)(s+1)-(2s^2+7s+4)}{(s+1)^2}\Big|_{s=-2} = \frac{(-8+7)(-2+1)-(8-14+4)}{(-2+1)^2} = 3,$$

therefore $F(s) = \frac{2s^2+7s+4}{(s+1)(s+2)^2} = \frac{-1}{s+1} + \frac{3}{s+2} + \frac{2}{(s+2)^2}$.

From entries #2 and #9 in Table 10.1 we obtain the inverse Laplace transform:

$$f(t) = \{-e^{-t} + 3e^{-2t} + 2te^{-2t}\} u(t).$$

If the function $F(s)$ has repeated roots in the denominator, then its form is:

$$F(s) = \frac{P(s)}{Q(s)} = \frac{P(s)}{(s-s_1)^k}. \tag{10.43}$$

For a k-times repeated root, the partial fraction expansion is given by:

$$F(s) = \frac{P(s)}{Q(s)} = \frac{C_1}{(s-s_1)^k} + \frac{C_2}{(s-s_1)^{k-1}} + \cdots + \frac{C_k}{(s-s_1)}. \tag{10.44}$$

Coefficients $C_1, C_2, \cdots C_k$ are obtained as follows:

$$C_1 = (s - s_1)^k F(s)\big|_{s=s_1},\tag{10.45}$$

$$C_2 = \frac{d}{ds}\{(s - s_1)^k F(s)\}\big|_{s=s_1},\tag{10.46}$$

$$C_3 = \frac{d^2}{ds^2}\{(s - s_1)^k F(s)\}\big|_{s=s_1}.\tag{10.47}$$

In general,

$$C_m = \frac{1}{(m-1)!} \frac{d^{m-1}}{ds^{m-1}}\{(s - s_1)^k F(s)\}\big|_{s=s_1} ; \quad m = 1, 2, \cdots, k.\tag{10.48}$$

With repeated roots the inverse Laplace transform is given by:

$$f(t) = \{C_k e^{s_1 t} + C_{k-1} t e^{s_1 t} + \cdots + C_2 t^{k-2} e^{s_1 t} + C_1 t^{k-1} e^{s_1 t}\} u(t)\tag{10.49}$$

For $k > 3$, finding a partial fraction expansion may require much work; it is advisable to use MATLAB® functions to determine the coefficients in such cases.

Shortcuts:

Although the Heaviside method of partial fraction expansion discussed above is popular, in some instances it is useful to find alternate way to arrive at the partial fraction expansion coefficients. Given an expression for $F(s)$, the equality holds for all values of s and it may be expeditious to evaluate both sides of the expression at a set of values of s that are not poles or zeros of the function $F(s)$. In particular, $s = 0$ and $s = \infty$ are excellent choices.

Example 10.7

Find the inverse Laplace transform of the following function:

$$F(s) = \frac{3s^2+5}{(s+2)(s^2+6s+25)}.$$

Solution: $F(s) = \dfrac{3s^2+5}{(s+2)(s^2+6s+25)} = \dfrac{C_1}{s+2} + \dfrac{C_{11}s+C_{12}}{s^2+6s+25}$,

$$C_1 = (s+2)F(s)\big|_{s=-2} = \dfrac{3s^2+5}{s^2+6s+25}\Big|_{s=-2} = 1,$$

$$\therefore \dfrac{3s^2+5}{(s+2)(s^2+6s+25)} = \dfrac{1}{s+2} + \dfrac{C_{11}s+C_{12}}{s^2+6s+25}.$$

Evaluating above equation on both sides at $s = 0$ and $s = 1$, we obtain

$$\dfrac{5}{50} = \dfrac{1}{2} + \dfrac{C_{12}}{25},$$

$$\dfrac{8}{(3)(32)} = \dfrac{1}{3} + \dfrac{C_{11}+C_{12}}{32}.$$

Solving the above two equations, we obtain $C_{12} = -10$ and $C_{11} = 2$. The partial fraction expansion is:

$$F(s) = \dfrac{3s^2+5}{(s+2)(s^2+6s+25)} = \dfrac{1}{s+2} + \dfrac{2s-10}{s^2+6s+25}.$$

Although we could use the previously-discussed inversion method for complex roots, we will make use of entries #11 and #12 in Table 10.1 to find the inverse:

$$F(s) = \dfrac{1}{s+2} + \dfrac{2s-10}{s^2+6s+25}$$

$$= \dfrac{1}{s+2} + 2\,\dfrac{(s+3)}{(s+3)^2+(4)^2} - 4\,\dfrac{(4)}{(s+3)^2+(4)^2}.$$

From Laplace transform table entries #11 and #12, using $a = -3$ and $\omega_0 = 4$, we obtain:

$$f(t) = \{e^{-2t} + 2e^{-3t}\cos(4t) - 4e^{-3t}\sin(4t)\}\,u(t)$$

10.2 Chapter Summary and Comments

- The Laplace Transform $F(s) = \mathcal{L}\{f(t)\}$ is a generalization of the Fourier Transform $F(\omega) = \mathcal{F}\{f(t)\}$.

- The s-domain is still the "frequency" domain, although $j\omega$ is generalized to be $s = \sigma + j\omega$ ($\sigma \in \mathcal{R}$). That results in the complex exponential signal $e^{(j\omega)t}$ (magnitude 1, angle ωt) becoming $e^{(\sigma+j\omega)t} = e^{\sigma t}e^{j\omega t}$ (magnitude $e^{\sigma t}$, angle ωt). Instead of decomposing real-valued time signals into a sum of constant-amplitude sinusoids (as does the CTFT), the Laplace Transform decomposes time signals into a sum of exponentially-decaying or growing sinusoids.

- Region of Convergence: the ROC is the range of $\sigma = \mathcal{R}e\{s\}$ where the integral defining $F(s) = \mathcal{L}\{f(t)\}$ converges. The ROC depends only on the real part of s ($= \sigma$).

- Bilateral Laplace Transform: $\mathcal{L}\{f(t)\} = \int_{-\infty}^{\infty} f(t)e^{-st}dt = F(s)$; ROC must be specified because more than one possibility may exist.

- Unilateral Laplace Transform: $\mathcal{L}\{f(t)u(t)\} = \int_{0^-}^{\infty} f(t)e^{-st}dt = F(s)$; there is only one ROC – the one corresponding to causal $f(t)$.

- Inverse Laplace Transform: $\mathcal{L}^{-1}\{F(s)\} = \frac{1}{2\pi j}\int_{\sigma_0-j\infty}^{\sigma_0+j\infty} F(s)e^{st}ds = f(t)$ ($\sigma_0 + j\omega \in$ ROC).

- Laplace Transform pairs: we focus on causal $f(t)$, since that is the case for impulse response of a causal system.

- $H(s)$ is said to be *rational* when it is the ratio of polynomials in s: $H(s) = P(s)/Q(s)$.

- Laplace transform pairs for the case of simple poles of rational $H(s)$:
  Causal: $e^{-at}u(t) \Leftrightarrow 1/(s+a)$, ROC: $\sigma = \mathcal{R}e\{s\} > \mathcal{R}e\{a\}$,
  Anticausal: $e^{-at}(u(t)-1) \Leftrightarrow 1/(s+a)$, ROC: $\sigma = \mathcal{R}e\{s\} < \mathcal{R}e\{a\}$.

- The ROC of $H(s) = \mathcal{L}\{f(t)\}$ for causal $f(t)$ is $Re\{s\} > \sigma_0$, where σ_0 is the right-most pole in $H(s)$.

- The ROC of $H(s) = \mathcal{L}\{f(t)\}$ for anticausal $f(t)$ is $Re\{s\} < \sigma_0$, where σ_0 is the left-most pole in $H(s)$.

- Laplace Transform properties are very much like the properties of the continuous-time Fourier transform (linearity, convolution in time is multiplication in s-domain, etc.).

- The inverse Laplace Transform of rational $H(s)$ can be done via table look-up, long division to obtain a proper rational function (if necessary) and then partial fraction expansion, or direct evaluation of the line integral expression.

- Lumped-element RLC linear circuit analysis gives finite-order constant-coefficient differential equations, which when transformed to the s-domain give rational transfer functions $H(s)$.

- Here are some advantages of the Laplace transform vs. continuous-time Fourier transform:

  a) $\mathcal{L}\{f(t)\}$ may converge when $\mathcal{F}\{f(t)\}$ does not, making it possible to analyze signals such as the ramp function,
  b) The Laplace transform makes it possible to analyze *unstable* systems when $\mathcal{L}\{h(t)\}$ exists but $\mathcal{F}\{h(t)\}$ does not,
  c) When $H(s)$ is rational, its poles give us information about system stability,
  d) When $H(s)$ is rational, its pole/zero locations give us information that is useful for designing and analyzing analog filters.

10.3 Homework Problems

P10.1 When $(s) = \frac{1}{s+0.5}$, what ROC corresponds to causal $f(t)$?

P10.2 When $F(s) = \frac{1}{s-2}$, what ROC corresponds to anticausal $f(t)$?

P10.3 Find $f(t)$ when $F(s) = \frac{1}{s-0.5}$ (ROC for causal $f(t)$).

P10.4 Find $f(t)$ when $F(s) = \frac{1}{s+2}$ (ROC for anticausal $f(t)$).

P10.5 Given: $F(s) = \frac{1}{s+0.5} + \frac{1}{s-2}$. What ROC's are possible?

P10.6 Given: $F(s) = \frac{2}{2s-0.5} - \frac{3}{3s+2}$. What ROC's are possible?

P10.7 Given: $F(s) = \frac{12s+2}{2s^2+3s+1}$. Find $f(t)$ when it is causal.

P10.8 Given: $H(s) = \frac{-2s}{s^2+2s+17}$. Draw the pole-zero diagram, with S.F.

P10.9 Find the Laplace transform of $f(t) = e^{-2(t-1)}u(t-1)$.

P10.10 Find the Laplace transform of $f(t) = e^{-5t}\cos(\pi t/4)\,u(t)$.

P10.11 Find the Laplace transform of $f(t) = \omega_0\{\cos(\omega_0 t)\,u(t) * u(t)\}$.

P10.12 Find the Laplace transform of $f(t) = \text{rect}\left(\frac{t}{6} - \frac{1}{2}\right)\cos(6t)$.

P10.13 Find causal impulse response $h(t)$, using the partial fraction expansion method, when $H(s) = \frac{-3s}{s^2+9s+20}$.

P10.14 Find causal impulse response $h(t)$, using the partial fraction expansion method, when $H(s) = \frac{s-1}{s^2+2s+17}$.

Chapter 11

Applications of Z-Domain Signal Processing

11.1 Introduction

As we learned in Ch. 9, because the Z transform is a generalization of the DTFT it has many of the same properties. One property both Z and Fourier transforms have is that convolution in time becomes multiplication in the transform domain. This is important since a linear shift-invariant system is completely described by its impulse response, which relates output and input signals via convolution. Therefore, the multiplicative operator known as the *transfer function* is important in both ω and z domains.

The z-domain makes discrete-time systems simpler to analyze and to design. Among its many benefits, transforming from time domain to the z-domain makes it possible to algebraically solve constant-coefficient linear difference equations. We use the z-domain to design digital filters, to test for digital filter stability, and to simplify mathematical notation by avoiding direct mention of $e^{j\omega}$. This chapter presents some interesting applications of z-domain representation and analysis.

11.2 Applications of Pole-Zero Analysis

In Ch. 7 we studied frequency response functions $H(e^{j\omega})$ that satisfy the general definitions of basic filter types. For example, a digital lowpass filter will have $H(e^{j\omega}) = 0$ at $\omega = \pm\pi$, and $H(e^{j\omega}) = 1$ at $\omega = 0$ (when normalized for peak unity gain). The independent variable of interest there was frequency variable ω. In this chapter, the z-domain transfer function

$H(z)$ permits us to describe each basic filter type in terms of its poles and zeros with respect to complex frequency variable z. An advantage of the z-domain description is that it provides additional insight into the implementation and stability of these filters.

11.2.1 *Poles and zeros of realizable systems*

Most engineering problems of interest involve rational transfer functions $(H(z) = N(z)/D(z))$. The coefficients of polynomials $N(z)$ and $D(z)$ are the same coefficients found in the time-domain difference equations describing that system. In practical situations, these coefficients will be real-valued, and so will be the system's impulse response $h(n)$. From the *complex conjugate root theorem* in mathematics, when $H(z)$ is rational with real coefficients, then the poles and zeros of $H(z)$ will either be real or will appear in complex-conjugate pairs.

Another obvious requirement for rational $H(z)$ to be realizable is that it have finite order. Therefore, most every linear discrete-time system we need to deal with may be analyzed by considering a finite number of poles and zeros that are real, or in complex-conjugate pairs.

11.2.2 *Frequency response from H(z)*

Before we discuss calculating frequency response $H(e^{j\omega})$ from pole and zero locations in the z-plane, let us not forget that the easiest way to find $H(e^{j\omega})$ is to substitute $z = e^{j\omega}$ into the expression for $H(z)$. For this to be correct, we must be sure that the unit circle (defined by $|z| = 1$, or $z = e^{j\omega}$) lies within the region of convergence of $H(z)$; otherwise, our expression for $H(z)$ is not valid at the points where we wish to evaluate it.

For example, given a causal system with transfer function:

$$H(z) = \frac{z}{z + 0.8}. \tag{11.1}$$

The region of convergence for a causal system is $|z| > |p|$, where p is the highest-magnitude pole of $H(z)$. In this example, we see there is one pole $p = -0.8$. The region of convergence is therefore $|z| > |-0.8|$, which contains the unit circle. Therefore, we may confidently state:

$$H(e^{j\omega}) = H(z)|_{z=e^{j\omega}} = \frac{e^{j\omega}}{e^{j\omega}+0.8} = \frac{1}{1+0.8e^{-j\omega}}. \quad (11.2)$$

To check the correctness of this result, recall Entry #1 in Table 9.2:

$$Z\{r^n u(n)\} = \frac{z}{z-r}, \quad \text{ROC: } |z| > |r|. \quad (11.3)$$

Using this Z transform pair to find $Z^{-1}\{H(z)\}$ in our example, we obtain $h(n) = (-0.8)^n u(n)$. Next, refer to Table 4.1, Entry #2:

$$\mathcal{F}\{a^n u(n)\} = \frac{1}{1-ae^{-j\omega}}, \quad |a| < 1. \quad (11.4)$$

Substituting $a = -0.8$ gives us $\mathcal{F}\{(-0.8)^n u(n)\} = \frac{1}{1+0.8e^{-j\omega}}$, which confirms our result in Eq. (11.2).

Here is an example where $H(e^{j\omega}) = H(z)|_{z=e^{j\omega}}$ does *not* work: given $(z) = \frac{z}{z-1}$, and the system is causal. We are tempted to say that the frequency response $(e^{j\omega}) = \frac{e^{j\omega}}{e^{j\omega}-1} = \frac{1}{1-e^{-j\omega}}$, but our conclusion is incorrect because the ROC, $|z| > 1$, does not include the unit circle. Note: $Z^{-1}\{\frac{z}{z-1}, ROC: |z| > 1\} = u(n)$ and $\mathcal{F}\{u(n)\} = \frac{1}{1-e^{-j\omega}} + \pi\delta_{2\pi}(\omega) \neq \frac{1}{1-e^{-j\omega}}$.

11.2.3 *Frequency response from pole/zero locations*

Consider the case where $H(z)$ is rational. Factoring its Mth order numerator and N^{th} order denominator polynomials gives

$$H(z) = S.F. \times \frac{(z-z_1)(z-z_2)\cdots(z-z_M)}{(z-p_1)(z-p_2)\cdots(z-p_N)} = S.F. \times \frac{\prod_{m=1}^{M}(z-z_m)}{\prod_{n=1}^{N}(z-p_n)}, \quad (11.5)$$

where *S.F.* is a constant (scale factor), numerator roots z_m are the zeros of $H(z)$, and denominator roots p_n are the poles of $H(z)$. Assume that the unit circle lies within the region of convergence for $H(z)$, so that frequency response $H(e^{j\omega})$ may be found as $H(z)|_{z=e^{j\omega}}$.

11.2.3.1 *Magnitude response*

From Eq. (11.5), the magnitude frequency response function $\left|H(e^{j\omega})\right|$ may be expressed in terms of the poles and zeros of $H(z)$ as:

$$\left|H(e^{j\omega})\right| = \left|S.F.\times \frac{\prod_{m=1}^{M}(z-z_m)}{\prod_{n=1}^{N}(z-p_n)}\right|_{z=e^{j\omega}}$$

$$= |S.F.|\frac{\prod_{m=1}^{M}|z-z_m|}{\prod_{n=1}^{N}|z-p_n|} \text{ when } z = e^{j\omega}. \tag{11.6}$$

Since $|z_a - z_b|$ is the distance between points z_a and z_b in the complex z-plane, we see that the magnitude of $H(e^{j\omega})$ at some frequency value $\omega = \omega_0$ is equal to $|S.F.|$ multiplied by the product of distances from each zero of $H(z)$ to the point $z_0 = e^{j\omega_0}$, divided by the product of distances from each pole of $H(z)$ to z_0. (Zeros and poles at ∞ do not directly enter into the calculation.) Thus, we may find $\left|H(e^{j\omega})\right|$ at some frequency ω, graphically, from a knowledge of the pole-zero plot of $H(z)$ and $S.F.$

Example 11.1

Given: $H(z) = \frac{-6(z-0.5)(z+1)}{(z+0.5)}$. Find $\left|H(e^{j\omega})\right|$ at frequency values $\omega = \{\pi/4, \pi/2, \pi, \text{ and } 2\pi\}$ using Eq. (11.6).

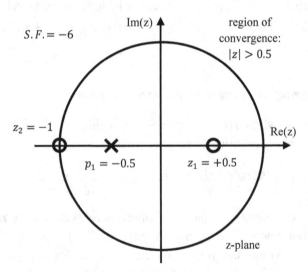

Figure 11.1. Pole-zero plot of $H(z) = -\frac{6(z-0.5)(z+1)}{z+0.5}$ from Example 11.1.

The graphical solutions for $\left|H\left(e^{j\omega}\right)\right|$, based on the pole and zero locations in the z-plane, are shown in Figs. 11.1 through 11.5.

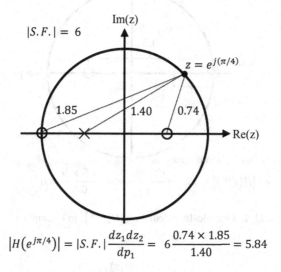

$|S.F.| = 6$

$\text{Im}(z)$

$z = e^{j(\pi/4)}$

1.85   1.40   0.74

$\text{Re}(z)$

$$\left|H\left(e^{j\pi/4}\right)\right| = |S.F.|\frac{dz_1 dz_2}{dp_1} = 6\frac{0.74 \times 1.85}{1.40} = 5.84$$

Figure 11.2. Graphically calculating $\left|H\left(e^{j\pi/4}\right)\right|$ in Example 11.1.

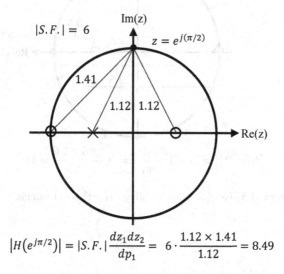

$|S.F.| = 6$

$\text{Im}(z)$

$z = e^{j(\pi/2)}$

1.41

1.12 1.12

$\text{Re}(z)$

$$\left|H\left(e^{j\pi/2}\right)\right| = |S.F.|\frac{dz_1 dz_2}{dp_1} = 6\cdot\frac{1.12 \times 1.41}{1.12} = 8.49$$

Figure 11.3. Graphically calculating $\left|H\left(e^{j\pi/2}\right)\right|$ in Example 11.1.

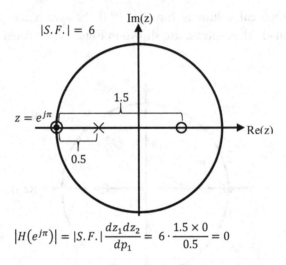

$$\left|H\left(e^{j\pi}\right)\right| = |S.F.| \frac{dz_1 dz_2}{dp_1} = 6 \cdot \frac{1.5 \times 0}{0.5} = 0$$

Figure 11.4. Graphically calculating $\left|H\left(e^{j\pi}\right)\right|$ in Example 11.1.

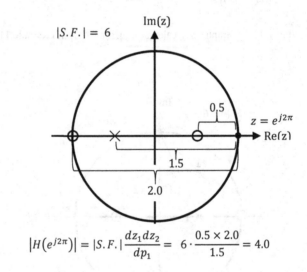

$$\left|H\left(e^{j2\pi}\right)\right| = |S.F.| \frac{dz_1 dz_2}{dp_1} = 6 \cdot \frac{0.5 \times 2.0}{1.5} = 4.0$$

Figure 11.5. Graphically calculating $\left|H\left(e^{j2\pi}\right)\right|$ in Example 11.1.

When plotting $\left|H(e^{j\omega})\right|$ on a logarithmic magnitude scale (e.g., as $10\log_{10}\left|H(e^{j\omega})\right|^2 = 20\log_{10}\left|H(e^{j\omega})\right|$ dB), the product of terms corresponding to each pole and zero becomes a sum:

$$20\log_{10}\left|H(e^{j\omega})\right| = 20\log_{10}\left\{|S.F.|\frac{\prod_{m=1}^{M}\left|e^{j\omega}-z_m\right|}{\prod_{n=1}^{N}\left|e^{j\omega}-p_n\right|}\right\}$$

$$= 20\log_{10}|S.F.| + \sum_{m=1}^{M}20\log_{10}\left|e^{j\omega}-z_m\right|$$

$$- \sum_{n=1}^{N}20\log_{10}\left|e^{j\omega}-p_n\right| \tag{11.7}$$

In this way, the contribution from each pole and zero (and the scale factor) is separated from the rest as an independent additive component in the log-magnitude plot of $H(e^{j\omega})$ vs ω.

Example 11.2
$H(z) = -6(z-0.5)(z+1)/(z+0.5)$, as in the previous example. Write MATLAB® code to show the additive contribution of each pole and zero in a plot of $\left|H(e^{j\omega})\right|^2$ (in dB) vs ω.

Solution:

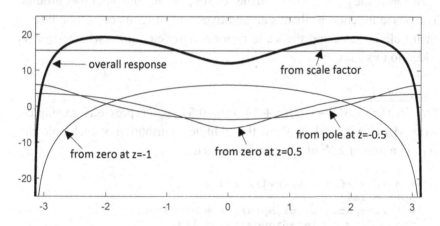

Figure 11.6. A graph of contributions to frequency magnitude response (dB) vs. ω from each zero, pole, and scale factor of $H(z)$ in Example 11.2.

Here is the MATLAB® code used to obtain the graph shown in Fig. 11.6:

```
% H(z) = -6(z-0.5)(z+1)/(z+0.5)
Npts = 1e4;
w = linspace(-pi,pi,Npts); z = exp(j*w);
H_SF_dB = 20*log10(abs(-6*ones(size(w))));
H_z1_dB = 20*log10(abs(z-0.5));
H_z2_dB = 20*log10(abs(z+1));
H_p1_dB = 20*log10(abs(1./(z+0.5)));
plot(w,H_SF_dB); hold on
plot(w,H_z1_dB)
plot(w,H_z2_dB)
plot(w,H_p1_dB)
plot(w,H_SF_dB + H_z1_dB + H_z2_dB + H_p1_dB);
axis([-pi pi -25 25])
```

11.2.3.2 *Phase response*

Based on Eq. (11.5), the phase function $\angle H(e^{j\omega})$ is also easily expressed in terms of the poles and zeros of $H(z)$:

$$\angle H(e^{j\omega}) = \angle S.F. + \sum_{m=1}^{M} \angle(e^{j\omega} - z_m) - \sum_{n=1}^{N} \angle(e^{j\omega} - p_n). \quad (11.8)$$

Note that the angles $\angle(e^{j\omega} - z_m)$ and $\angle(e^{j\omega} - p_n)$ are of vectors drawn *from* the point $z = e^{j\omega}$ *to* the zero or pole, not the other way around. The phase function is always an additive sum[a] of terms corresponding to each pole or zero, and the scale factor's sign determines whether $\pm\pi$ is added to this sum.

Example 11.3
For $H(z) = -6(z - 0.5)(z + 1)/(z + 0.5)$ as in previous examples, write MATLAB® code to show the additive contribution of each pole and zero in a plot of $\angle H(e^{j\omega})$ (in radians) vs ω.

```
% H(z) = -6(z-0.5)(z+1)/(z+0.5)
Npts = 1e4;
w = linspace(-pi,pi,Npts); z = exp(j*w);
theta_SF = angle(-6*ones(size(w)));
theta_z1 = angle(z-0.5);
theta_z2 = angle(z+1);
```

---

[a] Phase angle is never plotted on a logarithmic scale.

```
theta_p1 = angle(1./(z+0.5));
plot(w,theta_SF); hold on
plot(w,theta_z1)
plot(w,theta_z2)
plot(w,theta_p1)
plot(w,theta_SF + theta_z1 + theta_z2 + theta_p1);
axis([-pi pi -pi 2*pi])
```

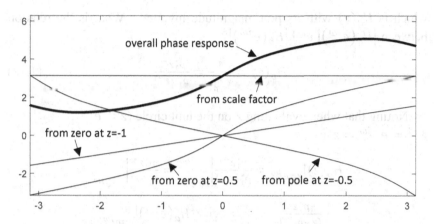

Figure 11.7. A graph of contributions to frequency phase response (rad/sec) vs. ω from each zero, pole, and scale factor of $H(z)$ in Example 11.3.

From Examples 11.2 and 11.3, we see that plots of magnitude and phase contributions from poles and zeros shows us the effect of each of these on overall frequency response. This understanding may be important when designing filters.

11.2.4 *Effect on $H(e^{j\omega})$ of reciprocating a pole*

It is interesting to see what happens to frequency response magnitude and phase contributions from a pole (or zero) that is reciprocated in magnitude (if inside the unit circle it is moved outside, and vice versa). To investigate this, consider the simple case:

$$H_a(z) = \frac{1}{z - p_1}. \tag{11.9}$$

Expressing complex constant p_1 in polar form, we obtain

$$H_a(z) = \frac{1}{z - r_1 e^{j\theta_1}}. \tag{11.10}$$

Next define

$$H_b(z) = \frac{1}{z - \frac{1}{r_1} e^{j\theta_1}} = \frac{1}{z - \frac{1}{r_1 e^{-j\theta_1}}} = \frac{1}{z - \frac{1}{p_1^*}}, \tag{11.11}$$

which is $H_a(z)$ with its pole magnitude inverted. What is the relation between $|H_a(e^{j\omega})|$ and $|H_b(e^{j\omega})|$?

$$H_b(z) = \frac{1}{z - \frac{1}{p_1^*}} = \frac{p_1^* z^{-1}}{p_1^* - z^{-1}}. \tag{11.12}$$

Noting that when evaluating z on the unit circle ($z = e^{j\omega}$), $z^{-1} = e^{-j\omega} = z^*$:

$$H_b(z)\big|_{z=e^{j\omega}} = \frac{p_1^* z^{-1}}{p_1^* - z^{-1}}\bigg|_{z=e^{j\omega}} = \frac{p_1^* z^*}{p_1^* - z^*}\bigg|_{z=e^{j\omega}}$$

$$= \frac{-p_1^* z^*}{(z - p_1)^*}\bigg|_{z=e^{j\omega}} = -p_1^* z^* H_a^*(e^{j\omega})\big|_{z=e^{j\omega}},$$

or
$$H_b(e^{j\omega}) = -p_1^* e^{-j\omega} H_a^*(e^{j\omega}). \tag{11.13}$$

Thus
$$|H_b(e^{j\omega})| = |-p_1^*||e^{-j\omega}||H_a(e^{j\omega})|$$

$$= |p_1||H_a(e^{j\omega})| = \text{const}_1 \times |H_a(e^{j\omega})|, \tag{11.14}$$

and
$$\angle H_b(e^{j\omega}) = \pi - \angle p_1 - \omega - \angle H_a(e^{j\omega})$$

$$= \text{const}_2 - \angle H_a(e^{j\omega}). \tag{11.15}$$

A similar result is obtained when a zero's magnitude is inverted: frequency response magnitude does not change except for a multiplicative constant, and the phase is subtracted from a constant. This may be used to advantage when designing all-pass filters, which are covered later in this chapter.

11.2.5 *System stability*

A linear, shift-invariant system is stable[b] if its transfer function $H(z)$ has region of convergence that includes the unit circle. That is, $H(z)$ must converge at $z = e^{j\omega_0}, \forall \omega_0$.

What are the practical consequences when a digital filter is unstable? In that case, for certain input sequences, the filter's output values will grow in amplitude until they exceed the largest numerical code available in the processor. This may even lead to physical danger in some applications (e.g., controlling the motion of a huge robotic arm).[c]

Let us consider two cases: (1) $H(z)$ is rational and the system is causal, and (2) $H(z)$ is rational and the system is anticausal.

11.2.5.1 *Causal systems*

For causal systems, the ROC of transfer function $H(z)$ is $|z| > |p|$, where p is the pole furthest from the origin. If a causal system is stable (ROC includes the unit circle) then all poles must lie inside the unit circle.

11.2.5.2 *Anticausal systems*

For anticausal systems, the ROC of transfer function $H(z)$ is $|z| < |p|$, where p is the pole closest to the origin. If an anticausal system is stable (ROC includes the unit circle) then all poles must lie outside the unit circle.

---

[b] In the sense that an input signal that is restricted to a finite range of amplitude values will result in an output signal that is also restricted to a finite range of amplitude values. This is called *Bounded Input, Bounded Output* (BIBO) stability.

[c] Unstable systems are not completely useless, however, because we can keep the output signal in check by applying a properly compensated input signal. This is a concept from the area of *control systems*, where feedback is used to control the output.

11.2.5.3 *Stabilizing an unstable causal system*

Let's say we are designing a causal digital filter network, and we arrive at a desired magnitude response $|H(e^{j\omega})|$ with some choice of $N(z)$ and $D(z)$ polynomials in transfer function $H(z) = N(z)/D(z)$. However, after plotting the roots of $D(z)$ we discover that some poles are outside of the unit circle making the filter unstable. How may we stabilize the system without changing its magnitude frequency response?

The solution is to reciprocate the magnitudes of those poles outside of the unit circle to move them to inside the unit circle.[d] From Section 11.2.4, and specifically Eq. (11.14), we learned that reciprocating the magnitude of a pole will not change $|H(e^{j\omega})|$ except for scaling it by a constant.[e] This constant may be compensated for by adjusting the scale factor of $H(z)$. The following example demonstrates this technique.

Example 11.4
A causal system with transfer function $H_a(z) = \dfrac{2z^2}{(z-1/2)(z-3/2)}$ is unstable due to one pole having magnitude > 1.

Find a different transfer function, $H_b(z)$, representing a stable causal system, where $|H_b(e^{j\omega})| = |H_a(e^{j\omega})|$. Confirm this in MATLAB®.

Solution:
Reciprocate the magnitude of the pole at $z = 3/2$ to give transfer function $H_b(z) = \dfrac{const. \times 2z^2}{(z-1/2)(z-2/3)}$. To match the gains of the two functions, choose a value of $z = e^{j\omega}$ where $H(e^{j\omega}) \neq 0$ and evaluate:

$$\left|H_a(z = e^{j\cdot 0} = 1)\right| = \left|\frac{2\cdot 1^2}{(1-1/2)(1-3/2)}\right| = \left|\frac{2}{(1/2)(-1/2)}\right| = 8$$

$$\therefore |H_b(1)| = \left|\frac{const. \times 2\cdot 1^2}{(1-1/2)(1-2/3)}\right| = \left|\frac{const. \times 2}{(1/2)(1/3)}\right| = |const. \times 12| = 8,$$

---

[d] If a pole lies *on* the unit circle, then this method will not stabilize the system.

[e] The phase response will change, but assume that is an acceptable price to pay for having a stable filter.

or const. $= \pm \dfrac{8}{12} = \pm \dfrac{2}{3}$ (choose 2/3): $H_b(z) = \dfrac{(4/3)\,z^2}{(z-1/2)(z-2/3)}$.

```
Npts = 1e4;
w = linspace(0,pi,Npts); z = exp(j*w);
Ha = (2*z.^2)./((z-1/2).*(z-3/2));
Hb = ((4/3)*z.^2)./((z-1/2).*(z-2/3));
subplot(2,1,1)
plot(w,abs(Ha)); axis([0 pi 0 8]); grid on
subplot(2,1,2)
plot(w,abs(Hb)); axis([0 pi 0 8]); grid on
```

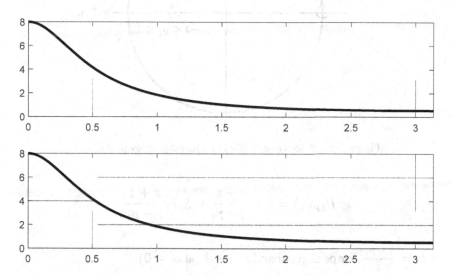

Figure 11.8. A graph of $|H_a(e^{j\omega})|$ and $|H_b(e^{j\omega})|$ vs. ω in Example 11.4, confirming that they are the same frequency magnitude response.

11.2.6 *Pole-zero plots of basic digital filters*

In this section, we describe various low-order digital filters in terms of their transfer functions, pole/zero locations and difference equations. The filters are stable and realizable: their poles lie inside the unit circle, and all poles/zeros are real or appear in complex conjugate pairs.

11.2.6.1 *Lowpass filter*

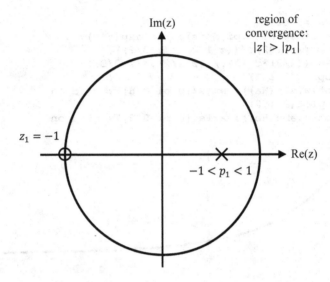

Figure 11.9. Pole-zero plot of a first-order digital lowpass filter.

$$H_{LP}(z) = S.F.\frac{z - z_1}{z - p_1} = S.F.\frac{z + 1}{z - p_1},$$

$S.F. = \frac{1 - p_1}{2}$ for peak passband gain = 1 (at $\omega = 0$):

$$H_{LP}(z) = \left(\frac{1 - p_1}{2}\right)\left(\frac{z + 1}{z - p_1}\right) = \left(\frac{1 - p_1}{2}\right)\left(\frac{1 + z^{-1}}{1 - p_1 z^{-1}}\right),$$

$$y(n) = \left(\frac{1 - p_1}{2}\right)x(n) + \left(\frac{1 - p_1}{2}\right)x(n - 1) + p_1 y(n - 1).$$

11.2.6.2 *Highpass filter*

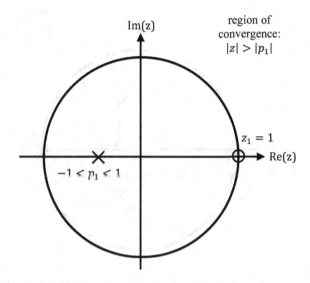

Figure 11.10. Pole-zero plot of a first-order digital highpass filter.

$$H_{HP}(z) = S.F.\frac{z - z_1}{z - p_1} = S.F.\frac{z - 1}{z - p_1},$$

$S.F. = \frac{1 + p_1}{2}$ for peak passband gain = 1 (at $\omega = \pi$):

$$H_{HP}(z) = \left(\frac{1 + p_1}{2}\right)\left(\frac{z - 1}{z - p_1}\right) = \left(\frac{1 + p_1}{2}\right)\left(\frac{1 - z^{-1}}{1 - p_1 z^{-1}}\right),$$

$$y(n) = \left(\frac{1 + p_1}{2}\right)x(n) - \left(\frac{1 + p_1}{2}\right)x(n - 1) + p_1 y(n - 1).$$

11.2.6.3 *Bandpass digital filter*

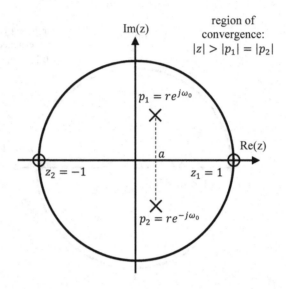

Figure 11.11. Pole-zero plot of a 2nd-order digital bandpass filter.

$$H_{BP}(z) = S.F. \frac{(z-1)(z+1)}{(z-p_1)(z-p_1^*)} = S.F. \frac{z^2-1}{z^2-2az+r^2},$$

$$(a = Re\{p_1\} = Re\{p_2\}, \quad r = |p_1| = |p_2|, \quad \omega_0 = \angle p_1)$$

$$\left|H_{BP}(e^{j\omega})\right| \text{ peaks at } \omega \cong \omega_0.$$

Because $H_{BP}(z) = c(1 - z^{-2})/(1 - 2az^{-1} + r^2 z^{-2})$,

$$y(n) = cx(n) - cx(n-2) + 2ay(n-1) - r^2 y(n-2).$$

Scale factor c is chosen to give desired gain at the peak frequency.

11.2.6.4 *Notch filter*

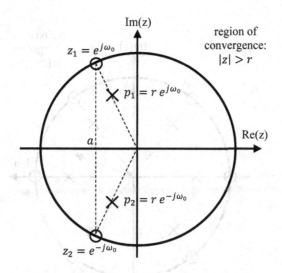

Figure 11.12. Pole-zero plot of a 2nd-order digital notch filter.

$$H_N(z) = S.F. \; \frac{(z - z_1)(z - z_1^*)}{(z - p_1)(z - p_1^*)} = S.F. \; \frac{z^2 - 2az + 1}{z^2 - 2raz + r^2},$$

$$(a = Re\{z_1\} = Re\{z_2\}, \;\; r = |p_1| = |p_2| < 1, \;\; \omega_0 = \angle p_1 = \angle z_1)$$

$\left|H_N\!\left(e^{j\omega}\right)\right| = 0$ at $\omega = \omega_0$.   A narrower notch results as $r \to 1$.

To obtain d.c. gain = 1, choose $S.F. = (1 - 2ra + r^2)/(2 - 2a)$.

Because $H_N(z) = c(1 - 2az^{-1} + z^{-2})/(1 - 2raz^{-1} + r^2z^{-2})$,

$$y(n) = c\,x(n) - 2ac\,x(n-1) + c\,x(n-2)$$
$$+2ra\,y(n-1) - r^2\,y(n-2).$$

11.2.6.5 *Comb filter*

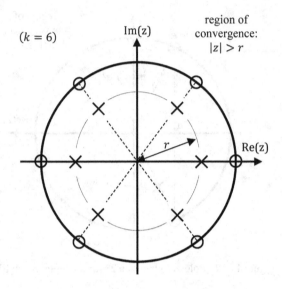

Figure 11.13. Pole-zero plot of a kth-order digital comb filter.

$$H_C(z) = S.F.\ \frac{z^k - 1}{z^k - r} = S.F.\ \prod_{n=0}^{k-1}\left(\frac{z - e^{jn(2\pi/k)}}{z - re^{jn(2\pi/k)}}\right),$$

$$(0 \leq r < 1)$$

$$\left|H_C(e^{j\omega})\right|_{\omega=2\pi n/k} = 0 \quad (n \in \mathbb{Z}).$$

$S.F. = (r+1)/2$ for peak gain magnitude = 1:

Because $H_C(z) = \left(\frac{r+1}{2}\right)\left(1 - z^{-k}\right)/\left(1 - rz^{-k}\right),$

$$y(n) = \left(\frac{r+1}{2}\right)x(n) - \left(\frac{r+1}{2}\right)x(n-k) + r\,y(n-k).$$

11.2.6.6 *Allpass filter (real pole and zero)*

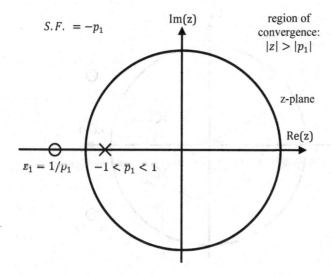

Figure 11.14. Pole-zero plot of a first-order digital allpass filter.

$$H_{AP_1}(z) = S.F. \, \frac{z - z_1}{z - p_1} = S.F. \, \frac{z - 1/p_1}{z - p_1},$$

$S.F. = -p_1$ for gain magnitude $= 1$ at $\forall \omega$:

$$H_{AP_1}(z) = -p_1 \, \frac{z - 1/p_1}{z - p_1} = \frac{-p_1 z + 1}{z - p_1} = \frac{-p_1 + z^{-1}}{1 - p_1 z^{-1}},$$

$$y(n) = -p_1 x(n) + x(n-1) + p_1 y(n-1).$$

11.2.6.7 *Allpass filter (complex conjugate poles and zeros)*

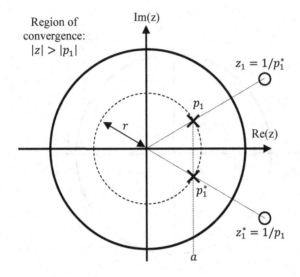

Figure 11.15. Pole-zero plot of a 2nd-order digital allpass filter having complex-conjugate poles and zeros.

$$H_{AP_2}(z) = S.F. \frac{(z - z_1)(z - z_1^*)}{(z - p_1)(z - p_1^*)} = S.F. \frac{(z - 1/p_1^*)(z - 1/p_1)}{(z - p_1)(z - p_1^*)},$$

$S.F. = |p_1|^2 = p_1 p_1^*$ for gain magnitude = 1 at $\forall \omega$:

$$H_{AP_2}(z) = p_1 p_1^* \frac{(z - 1/p_1^*)(z - 1/p_1)}{(z - p_1)(z - p_1^*)} = \frac{r^2 - 2az^{-1} + z^{-2}}{1 - 2az^{-1} + r^2 z^{-2}},$$

$$y(n) = r^2 x(n) - 2a\, x(n - 1) + x(n - 2)$$
$$+ 2a\, y(n - 1) - r^2 y(n - 2).$$

11.2.7 *Minimum-phase system*

A rational transfer function $H(z)$ is said to represent a *minimum-phase* system when all poles and zeros lie inside the unit circle.

Consider a causal system that is minimum-phase. It must be stable because all poles lie inside the unit circle. If we reciprocate $H(z)$ of this system to obtain $H^{-1}(z) = 1/H(z)$, this new system will also be stable because the zeros of $H(z)$ are the poles of $H^{-1}(z)$ that also lie within the unit circle. Therefore, a minimum-phase system can filter $x(n)$ to give $y(n)$, and an inverse system exists that can filter $y(n)$ to recover $x(n)$. That is to say, $h(n) * h^{-1}(n) = \delta(n)$, where both $h(n)$ and $h^{-1}(n)$ are causal and represent stable systems.

Recall, from Section 11.2.4 (Eq. (11.14)), that reciprocating the magnitude of a pole or zero will not change $|H(e^{j\omega})|$ except for scaling it by a constant, and this constant can be compensated for by changing the scaling factor of $H(z)$. Therefore, we may transform a non-minimum phase system into a minimum phase system, without affecting $|H(e^{j\omega})|$, by reciprocating the magnitudes of any poles/zeros that lie outside the unit circle and then adjusting the scale factor.

11.2.8 *Digital filter design based on analog prototypes*

Many analog filter designs already exist; for a given filter order and some other constraints, each has been designed to have an optimized frequency response. To tap the wealth of this knowledge in classical analog filter designs, in this section we consider how to convert an s-domain transfer function for use in the z-domain. To avoid any confusion, let $H_c(s)$ represent a *continuous*-time filter's s-domain transfer function, and let $H_d(z)$ represent a *discrete*-time filter's z-domain transfer function.

From Chs. 6 and 7, we know that $H_d(e^{j\omega})$, the frequency response of a digital filter, is periodic having period $\omega_s = 2\pi$ when the assumed spacing of its time samples is $T_s = 1$ second. In general, for an arbitrary value of T_s, the spectrum $H_d(e^{j\omega T_s})$ has period $\omega_s = 2\pi/T_s$ sec. So, we may express a digital filter's frequency response $H_d(e^{j\omega T_s})$ as a convolution product of some analog filter spectrum $X_c(\omega)$ with a periodic impulse train:

$$H_d(e^{j\omega T_s}) = X_c(\omega) * \delta_{\omega_s}(\omega) = \sum_{k=-\infty}^{\infty} X_c(\omega - k\omega_s). \qquad (11.16)$$

If the bandwidth of $X_c(\omega)$ exceeds $\omega_s/2$, then partial overlap (aliasing) of frequency-shifted copies of $X_c(\omega)$ will occur.

We will overview two methods of synthesizing a digital filter having frequency response $H_d(e^{j\omega T_s})$ based on the frequency response $H_c(\omega)$ of a reference analog filter. The first method chooses $X_c(\omega) = H_c(\omega)$, which may result in aliasing because $H_c(\omega)$ has no bandwidth restrictions (*impulse-invariant transformation* method). The second method chooses $X_c(\omega)$ to be a frequency-compressed version of $H_c(\omega)$ having bandwidth $= \omega_s/2$ to guarantee that aliasing does not occur (*bilinear transformation* method). In both methods $H_c(s)$ must be rational, and poles of $H_c(s)$ are mapped to become poles of $H_d(z)$.

11.2.8.1 *Impulse-invariant transformation*

Using this method, we define our digital filter's frequency response in terms of the reference analog filter's frequency response as follows:

$$H_d(e^{j\omega T_s}) = H_c(\omega) * \delta_{\omega_s}(\omega). \qquad (11.17)$$

Taking the inverse CTFT we obtain

$$\mathcal{F}^{-1}\{H_c(\omega) * \delta_{\omega_s}(\omega)\} = 2\pi\left\{h_c(t) \times \frac{1}{\omega_s}\delta_{T_s}(t)\right\} = T_s h_c(t)\delta_{T_s}(t),$$

which is a scaled and sampled version of $h_c(t)$. Therefore, we conclude that our digital filter's impulse response will be the same as a sampled version of the analog filter's impulse response

$$h_d(n) = T_s h_c(nT_s), \qquad (11.18)$$

which is where the name "impulse-invariant" comes from.

Begin with a partial fraction expansion of $H_c(s)$ (assumed to have N simple poles):

$$H_c(s) = \sum_{k=1}^{N} \frac{C_k}{s - s_k}.$$

Take the inverse Laplace transform of both sides:

$$h_c(t) = \sum_{k=1}^{N} C_k e^{s_k t} u(t). \tag{11.19}$$

Obtain sequence $h_d(n)$ by scaling and sampling $h_c(t)$:

$$h_d(n) = T_s h_c(t)|_{t=nT_s} = T_s \sum_{k=1}^{N} C_k e^{s_k n T_s} u(nT_s)$$

$$= T_s \sum_{k=1}^{N} C_k (e^{s_k T_s})^n u(n). \tag{11.20}$$

Take the Z transform of $h_d(n)$:

$$H_d(z) = T_s \sum_{k=1}^{N} \frac{C_k z}{z - e^{s_k T_s}}. \tag{11.21}$$

Comparing $H_c(s)$ to $H_d(z)$, we see that there was a one-to-one mapping of poles from the s-domain to the z-domain:

$$s = s_k \text{ (s-domain)} \rightarrow z_k = e^{s_k T_s} \text{ (z-domain)}.$$

We then algebraically combine the terms to obtain a z-domain transfer function $H_d(z) = N(z)/D(z)$. $D(z)$ will have N roots that were mapped from the N poles of $H_c(s)$. There is no equivalent mapping of zeros in $H_c(s)$ to roots in $N(z)$. For the impulse-invariant transformation method to work well, aliasing must be minimized by choosing an analog filter having band-limited frequency response (e.g., the method is not suitable for highpass filter types) and by choosing a sufficiently high value of sampling frequency $\omega_s = 2\pi/T_s$.

Example 11.5
Given 3rd-order Butterworth analog lowpass filter transfer function:

$$H_c(s) = \frac{1}{s^3 + 2s^2 + 2s + 1}.$$

Design a digital filter by letting $h_d(n) = T_s h_c(nT_s)$, where $T_s = 1.1$ sec. Use MATLAB® to plot $|H_c(\omega)|$ and $\left|H_d(e^{j\omega T_s})\right|$ on the same graph. Also, plot a pole-zero diagram for the digital filter you have designed.

Express transfer function $H_c(s)$ as a partial fraction expansion:

$$\frac{1}{s^3+2s^2+2s+1} = \frac{1}{s-(-1)} + \frac{-0.5000 - j0.2887}{s-(-0.5000 + j0.8660)} + \frac{-0.5000 + j0.2887}{s-(-0.5000 - j0.8660)}$$

Using $T_s = 1.1$ sec, we use Eq. (11.21) to obtain

$$H_d(z) = \frac{1.1\,z}{z-0.3329} + \frac{(-0.5500 - j0.3175)z}{z-(0.3344 + j0.4702)} + \frac{(-0.5500 + j0.3175)z}{z-(0.3344 - j0.4702)}.$$

Finally, converting $H_d(z)$ to a ratio of polynomials gives:

$$H_d(z) = \frac{0.5386z + 0.2043}{z^3-0.4767z^2+0.2797z-0.0498}.$$

Here is the MATLAB® code used to calculate and generate plots:

```
Ts = 1.5;
Bs = [0 0 0 1]; As = [1 2 2 1];
[Rs,Ps,K] = residue(Bs,As);
[Bz,Az] = residue(Ts*Rs,exp(Ps*Ts),K);
Npts = 1e4;
w = linspace(-2*pi/Ts,2*pi/Ts,Npts);
s = j*w;
Hs = polyval(Bs,s)./polyval(As,s);
plot(w,abs(Hs),':'); hold on
z = exp(j*w*Ts);
Hz = polyval(Bz,z)./polyval(Az,z);
plot(w,abs(Hz))
axis([min(w) max(w) 0 1.1])
```

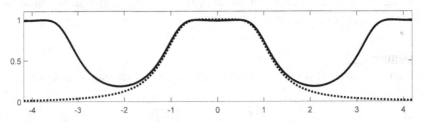

Figure 11.16. Frequency magnitude response plots, vs. ω, of analog reference filter (dotted line) and digital filter (solid line) created using impulse-invariant transformation in Example 11.5.

```
N = [0, 0, 0.5386, 0.2043];
D = [1.0000, -0.4767, 0.2797, -0.0498];
zplane(N,D)
```

$$S.F.= 0.539$$

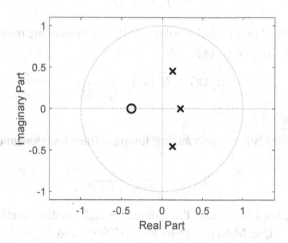

Figure 11.17. Pole-zero plot for the digital filter created using the impulse-invariant transformation in Example 11.5.

11.2.8.2 *Bilinear transformation*

Using this method, we define our digital filter's frequency response in terms of the reference analog filter's frequency response as follows:

$$H_d\left(e^{j\omega T_s}\right) = H_c\left(\frac{2}{T_s}\tan^{-1}\left(\frac{\omega T_s}{2}\right)\right) * \delta_{\omega_s}(\omega). \qquad (11.22)$$

Function $(2/T_s)\tan^{-1}(\omega T_s/2)$ maps frequency range $-\infty \leq \omega \leq \infty$ to $-\omega_s/2 \leq \omega \leq \omega_s/2$, so that aliasing cannot occur in Eq. (11.22). This warping is nonlinear and thus changes the shape of the filter's frequency response, more so at higher frequencies. However, useful digital filters result, which are usually preferred to those obtained via impulse invariant transformation due to the lack of aliasing. The bilinear transformation method can be used to transform any analog filter type, including highpass.

It may be shown [Jackson] that this transformation is done by replacing complex variable s in $H_c(s)$ with

$$s = \frac{2}{T_s}\left(\frac{z-1}{z+1}\right),$$

which is a ratio of two 1st-order polynomials in z (where the transformation gets its name), to give $H_d(z)$:

$$H_d(z) = H_c(s)\big|_{s=\frac{2}{T_s}\left(\frac{z-1}{z+1}\right)}. \tag{11.23}$$

Example 11.6
Given 3rd-order Butterworth analog lowpass filter transfer function, as in Example 11.5:

$$H_c(s) = \frac{1}{s^3+2s^2+2s+1}.$$

Design a digital filter using the bilinear transformation method, where $T_s = 1.1$ sec. Use MATLAB® to plot $|H_c(\omega)|$ and $|H_d(e^{j\omega T_s})|$ on the same graph. Also, plot a pole-zero diagram for the digital filter you have designed.

By substituting $s = (2/T_s)(z-1)/(z+1)$ into the given expression for $H_c(s)$, we obtain:

$$H_d(z) = \frac{0.0579z^3 + 0.1738z^2 + 0.1738z + 0.0579}{z^3 - 1.0434z^2 + 0.6248z - 0.1179}.$$

Here is the MATLAB® code used to calculate and generate plots:

```
Ts = 1.5;
Bs = [0 0 0 1]; As = [1 2 2 1];
[Bz,Az] = bilinear(Bs,As,1/Ts);
Npts = 1e4;
w = linspace(-2*pi/Ts,2*pi/Ts,Npts);
s = j*w;
Hs = polyval(Bs,s)./polyval(As,s);
plot(w,abs(Hs),':'); hold on
z = exp(j*w*Ts);
Hz = polyval(Bz,z)./polyval(Az,z);
plot(w,abs(Hz))
axis([min(w) max(w) 0 1.1])
```

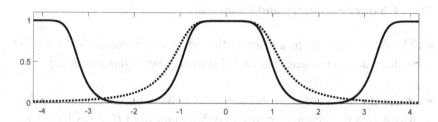

Figure 11.18. Frequency magnitude response plots, vs. ω, of analog reference filter (dotted line) and digital filter (solid line) created using the bilinear transformation in Example 11.6.

Here is the pole-zero diagram of $H_d(z)$ that was created using MATLAB® function `zplane`:

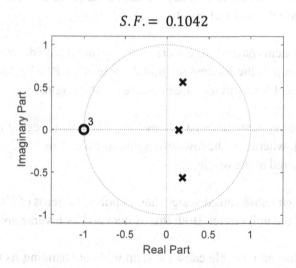

Figure 11.19. Pole-zero plot for the digital filter created using the bilinear transformation in Example 11.6.

11.3   Chapter Summary and Comments

- The Z transform maps a discrete-time system's impulse response $h(n)$ to the z-domain, where it is called *system transfer function* $H(z)$.

- When the region of convergence of $H(z)$ includes the unit circle, then a system's frequency response may be found from $H(z)$ as $H(e^{j\omega}) = H(z)|_{z=e^{j\omega}}$.

- For a system to be *stable*, the region of convergence of $H(z)$ must include the unit circle.

- In most practical applications $H(z)$ will be a rational function: $N(z)/D(z)$. The roots of $N(z)$ are called the *zeros* of $H(z)$, and roots of $D(z)$ are the *poles* of $H(z)$.

- Causal systems have ROC covering the z-plane region defined by $|z| > |p|$, where p is the highest-magnitude pole. This ROC is the entire z-plane except for a circular area centered at the origin.

- Anticausal systems have ROC covering the z-plane region defined by $|z| < |p|$, where p is the lowest-magnitude pole. This ROC is a circular area centered at the origin.

- A *minimum phase* causal system has all poles and zeros of $H(z)$ located inside of the unit circle. Both this system and its inverse are stable.

- To stabilize an unstable causal system without changing its magnitude frequency response, any poles that are outside the unit circle should be reciprocated in magnitude and scale factor adjusted accordingly.

- A system's frequency magnitude (dB) and phase responses can be decomposed into additive contributions from poles and zeros in $H(z)$.

- Pole-zero diagrams contain all the information describing $H(z)$ when it is rational. These diagrams show the locations of finite-magnitude poles and zeros in the z-plane, and specify a scale factor.

- This chapter presents various low-order digital filters (LP, HP, BP, BE, Comb, AP) in terms of their z-domain transfer functions, pole-zero diagrams, and difference equations.

- IIR digital filters may be designed by mapping s-domain poles from classical analog filters to the z-plane. This chapter overviews two ways to do this: impulse invariant method, bilinear transformation method.

- The impulse invariant method of digital filter design is equivalent to sampling an analog filter's impulse response $h(t)$ to obtain the digital filter's impulse response $h(n)$. Aliasing distortion may be a problem with this approach.

- The bilinear transformation method of digital filter design has no aliasing, but nonlinearly distorts the analog filter frequency response function (especially at high frequencies) to become one period of the digital filter's frequency response.

11.4 Homework Problems

P11.1 When $H(z) = \frac{z-1}{9z^2+4}$ is the transfer function of a causal system, write an expression for the frequency response of that system.

P11.2 When $(z) = \frac{z}{z^2+9}$, what ROC corresponds to a stable system?

P11.3 Find $f(n)$ when $F(z) = \frac{z}{z-0.9}$ (ROC for stable system).

P11.4 Given: $H(z) = \frac{12z^2+2z}{4z^2-1}$. Find $h(n)$ when the system is stable.

P11.5 Given: $H(z) = \frac{z^2+1}{z^2-1}$. Is this system stable? Justify your answer.

P11.6 Given: $H_a(z) = \frac{z^2-z}{2z^2-3z-2}$, the transfer function of an unstable causal system. Find the transfer function of a *stable* causal system, $H_b(z)$, such that $|H_b(e^{j\omega})| = |H_a(e^{j\omega})|$.

P11.7 Given: $H_a(z) = \frac{z^2-1}{2z^2+5z-3}$, the transfer function of an unstable causal system. Find the transfer function of a *stable* causal system, $H_b(z)$, such that $|H_b(e^{j\omega})| = |H_a(e^{j\omega})|$. In MATLAB®, plot both systems' frequency magnitude responses over the frequency range $0 \le \omega \le \pi$ rad/sec.

P11.8 Given: $H_a(z) = \frac{z^2-2z}{2z^2-3z-2}$, the transfer function of a system that is not minimum-phase. Find the transfer function of minimum-phase system $H_b(z)$, such that $|H_b(e^{j\omega})| = |H_a(e^{j\omega})|$. In MATLAB®, plot both systems' frequency magnitude *and phase* responses over frequency range $0 \le \omega \le \pi$ rad/sec.

P11.9 When $H(z) = N(z)/D(z)$ is the transfer function of a stable causal system, why must the degree of $N(z) \le$ degree of $D(z)$? Hint: the number of poles is always equal to the number of zeros when those at infinity are included.

P11.10 Label each of the following pole/zero plots in the z-plane as being that of a LP, HP, BP, BE or AP filter:

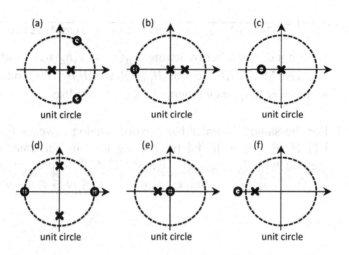

P11.11 Which pole-zero plot in problem P11.10 corresponds to a system having minimum-phase?

P11.12 By estimating pole and zero coordinates from problem P11.10 (d), use MATLAB® to plot the frequency magnitude response over frequency range $0 \leq \omega \leq \pi$ rad/sec.

P11.13 Design a first-order lowpass filter having maximum passband gain = 2.0 and cutoff frequency = $\pi/2$ rad/sec. Write this filter's difference equation, sketch the network diagram, and sketch its pole-zero diagram.

P11.14 Design a first-order highpass filter having maximum passband gain = −3.0 and cutoff frequency = $\pi/4$ rad/sec. Write this filter's transfer function, and use MATLAB® to plot its frequency magnitude response over frequency range $0 \leq \omega \leq \pi$ rad/sec.

P11.15 Design a 10th-order comb filter having maximum gain = 1. Write this filter's transfer function, and use MATLAB® to plot its frequency magnitude response over $0 \leq \omega \leq \pi$ rad/sec.

P11.16 An 8th-order Butterworth analog lowpass filter, with cutoff frequency at 1 rad/sec, has transfer function:

$$H_c(s) = \frac{1}{s^8 + 5.13s^7 + 13.14s^6 + 21.85s^5 + 25.69s^4 + 21.85s^3 + 13.14s^2 + 5.13s + 1}.$$

Design a digital filter by letting $h_d(n) = T_s h_c(nT_s)$ with $T_s = 2$ sec. Use MATLAB® to plot $|H_c(\omega)|$ and $|H_d(e^{j\omega T_s})|$ on the same graph over frequency range $-\pi \leq \omega \leq \pi$ rad/sec.

P11.17 For the same 8th-order Butterworth analog lowpass filter as in P11.16, design a digital filter using the bilinear transformation method with $T_s = 2$ sec. Use MATLAB® to plot $|H_c(\omega)|$ and $|H_d(e^{j\omega T_s})|$ on the same graph over $-\pi \leq \omega \leq \pi$ rad/sec.

Chapter 12

Applications of S-Domain
Signal Processing

12.1 Introduction

As we learned in Ch. 10, because the Laplace transform is a generalization of the CTFT it has many of the same properties. One property both Laplace and Fourier transforms have is that convolution in time becomes multiplication in the transform domain. This is important since a linear time-invariant system is completely described by its impulse response, which relates output and input signals via convolution. Therefore, the multiplicative operator known as the *transfer function* is important in both ω and s domains.

The s-domain makes continuous-time systems simpler to analyze and to design. Among its many benefits, transforming from time domain to the s-domain makes it possible to algebraically solve constant-coefficient linear differential equations. We use the s-domain to design analog filters, to test for filter stability, and to simplify mathematical notation by avoiding direct mention of $j\omega$. Circuits can be easily analyzed in the s-domain, and we discuss methods to consider nonzero initial conditions. This chapter presents some interesting applications of s-domain representation and analysis. But first, let us take another look at linear, time-invariant continuous systems and how they may be characterized in the s-domain.

463

12.2 Linear System Analysis in the S-Domain

12.2.1 *Linear time-invariant continuous system*

Let us represent a linear, time-invariant continuous system as a black box with one input and one output. Such a system is completely described by its response, $h(t)$, to an impulse input. Instead of applying an impulse to the system input, the system's impulse response $h(t)$ may also be determined by solving the differential equation modeling the system. We will not be concerned with that aspect for the moment. We will assume that we have knowledge of the impulse response. A block diagram of a system having impulse response $h(t)$, input $f(t)$, and output $g(t)$ is shown below:

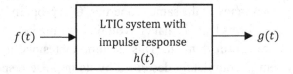

Figure 12.1. Linear system having output signal $g(t) = f(t) * h(t)$.

In a previous chapter, we discussed the significance of the impulse function. In the case of the Laplace transform, a generalization of Fourier transforms, an impulse function may be viewed as consisting of all *complex* frequency components e^{st}. This is easily recognized when we consider the inverse Laplace transform expression for an impulse:

$$\delta(t) = \frac{1}{2\pi j} \int_{\sigma_0 - j\infty}^{\sigma_0 + j\infty} 1\, e^{st} ds. \qquad (12.1)$$

Hence the output in response to an impulse at the input will have information about the behavior of the system to all frequencies. That is to say, $\mathcal{L}\{h(t)\} = H(s)$ is the output frequency response (coefficients) for all frequencies when the input frequency coefficients are all unity.

The definition of linearity was previously discussed along with the definition of a linear system. For a linear system, we can readily determine the relationship between the input $f(t)$ and output $g(t)$ of the system from the knowledge of $h(t)$ and/or $H(s)$. Since $F(s)$ represents the input coefficients for each frequency, and $f(t)$ is a linearly weighted (by $H(s)$) combination of all frequencies, the output frequency coefficients should be the product of $F(s)$ and $H(s)$ and the output $g(t)$ would be a linear combination of all frequencies in the product $G(s) = F(s)H(s)$.

Therefore, we may conclude that

$$g(t) \Longleftrightarrow G(s) = F(s) \times H(s). \tag{12.2}$$

And from the convolution property of Laplace transforms, we have

$$g(t) = \mathcal{L}^{-1}\{G(s)\} = f(t) * h(t). \tag{12.3}$$

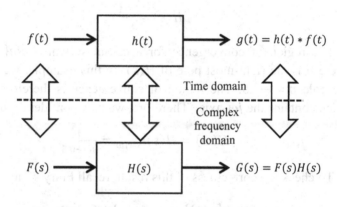

Figure 12.2. Linear system relationships in time and in complex frequency domains.

Figure 12.2 represents this process of determining output of a system with impulse response $h(t)$ to any input $f(t)$, and is analogous to that for the case of Fourier transform. With the knowledge of impulse response one could determine the output of a linear system via convolution process in the time domain OR if it is more advantageous, transform to frequency domain the input and impulse response and then convert the product to time domain using inverse transform.

12.2.2 *Frequency response from H(s)*

We mentioned that the Laplace transform is a generalization of the Fourier transform. So how then may we extract $H(\omega) = \mathcal{F}\{h(t)\}$ from the information present in $H(s) = \mathcal{L}\{h(t)\}$? One way is to recover $h(t)$ from $H(s)$ using the inverse Laplace transform, followed by calculating the Fourier transform to obtain $H(\omega)$.[a] A much easier way to find frequency response $H(\omega)$ is to substitute $s = j\omega$ into the expression for $H(s)$.[b] For this to be correct, we must be sure that the $j\omega$-axis (defined by $Re\{s\} = \sigma = 0$) lies within the region of convergence of $H(s)$; otherwise, our expression for $H(s)$ is not valid at the points where we wish to evaluate it.

For example, given a causal system with transfer function:

$$H(s) = \frac{1}{s+0.5}. \tag{12.4}$$

The region of convergence for a causal system is $Re\{s\} > Re\{p\}$, where p is the right-most pole of $H(s)$. In this example, we see there is one pole $p = -0.5$. The region of convergence is therefore $\sigma > -0.5$, which contains the $j\omega$-axis. Therefore, we may confidently state:

$$H(\omega) = H(s)|_{s=j\omega} = \frac{1}{j\omega+0.5}. \tag{12.5}$$

To check the correctness of this result, recall Entry #2 in Table 10.1:

$$\mathcal{L}\{e^{-at}u(t)\} = \frac{1}{s+a}, \quad \text{ROC: } \sigma > -a. \tag{12.6}$$

Using this Laplace transform pair to find $\mathcal{L}^{-1}\{H(s)\}$ in our example, we obtain $h(t) = e^{-0.5t}u(t)$. Next, refer to Table 5.1, Entry #2:

$$\mathcal{F}\{e^{-bt}u(t)\} = \frac{1}{b+j\omega}, \quad b > 0. \tag{12.7}$$

Substituting $b = 0.5$ gives us $\mathcal{F}\{e^{-0.5t}u(t)\} = \frac{1}{0.5+j\omega}$, which confirms our result in Eq. (12.5).

---

[a] Assuming that $\mathcal{F}\{h(t)\}$ exists.

[b] For this reason, $H(s)|_{s=j\omega} = $ "$H(j\omega)$" is often used to denote $H(\omega)$.

Here is an example where $H(\omega) = H(s)|_{s=j\omega}$ does *not* work: given $(s) = \frac{1}{s}$, and the system is causal. We are tempted to say that the frequency response $(\omega) = \frac{1}{j\omega}$, but our conclusion is incorrect because the ROC, $\sigma > 0$, does not include the $j\omega$ axis: $\mathcal{L}^{-1}\left\{\frac{1}{s}, ROC: \sigma > 0\right\} = u(t)$ and $\mathcal{F}\{u(t)\} = \pi\delta(\omega) + \frac{1}{j\omega} \neq \frac{1}{j\omega}$.

As discussed previously, the transfer function relates system output to an input complex exponential signal e^{st} with zero initial conditions. Then it is not surprising to realize that if we let $s = j\omega$, with ω being the sinusoidal signal frequency of interest, we would have obtained the phasor sinusoidal steady state response to a unit input. We will illustrate this with a circuit example.

Example 12.1
Determine the sinusoidal steady state voltage across the capacitor in Fig. 12.3 when the input voltage source is $V_i = 12\cos(10t)\,u(t)$.

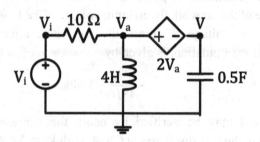

Figure 12.3. Circuit in Example 12.1.

To solve this network for the voltage across the capacitor, the best procedure is to write a generalized (super) node equation at node V_a. First, we convert the network to s-domain as shown in Fig. 12.4. We see that:

$$\frac{V_a - V_i}{10} + \frac{V_a}{4s} + \frac{V}{2/s} = 0, \tag{12.8}$$

$$V_a = 2V_a + V \text{ and } \therefore V = -V_a. \tag{12.9}$$

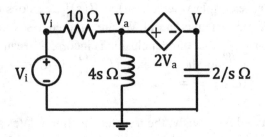

Figure 12.4. S-domain equivalent circuit in Example 12.1.

Substituting Eq. (12.9) into Eq. (12.8), and after some algebraic manipulation, we obtain

$$V\left(\frac{s}{2} - \frac{1}{4s} - \frac{1}{10}\right) = \frac{V_i}{10}, \tag{12.10}$$

$$\therefore V = \frac{2s}{10s^2 - 2s - 5} V_i. \tag{12.11}$$

We now substitute $s = j10$ and $V_i = 12\ V$ to obtain the steady state response of the network for an input voltage of $12\ V$ @ $\omega = 10$ rad/s. The output voltage V across the capacitor, after some complex algebraic manipulation, is given by

$$V = -\frac{48}{4 - j201} \text{ Volts.} \tag{12.12}$$

This result may be verified by using the sinusoidal steady state analysis technique discussed in Ch. 8, with $V_i = 12\ V$ @ 10 rad/s.

12.3 Applications of Pole-Zero Analysis

In Ch. 8 we studied frequency response functions $H(\omega)$ that satisfy the general definitions of basic filter types. For example, a lowpass filter will have $H(\omega) = 0$ at $\omega = \infty$, and $H(\omega) = 1$ at $\omega = 0$ (when normalized for peak unity gain). The independent variable of interest there was frequency variable ω. In this chapter, the s-domain transfer function $H(s)$ permits us to describe each basic filter type in terms of its poles and zeros with respect to complex frequency variable s. An advantage of the s-domain description is that it provides additional insight into the implementation and stability of these filters.

12.3.1 *Poles and zeros of realizable systems*

Most engineering problems of interest involve rational transfer functions $(H(s) = P(s)/Q(s))$. The coefficients of polynomials $P(s)$ and $Q(s)$ are the same coefficients found in the time-domain differential equations describing that system. In practical situations, these coefficients will be real-valued, and so will be the system's impulse response $h(t)$. From the *complex conjugate root theorem* in mathematics, when $H(s)$ is rational with real coefficients, then the poles and zeros of $H(s)$ will either be real or will appear in complex-conjugate pairs.

Another obvious requirement for rational $H(s)$ to be realizable is that it have finite order. Therefore, most every linear continuous-time system we need to deal with may be analyzed by considering a finite number of poles and zeros that are real, or in complex-conjugate pairs.

12.3.2 *Frequency response from pole/zero locations*

Consider the case where $H(s)$ is rational. Factoring its Mth order numerator and N^{th} order denominator polynomials gives

$$H(s) = S.F.\times \frac{(s-z_1)(s-z_2)\cdots(s-z_M)}{(s-p_1)(s-p_2)\cdots(s-p_N)} = S.F.\times \frac{\prod_{m=1}^{M}(s-z_m)}{\prod_{n=1}^{N}(s-p_n)}, \qquad (12.13)$$

where $S.F.$ is a constant (scale factor), numerator roots z_m are the zeros of $H(s)$, and denominator roots p_n are the poles of $H(s)$. Assume that the $j\omega$ axis lies within the region of convergence for $H(s)$, so that frequency response $H(\omega)$ may be found as $H(s)|_{s=j\omega}$.

12.3.2.1 *Magnitude response*

From Eq. (12.13), the magnitude frequency response function $|H(\omega)|$ may be expressed in terms of the poles and zeros of $H(s)$ as:

$$|H(\omega)| = \left|S.F.\times \frac{\prod_{m=1}^{M}(s-z_m)}{\prod_{n=1}^{N}(s-p_n)}\right|_{s=j\omega}$$

$$= |S.F.| \frac{\prod_{m=1}^{M}|s-z_m|}{\prod_{n=1}^{N}|s-p_n|} \quad \text{when } s = j\omega. \qquad (12.14)$$

Since $|s_a - s_b|$ is the distance between points s_a and s_b in the complex s-plane, we see that the magnitude of $H(\omega)$ at some frequency value $\omega = \omega_0$ is equal to $|S.F.|$ multiplied by the product of distances from each zero of $H(s)$ to the point $s_0 = j\omega_0$, divided by the product of distances from each pole of $H(s)$ to s_0. (Zeros and poles at ∞ do not directly enter into the calculation.) Thus, we may graphically find $|H(\omega)|$ at some frequency ω from knowledge of the pole-zero plot of $H(s)$ and its scale factor.

Example 12.2
Given: $H(s) = -6(s - 0.5)(s + 1)/(s + 0.5)$. Using Eq. (12.14), find $|H(\omega)|$ at frequency values $\omega = \{-0.5, 0, 0.5, \text{ and } 1\}$.

The graphical solutions for $|H(\omega)|$, based on the pole and zero locations in the s-plane, are shown in Figs. 12.5 through 12.9.

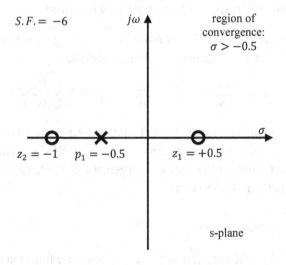

Figure 12.5. Pole-zero plot of $H(s) = \frac{-6(s-0.5)(s+1)}{s+0.5}$ from Example 12.2.

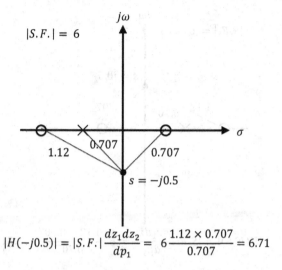

$$|H(-j0.5)| = |S.F.|\frac{dz_1 dz_2}{dp_1} = 6\frac{1.12 \times 0.707}{0.707} = 6.71$$

Figure 12.6. Graphically calculating $|H(-j0.5)|$ in Example 12.2.

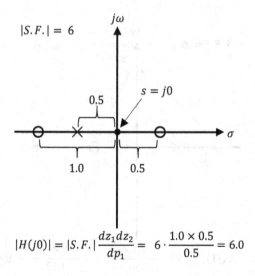

$$|H(j0)| = |S.F.|\frac{dz_1 dz_2}{dp_1} = 6 \cdot \frac{1.0 \times 0.5}{0.5} = 6.0$$

Figure 12.7. Graphically calculating $|H(j0)|$ in Example 12.2.

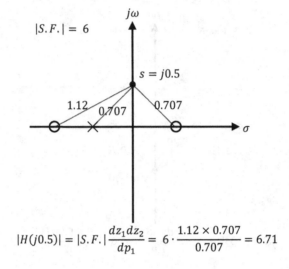

$$|H(j0.5)| = |S.F.| \frac{dz_1 dz_2}{dp_1} = 6 \cdot \frac{1.12 \times 0.707}{0.707} = 6.71$$

Figure 12.8. Graphically calculating $|H(j0.5)|$ in Example 12.2.

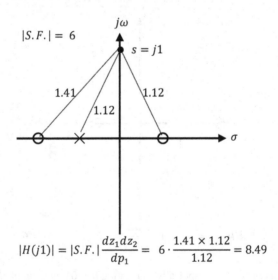

$$|H(j1)| = |S.F.| \frac{dz_1 dz_2}{dp_1} = 6 \cdot \frac{1.41 \times 1.12}{1.12} = 8.49$$

Figure 12.9. Graphically calculating $|H(j1)|$ in Example 12.2.

When plotting $|H(\omega)|$ on a logarithmic magnitude scale (e.g., as $10\log_{10}|H(\omega)|^2 = 20\log_{10}|H(\omega)|$ dB), the product of terms corresponding to each pole and zero becomes a sum:

$$20\log_{10}|H(\omega)| = 20\log_{10}\left\{|S.F.|\frac{\prod_{m=1}^{M}|j\omega-z_m|}{\prod_{n=1}^{N}|j\omega-p_n|}\right\}$$

$$= 20\log_{10}|S.F.| + \sum_{m=1}^{M}20\log_{10}|j\omega - z_m|$$

$$- \sum_{n=1}^{N}20\log_{10}|j\omega - p_n|. \qquad (12.15)$$

In this way, the contribution from each pole and zero (and the scale factor) is separated from the rest as an independent additive component in the log-magnitude plot of $H(\omega)$ vs ω.

Example 12.3
$H(s) = -6(s - 0.5)(s + 1)/(s + 0.5)$ as in the previous example. Write MATLAB® code to show the additive contribution of each pole and zero in a plot of $|H(\omega)|^2$ (in dB) vs ω.

Solution:

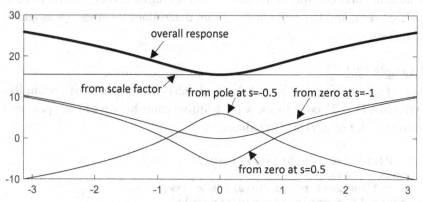

Figure 12.10. A graph of contributions to frequency magnitude response (dB) vs. ω from each zero, pole, and scale factor of $H(s)$ in Example 12.3.

The following MATLAB® code was used to draw the graph shown in Fig. 12.10:

```
% H(s)  =  -6(s-0.5)(s+1)/(s+0.5)
Npts = 1e4;
w = linspace(-pi,pi,Npts); s = j*w;
H_SF_dB = 20*log10(abs(-6*ones(size(w))));
H_z1_dB = 20*log10(abs(s-0.5));
H_z2_dB = 20*log10(abs(s+1));
H_p1_dB = 20*log10(abs(1./(s+0.5)));
plot(w,H_SF_dB); hold on
plot(w,H_z1_dB)
plot(w,H_z2_dB)
plot(w,H_p1_dB)
plot(w,H_SF_dB + H_z1_dB + H_z2_dB + H_p1_dB);
axis([-pi pi -10 30])
```

12.3.2.2 *Phase response*

Based on Eq. (12.13), the phase function $\angle H(\omega)$ is also easily expressed in terms of the poles and zeros of $H(s)$:

$$\angle H(\omega) = \angle S.F. + \sum_{m=1}^{M} \angle (j\omega - z_m) - \sum_{n=1}^{N} \angle (j\omega - p_n). \quad (12.16)$$

Note that the angles $\angle (j\omega - z_m)$ and $\angle (j\omega - p_n)$ are of vectors drawn *from* the point $s = j\omega$ *to* the zero or pole, not the other way around. The phase function is always an additive sum[c] of terms corresponding to each pole or zero, and the scale factor's sign determines whether $\pm\pi$ is added to this sum.

<u>Example 12.4</u>
For $H(z) = -6(s - 0.5)(s + 1)/(s + 0.5)$ as in previous examples, write MATLAB® code to show the additive contribution of each pole and zero in a plot of $\angle H(\omega)$ (in radians) vs ω.

```
% H(z)  =  -6(s-0.5)(s+1)/(s+0.5)
Npts = 1e4;
w = linspace(-pi,pi,Npts); s = j*w;
theta_SF = angle(-6*ones(size(w)));
theta_z1 = angle(s-0.5);
theta_z1 = unwrap(theta_z1);
theta_z1(theta_z1>0) = theta_z1(theta_z1>0)-2*pi;
theta_z2 = angle(s+1);
theta_p1 = angle(1./(s+0.5));
```

---

[c] Phase angle is never plotted on a logarithmic scale.

```
plot(w,theta_SF); hold on;axis([-pi pi -2*pi 2*pi]); pause
plot(w,theta_z1); pause
plot(w,theta_z2); pause
plot(w,theta_p1); pause
plot(w,theta_SF + theta_z1 + theta_z2 + theta_p1);
axis([-pi pi -2*pi 2*pi]); figure(gcf)
```

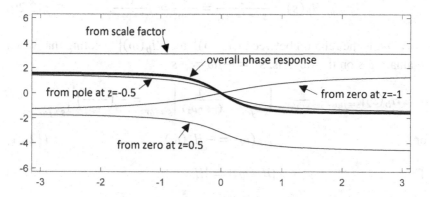

Figure 12.11. A graph of contributions to frequency phase response (rad) vs. ω from each zero, pole, and scale factor of $H(s)$ in Example 12.4.

From Examples 12.3 and 12.4, we see that plots of magnitude and phase contributions from poles and zeros shows us the effect of each of these on overall frequency response. This understanding may be important when designing filters.

12.3.3 *Effect on H(ω) of mirroring a pole about the jω axis*

It is interesting to see what happens to frequency response magnitude and phase contributions from a pole (or zero) that is flipped about the $j\omega$ axis by negating its real part. To investigate this, consider the simple case:

$$H_a(s) = \frac{1}{s - p_1}.$$ (12.17)

Expressing complex constant p_1 in rectangular form, we obtain

$$H_a(s) = \frac{1}{s - (\sigma_1 + j\omega_1)} = \frac{1}{s - \sigma_1 - j\omega_1}.$$ (12.18)

When the pole is flipped about the $j\omega$ axis, σ_1 becomes $-\sigma_1$:

$$H_b(s) = \frac{1}{s+\sigma_1-j\omega_1}, \tag{12.19}$$

which is $H_a(s)$ with its pole negated and conjugated:

$$H_b(s) = \frac{1}{s+\sigma_1-j\omega_1} = \frac{1}{s+p_1^*} = \frac{1}{(s^*+p_1)^*}. \tag{12.20}$$

What is the relation between $|H_a(\omega)|$ and $|H_b(\omega)|$? Noting that when evaluating s on the $j\omega$ axis, $s^* = -j\omega = -s$:

$$H_b(s)|_{s=j\omega} = \frac{1}{(s^*+p_1)^*}\bigg|_{s=j\omega} = \frac{1}{(-s+p_1)^*}\bigg|_{s=j\omega} = -\left(\frac{1}{s-p_1}\right)^*\bigg|_{s=j\omega},$$

or
$$H_b(\omega) = -H_a^*(\omega). \tag{12.21}$$

Thus
$$|H_b(\omega)| = |H_a(\omega)|, \tag{12.22}$$

and, since $\angle H_a^*(\omega) = -\angle H_a(\omega)$,

$$\angle H_b(\omega) = \pi - \angle H_a(\omega). \tag{12.23}$$

A similar result is obtained when a zero is flipped about the $j\omega$ axis: frequency response magnitude does not change except for a multiplicative constant, and the phase is subtracted from a constant.

12.3.4 *System stability*

A linear, shift-invariant system is stable[d] if its transfer function $H(s)$ has region of convergence that includes the $j\omega$ axis. That is, $H(s)$ must converge at $s = j\omega_0, \forall\omega_0$.

What are the practical consequences when an analog filter is unstable? In that case, for certain input signals, the filter's output signal grows in

---

[d] In the sense that an input signal that is restricted to a finite range of amplitude values will result in an output signal that is also restricted to a finite range of amplitude values. This is called *Bounded Input, Bounded Output* (BIBO) stability.

amplitude until it reaches the largest electrical voltage or current possible in the electronic circuit. Inevitably nonlinear behavior results, which can even lead to physical danger in some applications (e.g., controlling the braking system in a transport device).[e]

Let us consider two cases: (1) $H(s)$ is rational and the system is causal, and (2) $H(s)$ is rational and the system is anticausal.

12.3.4.1 *Causal systems*

For causal systems, the ROC of transfer function $H(s)$ is $\sigma > Re\{p\}$, where p is the right-most pole. If a causal system is stable (ROC includes the $j\omega$ axis) then all poles must lie to the left of the $j\omega$ axis in the s-plane.

12.3.4.2 *Anticausal systems*

For anticausal systems, the ROC of transfer function $H(s)$ is $\sigma < Re\{p\}$, where p is the left-most pole. If an anticausal system is stable (ROC includes the $j\omega$ axis) then all poles must lie to the right of the $j\omega$ axis in the s-plane.

12.3.4.3 *Stabilizing an unstable causal system*

Let's say we are designing a causal analog filter circuit, and we arrive at a desired magnitude response $|H(\omega)|$ with some choice of $P(s)$ and $Q(s)$ polynomials in transfer function $H(s) = P(s)/Q(s)$. However, after plotting the roots of $Q(s)$ we discover that some poles are to the right of the $j\omega$ axis making the filter unstable. How may we stabilize the system without changing its magnitude frequency response?

The solution is to negate the real parts of those poles that lie in the right-half s-plane so that they move to the left-half s-plane. From Section

---

[e] Unstable systems are not completely useless, however, because we can keep the output signal in check by applying a properly compensated input signal. This is a concept from the area of *control systems*, where feedback is used to control the output.

12.3.4, and specifically Eq. (12.22), we learned that mirroring a pole about the $j\omega$ axis will not change $|H(\omega)|$.[f] The following example demonstrates this technique.[g]

Example 12.5

A causal system with $H_a(s) = 2s/((s + 1/2)(s - 3/2))$ is unstable due to one pole lying in the right-half s-plane. Find a different transfer function, $H_b(s)$, representing a stable causal system, where $|H_b(\omega)| = |H_a(\omega)|$. Plot $|H_a(\omega)|$, $|H_b(\omega)|$, $\angle H_a(\omega)$ and $\angle H_b(\omega)$ in MATLAB®.

Negate the real part of the pole at $s = 3/2$ to give transfer function $H_b(s) = 2s/((s + 1/2)(s + 3/2))$.

```
Npts = 1e4; w = logspace(-2,2,Npts); s = j*w;
Ha = 2*s./((s+1/2).*(s-3/2));
Hb = 2*s./((s+1/2).*(s+3/2));
subplot(2,1,1); semilogx(w,abs(Ha))
axis([min(w) max(w) 0 1]); grid on
subplot(2,1,2); semilogx(w,abs(Hb));
axis([min(w) max(w) 0 1]); grid on
```

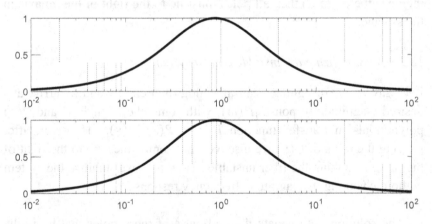

Figure 12.12 (a). A graph of $|H_a(\omega)|$ and $|H_b(\omega)|$ vs. ω in Example 12.5, confirming that they are the same frequency magnitude response.

---

[f] The phase response will change, but assume that is an acceptable price to pay for having a stable filter.

[g] If a pole lies *on* the $j\omega$ axis, then this method will not stabilize the system.

```
figure
subplot(2,1,1);
semilogx(w,angle(Ha))
grid on
subplot(2,1,2);
semilogx(w,angle(Hb));
grid on
```

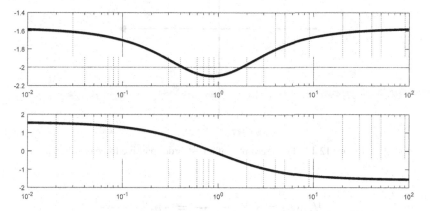

Figure 12.12 (b). A graph of $\angle H_a(\omega)$ and $\angle H_b(\omega)$ vs. ω in Example 12.5, showing how frequency phase response has changed as a result of flipping the pole locations about the $j\omega$ axis.

12.3.5 *Pole-zero plots of basic analog filters*

This section describes various low-order analog filters in terms of their transfer functions, pole/zero locations and RLC passive ladder circuits. The filters are stable and realizable: their poles lie in the left-half s-plane, and all poles/zeros are real or appear in complex conjugate pairs. Useful information regarding the bandwidth and passband gain is provided. You may use these diagrams as a guide for quickly designing a simple passive analog filter that is guaranteed to be stable.[h]

---

[h] Practical passive filters have some losses due to heat dissipation, and in the absence of an external source of power are guaranteed to be stable.

12.3.5.1 *Lowpass filter*

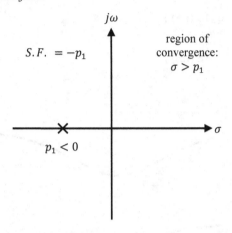

Figure 12.13. Pole-zero plot of a first-order passive lowpass filter.

$$H_{LP}(s) = S.F. \cdot \frac{1}{s - p_1} = -p_1 \left(\frac{1}{s - p_1} \right)$$

Peak passband gain = 1 at $\omega = 0$

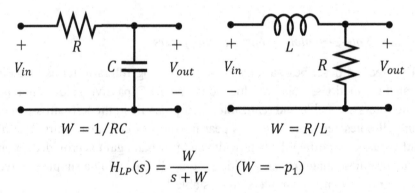

$$H_{LP}(s) = \frac{W}{s + W} \qquad (W = -p_1)$$

−3 dB Lowpass filter passband width = W rad/sec

12.3.5.2 *Highpass filter*

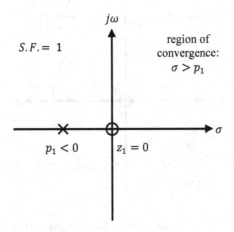

Figure 12.14. Pole-zero plot of a first-order passive highpass filter.

$$H_{HP}(s) = S.F. \frac{s - z_1}{s - p_1} = 1\left(\frac{s - 0}{s - p_1}\right)$$

Peak passband gain = 1 at $\omega = \infty$

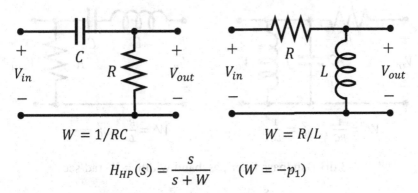

$W = 1/RC$ $W = R/L$

$$H_{HP}(s) = \frac{s}{s + W} \qquad (W = -p_1)$$

–3 dB Highpass filter stopband width = W rad/sec

12.3.5.3 *Bandpass filter*

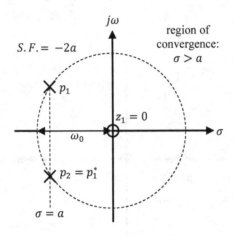

Figure 12.15. Pole-zero plot of a second-order passive bandpass filter.

$$H_{BP}(s) = S.F.\frac{(s - z_1)}{(s - p_1)(s - p_2)} = \frac{-2a(s - 0)}{s^2 - 2as + \omega_0{}^2}$$

$$a = Re\{p_1\}, \quad \omega_0 = |p_1|, \quad \text{peak gain} = 1 \text{ at } \omega = \omega_0$$

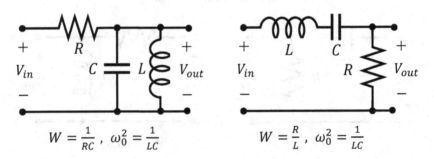

$$W = \frac{1}{RC}, \quad \omega_0^2 = \frac{1}{LC}$$

$$W = \frac{R}{L}, \quad \omega_0^2 = \frac{1}{LC}$$

−3 dB Bandpass filter passband width = W rad/sec

$$H_{BP}(s) = \frac{s}{s + \left(\dfrac{s^2 + \omega_0^2}{W}\right)} \qquad (W = -2a = -2Re\{p_1\})$$

12.3.5.4 *Notch (band-elimination) filter*

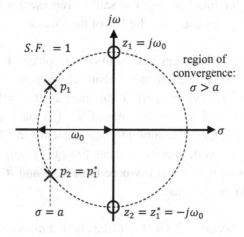

Figure 12.16. Pole-zero plot of a second-order passive notch filter.

$$H_N(s) = S.F.\frac{(s - z_1)(s - z_2)}{(s - p_1)(s - p_2)} = \frac{s^2 + \omega_0^2}{s^2 - 2as + \omega_0^2}$$

$$a = Re\{p_1\}, \quad \omega_0 = |p_1|, \quad \text{peak gain} = 1 \text{ at } \omega = \{0, \infty\}$$

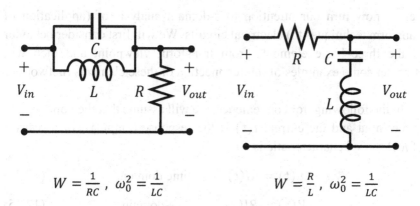

$$W = \frac{1}{RC}, \quad \omega_0^2 = \frac{1}{LC} \qquad\qquad W = \frac{R}{L}, \quad \omega_0^2 = \frac{1}{LC}$$

–3 dB Notch filter stopband width = W rad/sec, notch is at $\omega = \omega_0$

$$(W = -2a = -2Re\{p_1\})$$

12.3.6 *Minimum-phase system*

A rational transfer function $H(s)$ is said to represent a *minimum-phase* system when all poles and zeros lie left of the $j\omega$ axis.

Consider a causal system that is minimum-phase. It must be stable because all poles are in the left-half s-plane. If we reciprocate $H(s)$ of this system to obtain $H^{-1}(s) = 1/H(s)$, this new system will also be stable because the zeros of $H(s)$ are the poles of $H^{-1}(s)$ that also lie in the left-half s-plane. Therefore, a minimum-phase system can filter $x(t)$ to give $y(t)$, and an inverse system exists that can filter $y(t)$ to recover $x(t)$. That is to say, $h(t) * h^{-1}(t) = \delta(t)$, where both $h(t)$ and $h^{-1}(t)$ are causal and represent stable systems.

Recall, from Section 12.3.4 (Eq. (12.22)), that mirroring a pole or zero about the $j\omega$ axis does not change $|H(\omega)|$. Therefore, we may transform a non-minimum phase system into a minimum phase system, without affecting $|H(\omega)|$, by negating the real parts of any poles/zeros that lie in the right-half s-plane.

12.4 Circuit Analysis in the S-Domain

Let us now turn our attention to s-domain analysis of (application of transform techniques to) electrical circuits. We will first consider behavior of the three basic elements from transform viewpoint and then will consider some examples of interconnection of these elements (networks).

In the following, for convenience, we will assume that the voltage $v(t)$ is the input and the current $i(t)$ is the response (output). For a resistor of R Ω the v-i characteristic is

$$v(t) = Ri(t) \qquad \text{time domain,} \qquad (12.24)$$

$$V(s) = RI(s) \qquad \text{s-domain,} \qquad (12.25)$$

$$\therefore H(s) = V(s)/I(s) = R \ \ \Omega. \qquad (12.26)$$

The transfer function $H(s)$ has the units of impedance (Ω) and thus, in sinusoidal steady state circuit analysis, it is termed impedance $Z(s)$. If one assumes that the current $i(t)$ is the input and voltage $v(t)$ is the output, the transfer function will have units of Siemens and it is the inverse of impedance – admittance, denoted by $Y(s)$. An important observation to be made is that the transfer function is the ratio of output to input under zero initial conditions.

Let us consider the energy storage element, inductance L Henrys. Following an analogous derivation, we have started with the V-I characteristic:

$$v(t) = L\frac{di(t)}{dt} \qquad \text{time domain,} \qquad (12.27)$$

$$V(s) = sL\, I(s) - Li(0^-) \qquad \text{s-domain.} \qquad (12.28)$$

If the initial condition ($i(0^-)$) is assumed zero, then we can define

$$\therefore H(s) = \frac{V(s)}{I(s)} = sL \;\; \Omega. \qquad (12.29)$$

Thus, the impedance of an inductor is familiar value $Z(s) = sL\; \Omega$. Note that we have used one of the Laplace transform properties derived in Section 10.1.5.6 to arrive at the s-domain relationship.

A capacitor element, another energy storage element, C Farads has the impedance $Z(s) = 1/sC \;\; \Omega$. This may be derived as follows:

$$i(t) = C\frac{dv(t)}{dt} \qquad \text{time domain,} \qquad (12.30)$$

$$I(\omega) = sCV(s) - Cv(0^-) \qquad \text{s-domain,} \qquad (12.31)$$

$$\therefore H(s) = \frac{V(s)}{I(s)} = 1/sC \;\; \Omega \;;\; \text{zero initial voltage } v(0^-). \qquad (12.32)$$

The above results are summarized in Table 12.1. We will use these relationships, together with mesh and node analysis concepts from circuit theory, to derive s-domain equivalents of electrical circuits.

Table 12.1. V-I characteristic of R, L, and C in time and s-domains.

Element	Time domain / s-domain V-I Characteristics	Impedance $= \dfrac{V(s)}{I(s)}$ (Ω)
Resistor	$v(t) = Ri(t)$ $V(s) = R\,I(s)$	$Z_R(s) = R$
Inductor	$v(t) = L\dfrac{di(t)}{dt}$ $V(s) = sL\,I(s) - Li(0^-)$	$Z_L(s) = sL$
Capacitor	$i(t) = C\dfrac{dv(t)}{dt}$ $I(s) = sCV(s) - cv(0^-)$	$Z_C(s) = \dfrac{1}{sC}$

S-domain equivalent circuits for the two energy storage elements with nonzero initial conditioned are derived from the s-domain V-I characteristics. This leads to inclusion of a voltage or current source for each element., which are shown in Figs. 12.18 and 12.19.

Example 12.6
Find the voltage transfer function of the C-R ladder in Fig. 12.17(a).

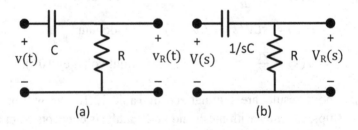

(a)   (b)

Figure 12.17. C-R ladder circuit in Example 12.6, and its s-domain equivalent.

Inductor with initial condition $i(0^-) \neq 0$:

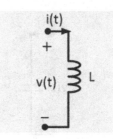

$$v(t) = L\frac{di}{dt}$$

s-domain Thevenin equivalent:

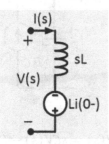

$$V(s) = sLI(s) - Li(0^-)$$

s-domain Norton equivalent:

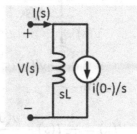

$$I(s) = \frac{V(s)}{sL} + \frac{i(0^-)}{s}$$

Figure 12.18. S-domain equivalent circuits for an inductor having nonzero initial conditions.

There are different approaches to finding the transfer function $H(s) = V_R(s)/V(s)$. They are analogous to the ones discussed for sinusoidal steady state analysis in Ch. 8. We will use the direct method of sinusoidal steady state analysis. First convert the element values into their s-domain impedances to obtain the transformed network shown in

Capacitor with initial condition $v(0^-) \neq 0$:

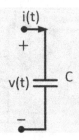

s-domain Thevenin equivalent: s-domain Norton equivalent:

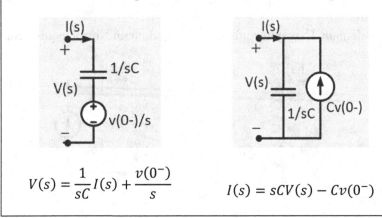

$$V(s) = \frac{1}{sC}I(s) + \frac{v(0^-)}{s} \qquad I(s) = sCV(s) - Cv(0^-)$$

Figure 12.19. S-domain equivalent circuits for a capacitor having nonzero initial conditions.

Fig. 12.17(b). Remember, when finding the transfer function, the initial conditions are assumed zero. We use series combination of impedances and voltage division to obtain the voltage transfer function as follows:

$$Z(s) = R + \frac{1}{sC} \ \ \Omega,$$

$$V_R(s) = \frac{R}{Z(s)}V(s) = \frac{sRC}{1+sRC} V(s),$$

$$\therefore H(s) = \frac{V_R(s)}{V(s)} = \frac{sRC}{1+sRC} = \frac{s}{s+\frac{1}{RC}}. \qquad (12.33)$$

This was an example of a first-order highpass filter. The transfer function has a pole at $s = -1/RC$ and zero at $s = 0$. The half-power frequency is $\omega_H = 1/RC$ rad/sec, and RC is called the time constant "τ" of the circuit. For a better understanding of filter design, the reader is referred to the many available textbooks, as well filter design handbooks, that deal with higher-order filters.

Recall that if the output is taken across the capacitor instead of the resistor, the output is that of a lowpass filter. For this case, it is easy to show that the transfer function

$$H(s) = \frac{V_C(s)}{V(s)} = \frac{1}{1+sRC}. \qquad (12.34)$$

Example 12.7

In the previous example described by Figs. 12.17 (a) and (b), replace the capacitor with an inductor and once again determine the circuit's voltage transfer function.

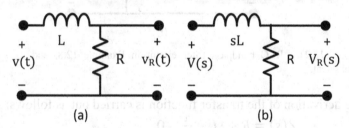

(a)   (b)

Figure 12.20. L-R ladder circuit in Example 12.7, and its s-domain equivalent.

We proceed with the analysis as in the previous example:

$$Z(s) = R + sL \ \Omega,$$

$$V_R(s) = \frac{R}{Z(s)} V(s) = \frac{R}{R+sL} V(s),$$

$$\therefore H(s) = \frac{V_R(s)}{V(s)} = \frac{R}{R+sL} = \frac{R/L}{s+R/L}. \qquad (12.35)$$

This was an example of a first-order lowpass filter. Note that the filter is defined by its pole at $= -R/L$. The half-power frequency $\omega_L = R/L$ rad/sec, and time constant $\tau = L/R$ sec.

Example 12.8
Consider the passive bandpass filter shown in Fig. 12.21(a), with its transformed version shown in 12.21(b). Determine its voltage transfer function.

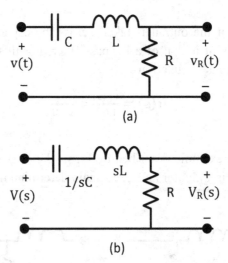

(a)

(b)

Figure 12.21.   RLC bandpass filter circuit in Example 12.8, and its s-domain equivalent.

The derivation of the transfer function is carried out as follows:

$$Z(s) = R + sL + \frac{1}{sC} \ \Omega,$$

$$V_R(s) = \frac{R}{Z(s)} V(s) = \frac{R}{R+sL+\frac{1}{sC}} V(s),$$

$$\therefore H(s) = \frac{V_R(s)}{V(s)} = \frac{sRC}{s^2LC+sRC+1} = \frac{s(R/L)}{s^2+s(R/L)+1/LC}. \qquad (12.36)$$

The transfer function of this band pass filter has zero at $s = 0$ and poles at $s = -R/(2L) \pm \sqrt{R^2C^2 - 4LC}/(2LC)$. Depending on the value of R, the poles may be either both real or both complex. In this circuit, if the $\{R\}$ and $\{L$ in series with $C\}$ branches are interchanged, the resulting circuit will have a notch filter transfer function characteristic.[i]

12.4.1 Transient Circuit Analysis

In this section, we will take up couple of simple examples to illustrate the power of s-domain analysis. The technique, coupled with node and mesh analysis, may be used to solve other complex circuit and system problems.

Example 12.9
Consider the circuit shown in Fig. 12.22. The switch in the network is closed and the circuit is in a steady state. At a reference time $t = 0$, the switch is opened. Find the current $i(t)$ that will flow in the inductor for $t \geq 0$.

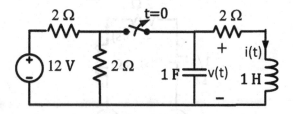

Figure 12.22. Circuit in Example 12.9.

Solution:
With the switch closed for a long time the circuit is in steady state. This allows us to determine the voltage across the capacitor and the current through the inductor before the switch opens at $t = 0$. To determine these initial conditions, we use the fact that for a dc source, the capacitor looks like an open circuit and the inductor looks like a short circuit. This leads to the following values:

$$v(0^-) = \frac{2||2}{2+2||2} \, 12 = \frac{1}{2+1} \, 12 = 4 \, V. \tag{12.37}$$

---

[i] $H_N(s) = \left(s^2 + \frac{1}{LC}\right) / \left(s^2 + s\left(\frac{R}{L}\right) + \frac{1}{LC}\right).$

$$i(0^-) = \frac{v(0^-)}{2} = 2\,A. \qquad (12.38)$$

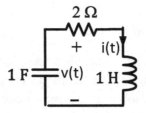

Figure 12.23. Circuit of interest in Example 12.9 for $t > 0$.

Figure 12.23 shows the circuit of interest when $t > 0$. Its s-domain version, shown in Fig. 12.24, has initial conditions incorporated using equivalent voltage sources (see Figs. 12.18 and 12.19):

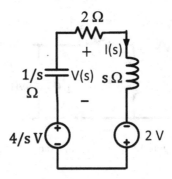

Figure 12.24. S-domain description of the circuit in Fig. 12.23 (Example 12.9).

We write a single mesh equation to analyze this circuit as follows:

$$\left(\frac{1}{s} + 2 + s\right)I(s) = 2 + \frac{4}{s}, \qquad (12.39)$$

$$I(s) = \frac{2(s+2)}{s^2 + 2s + 1} = \frac{2(s+2)}{(s+1)^2}. \qquad (12.40)$$

Using partial fraction expansion,

$$I(s) = \frac{2}{s+1} + \frac{2}{(s+1)^2}, \qquad (12.41)$$

and therefore

$$i(t) = [2e^{-t} + 2te^{-t}] u(t) \, A. \tag{12.42}$$

Example 12.10

Assuming zero initial conditions and $i(t) = 5e^{-2t}u(t) \, A$, determine $v(t)$ and $i_0(t)$ for $t > 0$ in the circuit below:

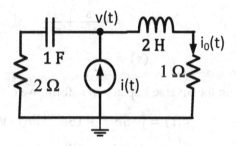

Figure 12.25. Circuit in Example 12.10.

We transform the network to s-domain. Since the initial conditions are zero, no additional sources are required. This circuit is shown in Fig. 12.26 where $I(s) = 5/(s + 2) \, A$.

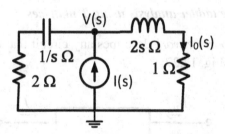

Figure 12.26. Circuit in Example 12.10, transformed to the s-domain.

We will use node analysis technique to solve for $V(s)$ and then we can readily determine the current $I_0(s)$ from it:

$$\frac{V(s)}{2+\frac{1}{s}} + \frac{V(s)}{2s+1} = I(s) = \frac{5}{s+2},$$

$$\therefore \frac{s+1}{2s+1} V(s) = \frac{5}{s+2}. \tag{12.43}$$

$$V(s) = \frac{5(2s+1)}{(s+1)(s+2)},$$

and
$$I_0(s) = \frac{V(s)}{2s+1} = \frac{5}{(s+1)(s+2)}. \tag{12.44}$$

Using partial fraction expansion, we have

$$V(s) = -\frac{5}{s+1} + \frac{15}{s+2},$$

and
$$I_0(s) = \frac{5}{s+1} - \frac{5}{s+2}. \tag{12.45}$$

After finding the inverse Laplace transform, we obtain

$$v(t) = [-5e^{-t} + 15e^{-2t}]u(t) \ V$$

and
$$i_0(t) = 5(e^{-t} - e^{-2t})u(t) \ A. \tag{12.46}$$

Note that we are now able to determine the instantaneous change in the current in the capacitor and the voltage across the inductor from $t = 0^-$ to $t = 0^+$. Verify that they are 5A and 10V, respectively.

12.4.2 *Passive ladder analysis using T matrices*

The term *two-port network* describes any circuit that may be put into the form shown in Fig. 12.27:

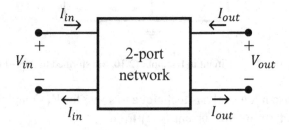

Figure 12.27.   Two-port network. Note that, by convention, currents are considered positive when entering the (+) terminal of each port.

When the circuit is linear and time-invariant, the relationship between externally applied voltage V_{in} and the resulting output voltage V_{out} is described in the s-domain as transfer function $H(s) = V_{out}(s)/V_{in}(s)$. Filter circuits made by concatenating series and shunt branches, or *ladder networks*, may be analyzed as two-port networks. The circuit shown in Fig. 12.28 is a ladder network.[j]

Transmission (or "T", sometimes called "ABCD") matrices are defined to relate $\{V_{in}, I_{in}\}$ to $\{V_{out}, I_{out}\}$ as follows:

$$V_{in} = A V_{out} - B I_{out},$$

$$I_{in} = C V_{out} - D I_{out}.$$

When expressed in matrix format,

$$\begin{bmatrix} V_{in} \\ I_{in} \end{bmatrix} = \begin{bmatrix} A & B \\ C & D \end{bmatrix} \begin{bmatrix} V_{out} \\ -I_{out} \end{bmatrix} = \overline{\mathbf{T}} \begin{bmatrix} V_{out} \\ -I_{out} \end{bmatrix}.$$

When working in the s-domain, coefficient B is an impedance, coefficient C is an admittance ($Y = 1/Z$), and coefficients A, D are dimensionless. T-matrix coefficients may be found using the following measures:

$$A = \frac{V_{in}}{V_{out}}\bigg|_{I_{out}=0} \qquad B = \frac{-V_{in}}{I_{out}}\bigg|_{V_{out}=0} \qquad C = \frac{I_{in}}{V_{out}}\bigg|_{I_{out}=0} \qquad D = \frac{-I_{in}}{I_{out}}\bigg|_{V_{out}=0}.$$

Of interest to us, for ladder analysis purposes, are these four properties of T-matrices:

(1) When two 2-port networks are placed in cascade, their overall T-matrix is the matrix product of their individual network T-matrices:
$$\overline{\mathbf{T}} = \overline{\mathbf{T}}_1 \times \overline{\mathbf{T}}_2$$

(2) The voltage transfer function of a 2-port network having no load termination ($I_{out} = 0$) is $H(s) = V_{out}(s)/V_{in}(s) = 1/A$.

---

[j] Usually the bottom terminal of the ladder is common to both input and output.

(3) The T-matrix of a two-port network having a series branch impedance $Z(s)$ is:

$$\overline{T} = \begin{bmatrix} 1 & Z(s) \\ 0 & 1 \end{bmatrix}$$

(4) The T-matrix of a two-port network having a shunt branch admittance $Y(s)$ is:

$$\overline{T} = \begin{bmatrix} 1 & 0 \\ Y(s) & 1 \end{bmatrix}$$

We demonstrate the application of T-matrix ladder analysis in the following example.

Example 12.11
Find an expression for $H(s) = V_{out}(s)/V_{in}(s)$, in terms of s and the unspecified component values, for the ladder circuit in Fig. 12.28 below:

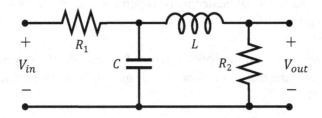

Figure 12.28. Passive ladder network in Example 12.11.

We note that the overall network is the cascade of a series branch (R_1), a shunt branch (C), a series branch (L), and a shunt branch (R_2). Therefore, the overall T-matrix will be the product of four T-matrices:

$$\overline{T} = \begin{bmatrix} 1 & Z_{R_1} \\ 0 & 1 \end{bmatrix} \begin{bmatrix} 1 & 0 \\ Y_C & 1 \end{bmatrix} \begin{bmatrix} 1 & Z_L \\ 0 & 1 \end{bmatrix} \begin{bmatrix} 1 & 0 \\ Y_{R_2} & 1 \end{bmatrix}.$$

We show the simplification process step-by-step:

$$\overline{T} = \underbrace{\begin{bmatrix} 1 & R_1 \\ 0 & 1 \end{bmatrix} \begin{bmatrix} 1 & 0 \\ sC & 1 \end{bmatrix}} \begin{bmatrix} 1 & sL \\ 0 & 1 \end{bmatrix} \begin{bmatrix} 1 & 0 \\ 1/R_2 & 1 \end{bmatrix}$$

$$= \begin{bmatrix} 1 + sR_1C & R_1 \\ sC & 1 \end{bmatrix} \underbrace{\begin{bmatrix} 1 & sL \\ 0 & 1 \end{bmatrix} \begin{bmatrix} 1 & 0 \\ 1/R_2 & 1 \end{bmatrix}}$$

$$= \begin{bmatrix} 1 + sR_1C & R_1 \\ sC & 1 \end{bmatrix} \begin{bmatrix} 1 + sL/R_2 & sL \\ 1/R_2 & 1 \end{bmatrix}.$$

Recall, we wish to find only coefficient A of the overall T-matrix, and thus calculating B, C, D is not necessary:

$$A = (1 + sR_1C)(1 + sL/R_2) + R_1/R_2$$

$$= 1 + sL/R_2 + sR_1C + s^2 R_1CL/R_2 + R_1/R_2$$

$$= s^2 \left(\frac{R_1CL}{R_2}\right) + s\left(\frac{L}{R_2} + R_1C\right) + \left(1 + \frac{R_1}{R_2}\right)$$

$$= \frac{R_1LC}{R_2}\left\{ s^2 + s\left(\frac{1}{R_1C} + \frac{R_2}{L}\right) + \left(\frac{R_1+R_2}{R_1LC}\right) \right\}.$$

This ladder circuit's voltage transfer function, therefore, is:

$$H(s) = \frac{V_{out}(s)}{V_{in}(s)} = \frac{1}{A} = \frac{\frac{R_2}{R_1LC}}{s^2 + s\left(\frac{1}{R_1C} + \frac{R_2}{L}\right) + \left(\frac{R_1+R_2}{R_1LC}\right)}.$$

Calculating $H(s)$ symbolically as we did is tedious; T-matrix analysis, however, is perfectly suited for matrix calculations using MATLAB®. The following code uses the T-matrix method to calculate and plot the frequency response $H(\omega)$ vs. ω for the same circuit:[k]

```
% Choose the following component values:
R1 = 1; C = 0.3; L = 2; R2 = 2;
```

---

[k] Recalculating the T-matrix at each value of ω is not the most efficient way of doing it, but this code runs fast anyway.

```
Npts = 1e4;
w = logspace(-2,2,Npts); % w from 0.01 to 100 rad/sec
H = zeros(1,Npts);
for n = 1:Npts
    s = j*w(n);
    T1 = [1, R1 ;  0,    1];
    T2 = [1, 0  ;  s*C,  1];
    T3 = [1, s*L;  0,    1];
    T4 = [1, 0  ;  1/R2, 1];
    T = T1*T2*T3*T4;
    H(n) = 1/T(1,1);   % A = T(1,1)
end
semilogx(w,abs(H))
grid on
```

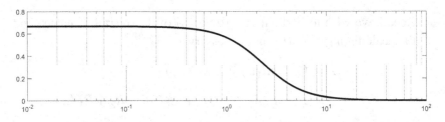

With this type of program, any number of branches in a ladder network may be easily included and analyzed.

12.5 Solution of Linear Differential Equations

In the previous chapter, we introduced Laplace transforms of several useful time-domain signals and considered how they may be represented as linear combinations of complex sinusoidal signals. In the engineering field, practical systems and devices are often modeled using linear components. Such mathematical models may be described with ordinary differential equations, or may be reduced to them. For time-invariant systems, the differential equation coefficients are constant since they depend on the parameters of the system. In some instances, the coefficients may be time-varying or the differential equations may be nonlinear. In those situations, it is common practice to develop an approximate (or piecewise) linear model. Here we will concern ourselves only with systems that may be modeled by constant coefficient linear differential equations.

Consider a general n^{th}-order ordinary differential equation with constant coefficients for response $y(t)$ with the input (forcing function) $f(t)$:

$$\frac{d^n y}{dt^n} + a_{n-1}\frac{d^{n-1}y}{dt^{n-1}} + a_{n-2}\frac{d^{n-2}y}{dt^{n-2}} + \cdots + a_1\frac{dy}{dt} + a_0 y(t)$$

$$= b_m\frac{d^m f}{dt^m} + b_{m-1}\frac{d^{m-1}f}{dt^{m-1}} + \cdots + b_1\frac{df}{dt} + b_0 f(t). \qquad (12.47)$$

Solve this equation using Laplace transform on both sides, under the assumption that $y(t)$ and $f(t)$ have Laplace transforms $Y(s)$ and $F(s)$, respectively. After some algebraic manipulation, the above equation reduces to

$$[s^n + a_{n-1}s^{n-1} + a_{n-2}s^{n-2} + \cdots a_1 s + a_0]Y(s)$$

$$= [b_m s^m + b_{m-1}s^{m-1} + \cdots + b_1 s + b_0]F(s). \qquad (12.48)$$

Here, we assumed all initial conditions on $y(t)$ at $t = 0^-$ are all zero. Consequently, according to the definition of transfer function, the ratio $Y(s)/F(s) = H(s)$. The system represented by the above differential equation has transfer function

$$H(s) = \frac{b_m s^m + b_{m-1}s^{m-1} + \cdots + b_1 s + b_0}{s^n + a_{n-1}s^{n-1} + a_{n-2}s^{n-2} + \cdots a_1 s + a_0}. \qquad (12.49)$$

The system impulse response $h(t) = \mathcal{L}^{-1}[H(s)]$.

One other important observation we should make is that from the knowledge of the transfer function, and thus by extension the impulse response, it is possible to reconstruct the differential equation governing the system. The transfer function directly helps determine not only the transient but also the sinusoidal steady state response of the system. Pole and zero locations of the transfer function play a key role in determining the stability and other aspects of a feedback control system.

Example 12.12
Consider a system described by the following differential equation. Determine its impulse response, transfer function and the zero-input response if $y(0^-) = 1$ and $dy(0^-)/dt = 0$.

$$\frac{d^2y(t)}{dt^2} + 5\frac{dy(t)}{dt} + 6y(t) = 2\frac{df(t)}{dt} + 3f(t). \qquad (12.50)$$

Taking the Laplace transform on both sides, we have

$$s^2Y(s) - sy(0^-) - \frac{dy(0^-)}{dt} + 5sY(s) - y(0^-) + 6Y(s)$$

$$= 2sF(s) + 3F(s). \qquad (12.51)$$

To find $H(s)$ and $h(t)$ we must assume the initial conditions are zero.

$$\therefore (s^2 + 5s + 6)Y(s) = (2s + 3)F(s) \qquad (12.52)$$

and
$$H(s) = \frac{Y(s)}{F(s)} = \frac{2s+3}{s^2+5s+6}. \qquad (12.53)$$

To find the impulse response we find the inverse Laplace transform of above transfer function using partial fraction expansion.

$$H(s) = \frac{2s+3}{s^2+5s+6} = \frac{2s+3}{(s+2)(s+3)} = -\frac{1}{s+2} + \frac{3}{s+3}, \qquad (12.54)$$

$$\therefore h(t) = [-e^{-2t} + 3e^{-3t}]u(t). \qquad (12.55)$$

To find the zero-input response, we set the right-hand side of the differential equation to zero (i.e., set $F(s) = 0$ in Eq. 12.51) and now include the initial conditions. Thus,

$$(s^2 + 5s + 6)Y(s) = s + 1, \qquad (12.56)$$

$$Y(s) = \frac{s+1}{s^2+5s+6} = -\frac{1}{s+2} + \frac{2}{s+3}, \qquad (12.57)$$

$$\therefore y_{ZIR}(t) = [-e^{-2t} + 2e^{-3t}]u(t). \qquad (12.58)$$

The system response to zero input is appropriately called $y_{ZIR}(t)$. It is also called a homogeneous solution to the differential equation.

Example 12.13
Determine the zero input response of the above system for the initial conditions $y(0^-) = 1$ and $dy(0^-)/dt = 1$.

We start with Eq. (12.51) and substitute $F(s) = 0$ and solve for $Y(s)$ to obtain

$$(s^2 + 5s + 6)Y(s) = s + 2, \tag{12.59}$$

$$Y(s) = \frac{s+2}{s^2+5s+6} = \frac{1}{s+3}, \tag{12.60}$$

$$\therefore y_{ZIR}(t) = e^{-3t}u(t). \tag{12.61}$$

We observe that with this choice of initial conditions, we do not measure the impact of the pole at $s = -2$ of the transfer function in the zero-input response!

12.6 Relation Between Transfer Function, Differential Equation, and State Equation

In this section, we show the significance of knowing the transfer function of a system. As discussed previously, in most cases the system's transfer function can be determined (within a constant) from a knowledge of poles and zeros.[1] For illustrative purposes, we will consider a 4th-order system whose general form of rational transfer function is given by:

$$H(s) = \frac{Y(s)}{F(s)} = \frac{b_3 s^3 + b_2 s^2 + b_1 s + b_0}{s^4 + a_3 s^3 + a_2 s^2 + a_1 s + a_0}. \tag{12.62}$$

12.6.1 *Differential equation from H(s)*

To obtain the corresponding differential equation we proceed as follows: from Eq. (12.62) we obtain

$$[s^4 + a_3 s^3 + a_2 s^2 + a_1 s + a_0] Y(s)$$

$$= [b_3 s^3 + b_2 s^2 + b_1 s + b_0] F(s). \tag{12.63}$$

Taking the inverse Laplace transform of the equation, term by term, and recalling that the initial conditions are all zero, we obtain the following differential equation:

---

[1] An exception to this case is when a pole and zero cancel each other (discussed in texts on control systems).

$$\frac{d^4y(t)}{dt^4} + a_3\frac{d^3y(t)}{dt^3} + a_2\frac{d^2y(t)}{dt^2} + a_1\frac{dy(t)}{dt} + a_0y(t)$$

$$= b_3\frac{d^3f(t)}{dt^3} + b_2\frac{d^2f(t)}{dt^2} + b_1\frac{df(t)}{dt} + b_0f(t). \qquad (12.64)$$

In this process, we have used the differentiation property of the Laplace transforms discussed previously in Ch. 10.

12.6.2 *State equations from H(s)*

To obtain a state equation form of system description we continue as discussed below. We shall consider only a single-input, single-output system. For that case, a standard form of state variable description of a system is:

State Equation: $\dot{X}(t) = [A]X(t) + [B]f(t),$ (12.65)

Output Equation: $y(t) = [C]X(t) + Df(t).$ (12.66)

Here $[A]$, $[B]$ and $[C]$ are appropriate-order matrices, and D is a scalar constant. Vector $X(t)$ is called a state variable and its length is determined by the order of the system. For the case of multiple-inputs and multiple-outputs, $y(t)$ and $f(t)$ are vectors and D is a matrix.

To relate the state variable description to a system's transfer function, we begin with the proper form of transfer function and introduce $X(s)$ in it as shown:

$$H(s) = \frac{Y(s)}{F(s)} = \frac{[b_3s^3+b_2s^2+b_1s+b_0]X(s)}{[s^4+a_3s^3+a_2s^2+a_1s+a_0]X(s)}. \qquad (12.67)$$

We now have

$$Y(s) = [b_3s^3 + b_2s^2 + b_1s + b_0]X(s), \qquad (12.68)$$

and $[s^4 + a_3s^3 + a_2s^2 + a_1s + a_0]X(s) = F(s). \qquad (12.69)$

Taking the inverse Laplace transform of Eq. (12.69) we obtain

$$\frac{d^4x(t)}{dt^4} + a_3\frac{d^3x(t)}{dt^3} + a_2\frac{d^2x(t)}{dt^2} + a_1\frac{dx(t)}{dt} + a_0x(t) = f(t). \qquad (12.70)$$

Define the following state variables:

$$x_1(t) = x(t),$$

$$x_2(t) = \frac{dx_1(t)}{dt},$$

$$x_3(t) = \frac{dx_2(t)}{dt},$$

$$x_4(t) = \frac{dx_3(t)}{dt}.$$

We may now write the following first-order differential equations involving variables $x_1(t)$, $x_2(t)$, $x_3(t)$ and $x_4(t)$ using 'dot' notation:

$$\dot{x}_1(t) = x_2(t),$$

$$\dot{x}_2(t) = x_3(t),$$

$$\dot{x}_3(t) = x_4(t).$$

And, from Eq. (12.70):

$$\dot{x}_4(t) = -a_0x_1(t) - a_1x_2(t) - a_2x_3(t) - a_3x_4(t) + f(t).$$

Converting these equations to a matrix form we obtain the state equation:

$$\begin{bmatrix} \dot{x}_1 \\ \dot{x}_2 \\ \dot{x}_3 \\ \dot{x}_4 \end{bmatrix} = \begin{bmatrix} 0 & 1 & 0 & 0 \\ 0 & 0 & 1 & 0 \\ 0 & 0 & 0 & 1 \\ -a_0 & -a_1 & -a_2 & -a_3 \end{bmatrix} \begin{bmatrix} x_1 \\ x_2 \\ x_3 \\ x_4 \end{bmatrix} + \begin{bmatrix} 0 \\ 0 \\ 0 \\ 1 \end{bmatrix} f(t) \qquad (12.71)$$

Proceeding similarly, from Eq. (12.68) we obtain the output equation:

$$y(t) = \begin{bmatrix} b_0 & b_1 & b_2 & b_3 \end{bmatrix} \begin{bmatrix} x_1 \\ x_2 \\ x_3 \\ x_4 \end{bmatrix} . \qquad (12.72)$$

One important aspect of deriving the state equations by this method is that there is no direct physical interpretation of the state variables. It is just a way to reformulate one M^{th}-order differential equation into a set of M simultaneous, first order equations to aid computer analysis in the absence of detailed knowledge of the system. It is possible to obtain a different set of state variables that may have a more physical relationship to the system's energy storage. Such state variables are interrelated.

12.7 Chapter Summary and Comments

- The Laplace transform maps an analog system's impulse response $h(t)$ to the s-domain, where it is called *system transfer function* $H(s)$.

- When the region of convergence of $H(s)$ includes the $j\omega$-axis, then a system's frequency response may be found from $H(s)$ as $H(\omega) = H(s)|_{s=j\omega}$.

- For a system to be *stable*, the region of convergence of $H(s)$ must include the $j\omega$-axis.

- In most practical applications $H(s)$ will be a rational function: $P(s)/Q(s)$. The roots of $P(s)$ are called the *zeros* of $H(s)$, and roots of $Q(s)$ are the *poles* of $H(s)$.

- Causal systems have ROC covering the s-plane region defined by the semi-plane $Re\{s\} > Re\{p\}$, where p is the right-most pole in the s-plane.

- Anticausal systems have ROC covering the s-plane region defined by the semi-plane $Re\{s\} < Re\{p\}$, where p is the left-most pole in the s-plane.

- A *minimum phase* causal system has all poles and zeros of $H(s)$ located left of the $j\omega$-axis. Both this system and its inverse are stable.

- To stabilize an unstable causal system without changing its magnitude frequency response, any poles that are right of the $j\omega$-axis should be mirrored about the $j\omega$-axis.

- A system's frequency magnitude (dB) and phase responses can be decomposed into additive contributions from poles and zeros in $H(s)$.

- Pole-zero diagrams contain all the information describing $H(s)$ when it is rational. These diagrams show the locations of finite-magnitude poles and zeros in the s-plane, and specify a scale factor.

- This chapter presents various 1st and 2nd-order digital filters (LP, HP, BP, BE) in terms of their s-domain transfer functions, pole-zero diagrams, and passive ladder circuit realizations.

- The Laplace transform simplifies frequency-dependent circuit analysis (i.e., circuits containing RLC components and sources). In this chapter, we analyze circuits in the s-domain including circuits having nonzero initial conditions.

- The Laplace transform is the standard method used to solve differential equations.

- An Mth-order differential equation may be reformulated as a coupled group of M first-order differential equations. These *state equations* are better suited for computer-aided analysis algorithms.

12.8 Homework Problems

P12.1 When $H(s) = \dfrac{s}{9s^2+2s+4}$ is the transfer function of a causal system, write an expression for the frequency response of that system.

P12.2 When $(s) = \dfrac{s^2-s+6}{s^2-s-6}$, what ROC corresponds to a stable system?

P12.3 Find $f(t)$ when $F(s) = \dfrac{1}{s-0.9}$ (ROC for stable system).

P12.4 Given: $H(s) = \dfrac{12s+2}{4s^2-1}$. Find $h(t)$ when the system is stable.

P12.5 Given: $H(s) = \dfrac{s^2-1}{s^2+1}$. Is this system stable? Justify your answer.

P12.6 Given: $H_a(s) = \dfrac{s-1}{s^2+2s-8}$, the transfer function of an unstable causal system. Find the transfer function of a stable causal system, $H_b(s)$, such that $|H_b(\omega)| = |H_a(\omega)|$.

P12.7 Given: $H_a(s) = \dfrac{-3}{s^2-2s-3}$, the transfer function of an unstable causal system. Find the transfer function of a stable causal system, $H_b(s)$, such that $|H_b(\omega)| = |H_a(\omega)|$. In MATLAB®, plot both $|H_a(\omega)|$ and $|H_b(\omega)|$ over range $0 \leq \omega \leq 10$ rad/sec.

P12.8 Given: $H_a(s) = \dfrac{s^2-2s-3}{4s^2-4s+101}$, the transfer function of a system that is not minimum-phase. Find the transfer function of minimum-phase system $H_b(s)$, such that $|H_b(\omega)| = |H_a(\omega)|$. In MATLAB®, plot $\angle H_a(\omega)$ and $\angle H_b(\omega)$ over $0 \leq \omega \leq 10$ rad/sec.

P12.9 When $H(s)$ is the transfer function of a causal system, why can't $H(s)$ have any poles at ∞?

P12.10 Is it possible for a notch filter to be a minimum-phase filter? Explain.

P12.11 Label each of the following pole/zero plots in the s-plane as being that of a LP, HP, BP, BE or AP filter:

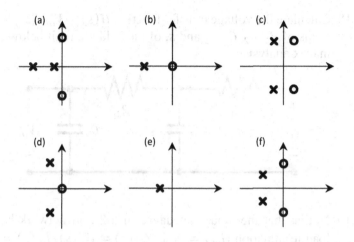

P12.12 Which pole-zero plot in problem P12.11 corresponds to a system having minimum-phase?

P12.13 Design an RC lowpass filter circuit having cutoff frequency = 10 rad/sec. Write this filter's transfer function $H(s)$, its frequency response function $H(\omega)$, and sketch its pole-zero diagram.

P12.14 Design and draw an RL highpass filter circuit having 5 rad/sec cutoff frequency. Write this filter's transfer function $H(s)$, and use MATLAB® to plot its frequency magnitude response over $10^{-2} \leq \omega \leq 10^2$ rad/sec on a log frequency scale.

P12.15 Design and draw an RLC bandpass filter circuit having peak $|H(\omega)|$ at 100 rad/sec. Write this filter's transfer function $H(s)$, and use MATLAB® to plot its frequency magnitude response *in dB* over $10^0 \leq \omega \leq 10^4$ rad/sec on a log frequency scale.

P12.16 Design and draw an RLC notch filter circuit having a notch at 1 rad/sec. Write this filter's s-domain transfer function $H(s)$, and write a simplified expression for $|H(\omega)|^2$.

P12.17 Find the impulse response $h(t)$ of an RC lowpass filter having cutoff frequency = 1 rad/sec.

P12.18 Calculate the voltage transfer function $H(s) = V_{out}(s)/V_{in}(s)$, in terms of R_1, R_2, C_1, C_2 and s, of the ladder circuit below using T-matrix analysis:

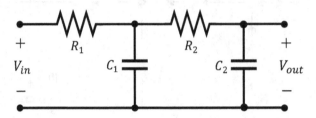

P12.19 Noting that the input impedance of a 2-port network having no load termination ($I_{out} = 0$) is $Z_{in}(s) = V_{in}(s)/I_{in}(s) = A/C$ in terms of T parameters, find an expression for $Z_{in}(s)$ for the ladder network shown in problem P12.18.

P12.20 Write MATLAB® code to plot the frequency phase response of the ladder network shown in problem P12.18, when $R_1 = R_2 = 1\ k\Omega$ and $C_1 = C_2 = 0.1$ uF, over $10^2 \le f \le 10^6$ Hz on a log frequency scale.

P12.21 For the circuit shown above in problem P12.18, assume that:
(a) $R_1 = R_2 = 1\ \Omega$ and $C_1 = C_2 = 0.5$ F;
(b) at times $t \ge 0$ the input port is shorted ($V_{in}(t) = 0$), and
(c) $V_{C1}(t = 0) = 1$ volt, and $V_{C2}(t = 0) = 2$ volts.
Find $V_{out}(t)$ for $t \ge 0$.

(*Note:* V_{C1} and V_{C2} are measured positive with respect to the common terminal of the ladder, just as V_{in} and V_{out} are.)

Appendix

Solved Homework Problems

Chapter 1

A1.1 Find the real part of $e^{j\pi/4}$.

> Answer: $Re(e^{j\pi/4}) = Re(\cos(\pi/4) + j\sin(\pi/4)) = \cos(\pi/4) = \sqrt{2}/2 = 0.707$.

A1.2 Find the imaginary part of $e^{-j\pi/2}$.

> Answer: $Im(e^{-j\pi/2}) = Re(\cos(\pi/2) - j\sin(\pi/2)) = -\sin(\pi/2) = -1$.

A1.3 Find the magnitude of $e^{j\theta}$ (for any real value of θ).

> Answer: $\left|e^{j\theta}\right|^2 = (e^{j\theta})(e^{j\theta})^* = e^{j\theta}e^{-j\theta} = e^{j\theta-j\theta} = e^0 = 1$, so $\left|e^{j\theta}\right| = \sqrt{1} = 1$.

A1.4 Find the magnitude of $Ae^{j\theta}$ (for any real values of A, θ).

> Answer: $\left|Ae^{j\theta}\right|^2 = (Ae^{j\theta})(Ae^{j\theta})^* = A^2e^{j\theta}e^{-j\theta} = A^2$, so $\left|Ae^{j\theta}\right| = \sqrt{A^2} = |A|$ (magnitude is always non-negative).

A1.5 Find the phase of $-5j$ in radians.

> Answer: When drawn on the complex plane as a vector of length 5 in the negative imaginary direction, its phase angle is $-\pi/2 = -1.571$ rad.

A1.6 Find the phase of -3 in degrees.

Answer: When drawn on the complex plane as a vector of length 3 in the negative real direction, its phase angle is $\pm\pi$ radians, or $\pm 180°$.

A1.7 Express $-2 + 2j$ in polar form.

Answer: $-2 + 2j = |-2 + 2j|e^{j\angle(-2+2j)} =$
$\sqrt{(-2)^2 + (2)^2}\ e^{\frac{j3\pi}{4}} = \sqrt{8} \angle 3\pi/4$ rad $= 2.828 \angle 135°$.

A1.8 Express $5e^{j\pi}$ in rectangular form.

Answer: $5e^{j\pi} = 5(\cos(\pi) + j\sin(\pi)) = -5 + j0$.

A1.9 Express $-e^{j7\pi}$ in rectangular form.

Answer: $-e^{j7\pi} = -(\cos(7\pi) + j\sin(7\pi)) = -(-1 + j0) = 1$.

A1.10 Find $(3 + 5j)(3 + 5j)^*$.

Answer: $(3 + 5j)(3 + 5j)^* = (3 + 5j)(3 - 5j) = 9 - 25j^2 = 9 + 25 = 34$.

A1.11 Write MATLAB® code to find the real part of $e^{j\pi/9}$.

```
>> real(exp(j*pi/9))
ans = 0.9397
```

A1.12 Write MATLAB® code to find the imaginary part of $e^{-j\pi/3}$.

```
>> imag(exp(-j*pi/3))
ans = -0.8660
```

A1.13 Write MATLAB® code to find the magnitude of $\pi e^{j45°}$.

```
>> abs(pi*exp(j*45*pi/180))
ans = 3.1416
```

A1.14 Write MATLAB® code to find the magnitude of $3e^{j\pi/8}$.

```
>> A = 3;
>> theta = pi/8;
```

```
>> abs(A*exp(j*theta))
ans = 3.0000
```

A1.15 Write MATLAB® code to find the magnitude of $-5e^{-6.3j}$.

```
>> A = -5;
>> theta = -6.3;
>> abs(A*exp(j*theta))
ans = 5
```

A1.16 Find the phase of $10 - 5j$ in degrees.

```
>> angle(10-5*j)*180/pi
ans = -26.5651
```

A1.17 Find the phase of -3 in radians.

```
>> angle(-3)
ans = 3.1416
```
(answer = $\pm\pi$ rad)

A1.18 Express $-4 + j$ in polar form.

```
>> abs(-4+j)
ans = 4.1231
>> angle(-4+j)
ans = 2.8966
```
(answer = $4.1231 \angle 2.8966$ rad)

A1.19 Express $4e^{-j\pi}$ in rectangular form.

```
>> 4*exp(-j*pi)
ans = -4.0000 - 0.0000i
```

A1.20 Express $e^{j\pi} + 1$ in rectangular form.

```
>> exp(j*pi)+1
ans = 0 + 1.2246e-16i
```
($= 0 + j0$ plus roundoff error)

A1.21 Express $(6 + 7j)(3 - 5j)^*$ in rectangular form.

```
>> (6+7*j)*conj(3-5*j)
ans = -17.0000 + 51.0000i
```

Chapter 2

A2.1 Simplify $\sum_{n=-\infty}^{\infty} \delta(n)x(n-2)$.

Solution:
$\sum_{n=-\infty}^{\infty} \delta(n)x(n-2) = \sum_{n=-\infty}^{\infty} \delta(n)x(0-2) =$
$x(-2)\sum_{n=-\infty}^{\infty} \delta(n) = x(-2)$.

A2.2 Simplify $\sum_{n=-\infty}^{\infty} \delta(n-1)y(n-2)$.

Solution:
$\sum_{n=-\infty}^{\infty} \delta(n-1)y(n-2) = \sum_{n=-\infty}^{\infty} \delta(n-1)y(1-2) =$
$y(-1)\sum_{n=-\infty}^{\infty} \delta(n-1) = y(-1)$.

A2.3 Find $y(26)$ when $(n) = 3\Delta_3(n) * \delta_5(n)$.

Solution:
$3\Delta_3(n) = \{1, 2, 3, 2, 1\}$ for $n = \{-2, -1, 0, 1, 2\}$; elsewhere this triangular pulse is zero. When convolved with an impulse train $\delta_5(n)$, copies of $3\Delta_3(n)$ are placed every 5 samples:
$y(n) = 3\Delta_3(n) * \delta_5(n) = 3\Delta_3(n) * \sum_{k=-\infty}^{\infty} \delta(n-k) =$
$\sum_{k=-\infty}^{\infty} 3\Delta_3(n-k)$. The sequence $y(n)$ is therefore periodic, having period 5 samples. Since $y(n) = y(n+5) = y(n+10)$, etc., $y(26) = y(26 - 5 \cdot 5) = y(1) = 2$.

A2.4 Is $w(n) = \cos(\pi n)$ periodic?
Answer: yes, $\cos(\omega_0 n)$ is periodic when $2\pi/\omega_0$ is an integer.

A2.5 State this in terms of impulse functions: $\text{rect}_2(n+1)\text{rect}_2(n-1)$

Solution:
$\text{rect}_2(n) = 1$ for $n = \{-2, -1, 0, 1, 2\}$; otherwise it is zero.
$\text{rect}_2(n+1) = 1$ for $n = \{-3, -2, -1, 0, 1\}$; otherwise it is zero.
$\text{rect}_2(n-1) = 1$ for $n = \{-1, 0, 1, 2, 3\}$; otherwise it is zero.

Therefore $\text{rect}_2(n+1)\text{rect}_2(n-1) = 1$ for $n = \{-1, 0, 1\}$; otherwise this product is zero. Stated in terms of impulse functions, $\text{rect}_2(n+1)\text{rect}_2(n-1) = \delta(n+1) + \delta(n) + \delta(n-1)$.

A2.6 State in terms of the product of two unit step functions: $\text{rect}_4(n)$
Answer: $\text{rect}_4(n) = u(n+4)u(-(n-4))$.

A2.7 Calculate the energy of $f(n) = -3j(5^{-n})u(n)$.

Solution:
Energy of $a^n u(n)$ is $\sum_{n=0}^{\infty}|a^n|^2 = \sum_{n=0}^{\infty}(a^2)^n = 1/(1-a^2)$ when $|a| < 1$. Also, energy of $cx(n)$ is equal to $|c|^2$ times the energy of $x(n)$. Therefore, energy of $f(n) = |-3j|^2 = 9$ times the energy of $(1/5)^n u(n) = 9/(1-0.2^2) = 9.375$. This is easily verified in MATLAB®: `sum(abs(-3*j*5.^(-[0:100]))).^2)`.

A2.8 What is the power of $u(n)$?
Solution: Power $= \lim_{M\to\infty} \frac{1}{2M+1} \sum_{n=-M}^{M}|u(n)|^2 =$
$\lim_{M\to\infty} \frac{1}{2M+1}\sum_{n=0}^{M}|1|^2 = \lim_{M\to\infty}\frac{M+1}{2M+1} = \lim_{M\to\infty}\frac{1+1/M}{2+1/M} = 0.5$.

A2.9 What is the energy of $\text{sgn}(n)$?
Solution: $\sum_{n=-\infty}^{\infty}|\text{sgn}(n)|^2 = \sum_{n=-\infty}^{\infty}|\pm 1|^2 = \infty$.

A2.10 What is the power of $\text{rect}_5(n)$?
Solution: Power $= 0$, since $\text{rect}_5(n)$ has finite energy $= 11$:
$\lim_{M\to\infty}\frac{1}{2M+1}(11) = 0$.

A2.11 Find the even part of $x(n) = \delta(n-2)$.
Solution: $x_e(n) = 0.5\{x(n) + x(-n)\} =$
$0.5\{\delta(n-2) + \delta(-n-2)\} = 0.5\{\delta(n-2) + \delta(-(n+2))\} =$
$= 0.5\{\delta(n-2) + \delta(n+2)\}$ since $\delta(n)$ is an even function.

A2.12 Simplify $\delta(n) * \text{sinc}(n-1)$:
Answer: $\text{sinc}(n-1)$.

A2.13 Simplify $\delta(n + 1) * \text{sinc}(n - 1)$:

Answer: $\text{sinc}(n)$.

A2.14 Periodic sequence $x_p(n)$ has period 3 samples, and its sample values over time span $0 \leq n \leq 8$ are $\{1, 1, 2, 1, 1, 2, 1, 1, 2\}$. Express $x_p(n)$ as a convolution product of impulse train with triangular pulse function.

Solution: $2\Delta_2(n) = \{1, 2, 1\}$ for $n = \{-1, 0, 1\}$, and is zero elsewhere. Since the peak occurs at $n = 0$, but in $x_p(n)$ the peak occurs at $n = 2$, there must be a delay of two samples. Thus,
$x_p(n) = 2\Delta_2(n - 2) * \delta_3(n) = 2\Delta_2(n) * \delta_3(n - 2)$.

Chapter 3

A3.1 $\text{sgn}(t - 2)|_{t=-3} = ?$ Answer: $\text{sgn}(-5) = -1$.

A3.2 $-u(t)|_{t=-2} = ?$ Answer: $-u(-2) = -0 = 0$.

A3.3 $-u(t + 2)|_{t=-3} = ?$ Answer: $-u(-1) = -0 = 0$.

A3.4 $\text{rect}\left(t - \frac{1}{2}\right)\Big|_{t=\frac{1}{2}} = ?$ Answer: $\text{rect}(0) = 1$.

A3.5 $\text{rect}\left(\frac{t-3}{2}\right)\Big|_{t=1} = ?$ Answer: $\text{rect}(-1) = 0$.

A3.6 $\text{rect}\left(-t - \frac{1}{2}\right)\Big|_{t=\frac{1}{2}} = ?$ Answer: $\text{rect}(-1) = 0$.

A3.7 $\text{rect}(2t - 2)|_{t=-1} = ?$ Answer: $\text{rect}(-4) = 0$.

A3.8 $-\text{rect}(1 - t/2)|_{t=2} = ?$ Answer: $-\text{rect}(0) = -1$.

A3.9 $\left(\text{rect}(t/2) - \text{rect}(t)\right)\Big|_{t=2} = ?$

Answer: $\text{rect}(1) - \text{rect}(2) = 0 - 0 = 0$.

A3.10 $\int_{-\infty}^{t} \delta\left(\tau + \frac{1}{2}\right) d\tau\Big|_{t=0} = ?$

Answer: $u(t + 1/2)|_{t=0} = u(1/2) = 1$.

A3.11 Express $\text{sgn}(t)$ as a function of $u(t)$.

Answer: $\text{sgn}(t) = 2u(t) - 1$.

A3.12 Express $\text{rect}(t)$ in terms of functions $\Delta(t)$ and $\text{sgn}(t)$.

Answer: $\text{rect}(t) = \text{sgn}(\Delta(t))$.

A3.13 Express $\Delta(t)$ as a function of only $r(t)$ and $u(t)$.

Answer: $\Delta(t) = 2r(t + 1/2)u(-t) + 2r(-t + 1/2)u(t)$

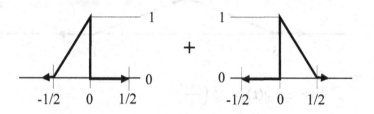

A3.14 Express $x(t)$ shown below as a sum of unit step functions at various time delays.

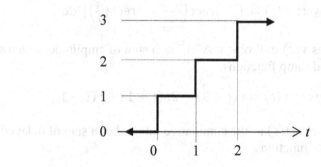

Answer:
$$x(t) = u(t) + u(t - 1) + u(t - 2).$$

A3.15 Express $x(t)$ in Problem A3.14 as a sum of amplitude-scaled and delayed rectangular pulse functions.

Answer: $x(t) = \text{rect}(t - 1/2) + 2\,\text{rect}(t - 3/2)$

$$+ \sum_{k=3}^{\infty} 3\,\text{rect}(t - (2k - 1)/2).$$

A3.16 Express $x(t)$ in Problem A3.14 as the cumulative integral of a sum of delayed impulse functions.

Answer: $x(t) = \int_{-\infty}^{t} \{\delta(\alpha) + \delta(\alpha - 1) + \delta(\alpha - 2)\}\, d\alpha.$

A3.17 Express $y(t)$ shown below as an amplitude-scaled and time-scaled triangular pulse function.

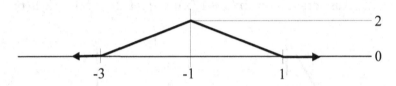

Answer: $y(t) = 2\Delta\left(\frac{t+1}{4}\right).$

A3.18 Express $y(t)$ in Problem A3.17 as a cumulative integral of time-scaled and delayed rectangular pulses.

Answer: $y(t) = \int_{-\infty}^{t} \left\{\text{rect}\left(\frac{\alpha+2}{2}\right) - \text{rect}\left(\frac{\alpha}{2}\right)\right\} d\alpha.$

A3.19 Express $y(t)$ in Problem A3.17 as a sum of amplitude-scaled and delayed ramp functions.

Answer: $y(t) = r(t + 3) - 2r(t + 1) + r(t - 1).$

A3.20 Express $\text{rect}(t)$ as the cumulative integral of a sum of delayed impulse functions.

Answer: $\text{rect}(t) = \int_{-\infty}^{t} \{\delta(\alpha + 1/2) - \delta(\alpha - 1/2)\}\, d\alpha.$

A3.21 Express $w(t)$ shown below as a product of triangular pulse and unit step functions.

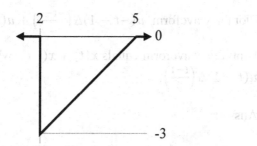

Answer: $w(t) = -3\Delta((t-2)/6)u(t-2)$.

A3.22 Express $w(t)$ in Problem A3.21 as the cumulated integral of the sum of impulse and rectangular pulse functions.

Answer: $w(t) = \int_{-\infty}^{t}\{-3\delta(\alpha - 2) + \text{rect}((\alpha - 3.5)/3)\}\,d\alpha$.

A3.23 Express the waveform below as the sum of two triangular pulses.

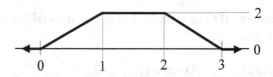

Answer: $2\Delta\left(\frac{t-1}{2}\right) + 2\Delta\left(\frac{t-2}{2}\right)$.

A3.24 Express the waveform below in terms of a time derivative of a sum of time-scaled and delayed triangular pulse functions.

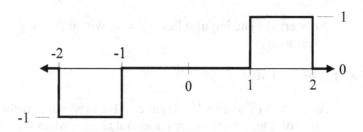

Answer: $\frac{d}{dt}\left\{-\Delta\left(\frac{t+1}{2}\right) - \Delta\left(\frac{t}{2}\right) - \Delta\left(\frac{t-1}{2}\right)\right\}$.

A3.25 Plot the waveform $u(-t-1)\Delta\left(\frac{-t-1}{2}\right) + u(t-1)\Delta\left(\frac{t-1}{2}\right)$.

Hint: this waveform equals $x(t) + x(-t)$, where $x(t) = u(t-1)\Delta\left(\frac{t-1}{2}\right)$.

Answer:

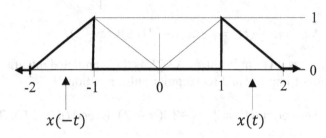

A3.26 $\int_{t_1}^{t_2} \delta(t)dt = ?$ Assume $t_1, t_2 < 0$.

Answer: 0 (The impulse lies at $t = 0$, outside the range of integration).

A3.27 $\int_{t_1}^{t_2} \delta(t)dt = ?$ Assume $t_1, t_2 > 0$.

Answer: 0 (The impulse lies at $t = 0$, outside the range of integration).

A3.28 $\int_{t_1}^{t_2} \delta(t)dt = ?$ Assume $t_1 < 0$, $t_2 > 0$.

Answer: 1 (The impulse lies at $t = 0$, within the range of integration).

A3.29 $\int_{-\infty}^{\infty} \delta(t-1)dt = ?$

Answer: 1 (The impulse lies at $t = 1$ sec, which is within the range of integration; its area does not change when it is shifted in time).

A3.30 $\int_{-\infty}^{\infty} \delta(t+1)dt = ?$

Answer: 1 (The impulse lies at $t = -1$ sec, which is within the range of integration; its area does not change when it is shifted in time).

A3.31 $\int_{-\infty}^{\infty}\{\delta(t+1) + \delta(t) + \delta(t-1)\}dt = ?$

Answer: 3 (The three impulses each with area 1, lie at times $t = \{-1, 0, +1\}$ sec).

A3.32 $\int_{-\infty}^{\infty}\{\delta(t+1) - \delta(t) + \delta(t-1)\}dt = ?$

Answer: 1 (There are three impulses, two having area 1 and one having area -1).

A3.33 $\int_{-\infty}^{\infty}\{2\delta(t+1) + \delta(t) - 2\delta(t-1)\}dt = ?$

Answer: 1 (The impulses have areas $2, 1, -2$).

A3.34 $\int_{-\infty}^{\infty}\delta(-t)dt = ?$

Answer: 1 (Time-reversing an impulse has no effect on its area. Also, in this specific case, $\delta(-t) = \delta(t)$).

A3.35 $\int_{-\infty}^{\infty}\delta(t/2)dt = ?$

Answer: 2 (Stretching the time scale by factor 2 doubles the impulse area, even though it is still infinitely tall and infinitesimally narrow.

A3.36 $\int_{-\infty}^{\infty}\delta(3t)dt = ?$

Answer: 1/3 (Compressing time scale by factor 3 scales the impulse area by 1/3).

A3.37 $\int_{-\infty}^{\infty}x(t)\delta(t-t_0)dt = ?$

Answer: $x(t_0)$ (This is the sifting property of the impulse function).

A3.38 $\int_{-\infty}^{\infty}x(t_0)\delta(t-t_0)dt = ?$

Answer: $x(t_0)$ (Treat $x(t_0)$ as a constant, move it outside of the integral.)

A3.39 $\int_{-\infty}^{\infty} x(t_1)\delta(t - t_0)dt = ?$ $(t_0 \neq t_1)$

Answer: $x(t_1)$ (treat $x(t_1)$ as a constant, move it outside of the integral; the information that $t_0 \neq t_1$ is superfluous.)

A3.40 $\int_{-\infty}^{\infty} \cos(t)\,\delta(t)dt = ?$

Answer: 1 ($\cos(t)\,\delta(t) = \cos(0)\,\delta(t) = \delta(t)$, area $= 1$.)

A3.41 $\int_{-\infty}^{\infty} \sin(t)\,\delta(t + \pi/2)dt = ?$

Answer: -1 ($\sin(t)\,\delta(t + \pi/2) = \sin(-\pi/2)\,\delta(t + \pi/2) = -\delta(t + \pi/2)$, which has area -1.)

A3.42 $\int_{-\infty}^{\infty}(t^2 - 4)\delta(t + 1)dt = ?$

Answer: -3 ($(t^2 - 4)\delta(t + 1) = ((-1)^2 - 4)\delta(t + 1) = -3\delta(t + 1)$).

A3.43 $\int_{-\infty}^{\infty} \delta_1(t)dt = ?$

Answer: ∞ (Impulse train $\delta_1(t) = \sum_{k=-\infty}^{\infty} \delta(t - k)$ has infinitely many impulses, each having area 1.)

A3.44 Sketch $\mathrm{sgn}(t - 2)$.

Answer:

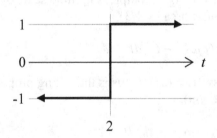

A3.45 Sketch $-u(t)$.

Answer:

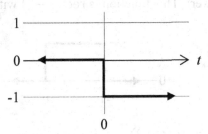

A3.46 Sketch $-u(t + 2)$.

Answer:

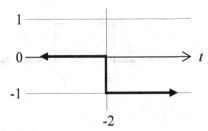

A3.47 Sketch $\text{rect}\left(t - \frac{1}{2}\right)$.

Answer:

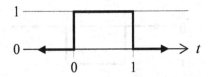

A3.48 Sketch $\text{rect}\left(\frac{t}{2} - \frac{1}{2}\right) = \text{rect}\left(\frac{t-1}{2}\right)$.

Answer: This function is $\text{rect}\left(\frac{t}{2}\right)$ with a 1-sec delay.

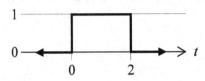

A3.49 Sketch $\text{rect}\left(-t - \frac{1}{2}\right)$.

Answer: This function is $\text{rect}\left(t - \frac{1}{2}\right)$ with time-reversal.

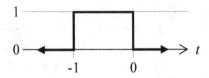

A3.50 Sketch $\text{rect}(2t - 2)$.

Answer: This is $\text{rect}(2(t - 1))$, or $\text{rect}(2t)$ with 1-sec delay.

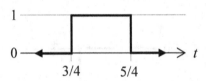

A3.51 Sketch $-\text{rect}\left(1 - \frac{t}{2}\right)$.

Answer: This is $-\text{rect}\left(\frac{-t+2}{2}\right)$, or $-\text{rect}\left(\frac{t}{2}\right)$ with a 2-sec advance and then time-reversal.

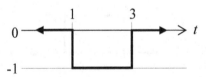

A3.52 Sketch $\text{rect}(t^2 - 4)$.

Answer: This function has pulse edges at $|t^2 - 4| = \frac{1}{2}$.
$t = \pm\sqrt{7/2} = \pm 1.87, \ \pm\sqrt{9/2} = \pm 2.12$ ($t = \pm 2$ lie within the pulses).

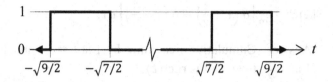

A3.53 Sketch $\text{rect}\left(t^2 - \frac{1}{4}\right)$.

Answer:
This function has pulse edges at $\left|t^2 - \frac{1}{4}\right| = \frac{1}{2}$.
$t = \pm\sqrt{3/4} = \pm 0.87$ ($t = 0$ lies within the pulse):

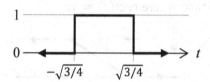

A3.54 Sketch $\text{rect}(t/2) - \text{rect}(t)$.

Answer: This is a pulse of width 2 minus a pulse of width 1:

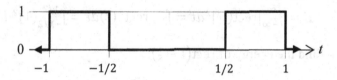

A3.55 Sketch $\int_{-\infty}^{t} \delta\left(\tau + \frac{1}{2}\right) d\tau$.

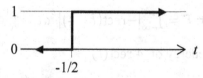

Answer: Cumulative area jumps by 1 at $= -\frac{1}{2}$; the result is $u\left(t + \frac{1}{2}\right)$.

A3.56 Sketch $\int_{-\infty}^{t} \left\{ \delta\left(\tau + \frac{1}{2}\right) - \delta\left(\tau - \frac{1}{2}\right) \right\} d\tau$.

Answer: Cumulative area jumps by 1 at $t = -\frac{1}{2}$ and drops by 1 at $= \frac{1}{2}$; the result is rect(t):

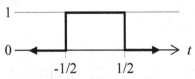

A3.57 Sketch $\int_{-\infty}^{t} \text{rect}(\tau) d\tau$.

Answer: Cumulative area linearly rises from 0 to 1 over the span of time where rect(t) = 1:

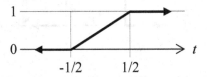

A3.58 Find the energy of rect(t).

Answer:
$$E = \int_{-\infty}^{\infty} |\text{rect}(t)|^2 dt = \int_{-\infty}^{\infty} \text{rect}^2(t) dt = \int_{-1/2}^{1/2} (1)^2 dt = 1.$$

A3.59 Find the energy of $\text{rect}\left(t - \frac{1}{2}\right)$.

Answer: $E = \int_{-\infty}^{\infty} \left|\text{rect}\left(t - \frac{1}{2}\right)\right|^2 dt = \int_{0}^{1} (1)^2 dt = 1$.

A3.60 Find the energy of $-\text{rect}\left(t - \frac{1}{2}\right)$.

Answer: $E = \int_{-\infty}^{\infty} \left|-\text{rect}\left(t - \frac{1}{2}\right)\right|^2 dt = \int_{0}^{1} (-1)^2 dt = 1$.

A3.61 Find the energy of $4\,\text{rect}(t)$.

Answer: $E = \int_{-\infty}^{\infty} |4\text{rect}(t)|^2 dt = \int_{-1/2}^{1/2} (4)^2 dt = 16$.

A3.62 Find the energy of $\Delta(t)$.

Answer: $E = \int_{-\infty}^{\infty} |\Delta(t)|^2 dt = \int_{-1/2}^{1/2} \Delta^2(t) dt = 2\int_0^{1/2} \Delta^2(t) dt$
$$= 2\int_0^{1/2}(1-2t)^2 dt = 2\int_0^{1/2}(1-4t+4t^2)dt$$
$$= 2\left(t - 2t^2 + \tfrac{4}{3}t^3\right)\Big|_{t=0}^{t=1/2} = 2\left(\tfrac{1}{2} - \tfrac{1}{2} + \tfrac{1}{6}\right) = 1/3.$$

A3.63 Find the energy of $\text{rect}(t/3)$.

Answer: $E = \int_{-\infty}^{\infty} |\text{rect}(t/3)|^2 dt = \int_{-3/2}^{3/2}(1)^2 dt = 3.$

A3.64 Find the energy of $\text{rect}\left(t + \tfrac{1}{2}\right) - \text{rect}\left(t - \tfrac{1}{2}\right)$.

Answer:

$E = \int_{-\infty}^{\infty}\left|\text{rect}\left(t + \tfrac{1}{2}\right) - \text{rect}\left(t - \tfrac{1}{2}\right)\right|^2 dt = \int_{-1}^{1}(\pm 1)^2 dt = 2.$

A3.65 Find the energy of $\text{rect}\left(t + \tfrac{1}{2}\right) + \text{rect}\left(t - \tfrac{1}{2}\right)$.

Answer:

$E = \int_{-\infty}^{\infty}\left|\text{rect}\left(t + \tfrac{1}{2}\right) + \text{rect}\left(t - \tfrac{1}{2}\right)\right|^2 dt = \int_{-1}^{1}(1)^2 dt = 2.$

A3.66 Find the energy of $\text{rect}\left(t + \tfrac{1}{4}\right) + \text{rect}(t - \tfrac{1}{4})$.

Answer:

$E = \int_{-\infty}^{\infty}\left|\text{rect}\left(t + \tfrac{1}{4}\right) + \text{rect}\left(t - \tfrac{1}{4}\right)\right|^2 dt = \int_{-3/4}^{-1/4}(1)^2 dt +$
$\int_{-1/4}^{1/4}(2)^2 dt + \int_{1/4}^{3/4}(1)^2 dt = \tfrac{1}{2} + \tfrac{4}{2} + \tfrac{1}{2} = 3.$

A3.67 Find the energy of $\sin(4t)$.

Answer: $E = \int_{-\infty}^{\infty} |\sin(4t)|^2 dt = \infty.$

A3.68 Find the power of $\text{sgn}(t)$.

Answer: $\lim\limits_{T\to\infty} \tfrac{1}{T}\int_{-T/2}^{T/2}|\text{sgn}(t)|^2 dt = \lim\limits_{T\to\infty} \tfrac{1}{T}\int_{-T/2}^{T/2}(1)^2 dt =$
$\lim\limits_{T\to\infty} \tfrac{1}{T}(T) = \lim\limits_{T\to\infty}(1) = 1.$

A3.69 Find the power of $u(t)$.

Answer: $\lim_{T\to\infty} \frac{1}{T} \int_{-T/2}^{T/2} |u(t)|^2 dt = \lim_{T\to\infty} \frac{1}{T} \int_0^{T/2} (1)^2 dt =$
$\lim_{T\to\infty} \frac{1}{T} \left(\frac{T}{2}\right) = \lim_{T\to\infty} \left(\frac{1}{2}\right) = \frac{1}{2}.$

A3.70 Find the power of $-u(t)$.

Answer: $\lim_{T\to\infty} \frac{1}{T} \int_{-T/2}^{T/2} |u(-t)|^2 dt = \lim_{T\to\infty} \frac{1}{T} \int_{-T/2}^0 (1)^2 dt =$
$\lim_{T\to\infty} \frac{1}{T} \left(\frac{T}{2}\right) = \lim_{T\to\infty} \left(\frac{1}{2}\right) = \frac{1}{2}.$

A3.71 Find the power of $\text{rect}\left(t - \frac{1}{2}\right)$.

Answer: $\lim_{T\to\infty} \frac{1}{T} \int_{-T/2}^{T/2} \left|\text{rect}\left(t - \frac{1}{2}\right)\right|^2 dt = \lim_{T\to\infty} \frac{1}{T} \int_0^1 (1)^2 dt =$
$\lim_{T\to\infty} \frac{1}{T} (1) = 0.$

A3.72 Find the power of $\cos(5\pi t)$.

Answer:

$P = \frac{1}{T_0} \int_{T_0} |\cos(5\pi t)|^2 dt = \frac{1}{T_0} \int_{T_0} \cos^2(5\pi t)\, dt =$
$\frac{1}{T_0} \int_{T_0} \left(\frac{1}{2} + \frac{1}{2}\cos(10\pi t)\right) dt \ = \frac{1}{T_0} \int_{T_0} \left(\frac{1}{2}\right) dt \ +$
$\frac{1}{T_0} \int_{T_0} \left(\frac{1}{2}\cos(10\pi t)\right) dt = \frac{1}{T_0} \int_{T_0} \left(\frac{1}{2}\right) dt = \frac{1}{T_0}\left(\frac{T_0}{2}\right) = 1/2.$
(true for any sinusoid having amplitude 1)

A3.73 Find the power of $\sin(5\pi t)$.

Answer: $P = 1/2$ (see previous problem).

A3.74 Find the power of $-3\cos(\pi t)$.

Answer: $P = \frac{|A|^2}{2} = \frac{|-3|^2}{2} = 4.5.$

A3.75 Sketch the even and odd parts of $\text{sgn}(t)$.

Answer:
$even\{\text{sgn}(t)\} = \frac{1}{2}\left(\text{sgn}(t) + \text{sgn}(-t)\right)$

$$= \tfrac{1}{2}\big(\mathrm{sgn}(t) - \mathrm{sgn}(t)\big) = 0$$

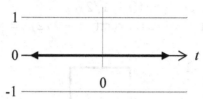

$$odd\{\mathrm{sgn}(t)\} = \tfrac{1}{2}\big(\mathrm{sgn}(t) - \mathrm{sgn}(-t)\big)$$
$$= \tfrac{1}{2}\big(\mathrm{sgn}(t) + \mathrm{sgn}(t)\big) = \mathrm{sgn}(t):$$

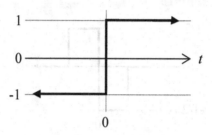

A3.76 Sketch the even and odd parts of $u(t)$.

Answer: $even\{u(t)\} = \tfrac{1}{2}\big(u(t) + u(-t)\big) = \tfrac{1}{2}:$

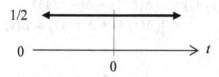

$$odd\{u(t)\} = \tfrac{1}{2}\big(u(t) - u(-t)\big) = \tfrac{1}{2}\mathrm{sgn}(t):$$

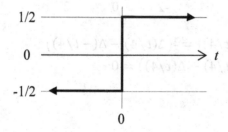

A3.77   Sketch the even and odd parts of $\text{rect}(t - 1/2)$.

Answer: $even\{\text{rect}(t - 1/2)\}$
$= \frac{1}{2}\left(\text{rect}(t - 1/2) + \text{rect}(-t - 1/2)\right) = \frac{1}{2}\text{rect}(t/2)$:

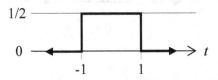

$odd\{\text{rect}(t - 1/2)\} = \frac{1}{2}\left(\text{rect}(t - 1/2) - \text{rect}(-t - 1/2)\right)$:

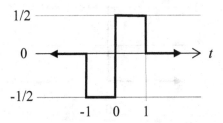

A3.78   Sketch the even and odd parts of $\Delta(t/4)$.

Answer:
$even\{\Delta(t/4)\} = \frac{1}{2}\left(\Delta(t/4) + \Delta(-t/4)\right)$
$\qquad\qquad\quad = \frac{1}{2}\left(\Delta(t/4) + \Delta(t/4)\right) = \Delta(t/4)$

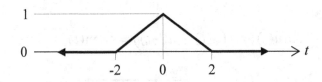

$odd\{\Delta(t/4)\} = \frac{1}{2}\left(\Delta(t/4) - \Delta(-t/4)\right)$
$= \frac{1}{2}\left(\Delta(t/4) - \Delta(t/4)\right) = 0$:

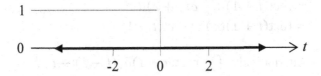

A3.79 Using the convolution integral, or by applying the properties of convolution, simplify $\delta(t - 1.5) * \text{rect}(t)$.

Solution: $\int_{-\infty}^{\infty} \delta(\tau - 1.5)\text{rect}(t - \tau)d\tau$
$= \int_{-\infty}^{\infty} \delta(\tau - 1.5)\text{rect}(t - 1.5)d\tau$
$= \text{rect}(t - 1.5) \int_{-\infty}^{\infty} \delta(\tau - 1.5)d\tau$
$= \text{rect}(t - 1.5)(1) = \text{rect}(t - 1.5)$.

Alternatively, $\int_{-\infty}^{\infty} \text{rect}(\tau)\delta((t - 1.5) - \tau)d\tau$
$= \int_{-\infty}^{\infty} \text{rect}(t - 1.5)\delta((t - 1.5) - \tau)d\tau$
$= \text{rect}(t - 1.5) \int_{-\infty}^{\infty} \delta((t - 1.5) - \tau)d\tau$
$= \text{rect}(t - 1.5) \int_{-\infty}^{\infty} \delta(\tau - (t - 1.5))d\tau$
$= \text{rect}(t - 1.5)(1) = \text{rect}(t - 1.5)$.

A3.80 Using the convolution integral, or by applying the properties of convolution, simplify
$\text{rect}(t) * \delta(t + 0.5) + \text{rect}(t) * \delta(t - 0.5)$.

Solution: Using the same method as in part (a) above,
$\text{rect}(t) * \delta(t + 0.5) = \text{rect}(t + 0.5)$ and
$\text{rect}(t) * \delta(t - 0.5) = \text{rect}(t - 0.5)$. The answer is then
$\text{rect}(t + 0.5) + \text{rect}(t - 0.5)$, which graphically may be
shown equal to $\text{rect}(t/2)$.

A3.81 Using the convolution integral, or by applying the properties of convolution, simplify $\delta(t + 2) * \text{rect}(t - 1)$.

Solution: $\int_{-\infty}^{\infty} \delta(\tau + 2)\text{rect}((t - 1) - \tau)d\tau$
$= \int_{-\infty}^{\infty} \delta(\tau + 2)\text{rect}((t - 1) - (-2))d\tau$
$= \int_{-\infty}^{\infty} \delta(\tau + 2)\text{rect}(t + 1)d\tau$

$$= \text{rect}(t + 1) \int_{-\infty}^{\infty} \delta(\tau + 2)d\tau$$
$$= \text{rect}(t + 1)(1) = \text{rect}(t + 1).$$

Alternatively, $\int_{-\infty}^{\infty} \text{rect}(\tau - 1)\delta((t + 2) - \tau)d\tau$
$$= \int_{-\infty}^{\infty} \text{rect}((t + 2) - 1)\delta((t + 2) - \tau)d\tau$$
$$= \int_{-\infty}^{\infty} \text{rect}(t + 1)\delta((t + 2) - \tau)d\tau$$
$$= \text{rect}(t + 1) \int_{-\infty}^{\infty} \delta((t + 2) - \tau)d\tau$$
$$= \text{rect}(t + 1) \int_{-\infty}^{\infty} \delta(\tau - (t + 2))d\tau$$
$$= \text{rect}(t + 1)(1) = \text{rect}(t + 1).$$

A3.82   Using the convolution integral, or by applying the properties of convolution, simplify $(u(t) * \delta(t + 1))(u(-t) * \delta(t - 1))$.

Solution: For any signal $f(t)$, $f(t) * \delta(t + 1) = f(t + 1)$ and $f(t) * \delta(t - 1) = f(t - 1)$. Therefore, $u(t) * \delta(t + 1) = u(t + 1)$ and $u(-t) * \delta(t - 1) = u(-(t - 1))$. The answer to this problem is the multiplicative product $u(t + 1)u(-(t - 1))$, which is equal to $\text{rect}(t/2)$.

A3.83   Using the convolution integral, or by applying the properties of convolution, simplify $\delta(t - 1) * \delta(t + 1)$.

Solution: $\delta(t - 1) * \delta(t + 1) = \delta((t + 1) - 1) = \delta(t)$.
We can also show this using the convolution integral:
$\int_{-\infty}^{\infty} \delta(\tau - 1)\delta((t + 1) - \tau)d\tau =$
$\int_{-\infty}^{\infty} \delta(\tau - 1)\delta((t + 1) - 1)d\tau = \int_{-\infty}^{\infty} \delta(\tau - 1)\delta(t)d\tau =$
$\delta(t) \int_{-\infty}^{\infty} \delta(\tau - 1)d\tau = \delta(t)(1) = \delta(t)$.

A3.84   Using the convolution integral, or by applying the properties of convolution, simplify $\delta(t - 1) * \text{rect}(t) * \delta(t + 1)$.

Solution: Since the commutative property holds for convolution, we may convolve these three terms in any order. $\delta(t - 1) * \text{rect}(t) = \text{rect}(t - 1)$, and then $\text{rect}(t - 1) * \delta(t + 1) = \text{rect}((t + 1) - 1) = \text{rect}(t)$.
Alternatively: $\text{rect}(t) * \delta(t + 1) = \text{rect}(t + 1)$, and then $\delta(t - 1) * \text{rect}(t + 1) = \text{rect}((t - 1) + 1) = \text{rect}(t)$.

Alternatively: $\delta(t-1) * \delta(t+1) = \delta(t)$, and then $\delta(t) *$ rect(t) = rect(t).

A3.85 Using the convolution integral, or by applying the properties of convolution, simplify rect$(t/2\pi) * \cos(t)$.

Solution: The rect pulse width is exactly one period of the cosine pulse; the area of their product, regardless of shift of one or the other (as calculated by the convolution integral), is zero.

A3.86 Graphically calculate the convolution product sgn$(t) *$ rect(t).

Solution: sgn$(t) *$ rect$(t) = \int_{-\infty}^{\infty}$ sgn(τ) rect$(t - \tau)d\tau =$ $\int_{-\infty}^{\infty}$ sgn(τ) rect$(-(\tau - t))d\tau$. We may solve this integral using a series of graphs that follow.

Plot of sgn(τ) vs. τ

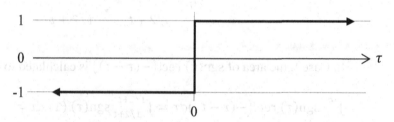

Plots of rect$(-(\tau - t))$ vs. τ

Case I: $1/2 + t \le 0$ $(t \le -1/2)$

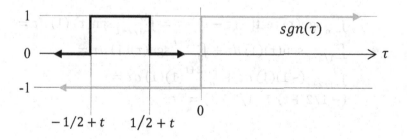

Case II: $-1/2 + t < 0$ *and* $1/2 + t > 0$ $(-1/2 < t < 1/2)$

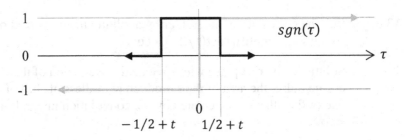

Case III: $-1/2 + t \geq 0$ $(t \geq 1/2)$

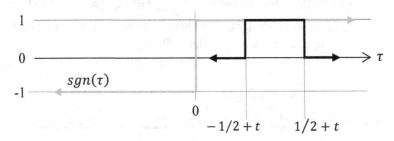

In Case I, the area of $\text{sgn}(\tau)\,\text{rect}\big(-(\tau - t)\big)$ is calculated to be:

$$\int_{-\infty}^{\infty} \text{sgn}(\tau)\,\text{rect}\big(-(\tau - t)\big)d\tau = \int_{-1/2+t}^{1/2+t} \text{sgn}(\tau)\,(1)\,d\tau =$$
$$\int_{-1/2+t}^{1/2+t}(-1)(1)\,d\tau = -1.$$

In Case II, the area of $\text{sgn}(\tau)\,\text{rect}\big(-(\tau - t)\big)$ is calculated to be:

$$\int_{-\infty}^{\infty} \text{sgn}(\tau)\,\text{rect}\big(-(\tau - t)\big)d\tau = \int_{-1/2+t}^{1/2+t} \text{sgn}(\tau)\,(1)\,d\tau =$$
$$\int_{-1/2+t}^{0} \text{sgn}(\tau)(1)\,d\tau + \int_{0}^{1/2+t} \text{sgn}(\tau)(1)\,d\tau =$$
$$\int_{-1/2+t}^{0}(-1)(1)\,d\tau + \int_{0}^{1/2+t}(1)(1)\,d\tau =$$
$$(-1/2 + t) + (1/2 + t) = 2t.$$

In Case III, the area of $\text{sgn}(\tau)\,\text{rect}\big(-(\tau - t)\big)$ is calculated to be:

$$\int_{-\infty}^{\infty} \text{sgn}(\tau)\,\text{rect}\big(-(\tau - t)\big)d\tau = \int_{-1/2+t}^{1/2+t} \text{sgn}(\tau)\,(1)\,d\tau =$$
$$\int_{-1/2+t}^{1/2+t}(1)(1)\,d\tau = 1.$$

Therefore, as plotted below,

$$\text{sgn}(t) * \text{rect}(t) = \begin{cases} -1, & t \leq -1/2; \\ 2t, & -1/2 < t < 1/2; \\ 1, & t \geq 1/2. \end{cases}$$

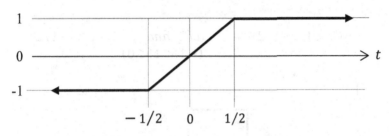

A3.87 Graphically calculate the convolution product $1.5\text{rect}(t) * \text{rect}(t)$.

Solution:
$1.5\text{rect}(t) * \text{rect}(t) = \int_{-\infty}^{\infty} 1.5\text{rect}(\tau)\text{rect}(t - \tau)d\tau =$
$\int_{-\infty}^{\infty} 1.5\text{rect}(\tau)\,\text{rect}(-(\tau - t))d\tau.$ We may solve this integral using a series of graphs that follow.

Plot of $1.5\text{rect}(\tau)$ vs. τ

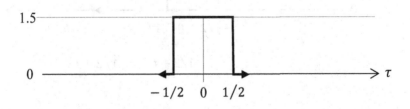

Plots of $\underline{rect(-(\tau - t))}$ vs. τ

Case I: $1/2 + t \leq -1/2 \ \ (t \leq -1)$

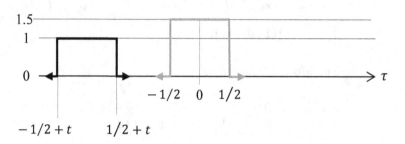

$-1/2 + t \qquad\qquad 1/2 + t$

Case II: $-1/2 + t < -1/2 \ \ and \ \ 1/2 + t > -1/2$
$(-1 < t < 0)$

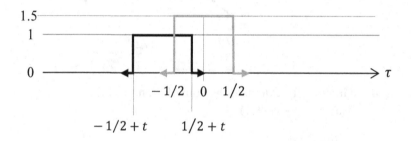

$-1/2 + t \qquad\qquad 1/2 + t$

Case III: $-1/2 + t < 1/2 \ \ and \ \ 1/2 + t > 1/2 \ \ (0 < t < 1):$

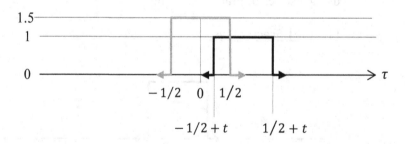

$-1/2 + t \qquad\qquad 1/2 + t$

Case IV: $-1/2 + t \geq 1/2$ $(t \geq 1)$

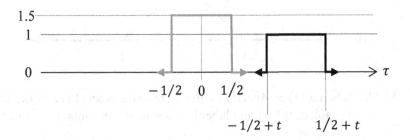

In Case I, the area of $1.5\text{rect}(\tau)\text{rect}(-(\tau - t))$ is calculated to be $\int_{-\infty}^{\infty} 1.5\text{rect}(\tau)\text{rect}(-(\tau - t)) \, d\tau = \int_{-1/2+t}^{1/2+t} (0)(1) \, d\tau = 0.$

In Case II, the area of $1.5\text{rect}(\tau)\text{rect}(-(\tau - t))$ is calculated to be $\int_{-\infty}^{\infty} 1.5\text{rect}(\tau)\text{rect}(-(\tau - t)) \, d\tau = \int_{-1/2}^{1/2+t} (1.5)(1) \, d\tau = 1.5(1 + t).$

In Case III, the area of $1.5\text{rect}(\tau)\text{rect}(-(\tau - t))$ is calculated to be $\int_{-\infty}^{\infty} 1.5\text{rect}(\tau)\text{rect}(-(\tau - t)) \, d\tau = \int_{-1/2+t}^{1/2} (1.5)(1) \, d\tau = 1.5(1 - t).$

In Case IV, the area of $1.5\text{rect}(\tau)\text{rect}(-(\tau - t))$ is calculated to be $\int_{-\infty}^{\infty} 1.5\text{rect}(\tau)\text{rect}(-(\tau - t)) \, d\tau = \int_{-1/2+t}^{1/2+t} (0)(1) \, d\tau = 0.$

Therefore, as plotted below,

$$1.5\text{rect}(t) * \text{rect}(t) = \begin{cases} 0, & t \leq -1; \\ 1.5(1 + t), & -1 < t \leq 0; \\ 1.5(1 - t), & 0 < t \leq 1; \\ 0, & t > 1. \end{cases}$$
$$= 1.5\Delta(t/2).$$

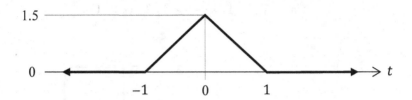

A3.88 When $w(t) = A\text{rect}(t) * B\text{rect}(t)$, what is $w(t)$ at $t = t_0$? Draw a graph of $w(t_0)$ and label it in terms of constants A, B, and t_0.
Answer:
Using the graphical method shown in Problem A3.89, one will obtain the following graph.

$w(t_0)$:

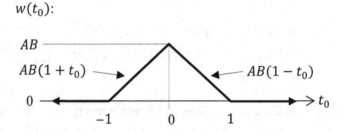

A3.89 When $w(t) = A\text{rect}(t) * B\text{rect}(t/C)$, what is $w(t)$ at $t = t_0$? Draw a graph of $w(t_0)$ and label it in terms of constants A, B, C, and t_0. Assume that $A, B, C > 1$.

Answer:
Using the graphical method shown in Problem A3.89, one will obtain the following graph.

$w(t_0)$:

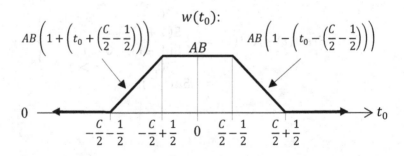

A3.90 When $w(t) = Au(t) * Bu(t)$, what is $w(t)$ at $t = t_0$? Draw a graph of $w(t_0)$ and label it in terms of constants A, B, and t_0.

Answer:
Using the graphical method shown in Problem A3.89, one will obtain the following graph.

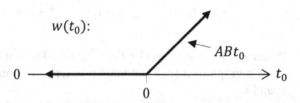

A3.91 When $w(t) = Ae^{-bt}u(t) * Bu(t)$, what is $w(t)$ at $t = t_0$? Draw a graph of $w(t_0)$ and label it in terms of constants A, B, b, and t_0.

Answer:
Using the graphical method shown in Problem A3.89, one will obtain the following graph.

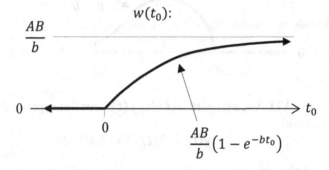

A3.92 When $w(t) = A\text{rect}(t - t_1) * B\text{rect}(t - t_2)$, find $w(t)$ at $t = t_0$. Draw a graph of $w(t_0)$ and label it in terms of constants A, B, t_1, t_2, and t_0.

Answer:
Using the graphical method shown in Problem A3.89, one will obtain the following graph.

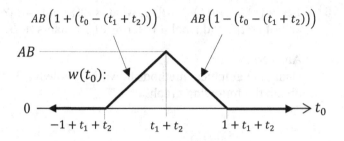

A3.93 When $w(t) = Ae^{-bt}u(t) * Be^{-ct}u(t)$, what is $w(t)$ at $t = t_0$?
Draw a graph of $w(t_0)$ and label it in terms of constants $A, B, b,$
$c,$ and t_0.

   Answer:
   Using the graphical method shown in Problem A3.89, one will
   obtain the following graph.

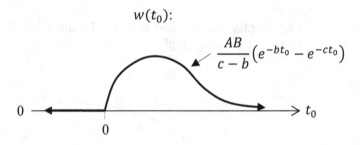

A3.94 In MATLAB, estimate and plot $y(t)$ over time span $-3 \leq t \leq 3$
   seconds:
$$y(t) = \Delta(t/2) * \text{rect}(t/2)$$

   Solution:
   Taking advantage of the fact that $\Delta(t/2) = \text{rect}(t) * \text{rect}(t)$,
   we will simplify the MATLAB code by working only with
   rectangular pulse functions:

$$y(t) = \text{rect}(t) * \text{rect}(t) * \text{rect}(t/2).$$

   The continuous-time convolution integral may be approximated using a summation of terms:

$$\int_{-\infty}^{\infty} f(\tau)\, g(t-\tau) d\tau = \lim_{\Delta\tau \to \infty} \sum_{k=-\infty}^{\infty} f(k\Delta\tau) g(t-k\Delta\tau)\Delta\tau$$

$$\approx \sum_{k=-\infty}^{\infty} f(k\Delta\tau) g(n\Delta\tau - k\Delta\tau)\Delta\tau,$$

where $\Delta\tau$ is small and samples of the result are found only at discrete times $t = n\Delta\tau$. The discrete-time convolution sum is defined by $f(n) * g(n) = \sum_{n=-\infty}^{\infty} f(k)g(n-k)$, which is the same as our final approximation to the convolution integral when labelling $"f(k)" = f(k\Delta\tau)$ and $"g(n-k)" = g\big((n-k)\Delta\tau\big)$ as samples of $f(t)$ and $g(t)$ taken every $\Delta\tau$ seconds, but for a scale factor of $\Delta\tau$. Thus, we may use MATLAB's built-in discrete convolution function **conv** to estimate the continuous-time convolution integral by scaling the result with constant $\Delta\tau$:

```
N = 5000; t1 = -1.5; t2 = 1.5;
% sample time at N pts between t1 and t2
t = linspace(t1,t2,N);
% this is the scaling factor delta_tau
dt = t(2)-t(1);
% get samples of f(t): rect pulse of width 1
f = zeros(size(t)); f(abs(t)<=0.5) = 1;
% get samples of g(t): rect pulse of width 2
g = zeros(size(t)); g(abs(t/2)<=0.5) = 1;
f1 = conv(f,f)*dt;
y = conv(f1,g)*dt;
% define output time samples for plotting
t_out = linspace(3*t1,3*t2,N+N-1+N-1);
plot(t_out,y)
xlabel('time (sec)')
% plot over time span [-3,+3] sec
axis([-3,3,0,1])
grid on
```

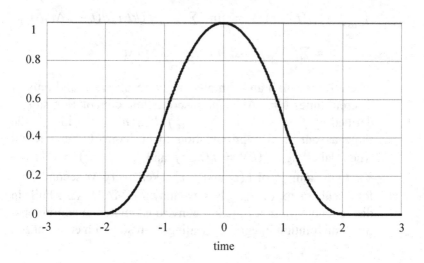

time

Chapter 4

A4.1 One period of periodic sequence $f_1(n)$ is equal to
$\{1, 0, 0, -1\}$ for $n = \{0, 1, 2, 3\}$. Find the period of
$F_p(k) = DFT\{f_1(n)\}$ over index range $k = \{0, 1, 2, 3\}$.

Answer:
$F_p(k) = \sum_{n=0}^{N-1} f_p(n)e^{-jk2\pi n/N} = \sum_{n=0}^{3} f_p(n)e^{-jk2\pi n/4}$. ∴
$F_p(0) = \sum_{n=0}^{3} f_p(n)e^0 = (1 + 0 + 0 - 1) = 0$. Similarly, we
find that $F_p(k) = \{0, \ 1 - j, \ 2, \ 1 + j\}$ over index range
$k = \{0, 1, 2, 3\}$.

A4.2 One period of periodic sequence $f_2(n)$ is equal to
$\{-1, 1, 0, 0\}$ for $n = \{0, 1, 2, 3\}$. Find the period of
$F_p(k) = DFT\{f_2(n)\}$ over index range $k = \{0, 1, 2, 3\}$.

Answer:
Using the same direct approach as in Problem A4.1, we find
that $F_p(k) = \{0, -1 - j, -2, -1 + j\}$ over index range
$k = \{0, 1, 2, 3\}$.

A4.3 One period of periodic sequence $f_3(n)$ is equal to $\{0, 1, 0, -1\}$ for $n = \{0, 1, 2, 3\}$. Find the period of $F_p(k) = DFT\{f_3(n)\}$ over index range $k = \{0, 1, 2, 3\}$.

Answer:
Using the same direct approach as in Problem A4.1, we find that $F_p(k) = \{0, -2j, 0, 2j\}$ over index range $k = \{0, 1, 2, 3\}$.

A4.4 One period of periodic sequence $f_4(n)$ is equal to $\{1, 1, 0, -2\}$ for $n = \{0, 1, 2, 3\}$. Find the period of $F_p(k) = DFT\{f_4(n)\}$ over index range $k = \{0, 1, 2, 3\}$,

Answer:
Using the same direct approach as in Problem A4.1, we find that $F_p(k) = \{0, 1 - 3j, 2, 1 + 3j\}$ over index range $k = \{0, 1, 2, 3\}$.

A4.5 Express $f_3(n)$ (Problem A4.3) as a linear combination of $f_1(n)$ and $f_2(n)$ (Problems A4.1 and A4.2), and show that the linearity property holds in determining $F_3(k)$ from $F_1(k)$ and $F_2(k)$.

Answer:
$f_3(n) = f_1(n) + f_2(n)$, therefore $F_3(k) = F_1(k) + F_2(k)$
$= \{0, -2j, 0, 2j\}$.

A4.6 Express $f_2(n)$ (Problem A4.2) as a time-shifted version of $f_1(n)$ (Problem A4.1), and show that the time-shifting property holds in determining $F_2(k)$ from $F_1(k)$.

Answer:
$f_2(n) = f_1(n - 1)$, therefore $F_2(k) = F_1(k)e^{j\omega(-1)}$
$= F_1(k)e^{j(k2\pi/N)(-1)} = F_1(k)e^{j(k2\pi/4)(-1)} = F_1(k)e^{-jk\pi/2}$
$= \{0e^{-j0\pi/2}, (1 - j)e^{-j1\pi/2}, 2e^{-j2\pi/2}, (1 + j)e^{-j3\pi/2}\}$
$= \{0, (1 - j)(-j), 2(-1), (1 + j)(j)\} = \{0, -1 - j, -2, -1 + j\}$.

A4.7 $\mathcal{F}\{\delta_1(n)\} = ?$

> Answer:
> $\mathcal{F}\{\delta_1(n)\} = \mathcal{F}\{1\} = 2\pi\delta_{2\pi}(\omega) = \delta_{2\pi}(\omega) * 2\pi\delta(\omega).$

A4.8 $\mathcal{F}\{e^{j2n}\} = ?$

> Answer:
> $\mathcal{F}\{e^{j2n}\} = 2\pi\,\delta_{2\pi}(\omega - 2) = \delta_{2\pi}(\omega) * 2\pi\delta(\omega - 2).$

A4.9 $\mathcal{F}\{\text{rect}_3(n) + \text{rect}_3(n - 4)\} = ?$

> Answer:
> $\mathcal{F}\{\text{rect}_3(n) + \text{rect}_3(n - 4)\}$
> $= 7\text{sinc}\left(\omega \cdot \frac{7}{2}\right) * \delta_{2\pi}(\omega) + e^{-j\omega 4}7\text{sinc}\left(\omega \cdot \frac{7}{2}\right) * \delta_{2\pi}(\omega)$
> $= \delta_{2\pi}(\omega) * \{7\text{sinc}(\omega \cdot \frac{7}{2})(1 + e^{-j\omega 4})\}.$

A4.10 $\mathcal{F}\{\text{sinc}(\pi n) * \text{sinc}(\pi n/2)\} = ?$

> Answer:
> $\mathcal{F}\{\text{sinc}(\pi n) * \text{sinc}(\pi n/2)\}$
> $= \{\text{rect}(\omega/2\pi) * \delta_{2\pi}(\omega)\} \times \{2\text{rect}(\omega/\pi) * \delta_{2\pi}(\omega)\}$
> $= \{\text{rect}(\omega/2\pi) \times 2\text{rect}(\omega/\pi)\} * \delta_{2\pi}(\omega)$
> $= \delta_{2\pi}(\omega) * 2\text{rect}(\omega/\pi).$

A4.11 $\mathcal{F}\{\Delta_4(n) \cos(n\pi/8)\} = ?$

> Answer:
> $\mathcal{F}\{\Delta_4(n) \cos(n\pi/8)\}$
> $= \{4\text{sinc}^2(2\omega) * \delta_{2\pi}(\omega)\} * \left\{\frac{1}{2}\delta(\omega + \pi/8) + \frac{1}{2}\delta(\omega - \pi/8)\right\}$
> $= \delta_{2\pi}(\omega) * \{2\text{sinc}^2(2(\omega + \pi/8)) + 2\text{sinc}^2(2(\omega - \pi/8))\}.$

A4.12 $\mathcal{F}^{-1}\{e^{-j5\omega} - e^{j4\omega} + 1\} = ?$

   Answer:
   $$\mathcal{F}^{-1}\{e^{-j5\omega} - e^{j4\omega} + 1\} = \delta(n - 5) - \delta(n + 4) + \delta(n).$$

A4.13 $\mathcal{F}^{-1}\{\cos(\omega) + \cos(2\omega) + \cos(3\omega)\} = ?$

   Answer:
   Since $\text{rect}_3(n) \Leftrightarrow 1 + 2\sum_{k=1}^{3}\cos(k\omega)$
   $= 1 + 2\{\cos(\omega) + \cos(2\omega) + \cos(3\omega)\}$,
   $\mathcal{F}^{-1}\{\cos(\omega) + \cos(2\omega) + \cos(3\omega)\}$
   $= \frac{1}{2}\{\text{rect}_3(n) - \delta(n)\}$
   $= \frac{1}{2}\delta(n \pm 1) + \frac{1}{2}\delta(n \pm 2) + \frac{1}{2}\delta(n \pm 3).$

A4.14 $\mathcal{F}^{-1}\{(1 + e^{j\omega})/(1 - e^{j\omega})\} = ?$

   Answer:
   $\mathcal{F}^{-1}\{(1 + e^{j\omega})/(1 - e^{j\omega})\}$
   $= \mathcal{F}^{-1}\left\{(1 + e^{-j\omega})/(1 - e^{-j\omega})\big|_{\omega \leftarrow -\omega}\right\}$
   $= \text{sgn}(n)|_{n \leftarrow -n} = \text{sgn}(-n) = -\text{sgn}(n).$

A4.15 $\mathcal{F}^{-1}\{1/(1 - e^{-j\omega})\} = ?$

   Answer:
   $\mathcal{F}^{-1}\{1/(1 - e^{-j\omega})\} = u(n) - \mathcal{F}^{-1}\{\pi\, \delta_{2\pi}(\omega)\}$
   $= u(n) - 1/2.$

A4.16 $\mathcal{F}^{-1}\{\sum_{k=-\infty}^{\infty}\Delta((\omega - k2\pi)/(\pi/10))\} = ?$

   Answer:
   $\mathcal{F}^{-1}\{\sum_{k=-\infty}^{\infty}\Delta((\omega - k2\pi)/(\pi/10))\}$
   $= \mathcal{F}^{-1}\{\Delta(\omega/(\pi/10)) * \delta_{2\pi}(\omega)\} = \frac{1}{40}\text{sinc}^2(n\pi/40).$

A4.17 Find the energy of sequence $\text{sinc}(n)$.

   Answer: Energy$\{\text{sinc}(n)\} = \sum_{n=-\infty}^{\infty}|\text{sinc}(n)|^2$
   $= \frac{1}{2\pi}\int_{-\pi}^{\pi}|\mathcal{F}\{\text{sinc}(n)\}|^2\, d\omega = \frac{1}{2\pi}\int_{-\pi}^{\pi}|\pi\, \text{rect}(\omega/2)|^2\, d\omega$

$$= \frac{1}{2\pi}\int_{-1}^{1}|\pi\,(1)|^2\,d\omega = \frac{1}{2\pi}\int_{-1}^{1}\pi^2\,d\omega = \frac{1}{2\pi}(2\pi^2) = \pi.$$

Chapter 5

A5.1   Find the Fourier Transform of $(1 + \cos(3t))$.

Answer:
$$\mathcal{F}\{1 + \cos(3t)\} = \mathcal{F}\{1\} + \mathcal{F}\{\cos(3t)\}$$
$$= 2\pi\delta(\omega) + \pi\delta(\omega + 3) + \pi\delta(\omega - 3).$$

A5.2   Find the Fourier Transform of $(\delta(t + 1) - 5\delta(t - 2))$.

Answer:
$$\mathcal{F}\{\delta(t + 1) - 5\delta(t - 2)\} = \mathcal{F}\{\delta(t + 1)\} - 5\mathcal{F}\{\delta(t - 2)\}$$
$$= e^{j\omega(1)}\mathcal{F}\{\delta(t)\} - 5e^{j\omega(-2)}\mathcal{F}\{\delta(t)\}$$
$$= e^{j\omega}(1) - 5e^{-2j\omega}(1) = e^{j\omega} - 5e^{-2j\omega}.$$

A5.3   Find the Fourier Transform of $(\delta_2(t) + \delta_2(t - 1))$:

Answer:
$$\mathcal{F}\{\delta_2(t) + \delta_2(t - 1)\} = \mathcal{F}\{\delta_2(t)\} + \mathcal{F}\{\delta_2(t - 1)\}$$
$$= \mathcal{F}\{\delta_2(t)\} + e^{j\omega(-1)}\mathcal{F}\{\delta_2(t)\} = \pi\delta_\pi(\omega) + e^{-j\omega}\pi\delta_\pi(\omega)$$
$$= \pi\delta_\pi(\omega)(1 + e^{-j\omega}). \text{ (This may be simplified to } 2\pi\delta_{2\pi}(\omega)$$
since $1 + e^{-j(k\pi)} = \{2, k \text{ even}; 0, k \text{ odd}\})$.

A5.4   Find the Fourier Transform of $u(t + 1)$.

Answer:
$$\mathcal{F}\{u(t + 1)\} = e^{j\omega(1)}\mathcal{F}\{u(t)\}$$
$$= e^{j\omega}(1/j\omega + \pi\delta(\omega)) = e^{j\omega}/j\omega + \pi e^{j\omega}\delta(\omega)$$
$$= e^{j\omega}/j\omega + \pi e^{j(0)}\delta(\omega) = e^{j\omega}/j\omega + \pi(1)\delta(\omega)$$
$$= e^{j\omega}/j\omega + \pi\delta(\omega).$$

A5.5 Find the Fourier Transform of $\delta(t/3)$.

Answer:
$$\mathcal{F}\{\delta(t/3)\} = \mathcal{F}\{3\delta(t)\} = 3\mathcal{F}\{\delta(t)\} = 3(1) = 3.$$

A5.6 Find the Fourier Transform of $2u(t) + u(-t) - 1$.

Answer:
$$\mathcal{F}\{2u(t) + u(-t) - 1\} = 2\mathcal{F}\{u(t)\} + \mathcal{F}\{u(-t)\} - \mathcal{F}\{1\}$$
$$= 2(1/j\omega + \pi\delta(\omega)) + (1/j(-\omega) + \pi\delta(-\omega)) - 2\pi\delta(\omega)$$
$$= 2/j\omega + 2\pi\delta(\omega) - 1/j\omega + \pi\delta(\omega) - 2\pi\delta(\omega)$$
$$= 1/j\omega + \pi\delta(\omega). \quad \text{(Note that } 2u(t) + u(-t) - 1 = u(t)\text{).}$$

A5.7 Find the inverse Fourier Transform of $F(\omega) = -4\operatorname{sinc}^2(\omega/4)$.

Answer:
$$\mathcal{F}^{-1}\{-4\operatorname{sinc}^2(\omega/4)\} = -4\,\mathcal{F}^{-1}\{\operatorname{sinc}^2(\omega/4)\} = -4\,\{2\Delta(t)\}$$
$$= -8\Delta(t).$$

A5.8 Find the inverse Fourier Transform of $\pi\delta(\omega - 1) + \pi\delta(\omega + 1)$.

Answer:
$$\mathcal{F}^{-1}\{\pi\delta(\omega - 1) + \pi\delta(\omega + 1)\}$$
$$= \pi\mathcal{F}^{-1}\{\delta(\omega - 1)\} + \pi\mathcal{F}^{-1}\{\delta(\omega + 1)\}$$
$$= \pi e^{-j(-1)t}\mathcal{F}^{-1}\{\delta(\omega);\} + \pi e^{-j(1)t}\mathcal{F}^{-1}\{\delta(\omega)\}$$
$$= \pi(e^{jt} + e^{-jt})\mathcal{F}^{-1}\{\delta(\omega)\} = \pi(e^{jt} + e^{-jt})(1/2\pi)$$
$$= (1/2)(e^{jt} + e^{-jt}) = \cos(t).$$

A5.9 Find the inverse Fourier Transform of $1/\omega$.

Answer:
$$\mathcal{F}^{-1}\left\{\frac{1}{\omega}\right\} = \mathcal{F}^{-1}\left\{\frac{j}{2} \cdot \frac{2}{j\omega}\right\} = \frac{j}{2}\mathcal{F}^{-1}\left\{\frac{2}{j\omega}\right\} = \frac{j}{2}\operatorname{sgn}(t).$$

A5.10 Find the Fourier Transform of $1/t$.

   Answer:
   From Problem A5.10 we know that $(j/2)\text{sgn}(t) \Leftrightarrow 1/\omega$.
   By the duality property of the Fourier Transform: If $f(t) \Leftrightarrow$
   $F(\omega)$, then $\mathcal{F}(t) \Leftrightarrow 2\pi f(-\omega)$. Therefore, $\mathcal{F}\{1/t\} =$
   $2\pi\big((j/2)\text{sgn}(-\omega)\big) = 2\pi\big(-(j/2)\text{sgn}(\omega)\big) = -j\pi\,\text{sgn}(\omega)$.

A5.11 Find the Fourier Transform of $\delta(t+1) + \delta(t-1)$.

   Answer:
   $$\mathcal{F}\{\delta(t+1) + \delta(t-1)\} = \mathcal{F}\{\delta(t+1)\} + \mathcal{F}\{\delta(t-1)\}$$
   $$= e^{j\omega(1)}\mathcal{F}\{\delta(t)\} + e^{j\omega(-1)}\mathcal{F}\{\delta(t)\}$$
   $$= e^{j\omega(1)}(1) + e^{j\omega(-1)}(1) = e^{j\omega} + e^{-j\omega} = 2\cos(\omega).$$

A5.12 Find the Fourier Transform of $\frac{1}{2}\sum_{k=-\infty}^{\infty}\delta(t-k\pi)$.

   Answer:
   $$\mathcal{F}\left\{\frac{1}{2}\sum_{k=-\infty}^{\infty}\delta(t-k\pi)\right\} = \mathcal{F}\left\{\frac{1}{2}\delta_{\pi}(t)\right\} = \frac{1}{2}\mathcal{F}\{\delta_{\pi}(t)\}$$
   $$= \frac{1}{2}(2\delta_2(\omega)) = \delta_2(\omega).$$

A5.13 Find the Fourier Transform of $\text{sgn}(2t)$.

   Answer:
   $$\mathcal{F}\{\text{sgn}(2t)\} = \frac{1}{|2|} \cdot \frac{2}{j(\omega/2)} = \frac{2}{j\omega}.\ \ \text{(Note that } \text{sgn}(2t) = \text{sgn}(t)).$$

A5.14 Find the Fourier Transform of $\text{sinc}^2(t)$.

   Answer:
   $$\mathcal{F}\{\text{sinc}^2(t)\} = \mathcal{F}\left\{\pi \cdot \frac{1}{\pi}\text{sinc}^2(t)\right\} = \pi\mathcal{F}\left\{\frac{1}{\pi}\text{sinc}^2(t)\right\}$$
   $$= \pi\Delta\left(\frac{\omega}{4}\right).$$

A5.15 Express the periodic signal $f_p(t)$ below as an Exponential Fourier Series:

Solution:

This periodic signal has period $T_0 = 4$ sec ($\omega_0 = \frac{2\pi}{T_0} = \frac{\pi}{2}$ rad/sec). The expression for $f_p(t) = p(t) * \delta_4(t)$, where pulse $p(t)$ defining one period is measured over time span $t = [-2, 2]$: $p(t) = \text{rect}(t/2) \Leftrightarrow 2\,\text{sinc}(\omega)$. Therefore, $D_k = (1/T_0)P(k\omega_0) = \frac{1}{4}(2\text{sinc}(k\omega_0)) = \frac{1}{2}\text{sinc}(k\pi/2)$. The final answer is then $f_p(t) = \sum_{k=-\infty}^{\infty} D_k e^{jk\omega_0 t}$

$= \sum_{k=-\infty}^{\infty} \frac{1}{2}\text{sinc}(k\pi/2)e^{jk(\pi/2)t}$.

A5.16 What is the energy of $\text{sinc}(t)$?

Answer: $\text{Energy}\{\text{sinc}(t)\} = \frac{1}{2\pi}\int_{-\infty}^{\infty}|\mathcal{F}\{\text{sinc}(t)\}|^2 d\omega$

$= \frac{1}{2\pi}\int_{-\infty}^{\infty}|\pi\,\text{rect}(\omega/2)|^2 d\omega = \frac{\pi^2}{2\pi}\int_{-\infty}^{\infty}|\text{rect}(\omega/2)|^2 d\omega$

$= \frac{\pi}{2}\int_{-\infty}^{\infty}\text{rect}(\omega/2)d\omega = \frac{\pi}{2}\int_{-1}^{1}(1)d\omega = \frac{\pi}{2}(2) = \pi.$

A5.17 What is the energy of $\text{sinc}^2(t)$?

Answer:

$\text{Energy}\{\text{sinc}^2(t)\} = \frac{1}{2\pi}\int_{-\infty}^{\infty}|\mathcal{F}\{\text{sinc}^2(t)\}|^2 d\omega$

$= \frac{1}{2\pi}\int_{-\infty}^{\infty}\left|\pi\Delta\left(\frac{\omega}{4}\right)\right|^2 d\omega = \frac{\pi^2}{2\pi}\int_{-\infty}^{\infty}\left|\Delta\left(\frac{\omega}{4}\right)\right|^2 d\omega$

$= \frac{\pi}{2}\int_{-2}^{2}\left(1-\frac{|\omega|}{2}\right)^2 d\omega = \pi\int_{0}^{2}\left(1-\frac{\omega}{2}\right)^2 d\omega$

$= \pi\int_{0}^{2}\left(1-\omega+\frac{\omega^2}{4}\right)d\omega = \pi\left(\omega-\frac{\omega^2}{2}+\frac{\omega^3}{12}\right)\Big|_{0}^{2}$

$$= \pi \left(2 - \frac{4}{2} + \frac{8}{12}\right)\Big|_0^2 = \frac{2\pi}{3}.$$

A5.18 What is the energy of $5j\ \text{rect}(t/4)$?

 Answer:

$$\text{Energy}\{5j\ \text{rect}(t/4)\} = \int_{-\infty}^{\infty}|5j\ \text{rect}(t/4)|^2 dt$$

$$= 25 \int_{-\infty}^{\infty}|\text{rect}(t/4)|^2 dt = 25 \int_{-2}^{2}|1|^2 dt = 25 \cdot 4 = 100.$$

A5.19 What is the power of $-6\sin(2t)$?

 Answer: $\text{Power}\{-6\sin(2t)\} = \frac{|-6|^2}{2} = 18.$

A5.20 What is the power of the signal whose spectrum is
$2\delta(\omega + 2\pi) - \delta(\omega) + j\delta(\omega - 2\pi)$?

 Answer: Since impulses are at different frequencies, each

 contributes power $= \left|\frac{Area}{2\pi}\right|^2$. The total power is then equal to

$$\left|\frac{2}{2\pi}\right|^2 + \left|\frac{-1}{2\pi}\right|^2 + \left|\frac{j}{2\pi}\right|^2 = \frac{4+1+1}{4\pi^2} = \frac{3}{2\pi^2}.$$

A5.21 Find the Fourier Transform of $\cos(2t) + j\sin(2t)$.

 Answer:

$$\mathcal{F}\{\cos(2t) + j\sin(2t)\} = \mathcal{F}\{\cos(2t)\} + j\mathcal{F}\{\sin(2t)\}$$

$$= \pi\big(\delta(\omega + 2) + \delta(\omega - 2)\big) + j(j\pi)\big(\delta(\omega + 2) - \delta(\omega - 2)\big)$$

$$= \pi\delta(\omega + 2) + \pi\delta(\omega - 2) - \pi\delta(\omega + 2) + \pi\delta(\omega - 2)$$

$$= 2\pi\delta(\omega - 2).$$

A5.22 Find the Fourier Transform of $\delta(t + 1) - \delta(t - 1)$.

 Answer:

$$\mathcal{F}\{\delta(t + 1) - \delta(t - 1)\} = \mathcal{F}\{\delta(t + 1)\} - \mathcal{F}\{\delta(t - 1)\}$$

$$= e^{j\omega(1)} - e^{j\omega(-1)} = e^{j\omega} - e^{-j\omega} = 2j\sin(\omega).$$

A5.23 Find the Fourier Transform of impulse train function $\frac{1}{\pi}\delta_1(t)$.

Answer:

$$\mathcal{F}\left\{\frac{1}{\pi}\delta_1(t)\right\} = \frac{1}{\pi}\mathcal{F}\{\delta_1(t)\} = \frac{1}{\pi}\left(2\pi\,\delta_{2\pi}(\omega)\right) = 2\delta_{2\pi}(\omega).$$

A5.24 Find the Fourier Transform of $u(t+1)$.

Answer:

$$\mathcal{F}\{u(t+1)\} = e^{j\omega(1)}\mathcal{F}\{u(t)\} = e^{j\omega}\left(1/j\omega + \pi\delta(\omega)\right)$$

$$= \frac{e^{j\omega}}{j\omega} + \pi e^{j\omega}\delta(\omega) = \frac{e^{j\omega}}{j\omega} + \pi e^{j(0)}\delta(\omega)$$

$$= \frac{e^{j\omega}}{j\omega} + \pi(1)\delta(\omega) = \frac{e^{j\omega}}{j\omega} + \pi\delta(\omega).$$

A5.25 Find the Fourier Transform of $f_6(t) = \delta(2t)$.

Answer:

$$\mathcal{F}\{\delta(2t)\} = \mathcal{F}\left\{\frac{1}{2}\delta(t)\right\} = \frac{1}{2}\mathcal{F}\{\delta(t)\} = \frac{1}{2}(1) = \frac{1}{2}.$$

A5.26 Find the Fourier Transform of $x(t) = 2u(t) + u(-t) - 1$.

Answer:

$$\mathcal{F}\{2u(t) + u(-t) - 1\} = 2\mathcal{F}\{u(t)\} + \mathcal{F}\{u(-t)\} - \mathcal{F}\{1\}$$

$$= 2\left(1/j\omega + \pi\delta(\omega)\right) + \left(1/j(-\omega) + \pi\delta(-\omega)\right) - 2\pi\delta(\omega)$$

$$= \frac{2}{j\omega} + 2\pi\delta(\omega) - \frac{1}{j\omega} + \pi\delta(\omega) - 2\pi\delta(\omega) = \frac{1}{j\omega} + \pi\delta(\omega).$$

A5.27 Find the inverse Fourier Transform of $F(\omega) = \text{sinc}^2(\omega/4)$.

Answer:

$$\mathcal{F}^{-1}\{\text{sinc}^2(\omega/4)\} = \mathcal{F}^{-1}\{2\text{sinc}^2(\omega/4)/2\}$$

$$= 2\mathcal{F}^{-1}\{\text{sinc}^2(\omega/4)/2\} = 2\Delta(t).$$

A5.28   Find the Fourier Transform of $\delta(t/2)$.

Answer:
$$\mathcal{F}\{\delta(t/2)\} = \mathcal{F}\{2\delta(t)\} = 2\mathcal{F}\{\delta(t)\} = 2(1) = 2.$$

A5.29   Find the Fourier Transform of $\text{sinc}^2(t)$.

Answer:
$$\mathcal{F}\{\text{sinc}^2(t)\} = \mathcal{F}\left\{\pi \cdot \frac{1}{\pi}\text{sinc}^2(t)\right\} = \pi\mathcal{F}\left\{\frac{1}{\pi}\text{sinc}^2(t)\right\}$$
$$= \pi\Delta\left(\frac{\omega}{4}\right).$$

A5.30   Find the Fourier Transform of $2u(t) - 1$.

Answer:
$$\mathcal{F}\{2u(t) - 1\} = 2\mathcal{F}\{u(t)\} - \mathcal{F}\{1\}$$
$$= 2\big(1/j\omega + \pi\delta(\omega)\big) - 2\pi\delta(\omega) = 2/j\omega.$$

A5.31   Find the inverse Fourier Transform of $\text{sinc}(\pi\omega)$.

Answer:
$$\mathcal{F}^{-1}\{\text{sinc}(\pi\omega)\} = \mathcal{F}^{-1}\left\{\frac{1}{2\pi} \cdot 2\pi\,\text{sinc}(\pi\omega)\right\}$$
$$= \frac{1}{2\pi}\mathcal{F}^{-1}\{2\pi\,\text{sinc}(\pi\omega)\} = \frac{1}{2\pi}\,\text{rect}\left(\frac{t}{2\pi}\right).$$

A5.32   Find the inverse Fourier Transform of
$2\pi\big(\delta(\omega - 1) + \delta(\omega + 2)\big)$.

Answer:
$$\mathcal{F}^{-1}\{2\pi\big(\delta(\omega - 1) + \delta(\omega + 2)\big)\}$$
$$= \mathcal{F}^{-1}\{2\pi\delta(\omega - 1)\} + \mathcal{F}^{-1}\{2\pi\delta(\omega + 2)\}$$
$$= e^{j(1)t}\mathcal{F}^{-1}\{2\pi\delta(\omega)\} + e^{j(-2)t}\mathcal{F}^{-1}\{2\pi\delta(\omega)\}$$
$$= e^{jt}(1) + e^{-j2t}(1) = e^{jt} + e^{-j2t}.$$

A5.33 Find the inverse Fourier Transform of $\frac{1}{1+\omega^2}$.

Answer:

$$\mathcal{F}^{-1}\left\{\frac{1}{1+\omega^2}\right\} = \mathcal{F}^{-1}\left\{\frac{1}{2}\cdot\frac{2}{1+\omega^2}\right\} = \frac{1}{2}\mathcal{F}^{-1}\left\{\frac{2}{1+\omega^2}\right\}$$
$$= \frac{1}{2}\left(e^{-|t|}\right) = \frac{1}{2}e^{-|t|}.$$

A5.34 Find the Fourier Transform of $\frac{1}{1+t^2}$.

Answer:

From Problem A5.34, we know that $\frac{1}{2}e^{-|t|} \Leftrightarrow \frac{1}{1+\omega^2}$.

Therefore, by the duality property of the Fourier Transform,

$$\frac{1}{1+t^2} \Leftrightarrow 2\pi\left(\frac{1}{2}e^{-|-\omega|}\right) = \pi e^{-|\omega|}.$$

A5.35 Assuming that $f_1(t) \Leftrightarrow F_1(\omega)$ and $f_2(t) \Leftrightarrow F_2(\omega)$, show that $f_1(t) * f_2(t) \Leftrightarrow F_1(\omega)F_2(\omega)$.

Solution:

$$f_1(t) * f_2(t) \Leftrightarrow \mathcal{F}\{f_1(t) * f_2(t)\}$$
$$= \int_{-\infty}^{\infty}\{f_1(t) * f_2(t)\}\,e^{-j\omega t}dt$$
$$= \int_{-\infty}^{\infty}\left\{\int_{-\infty}^{\infty}f_1(\tau)f_2(t-\tau)d\tau\right\}e^{-j\omega t}dt$$
$$= \int_{-\infty}^{\infty}f_1(\tau)\left\{\int_{-\infty}^{\infty}f_2(t-\tau)\,e^{-j\omega t}dt\right\}d\tau$$
$$= \int_{-\infty}^{\infty}f_1(\tau)\,\mathcal{F}\{f_2(t-\tau)\}\,d\tau = \int_{-\infty}^{\infty}f_1(\tau)e^{-j\omega\tau}\mathcal{F}\{f_2(t)\}d\tau$$
$$= \int_{-\infty}^{\infty}f_1(\tau)e^{-j\omega\tau}F_2(\omega)\,d\tau = F_2(\omega)\int_{-\infty}^{\infty}f_1(\tau)e^{-j\omega\tau}d\tau$$
$$= F_2(\omega)F_1(\omega) = F_1(\omega)F_2(\omega).$$

A5.36 Show that $a_1\text{sinc}(b_1t) * a_2\text{sinc}(b_2t) = a_3\text{sinc}(b_3t)$, where $\{a_1, a_2, a_3, b_1, b_2, b_3\}$ are real constants. Find expressions for $\{a_3, b_3\}$ in terms of $\{a_1, a_2, b_1, b_2\}$.

Solution:

From Table 5.1, $\frac{W}{\pi}\text{sinc}(Wt) \Leftrightarrow \text{rect}\left(\frac{\omega}{2W}\right)$. Therefore,

$\text{sinc}(Wt) \Leftrightarrow \frac{\pi}{W}\text{rect}\left(\frac{\omega}{2W}\right)$. It then follows that:

$a_1\text{sinc}(b_1 t) \Leftrightarrow \frac{a_1\pi}{b_1}\text{rect}\left(\frac{\omega}{2b_1}\right)$, $a_2\text{sinc}(b_2 t) \Leftrightarrow \frac{a_2\pi}{b_2}\text{rect}\left(\frac{\omega}{2b_2}\right)$.

$\mathcal{F}\{a_1\text{sinc}(b_1 t) * a_2\text{sinc}(b_2 t)\} = \frac{a_1\pi}{b_1}\text{rect}\left(\frac{\omega}{2b_1}\right) \times$

$\frac{a_2\pi}{b_2}\text{rect}\left(\frac{\omega}{2b_2}\right) = \frac{a_1 a_2\pi^2}{b_1 b_2}\text{rect}\left(\frac{\omega}{2\min\{b_1,b_2\}}\right)$, because the product

of two rectangular pulses is the narrower of the two. Taking
the inverse Fourier Transform we obtain: $\{c\,\text{rect}(\omega/d)\} =$

$\frac{c(d/2)}{\pi}\text{sinc}\big((d/2)t\big) = a_3\text{sinc}(b_3 t)$, where $a_3 = \frac{c}{\pi}\frac{d}{2} =$

$\frac{a_1 a_2\pi}{b_1 b_2}\min\{b_1, b_2\}$ and $b_3 = \frac{d}{2} = \min\{b_1, b_2\}$.

A5.37 Express the periodic signal $f_p(t)$ below as an Exponential
Fourier Series:

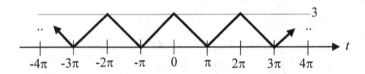

Solution:

This periodic signal has period $T_0 = 2\pi$ sec ($\omega_0 = 2\pi/T_0 = 1$
rad/sec). The expression for $f_p(t) = p(t) * \delta_{2\pi}(t)$, where
pulse $p(t)$ defining one period is over time span $t = [-\pi, \pi]$:

$p(t) = 3\Delta\left(\frac{t}{2\pi}\right) \Leftrightarrow 3\pi\text{sinc}^2\left(\frac{\omega\pi}{2}\right)$. Therefore,

$D_k = \frac{1}{T_0}P(k\omega_0) = \frac{1}{2\pi}\left(3\pi\text{sinc}^2\left(\frac{k\omega_0\pi}{2}\right)\right) = \frac{3}{2}\text{sinc}^2\left(\frac{k\pi}{2}\right)$.

The final answer is then: $f_p(t) = \sum_{k=-\infty}^{\infty} D_k e^{jk\omega_0 t}$,

or $f_p(t) = \sum_{k=-\infty}^{\infty}\big((3/2)\text{sinc}^2(k\pi/2)\big)e^{jkt}$.

A5.38 Plot, using MATLAB, the Exponential Fourier Series expression found in Problem A5.38, using 10,000 points linearly spaced over time span $[-4\pi, 4\pi]$ sec.

$$f_p(t) = \sum_{k=-\infty}^{\infty}\left(\frac{3}{2}\operatorname{sinc}^2\left(\frac{k\pi}{2}\right)\right)e^{jkt}$$

```
N = 10000;
t = linspace(-4*pi,4*pi,N);
f = ones(1,N) * 3/2;   % k = 0
% Note: MATLAB function sinc(x) returns sin(pi*x)/(pi*x)
for k = 1:100   % stop at k = 100
        f = f + (3/2)*(sinc(k/2).^2).*exp(j*k*t);    % k > 0
        f = f + (3/2)*(sinc(-/2).^2).*exp(-j*k*t);   % k < 0
    end
plot(t/pi,f)
xlabel('t/\pi in sec')
```

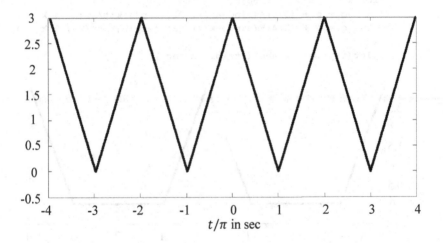

A5.39 Express the periodic signal $g_p(t)$ below as an Exponential Fourier Series:

Solution:
This periodic signal has period $T_0 = 4\pi$ sec ($\omega_0 = 2\pi/T_0 = 0.5$ rad/sec). The expression for $g_p(t) = p(t) * \delta_{4\pi}(t)$, where pulse $p(t)$ defining one period is measured over time span $t =$

$[-2\pi, 2\pi]$: $p(t) = 3\Delta(t/2\pi) \Leftrightarrow 3\pi\text{sinc}^2(\omega\pi/2)$.
Therefore, $D_k = (1/T_0)P(k\omega_0) =$
$(1/4\pi)(3\pi\text{sinc}^2(k\omega_0\pi/2)) = (3/4)\text{sinc}^2(k\pi/4)$.
The final answer is then: $g_p(t) = \sum_{k=-\infty}^{\infty} D_k e^{jk\omega_0 t}$,
or $g_p(t) = \sum_{k=-\infty}^{\infty}((3/4)\text{sinc}^2(k\pi/4))e^{jkt/2}$.

A5.40   Plot, using MATLAB, the Exponential Fourier Series expression found in Problem A5.40, using 10,000 points linearly spaced over time span $[-4\pi, 4\pi]$ sec.

$$g_p(t) = \sum_{k=-\infty}^{\infty}\left(\frac{3}{4}\text{sinc}^2\left(\frac{k\pi}{4}\right)\right)e^{jkt/2}.$$

```
N = 10000; t = linspace(-4*pi,4*pi,N);
g = ones(1,N) * 3/4;     % k = 0
% Note: MATLAB function sinc(x) returns sin(pi*x)/(pi*x)
for k = 1:100  % stop at k = 100
      g = g + (3/4)*(sinc(k/4).^2).*exp(j*k*t/2);    % k > 0
      g = g + (3/4)*(sinc(-k/4).^2).*exp(-j*k*t/2);  % k < 0
end
plot(t/pi,g); xlabel('t/\pi in sec')
```

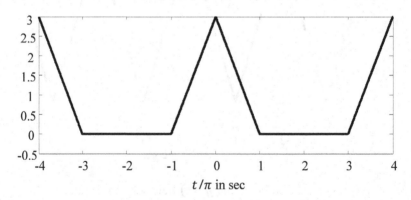

t/π in sec

A5.41   Express the periodic signal $h_p(t)$ below as an Exponential Fourier Series:

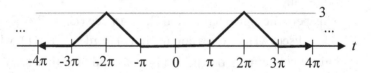

Solution:

This periodic signal has period $T_0 = 4\pi$ sec ($\omega_0 = 2\pi/T_0$ $= 0.5$ rad/sec). The expression for $h_p(t) = p(t) * \delta_{4\pi}(t)$, where pulse $p(t)$ defining one period is measured over time span $t = [0, 4\pi]$: $p(t) = 3\Delta((t - 2\pi)/2\pi) \Leftrightarrow$ $3\pi e^{-j\omega(2\pi)}\text{sinc}^2(\omega\pi/2)$. Therefore, $D_k = (1/T_0)P(k\omega_0)$ $= (1/4\pi)(3\pi e^{-j(k\omega_0)2\pi}\text{sinc}^2(k\omega_0\pi/2))$ $= (3/4)e^{-jk\pi}\text{sinc}^2(k\pi/4)$. The final answer is then: $h_p(t) = \sum_{k=-\infty}^{\infty} D_k e^{jk\omega_0 t}$, or $h_p(t) =$ $\sum_{k=-\infty}^{\infty} \left(\frac{3}{4} e^{-jk\pi}\text{sinc}^2\left(\frac{k\pi}{4}\right)\right) e^{jkt/2}$.

A5.42 Plot, using MATLAB, the Exponential Fourier Series expression found in Problem A5.42, using 10,000 points linearly spaced over time span $[-4\pi, 4\pi]$ sec.

$$h_p(t) = \sum_{k=-\infty}^{\infty} \left(\frac{3}{4} e^{-jk\pi}\text{sinc}^2\left(\frac{k\pi}{4}\right)\right) e^{jkt/2}.$$

```
N = 10000;
t = linspace(-4*pi,4*pi,N);
% Note: MATLAB function sinc(x) returns sin(pi*x)/(pi*x)
h = ones(1,N) * 3/4;        % k = 0
for k = 1:100               % stop at k = 100
    % k > 0
    h = h+(3/4)*exp(-j*k*pi).*(sinc(k/4).^2).*exp(j*k*t/2);
    % k < 0
    h = h+(3/4)*exp(j*k*pi).*(sinc(-k/4).^2).*exp(-j*k*t/2);
    end
plot(t/pi,h)
xlabel('t/\pi in sec')
```

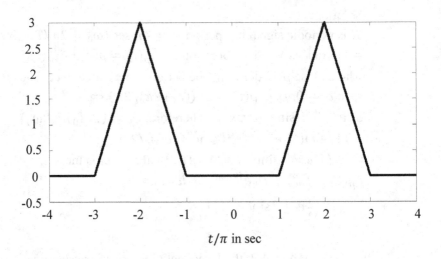

$$t/\pi \text{ in sec}$$

A5.43   Express the periodic signal $f_p(t)$ shown below as a
        Trigonometric Fourier Series:

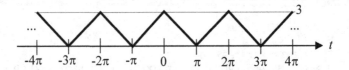

Solution:
From the solution to Problem A5.43 we know that for this
waveform, the Exponential Fourier Series coefficients in the
expression $f_p(t) = \sum_{k=-\infty}^{\infty} D_k e^{jk\omega_0 t}$ ($\omega_0 = 1$ rad/sec) are
$D_k = \frac{3}{2}\text{sinc}^2(k\pi/2)$. The Trigonometric Fourier Series
coefficients are: $a_0 = D_0 = \frac{3}{2}\text{sinc}^2(0) = \frac{3}{2}$; $a_k = D_k + D_{-k}$
$= \frac{3}{2}\text{sinc}^2(k\pi/2) + \frac{3}{2}\text{sinc}^2(-k\pi/2) = 3\text{sinc}^2(k\pi/2)$, $k \geq 1$;
$b_k = j(D_k - D_{-k}) = \frac{3j}{2}\text{sinc}^2(k\pi/2) - \frac{3j}{2}\text{sinc}^2(-k\pi/2) = 0$,
$k \geq 1$. The expression for $f_p(t)$ as a Trigonometric Fourier

Series is then: $f_p(t) = \frac{3}{2} + \sum_{k=1}^{\infty} 3\text{sinc}^2(k\pi/2)\cos(k\omega_0 t)$, where $\omega_0 = 1$ rad/sec.

A5.44 Plot, using MATLAB, the Trigonometric Fourier Series expression found in Problem A5.44, using 10,000 points linearly spaced over time span $[-4\pi, 4\pi]$ sec.

$$f_p(t) = \frac{3}{2} + \sum_{k=1}^{\infty} 3\text{sinc}^2\left(\frac{k\pi}{2}\right)\cos(kt)$$

```
N = 10000;
t = linspace(-4*pi,4*pi,N);
% Note: MATLAB function sinc(x) returns sin(pi*x)/(pi*x)
f = ones(1,N) * 3/2;   % k = 0
for k = 1:100          % stop at k = 100
     f = f + 3*(sinc(k/2).^2).*cos(k*t);   % k > 0
  end
plot(t/pi,f)
xlabel('t/\pi in sec')
```

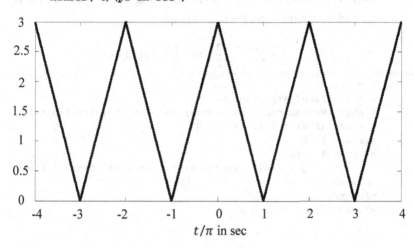

A5.45 Express the periodic signal $g_p(t)$ that is shown below as a *compact* Trigonometric Fourier Series:

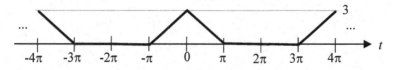

Solution:

From Problem A5.45 solution we know that for this waveform, the Exponential Fourier Series coefficients in the expression $g_p(t) = \sum_{k=-\infty}^{\infty} D_k e^{jk\omega_0 t}$ ($\omega_0 = 0.5$ rad/sec) are $D_k = (3/4)\text{sinc}^2(k\pi/4)$. The Compact Trigonometric Fourier Series coefficients are $C_0 = D_0 = \frac{3}{4}\text{sinc}^2(0) = \frac{3}{4}$; $C_k = 2|D_k| = 2\left|\frac{3}{4}\text{sinc}^2\left(\frac{k\pi}{4}\right)\right| = \frac{3}{2}\text{sinc}^2\left(\frac{k\pi}{4}\right)$, $k \geq 1$; $\theta_k = \angle D_k = 0, k \geq 1$. The expression for $g_p(t)$ as a compact Trigonometric Fourier Series is then $g_p(t) = \frac{3}{4} + \sum_{k=-\infty}^{\infty} \frac{3}{2}\text{sinc}^2\left(\frac{k\pi}{4}\right)\cos(k\omega_0 t + 0)$

where $\omega_0 = 0.5$ rad/sec.

A5.46 Plot in MATLAB the Compact Trigonometric Fourier Series expression found in Problem A5.46, using 10,000 points linearly spaced over time span $[-4\pi, 4\pi]$ sec.

$$g_p(t) = \frac{3}{4} + \sum_{k=-\infty}^{\infty} \frac{3}{2}\text{sinc}^2\left(\frac{k\pi}{4}\right)\cos(kt/2).$$

```
N = 10000;
t = linspace(-4*pi,4*pi,N);
% Note: MATLAB function sinc(x) returns sin(pi*x)/(pi*x)
g = ones(1,N) * 3/4;    % k = 0
for k = 1:100                                              %
stop at k = 100
        g = g + (3/2)*(sinc(k/4).^2).*cos(k*0.5*t); % k > 0
    end
plot(t/pi,g)
xlabel('t/\pi in sec')
```

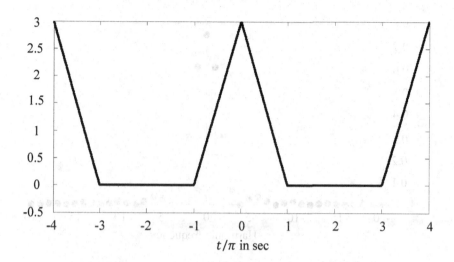

t/π in sec

A5.47 Using MATLAB plot $|D_k|$ vs. k, for $-20 \leq k \leq 20$, in the
Exponential Fourier Series for $h_p(t)$ shown below:

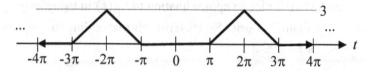

Solution:
From Problem A5.42 we know that, for this waveform,
the Exponential Fourier Series coefficients in the expression
$f_p(t) = \sum_{k=-\infty}^{\infty} D_k e^{jk\omega_0 t}$ ($\omega_0 = 1$ rad/sec) are
$D_k = (3/4)e^{-jk\pi}\text{sinc}^2(k\pi/4)$.

Here is the MATLAB code to plot $|D_k|$ vs. k:
```
k = -20:20;
% MATLAB function sinc(x) returns sin(pi*x)/(pi*x)
Dk = (3/4)*exp(-j*k*pi).*(sinc(k/4).^2);
stem(k,abs(Dk),'filled')
xlabel('Harmonic frequency index k')
ylabel('|D_k|')
```

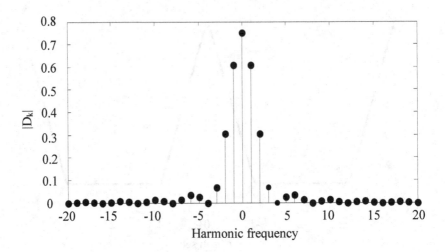

A5.48   When we look at the plot of $|D_k|$ vs. k in the solution to Problem A5.48 above, much of the signal power is concentrated at zero frequency and the first three harmonics ($-3 \leq k \leq 3$). Using MATLAB, plot an approximation to $h_p(t)$ in using only Exponential Fourier Series terms corresponding to frequencies $-3 \leq k\omega_0 \leq 3$.

Solution:

$$h_p(t) \approx \sum_{k=-3}^{3} D_k e^{jk\omega_0 t} = \sum_{k=-3}^{3} \left\{ \frac{3}{4} e^{-jk\pi} \text{sinc}^2\left(\frac{k\pi}{4}\right) \right\} e^{jkt}$$

```
N = 10000;
t = linspace(-4*pi,4*pi,N);
% MATLAB function sinc(x) returns sin(pi*x)/(pi*x)
h = ones(1,N) * 3/4;   %  k = 0
for k = 1:3
    %  k > 0
    h=h+(3/4)*exp(-j*k*pi).*(sinc(k/4).^2).*exp(j*k*t/2);
    %  k < 0
    h=h+(3/4)*exp(j*k*pi).*(sinc(-k/4).^2).*exp(-j*k*t/2);
    end
plot(t/pi,h)
xlabel('t/\pi in sec')
```

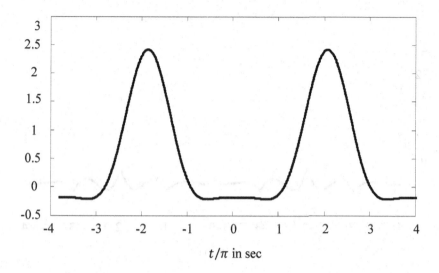

t/π in sec

A5.49   In Problem A5.49 you were asked to approximate periodic signal
$h_p(t)$ based on a few of its harmonics in the Exponential Fourier
Series. Calculate the approximation error of that result, over the
same spans of time and amplitude.

Solution:

$e_p(t) = h_p(t) - \sum_{k=-3}^{3} D_k e^{jk\omega_0 t} = \sum_{k=-\infty}^{\infty} D_k e^{jk\omega_0 t} - \sum_{k=-3}^{3} D_k e^{jk\omega_0 t} = \sum_{k=-\infty}^{-4} D_k e^{jk\omega_0 t} + \sum_{k=-4}^{\infty} D_k e^{jk\omega_0 t}.$

```
N = 10000;
t = linspace(-4*pi,4*pi,N);
% Note: MATLAB function sinc(x) returns sin(pi*x)/(pi*x)
e = zeros(1,N);
for k = 4:100   % stop at k = 100
    % k > +3
    e=e+(3/4)*exp(-j*k*pi).*(sinc(k/4).^2).*exp(j*k*t/2);
    % k < -3
    e=e+(3/4)*exp(j*k*pi).*(sinc(-k/4).^2).*exp(-j*k*t/2);
    end
plot(t/pi,e)
axis([-4, 4, -0.5, 3])
xlabel('t/\pi in sec')
```

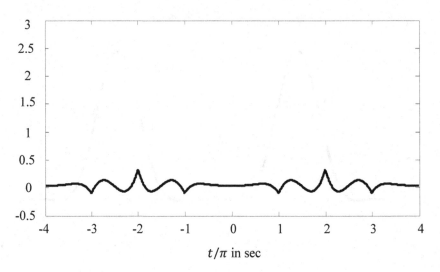

$$t/\pi \text{ in sec}$$

A5.50   Calculate the energy of $-2\text{sinc}(5t)$ in either the time or in the frequency domain (whichever is easiest):

Solution:

Energy$\{-2\text{sinc}(5t)\}$

$= \frac{1}{2\pi}\int_{-\infty}^{\infty}|\mathcal{F}\{-2\text{sinc}(5t)\}|^2 d\omega$

$= \frac{1}{2\pi}\int_{-\infty}^{\infty}|-2\mathcal{F}\{\text{sinc}(5t)\}|^2 d\omega$

$= \frac{1}{2\pi}\int_{-\infty}^{\infty}\left|-2\left\{\frac{\pi}{5}\text{rect}\left(\frac{\omega}{10}\right)\right\}\right|^2 d\omega$

$= \frac{1}{2\pi}\int_{-\infty}^{\infty}\frac{4\pi^2}{25}\left|\text{rect}\left(\frac{\omega}{10}\right)\right|^2 d\omega$

$= \frac{1}{2\pi}\cdot\frac{4\pi^2}{25}\int_{-5}^{5}|1|^2 d\omega = \frac{1}{2\pi}\cdot\frac{4\pi^2}{25}\cdot 10 = \frac{4\pi}{5}.$

Alternatively: We know that $\text{sinc}(t)$ has energy $= \pi$. A time-compressed $\text{sinc}(5t)$ will have energy $\pi/5$. Therefore, $-2\text{sinc}(5t)$ will have energy $= |-2|^2\frac{\pi}{5} = \frac{4\pi}{5}$.

A5.51   Calculate the energy of $5j\,\text{rect}(t/4)$ in either the time or in the frequency domain (whichever is easiest):

Solution:

Energy$\{5j\,\text{rect}(t/4)\} = \int_{-\infty}^{\infty}|5j\,\text{rect}(t/4)|^2 dt$

$= 25\int_{-\infty}^{\infty}|\text{rect}(t/4)|^2 dt = 25\int_{-2}^{2}|1|^2 dt = 25\cdot 4 = 100.$

Alternatively, we know that $\text{rect}(t)$ has energy $= 1$. A time-stretched $\text{rect}(t/4)$ will have energy 4. Therefore, $5j\,\text{rect}(t/4)$ will have energy $= |5j|^2 4 = 25\cdot 4 = 100.$

A5.52 Calculate the energy of $\text{sinc}(5t)\cos(15t)$ in either the time or in the frequency domain (whichever is easiest):

Solution:

Energy$\{\text{sinc}(5t)\cos(15t)\}$

$= \frac{1}{2\pi}\int_{-\infty}^{\infty}|\mathcal{F}\{\text{sinc}(5t)\cos(15t)\}|^2 d\omega.$

Since $\text{sinc}(5t) \Leftrightarrow \frac{\pi}{5}\text{rect}\left(\frac{\omega}{10}\right)$, $\text{sinc}(5t)\cos(15t) \Leftrightarrow$

$\frac{1}{2}\left\{\frac{\pi}{5}\text{rect}\left(\frac{\omega+15}{10}\right) + \frac{\pi}{5}\text{rect}\left(\frac{\omega-15}{10}\right)\right\}$. Therefore the energy equals

$\frac{1}{2\pi}\int_{-\infty}^{\infty}\left|\frac{1}{2}\left\{\frac{\pi}{5}\text{rect}\left(\frac{\omega+15}{10}\right) + \frac{\pi}{5}\text{rect}\left(\frac{\omega-15}{10}\right)\right\}\right|^2 d\omega$

$= \frac{1}{2\pi}\int_{-\infty}^{\infty}\frac{\pi^2}{100}\left|\text{rect}\left(\frac{\omega+15}{10}\right) + \text{rect}\left(\frac{\omega-15}{10}\right)\right|^2 d\omega$

$= \frac{\pi}{200}\int_{-\infty}^{\infty}\left(\text{rect}\left(\frac{\omega+15}{10}\right) + \text{rect}\left(\frac{\omega-15}{10}\right)\right) d\omega$

$= \frac{\pi}{200}\left(\int_{-20}^{-10}(1)\,d\omega + \int_{10}^{20}(1)\,d\omega\right)$

$= \frac{\pi}{200}(10+10) = \frac{\pi}{10}.$

Alternatively: we know that that $\text{sinc}(t)$ has energy $= \pi$. A time-compressed $\text{sinc}(5t)$ will have energy $= \pi/5$. When multiplied by a cosine, its energy/power are cut by factor 2 (if there is no overlap between the resulting two copies of the spectrum that are shifted left and right); therefore, $\text{sinc}(5t)\cos(15t)$ has energy $= \pi/10$.

A5.53 Calculate the power of $2\cos(2t + \pi/3)$ in either the time or in the frequency domain (whichever is easiest):

Solution:
Power $\{2\cos(2t + \pi/3)\} = |2|^2/2 = 2$.

A5.54 Calculate the power of $\mathrm{sinc}(10t)\cos(8t)$ in either the time or in the frequency domain (whichever is easiest):

Solution:
Power$\{\mathrm{sinc}(10t)\cos(8t)\} = 0$, since this is an energy signal. (There are no impulses in the spectrum, and $|F(\omega)|^2$ has finite area).

A5.55 Calculate the power of $\mathrm{sinc}(10t) * \delta_1(t)$ in either the time or in the frequency domain (whichever is easiest).

Solution:
Power $\{\mathrm{sinc}(10t) * \delta_1(t)\}$ may be found in the frequency domain: $\mathrm{sinc}(10t) * \delta_1(t) \Leftrightarrow \mathcal{F}\{\mathrm{sinc}(10t)\} \cdot \mathcal{F}\{\delta_1(t)\}$
$= (\pi/10)\mathrm{rect}(\omega/20) \cdot 2\pi\,\delta_{2\pi}(\omega) = \pi^2/5 \times \{\text{sum of those impulses in } \delta_{2\pi}(\omega) \text{ that lie within } -10 \le \omega \le 10\}$
$= (\pi^2/5)\{\delta(\omega + 2\pi) + \delta(\omega) + \delta(\omega - 2\pi)\}$. These three impulses each contribute power $|\mathrm{Area}/2\pi|^2 = \left|(\pi^2/5)/2\pi\right|^2$
$= \pi^2/100$, so that the total power is equal to $3\pi^2/100$.

A5.56 Calculate the power of $\mathrm{rect}(t) * \delta_1(t)$ in either the time or in the frequency domain (whichever is easiest):

Solution:|
Power $\{\mathrm{rect}(t) * \delta_1(t)\}$ may be calculated in either the time or the frequency domain. In the time domain, one can show that $\mathrm{rect}(t) * \delta_1(t) = 1$. The power is the average value of $|1|^2$, which equals 1. In the frequency domain, $\mathrm{rect}(t) * \delta_1(t) \Leftrightarrow$

$\mathcal{F}\{rect(t)\} \cdot \mathcal{F}\{\delta_1(t)\} = sinc(\omega/2) \cdot 2\pi \, \delta_{2\pi}(\omega)$. The function $sinc(\omega/2)$ has zero-crossings at every integer multiple of 2π except at $\omega = 0$, so the product of these two functions gives only the impulse at $\omega = 0$: $2\pi \, \delta(\omega)$. The power is then found as $|Area/2\pi|^2 = |2\pi/2\pi|^2 = 1$.

A5.57 Find the area of $sinc^2(t)$ using Parseval's Theorem.

Solution:
$$\int_{-\infty}^{\infty} sinc^2(t)dt = \int_{-\infty}^{\infty} |sinc(t)|^2 dt = Energy\{sinc(t)\}$$
$$= \frac{1}{2\pi}\int_{-\infty}^{\infty}|\mathcal{F}\{sinc(t)\}|^2 d\omega = \frac{1}{2\pi}\int_{-\infty}^{\infty}\left|\pi \, rect\left(\frac{\omega}{2}\right)\right|^2 d\omega$$
$$= \frac{\pi^2}{2\pi}\int_{-\infty}^{\infty}\left|rect\left(\frac{\omega}{2}\right)\right|^2 d\omega = \frac{\pi}{2}\int_{-\infty}^{\infty} rect\left(\frac{\omega}{2}\right) d\omega$$
$$= \frac{\pi}{2}\int_{-1}^{1}(1)d\omega = \frac{\pi}{2}(2) = \pi.$$

Chapter 6

A6.1 Calculate the Fourier Transform of $x(t) = \frac{1}{\pi}\cos(3t)$ and plot that spectrum over frequency range $-10 \leq \omega \leq 10$ rad/sec.

Answer:
$$\frac{1}{\pi}\cos(3t) \Leftrightarrow \delta(\omega + 3) + \delta(\omega - 3).$$

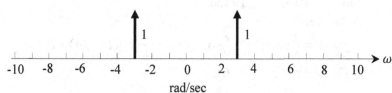

A6.2 Calculate the Fourier Transform of $x_s(t) = \frac{1}{\pi}\cos(3t) \cdot \frac{2\pi}{5}\delta_{\frac{2\pi}{5}}(t)$ and plot that spectrum over frequency range $-10 \leq \omega \leq 10$ rad/sec.

Answer:

$$\frac{1}{\pi}\cos(3t) \cdot \frac{2\pi}{5}\delta_{\frac{2\pi}{5}}(t) \Leftrightarrow \{\delta(\omega + 3) + \delta(\omega - 3)\} * \delta_5(\omega).$$

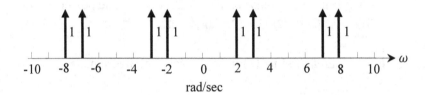

A6.3 Calculate the Fourier Transform of $y(t) = \frac{2}{\pi}\mathrm{sinc}(2t)$ and plot that spectrum over frequency range $-10 \leq \omega \leq 10$ rad/sec.

Answer: $y(t) = \frac{2}{\pi}\mathrm{sinc}(2t) \Leftrightarrow \mathrm{rect}\left(\frac{\omega}{4}\right)$.

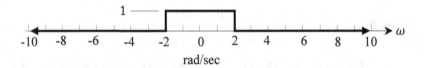

A6.4 Calculate the Fourier Transform of $y_s(t) = \frac{2}{\pi}\mathrm{sinc}(2t) \cdot \frac{2\pi}{6}\delta_{2\pi/6}(t)$ and plot that spectrum over frequency range $-10 \leq \omega \leq 10$ rad/sec.

Answer:

$$y_s(t) = \frac{2}{\pi}\mathrm{sinc}(2t) \cdot \frac{2\pi}{6}\delta_{\frac{2\pi}{6}}(t) \Leftrightarrow \frac{1}{2\pi}\left(\mathrm{rect}(\omega/4) * \delta_6(\omega)\right)$$

$(T_s = 2\pi/6, \quad \omega_s = 6).$

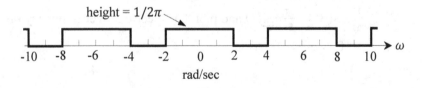

A6.5   Calculate the Fourier Transform of $w(t) = \frac{1}{2\pi}\text{sinc}^2(t/2)$ and plot that spectrum over frequency range $-10 \leq \omega \leq 10$ rad/sec.

Answer:
$$w(t) = \frac{1}{2\pi}\text{sinc}^2(t/2) \Leftrightarrow \Delta(\omega/2).$$

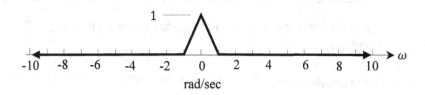

A6.6   Calculate the Fourier Transform of $w_s(t) =$ $\frac{1}{2\pi}\text{sinc}^2(t/2) \cdot \pi\delta_\pi(t)$ and plot that spectrum over frequency range $-10 \leq \omega \leq 10$ rad/sec.

Answer:
$$w_s(t) = \frac{1}{2\pi}\text{sinc}^2(t/2) \cdot \pi\delta_\pi(t) \Leftrightarrow \frac{1}{2\pi}\left(\Delta(\omega/2) * \delta_2(\omega)\right)$$
$$(T_s = \pi, \quad \omega_s = 2).$$

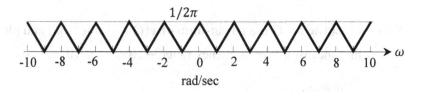

A6.7   Calculate the Fourier Transform of $f(t) = \frac{1}{\pi}\cos(7t)$ and plot that spectrum over frequency range $-10 \leq \omega \leq 10$ rad/sec.

Answer:
$$f(t) = \frac{1}{\pi}\cos(7t) \Leftrightarrow \delta(\omega \pm 7).$$

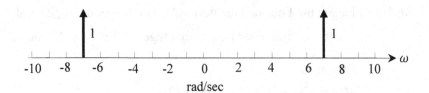

A6.8   Calculate the Fourier Transform of $f_s(t) = \frac{1}{\pi}\cos(7t) \cdot \frac{2\pi}{5}\delta_{\frac{2\pi}{5}}(t)$
and plot that spectrum over frequency range $-10 \leq \omega \leq 10$
rad/sec.

     Answer:

$$f_s(t) = \frac{1}{\pi}\cos(7t) \cdot \frac{2\pi}{5}\delta_{\frac{2\pi}{5}}(t) \Leftrightarrow \delta(\omega \pm 7) * \delta_5(\omega)$$

$$(T_s = 2\pi/5, \quad \omega_s = 5).$$

all impulse areas = 1

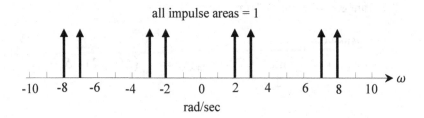

A6.9   Calculate the Fourier Transform of $x(t) = \frac{3}{\pi}\text{sinc}^2(3t)$ and plot
that spectrum over frequency range $-10 \leq \omega \leq 10$ rad/sec.

     Answer:

$$x(t) = \frac{3}{\pi}\text{sinc}^2(3t) \Leftrightarrow \Delta(\omega/12).$$

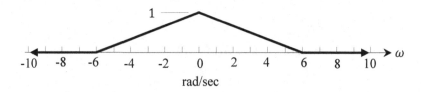

A6.10   Calculate the Fourier Transform of $x_{s1}(t) = x(t) \cdot \frac{\pi}{6} \delta_{\pi/6}(t)$ and

plot that spectrum over frequency range $-10 \leq \omega \leq 10$ rad/sec. (Assume $x(t)$ from Problem A6.9).

Answer:

$$x_{s1}(t) = x(t) \cdot \frac{\pi}{6} \delta_{\frac{\pi}{6}}(t) \Leftrightarrow \Delta(\omega/12) * \delta_{12}(\omega)$$
$$= X_{s1}(\omega) * \delta_{12}(\omega)$$

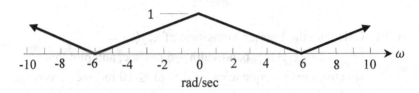

A6.11   Calculate the Fourier Transform of $x_{r1}(t) = $ Ideal LPF$\{x_{s1}(t)\}$, bandwidth = 6 rad/sec, and plot that spectrum over frequencies $-10 \leq \omega \leq 10$ rad/sec. (Assume $x_{s1}(t)$ from Problem A6.10.)

Answer:
The spectrum of $x_{s1}(t)$ shown above is a periodically-extended version of the spectrum of $x(t)$ in Problem A6.9. An ideal lowpass filter passing only frequencies $-12 \leq \omega \leq 12$ rad/sec exactly recovers $x(t)$ from $x_{s1}(t)$, as shown below.

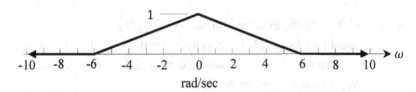

A6.12   Calculate the Fourier Transform of $x_{s2}(t) = x(t) \cdot \frac{\pi}{5} \delta_{\pi/5}(t)$ and plot that spectrum over frequency range $-10 \leq \omega \leq 10$ rad/sec. (Assume $x(t)$ from Problem A6.9.)

Answer:

$$x_{s2}(t) = x(t) \cdot \frac{\pi}{5}\delta_{\pi/5}(t) \Leftrightarrow \Delta(\omega/12) * \delta_{10}(\omega)$$

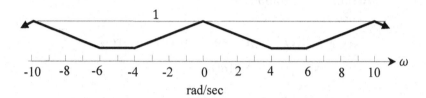

A6.13 Calculate the Fourier Transform of $x_{r2}(t) =$
Ideal LPF$\{x_{s2}(t)\}$, bandwidth = 6 rad/sec, and plot that
spectrum over frequencies $-10 \le \omega \le 10$ rad/sec. (Assume
$x_{s2}(t)$ from Problem A6.12).

Answer:
Referring to the plot in Problem A6.12, its lowpass-filtered
version is graphed below.

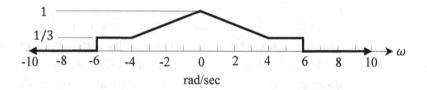

A6.14 Calculate the Fourier Transform of $x_{r3}(t) =$
Ideal LPF$\{x_{s2}(t)\}$, bandwidth = 4 rad/sec, and plot that
spectrum over frequencies $-10 \le \omega \le 10$ rad/sec. (Assume
$x_{s2}(t)$ from Problem A6.12).
Answer:

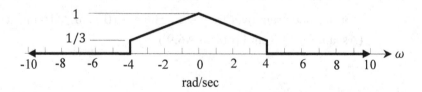

A6.15　Calculate the Fourier Transform of $y(t) = $ Ideal LPF $\{x(t)\}$,
bandwidth = 5 rad/sec, and plot that spectrum over frequencies
$-10 \leq \omega \leq 10$ rad/sec.　(Assume $x(t)$ from Problem A6.9).

Answer:
$Y(\omega)$ is plotted below:

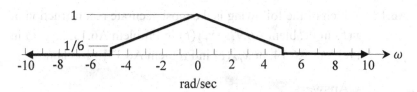

A6.16　Calculate the Fourier Transform of $y_s(t) = y(t) \cdot \frac{\pi}{5}\delta_{\pi/5}(t)$, and
plot that spectrum over frequency range $-10 \leq \omega \leq 10$ rad/sec.
(Assume $y(t)$ from Problem A6.15).

Answer:
$$y_s(t) = y(t) \cdot \frac{\pi}{5}\delta_{\frac{\pi}{5}}(t) \Leftrightarrow Y(\omega) * \delta_{10}(\omega) = Y_s(\omega).$$

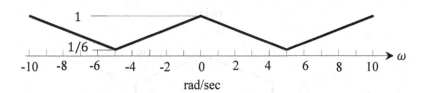

A6.17　Calculate the Fourier Transform of $x_{r4}(t) = $ Ideal LPF$\{y_s(t)\}$,
bandwidth = 5 rad/sec, and plot that spectrum over frequency
range $-10 \leq \omega \leq 10$ rad/sec.　(Assume $y_s(t)$ from Problem
A6.16).

Answer:
$X_{r4}(\omega)$ is plotted below.

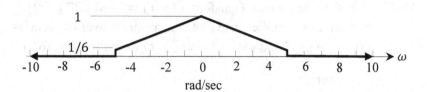

A6.18 Which of the following is the most accurate reconstruction of $x(t)$ in Problem A6.9?: {$x_{r2}(t)$ in Problem A6.13, $x_{r3}(t)$ in Problem A6.14, or $x_{r4}(t)$ in Problem A6.17}. Explain.

Answer:
$X_{r4}(\omega)$ best represents $X(\omega)$, in the sense that these two spectra are identical over 0-5 rad/sec frequency range, whereas the other two match $X(\omega)$ only over 0-4 rad/sec.

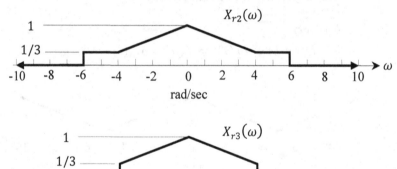

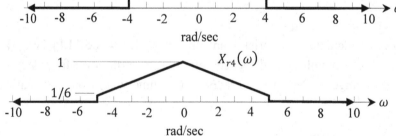

A6.19 Calculate the Fourier Transform of $w_s(t) = w(t) \cdot \frac{2\pi}{9} \delta_{2\pi/9}(t)$,

where $w(t) = \frac{1}{2\pi} \text{sinc}^2(t/2)$, and plot that spectrum over

frequency range $-10 \le \omega \le 10$ rad/sec.

Answer:

From Problem A6.5, $w(t) = \frac{1}{2\pi} \text{sinc}^2(t/2) \Leftrightarrow \Delta(\omega/2)$.

Therefore $w_s(t) = w(t) \cdot \frac{2\pi}{9} \delta_{\frac{2\pi}{9}}(t) \Leftrightarrow \Delta(\omega/2) * \delta_9(\omega)$.

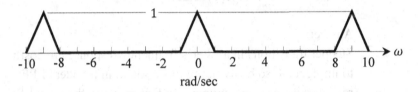

A6.20 Given: $h(t) \Leftrightarrow H(\omega) = \Delta\left(\frac{\omega+4}{8}\right) + \Delta\left(\frac{\omega}{8}\right) + \Delta\left(\frac{\omega-4}{8}\right)$. Plot

$H(\omega)$ over frequency range $-10 \le \omega \le 10$ rad/sec.

Answer:

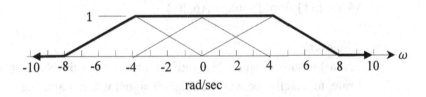

A6.21 Calculate the Fourier Transform of $w_r(t) = w_s(t) * h(t) \Leftrightarrow$

$W_s(\omega)H(\omega)$ and plot that spectrum over $-10 \le \omega \le 10$ rad/sec.

(Assume $w_s(t)$ from Problem A6.19, $h(t)$ from Problem A6.20.)

Answer:

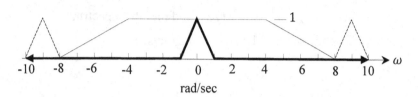

rad/sec

A6.22   What is the advantage of using $H(\omega)$ to recover $w(t)$ instead of using an ideal lowpass filter? (Assume $w_s(t)$ from Problem A6.19, $h(t)$ from Problem A6.20).

Answer:
Steeper transitions in a filter's frequency response are costlier to implement, so by avoiding those found in the ideal LPF response we can save money. In fact, because the ideal LPF has instantaneous jumps in its frequency response, the ideal LPF is not realizable.

A6.23   What is the disadvantage of using $H(\omega)$ to recover $w(t)$ instead of using an ideal lowpass filter? (Assume $w_s(t)$ from Problem A6.19, $h(t)$ from Problem A6.20.)

Answer:
$H(\omega)$ requires that $\omega_s > 9$ rad/sec, or 4.5 times the Nyquist rate, to exactly recover the original signal without aliasing distortion; a higher sampling rate is undesirable due to more samples/second that need to be stored or transmitted.

Chapter 7

A7.1   Write an expression for the frequency response of an ideal digital lowpass filter having bandwidth $= \pi/3$ rad/sec.

Answer: $H_{LPF}(e^{j\omega}) = \text{rect}\left(\dfrac{\omega}{2\omega_{max}}\right) * \delta_{2\pi}(\omega)$

$$= \text{rect}\left(\dfrac{\omega}{2(\pi/3)}\right) * \delta_{2\pi}(\omega).$$

A7.2 Write an expression for the frequency response of an ideal digital highpass filter having cutoff frequency $= \pi/2$ rad/sec.

Answer: $H_{HPF}(e^{j\omega}) = 1 - \text{rect}\left(\dfrac{\omega}{2\omega_{min}}\right) * \delta_{2\pi}(\omega)$

$$= 1 - \text{rect}\left(\dfrac{\omega}{\pi}\right) * \delta_{2\pi}(\omega).$$

A7.3 Write an expression for the frequency response of an ideal digital bandpass filter having passband edge frequencies $\pi/4$ and $3\pi/4$ rad/sec.

Answer:

$$H_{BPF}(e^{j\omega}) = \left\{\text{rect}\left(\dfrac{\omega}{2\omega_{max}}\right) - \text{rect}\left(\dfrac{\omega}{2\omega_{min}}\right)\right\} * \delta_{2\pi}(\omega)$$

$$= \left\{\text{rect}\left(\dfrac{\omega}{2(3\pi/4)}\right) - \text{rect}\left(\dfrac{\omega}{2(\pi/4)}\right)\right\} * \delta_{2\pi}(\omega)$$

A7.4 Design and sketch a causal lowpass digital filter network having bandwidth $\pi/4$ rad/sec. Label input/output signals and function of every block.

Answer: (using the design formulas given in Chapter 7)

$$BW = \pi/4 = \omega_0$$

$$C_1 = \dfrac{\sin(\omega_0/2)}{\cos(\omega_0/2)+\sin(\omega_0/2)} = 0.2929,$$

$$\alpha = \dfrac{\cos(\omega_0/2)-\sin(\omega_0/2)}{\cos(\omega_0/2)+\sin(\omega_0/2)} = 0.4142.$$

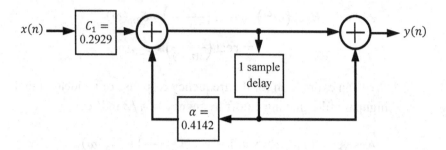

A7.5 Write the causal difference equation corresponding to the digital filter network in Problem A7.4.

Answer: Based on the network diagram shown in Problem A7.4, we can determine the filter's difference equation by labelling the signal coming out of the delay unit "$w(n)$". Then, we may write:

$$y(n) = w(n) + \alpha w(n) + C_1 x(n),$$

$$w(n) = \alpha w(n-1) + C_1 x(n-1).$$

Solving for $w(n)$ in the first equation we obtain:

$$w(n) = \frac{1}{\alpha+1}(y(n) - C_1 x(n)) = \frac{1}{\alpha+1}y(n) - \frac{C_1}{\alpha+1}x(n).$$

Then, substituting this expression into the second equation we obtain $y(n) = C_1 x(n) + C_1 x(n-1) + \alpha y(n-1)$, or

$$y(n) = 0.2929x(n) + 0.2929x(n-1) + 0.4142y(n-1).$$

A7.6 Write an expression for $H(e^{j\omega}) = Y(e^{j\omega})/X(e^{j\omega})$ corresponding to the digital filter network in Problem A7.4.

Answer: From Chapter 7,

$$H_{LPF}(e^{j\omega}) = C_1 \frac{1+e^{-j\omega}}{1-\alpha e^{-j\omega}} = 0.2929\left(\frac{1+e^{-j\omega}}{1-0.4142e^{-j\omega}}\right).$$

A7.7 Plot $\left|H(e^{j\omega})\right|^2$ in dB vs. normalized frequency ω/ω_s, over frequency range $0 \le \omega/\omega_s \le 0.5$, corresponding to the digital filter network in Problem A7.4.

Answer: We use the expression for $H(e^{j\omega})$ found in Problem A7.6 to find $|H(e^{j\omega})|^2$ in dB using MATLAB code shown below.

```
Npts = 1e4;
wn = linspace(0,0.5,Npts);
w = 2*pi*wn;
H = 0.2929*((1+exp(-j*w))./(1-0.4142*exp(-j*w)));
plot(wn,20*log10(abs(H)))
% change vertical axis for better-looking plot
axis([0 0.5 -50 10])
xlabel('normalized frequency \omega/\omega_s')
ylabel('|H(e^j^\omega)|^2 (dB)')
title(['H(e^j^\omega) = ', ...
    '0.2929*(1+e^-^j^\omega)/(1-0.4142e^-^j^\omega)'])
grid
```

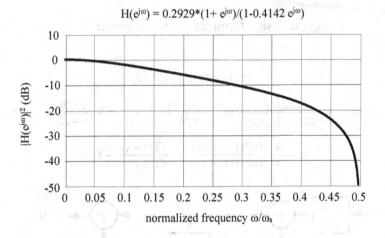

$$H(e^{j\omega}) = 0.2929*(1+ e^{j\omega})/(1-0.4142\ e^{j\omega})$$

A7.8 Draw a signal flow graph that implements difference equation
$$y(n) = x(n) + 2x(n-1) + 3x(n-2) - 5y(n-2).$$

Answer:

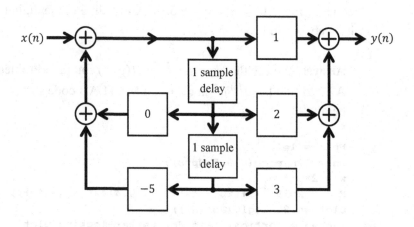

A7.9   Design a simple causal highpass digital filter network having
$\omega_{-3dB} = \pi/2$ rad/sec. Sketch your network. Label input signal
$x(n)$, output signal $y(n)$, and the function of every block.

Answer:

Using the design formulas given in Chapter 7:

$$\omega_{-3dB} = \omega_0 = \pi/2,$$

$$C_2 = \frac{\cos(\omega_0/2)}{\cos(\omega_0/2)+\sin(\omega_0/2)} = \frac{\cos(\pi/4)}{\cos(\pi/4)+\sin(\pi/4)} = \frac{\sqrt{2}/2}{\sqrt{2}/2+\sqrt{2}/2} = 0.5,$$

$$\alpha = \frac{\cos(\pi/4)-\sin(\pi/4)}{\cos(\pi/4)+\sin(\pi/4)} = \frac{\sqrt{2}/2 - \sqrt{2}/2}{\sqrt{2}/2 + \sqrt{2}/2} = 0.$$

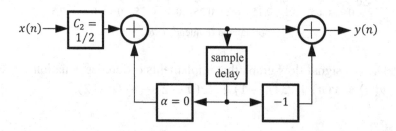

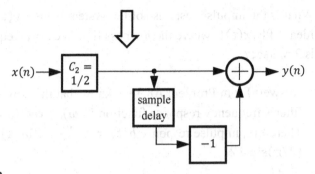

Chapter 8

A8.1 What is the impulse response of the system having input/output relation $y(t) = -6x(t)$?

Answer: Because $y(t) = x(t) * h(t)$, then it must be true that $h(t) = -6\delta(t)$. We could have solved this problem more systematically in the frequency domain: since $y(t) = -6x(t)$, or $Y(\omega) = -6X(\omega)$, then $H(\omega) = Y(\omega)/X(\omega) = -6$ and $h(t) = \mathcal{F}^{-1}\{-6\} = -6\delta(t)$.

A8.2 What is the impulse response of the system having input/output relation $(t) = x(t + 2)$?

Answer: From $y(t) = x(t + 2)$ we deduce that $h(t) = \delta(t + 2)$.

A8.3 What is the frequency response of the system having input/output relation $y(t) = \text{Ideal LPF}\{x(t)\}$, where the lowpass filter's cutoff frequency is 2 rad/sec?

Answer: The lowpass filter bandwidth is 2 rad/sec (the highest frequency in the passband minus the lowest frequency in the passband, over positive frequencies only), and therefore $H(\omega) = \text{rect}(\omega/4)$.

A8.4   What is the impulse response of the system having $y(t) =$ Ideal LPF$\{x(t)\}$, where the lowpass filter's cutoff frequency is 2 rad/sec?

   Answer: From Problem A8.3 we know that this lowpass filter's frequency response function $H(\omega) = \text{rect}(\omega/4)$. Therefore, impulse response $h(t) = \mathcal{F}^{-1}\{\text{rect}(\omega/4)\} = (2/\pi)\text{sinc}(2t)$.

A8.5   What is the frequency response of the system having $y(t) =$ Ideal HPF$\{x(t)\}$, passing all frequencies above 2 rad/sec?

   Answer: The ideal highpass filter frequency response is $H_{HPF}(\omega) = 1 - H_{LPF}(\omega)$, where $H_{LPF}(\omega)$ is that of an ideal lowpass filter having 2 rad/sec bandwidth. Therefore, referring to Problem A8.3, $H(\omega) = 1 - \text{rect}(\omega/4)$.

A8.6   What is the impulse response of the system having $y(t) =$ Ideal HPF$\{x(t)\}$, passing all frequencies above 2 rad/sec?

   Answer: From Problem A8.5 we know that this highpass filter's frequency response function $H(\omega) = 1 - \text{rect}(\omega/4)$. Therefore, $h(t) = \mathcal{F}^{-1}\{1 - \text{rect}(\omega/4)\}$
   $= \delta(t) - (2/\pi)\text{sinc}(2t)$.

A8.7   What is the frequency response of the system having $y(t) =$ Ideal BPF$\{x(t)\}$, having passband between 1 and 2 rad/sec?

   Answer: An ideal bandpass filter frequency response may be formulated by subtracting a narrower rectangular pulse from a wider rectangular pulse in the frequency domain. Following that approach, $H(\omega) = \text{rect}(\omega/4) - \text{rect}(\omega/2)$.

A8.8 What is the impulse response of the system having $y(t) =$ Ideal BPF$\{x(t)\}$, having passband between 1 and 2 rad/sec?

Answer: From Problem A8.7 we know that this bandpass filter's frequency response function $H(\omega) = \text{rect}(\omega/4) - \text{rect}(\omega/2)$. Therefore $h(t) = \mathcal{F}^{-1}\{\text{rect}(\omega/4) - \text{rect}(\omega/2)\} = (2/\pi)\text{sinc}(2t) - (1/\pi)\text{sinc}(t)$.

A8.9 Calculate the frequency response function $H(\omega) = V_{out}(\omega)/V_{in}(\omega)$ of the circuit below:

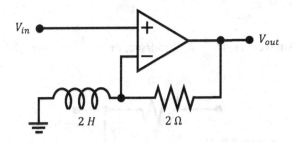

Answer:
$$H(\omega) = 1 + \frac{Z_2}{Z_1} = 1 + \frac{2}{j2\omega} = 1 + \frac{1}{j\omega} = 1 - j\left(\frac{1}{\omega}\right).$$

A8.10 Design an operational amplifier circuit that realizes a filter with frequency response $H(\omega) = -2/(1/j\omega + j2\omega)$. Draw your circuit and label every component with its value.

Answer: Because of the minus sign in $H(\omega)$, use the inverting op-amp circuit having $H(\omega) = -Z_2/Z_1$. One possibility is that Z_2 is a 2Ω resistor, and Z_1 is a $1\,F$ capacitor in series with a $2H$ inductor:

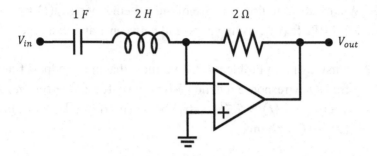

A8.11   Design a simple lowpass filter circuit having half-power frequency = 2 rad/sec. One of your components should be a 5 F capacitor.

Answer: $W = \dfrac{1}{RC} = 2$ rad/sec, or $= \dfrac{1}{CW} = \dfrac{1}{5 \cdot 2} = 0.1 \, \Omega$:

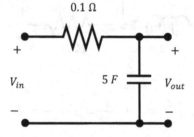

A8.12   Design a simple bandpass filter having center frequency = 10 rad/sec and bandwidth = 1 rad/sec. Your components should include a 1 F capacitor.

Answer:

(Possible circuit #1) $\omega_0^2 = \dfrac{1}{LC} = 10^2 = 100$, or $L = \dfrac{1}{C\omega_0^2}$

$= \dfrac{1}{1 \cdot 100} = 0.01 \, H$. $W = \dfrac{1}{RC} = 1$, or $R = \dfrac{1}{C} = 1 \, \Omega$.

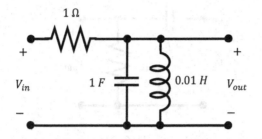

(Possible circuit #2) $\omega_0^2 = \dfrac{1}{LC} = 10^2 = 100$, or $L = \dfrac{1}{C\omega_0^2}$

$= \dfrac{1}{1 \cdot 100} = 0.01\ H$. $W = \dfrac{R}{L} = 1$, or $R = LW = 0.01\ \Omega$.

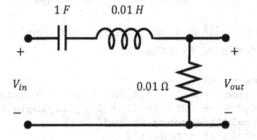

A8.13 Design a simple highpass analog filter circuit using two components: R, C. Choose any value of R in the range $1\Omega \le R \le 10\Omega$.
The filter should have a half-power frequency equal to 5 rad/sec.

a) Sketch your circuit. Label input/output signals and
component values.

Answer: To make $W = \omega_{-3dB} = \dfrac{1}{RC} = 5$ rad/sec, this
solution has $R = 1\ \Omega, C = 1/5$ F.

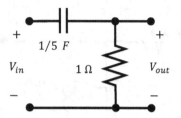

b) Write an expression for $|H(\omega)|^2$, your circuit's magnitude-squared transfer function, in terms of ω.

Answer:

$$|H_{HP}(\omega)|^2 = \left|\frac{Z_R}{Z_R+Z_C}\right|^2 = \left|\frac{1}{1+5/j\omega}\right|^2 = \left|\frac{j\omega}{j\omega+5}\right|^2 = \frac{\omega^2}{\omega^2+25}$$

c) Plot $|H(\omega)|^2$ vs. ω using the log-log method, over the range $1/25 \le \omega \le 25$ rad/sec.

```
Npts = 1e4;
w = logspace(log10(1/25),log10(25),Npts);
Hmag2 = (w.^2)./(w.^2 + 25); loglog(w,Hmag2);
xlabel('frequency in rad/sec')
ylabel('|H(\omega)|^2')
title('|H(\omega)|^2 = \omega^2/(\omega^2+25)')
grid on
```

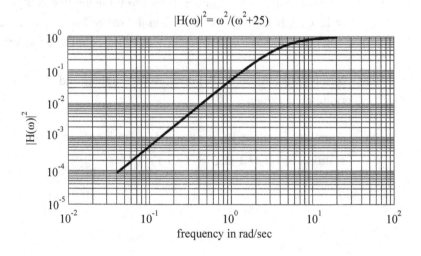

A8.14 Design a simple bandpass analog filter circuit using three components: resistor, inductor and capacitor. Choose any value of resistance in the range $1k\Omega \leq R \leq 10k\Omega$. The filter should have 5 kHz bandwidth and 100 kHz passband center frequency.

a) Sketch your circuit. Label input/output signals and component values.

Answer:
To make W = BPF bandwidth $= 1/RC = 2\pi \cdot 5\text{kHz}$, this solution has $R = 1\ k\Omega$, $C = 32$ nF. Then $L = 1/(\omega_0^2 C)$ $= 1/\left((2\pi \cdot 100\text{kHz})^2(32 \times 10^{-9})\right) = 80$ uH.

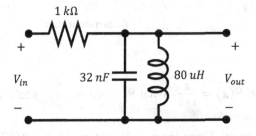

b) Plot $|H(f)|^2$ in dB vs. f on a log scale, over the range $1\ \text{kHz} \leq f \leq 10\ \text{MHz}$.

Answer:
$$|H_{BP}(\omega)|^2 = \frac{\omega^2}{\omega^2+\left((\omega_0^2-\omega^2)/W\right)^2}, \text{ where } \omega_0 = 2\pi \cdot 100\text{kHz},$$
$W = 2\pi \cdot 5\text{kHz}.$

MATLAB code:

```
Npts = 1e4;
f = logspace(log10(1e3),log10(10e6),1e4);
w = 2*pi*f;
w0 = 100e3 * 2 * pi;
W  =   5e3 * 2 * pi;
Hmag2 = (w.^2)./(w.^2 + ((w0^2-w.^2)/W).^2);
semilogx(f,10*log10(Hmag2))
```

```
xlabel('frequency in Hz')
ylabel('|H(f)|^2 (in dB)')
grid on
```

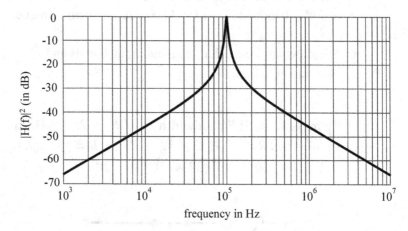

Chapter 9

A9.1   When $F(z) = \frac{z}{z+2}$, what ROC corresponds to causal $f(n)$?

       Answer: Since the pole is at $z = -2$, the ROC for causal $f(n)$ is: $|z| > |-2| = 2$.

A9.2   Given: $H(z) = \frac{3z}{3z+1} - \frac{4z}{2z-1}$

     a)   What regions of convergence are possible for this $H(z)$?
         (Specify ROC_1: $r_1 < |z| < r_2$, ROC_2: $r_3 < |z| < r_4$, etc.)

     Answer:

$$H(z) = \frac{3z}{3z+1} - \frac{4z}{2z-1} = \frac{z}{z+1/3} - \frac{2z}{z-1/2} = \frac{-(z-0)(z+7/6)}{(z+1/3)(z-1/2)}.$$

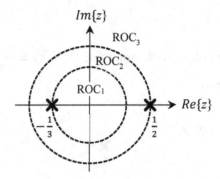

Therefore, the pole values are $\{-1/3,\ 1/2\}$ and their magnitudes define the boundaries of ROCs:

$$ROC_1:\quad 0 \le |z| < 1/3,\quad ROC_2:\quad 1/3 < |z| < 1/2$$

$$ROC_3:\quad 1/2 < |z| \le \infty$$

b) Write an expression for $h(n) = \mathcal{L}^{-1}\{H(z)\}$ for each region of convergence found in (a).

Answer:

$ROC_1:\quad h_1(n) = (-1/3)^n\{u(n) - 1\} - 2(1/2)^n\{u(n) - 1\},$

$ROC_2:\quad h_2(n) = (-1/3)^n\, u(n) \qquad\quad\ - 2(1/2)^n\{u(n) - 1\},$

$ROC_3:\quad h_3(n) = (-1/3)^n\, u(n) \qquad\quad\ - 2(1/2)^n\, u(n).$

c) Plot a pole-zero diagram for $H(z)$.

Answer: From (b), $H(z) = (-1)\dfrac{(z-0)(z+7/6)}{(z+1/3)(z-1/2)}.$

$S.F. = -1 \qquad\qquad\qquad Im\{z\}$

Chapter 10

A10.1 Plot a pole-zero diagram for: $\quad H(s) = \dfrac{4s(s-1)}{2(s+1+2j)(s+1-2j)}$

Answer:

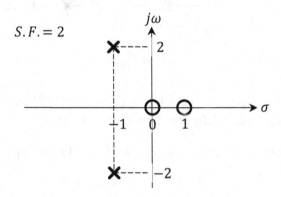

A10.2 Find an expression for $H(s)$ from its pole-zero diagram below (do not specify ROC):

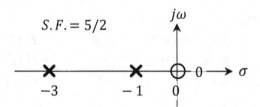

Answer:

$$H(s) = \frac{5}{2}\frac{s}{(s+3)(s+1)}.$$

A10.3 Given: $H(s) = \dfrac{2}{s^2+6s+34}$, and it is known that $h(t) = \mathcal{L}^{-1}\{H(s)\}$ is causal. Find $h(t)$.

Solution:

$$H(s) = \frac{2}{s^2+6s+34} = \frac{2}{(s+3)^2+25} = \frac{\frac{2}{5}(5)}{(s+3)^2+25}.$$

Since, from Table 9.1, we know the causal Laplace transform pair $e^{at}\sin(\omega_0 t)u(t) \Leftrightarrow \frac{\omega_0}{(s-a)^2+\omega_0^2}$, we obtain:

$$h(t) = \tfrac{2}{5}e^{-3t}\sin(5t)\,u(t).$$

A10.4 Find the inverse bilateral Laplace transform of transfer function

$$(s) = \frac{2s+5}{(s+2)(s+3)}, \quad -3 < \sigma < -2.$$

Solution:
Using partial fraction expansion we obtain

$$\frac{2s+5}{(s+2)(s+3)} = \frac{1}{s+2} + \frac{1}{s+3}.$$

For region of convergence $-3 < \sigma < -2$, the term $\frac{1}{s+3}$ corresponds to a causal time function and the term $\frac{1}{s+2}$ corresponds to an anticausal time function:

$$\mathcal{L}^{-1}\left\{\frac{1}{s+3}\right\} = e^{-3t}u(t),$$

$$\mathcal{L}^{-1}\left\{\frac{1}{s+2}\right\} = e^{-2t}\{u(t) - 1\} = -e^{-2t}u(-t),$$

From this it follows that $h(t) = e^{-3t}u(t) - e^{-2t}u(-t)$.

A10.5 Find the inverse bilateral Laplace transform of transfer function

$$(s) = \frac{2s-5}{(s-2)(s-3)}, \quad 2 < \sigma < 3.$$

Solution: Using partial fraction expansion we obtain

$$\frac{2s-5}{(s-2)(s-3)} = \frac{1}{s-2} + \frac{1}{s-3}.$$

For region of convergence $2 < \sigma < 3$, the term $\frac{1}{s-2}$

corresponds to a causal time function and the term $\frac{1}{s-3}$ corresponds to an anticausal time function:

$$\mathcal{L}^{-1}\left\{\frac{1}{s-2}\right\} = e^{2t}u(t),$$

$$\mathcal{L}^{-1}\left\{\frac{1}{s-3}\right\} = e^{3t}\{u(t) - 1\} = -e^{3t}u(-t),$$

From this it follows that $h(t) = e^{2t}u(t) - e^{3t}u(-t)$.

A10.6 Find the inverse bilateral Laplace transform of transfer function
$$(s) = \frac{2s+3}{(s+1)(s+2)}, \ \sigma > -1.$$

Solution: Using partial fraction expansion we obtain

$$\frac{2s+3}{(s\mp1)(s+2)} = \frac{1}{s+1} + \frac{1}{s+2}.$$

For region of convergence $\sigma > -1$, both terms $\frac{1}{s+1}$ and $\frac{1}{s+2}$ correspond to causal time functions:

$$\mathcal{L}^{-1}\left\{\frac{1}{s+1}\right\} = e^{-t}u(t),$$

$$\mathcal{L}^{-1}\left\{\frac{1}{s+2}\right\} = e^{-2t}u(t),$$

From this it follows that $h(t) = \{e^{-t} + e^{-2t}\}u(t)$.

A10.7 Find the inverse bilateral Laplace transform of transfer function
$$(s) = \frac{2s+3}{(s+1)(s+2)}, \ \sigma < -2.$$

Solution: Using partial fraction expansion we obtain
$$\frac{2s+3}{(s\mp1)(s+2)} = \frac{1}{s+1} + \frac{1}{s+2}.$$

For region of convergence $\sigma < -2$, both terms $\frac{1}{s+1}$ and $\frac{1}{s+2}$ correspond to anticausal time functions:

$$\mathcal{L}^{-1}\left\{\frac{1}{s+1}\right\} = e^{-t}\{u(t) - 1\} = -e^{-t}u(-t),$$

$$\mathcal{L}^{-1}\left\{\frac{1}{s+2}\right\} = e^{-2t}\{u(t) - 1\} = -e^{-2t}u(-t),$$

From this it follows that $h(t) = \{-e^{-t} - e^{-2t}\}u(-t)$.

A10.8 Find the inverse bilateral Laplace transform of transfer function

$$(s) = \frac{3s^2 - 2s - 17}{(s+1)(s+3)(s-5)}, \quad -1 < \sigma < 5.$$

Solution: Using partial fraction expansion we obtain

$$\frac{3s^2 - 2s - 17}{(s+1)(s+3)(s-5)} = \frac{1}{s+1} + \frac{1}{s+3} + \frac{1}{s-5}.$$

For region of convergence $-1 < \sigma < 5$, the terms $\frac{1}{s+1}$ and $\frac{1}{s+3}$ correspond to causal time functions and the term $\frac{1}{s-5}$ corresponds to an anticausal time function:

$$\mathcal{L}^{-1}\left\{\frac{1}{s+1}\right\} = e^{-t}u(t),$$

$$\mathcal{L}^{-1}\left\{\frac{1}{s+3}\right\} = e^{-3t}u(t),$$

$$\mathcal{L}^{-1}\left\{\frac{1}{s-5}\right\} = e^{5t}\{u(t) - 1\} = -e^{5t}u(-t),$$

From this it follows that $h(t) = \{e^{-t} + e^{-3t}\}u(t) - e^{5t}u(-t)$.

A10.9 Repeat Problem A10.8, this time assuming a region of convergence corresponding to a causal signal in the time domain.

Solution:

Recall that $H(s) = \frac{3s^2 - 2s - 17}{(s+1)(s+3)(s-5)} = \frac{1}{s+1} + \frac{1}{s+3} + \frac{1}{s-5}$. When an inverse Laplace transform exists such that the time function

$h(t)$ is causal, every term in the partial fraction expansion is causal:[a]

$$\mathcal{L}^{-1}\left\{\frac{1}{s+1}\right\} = e^{-t}u(t)$$

$$\mathcal{L}^{-1}\left\{\frac{1}{s+3}\right\} = e^{-3t}u(t)$$

$$\mathcal{L}^{-1}\left\{\frac{1}{s-5}\right\} = e^{5t}u(t)$$

From this it follows that $h(t) = \{e^{-t} + e^{-3t} + e^{5t}\}u(t)$.

A10.10 Repeat Problem A10.8, this time assuming a region of convergence corresponding to an anticausal signal in the time domain.

Solution:

Recall that $H(s) = \dfrac{3s^2 - 2s - 17}{(s+1)(s+3)(s-5)} = \dfrac{1}{s+1} + \dfrac{1}{s+3} + \dfrac{1}{s-5}$. When an inverse Laplace transform exists such that the time function $h(t)$ is anticausal, every term in the partial fraction expansion is anticausal:[b]

$$\mathcal{L}^{-1}\left\{\frac{1}{s+1}\right\} = e^{-t}\{u(t) - 1\} = -e^{-t}u(-t)$$

$$\mathcal{L}^{-1}\left\{\frac{1}{s+3}\right\} = e^{-3t}\{u(t) - 1\} = -e^{-3t}u(-t)$$

$$\mathcal{L}^{-1}\left\{\frac{1}{s-5}\right\} = e^{5t}\{u(t) - 1\} = -e^{5t}u(-t)$$

From this it follows that $h(t) = \{-e^{-t} - e^{-3t} - e^{5t}\}u(-t)$.

A10.11 Repeat Problem A10.8, this time assuming a region of convergence corresponding to a stable system.

---

[a] The region of convergence is $(\sigma > -1) \cap (\sigma > -3) \cap (\sigma > 5) = (\sigma > 5)$.
[b] The region of convergence is $(\sigma < -1) \cap (\sigma < -3) \cap (\sigma < 5) = (\sigma < -1)$.

Solution:

Recall that $H(s) = \frac{3s^2-2s-17}{(s+1)(s+3)(s-5)} = \frac{1}{s+3} + \frac{1}{s+1} + \frac{1}{s-5}$, which

has pole values $s = \{-3, -1, +5\}$. When an inverse Laplace transform exists such that the time function $h(t)$ is the impulse response of a stable system, the region of convergence must contain the $j\omega$ axis ($\sigma = 0$). Thus the terms corresponding to poles at $s = \{-3, -1\}$ must be causal and the term corresponding to a pole at $s = 5$ must be anticausal:[c]

$$\mathcal{L}^{-1}\left\{\frac{1}{s+1}\right\} = e^{-t}u(t), \quad \mathcal{L}^{-1}\left\{\frac{1}{s+3}\right\} = e^{-3t}u(t);$$

$$\mathcal{L}^{-1}\left\{\frac{1}{s-5}\right\} = e^{5t}\{u(t) - 1\} = -e^{5t}u(-t).$$

Therefore, $h(t) = \{e^{-t} + e^{-3t}\}u(t) - e^{5t}u(-t)$.

A10.12 Given: $H(s) = (3s^2 + 14s - 2)/(s^3 + s^2 - 2s)$

a) Plot a pole-zero diagram for $H(s)$

Answer: Factor the numerator and denominator polynomials to obtain

$$H(s) = \frac{3s^2+14s-2}{s^3+s^2-2s} = 3\frac{(s-z_1)(s-z_2)}{(s-p_1)(s-p_2)(s-p_3)} = 3\frac{(s+4.8)(s-0.14)}{(s+2)(s-1)(s-0)}.$$

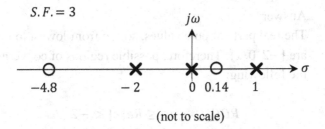

$S.F. = 3$

(not to scale)

---

[c] The region of convergence is $(\sigma > -1) \cap (\sigma > -3) \cap (\sigma < 5) = (-1 < \sigma < 5)$.

b) Express $H(s)$ in the form of a partial fraction expansion.

Answer:

$$H(s) = \frac{3s^2+14s-2}{(s+2)(s-1)(s-0)} = \frac{A_1}{s+2} + \frac{A_2}{s-1} + \frac{A_3}{s}$$

$$A_1 = H(s)(s+2)\big|_{s=-2} = \frac{3s^2+14s-2}{(s-1)s}\bigg|_{s=-2}$$

$$= \frac{3(-2)^2+14(-2)-2}{((-2)-1)(-2)} = \frac{-18}{6} = -3$$

$$A_2 = H(s)(s-1)\big|_{s=1} = \frac{3s^2+14s-2}{(s+2)s}\bigg|_{s=1}$$

$$= \frac{3(1)^2+14(1)-2}{((1)+2)(1)} = \frac{15}{3} = 5$$

$$A_3 = H(s)(s)\big|_{s=0} = \frac{3s^2+14s-2}{(s+2)(s-1)}\bigg|_{s=0}$$

$$= \frac{3(0)^2+14(0)-2}{((0)+2)((0)-1)} = \frac{-2}{-2} = 1$$

$$\therefore H(s) = \frac{-3}{s+2} + \frac{5}{s-1} + \frac{1}{s}.$$

c) What regions of convergence are possible for these pole locations?

Answer:
The real parts of pole values, sorted from lowest to highest, are $\{-2, 0, 1\}$. Therefore, possible regions of convergence are the following:

$$ROC_1: \quad -\infty \le Re\{s\} < -2$$
$$ROC_2: \quad -2 < Re\{s\} < 0$$
$$ROC_3: \quad 0 < Re\{s\} < 1$$
$$ROC_4: \quad 1 < Re\{s\} \le \infty$$

d) Write an expression for $h(t) = \mathcal{L}^{-1}\{H(s)\}$ for each region of convergence in (c).

Answer:
$$h_1(t) = -3e^{-2t}(u(t)-1) + (u(t)-1) + 5e^t(u(t)-1),$$
$$h_2(t) = -3e^{-2t}u(t) \qquad + (u(t)-1) + 5e^t(u(t)-1),$$
$$h_3(t) = -3e^{-2t}u(t) \qquad + u(t) \qquad + 5e^t(u(t)-1),$$
$$h_4(t) = -3e^{-2t}u(t) \qquad + u(t) \qquad + 5e^tu(t).$$

Chapter 11

A11.1 Draw a discrete-time network that realizes this transfer function when it is causal. (Hint: Rewrite the transfer function in terms of z^{-1} instead of z.)
$$H(z) = \frac{3z}{3z+1} - \frac{4z}{2z-1}.$$

Solution:

$$H(z) = \frac{3z}{3z+1} - \frac{4z}{2z-1} = \frac{-6z^2-7z}{6z^2-z-1} = \frac{-6-7z^{-1}}{6-z^{-1}-z^{-2}} = \frac{-1-\frac{7}{6}z^{-1}}{1-\frac{1}{6}z^{-1}-\frac{1}{6}z^{-2}}$$

$$= Y(z)/X(z).$$

$$Y(z)\left\{1 - \tfrac{1}{6}z^{-1} - \tfrac{1}{6}z^{-2}\right\} = X(z)\left\{-1 - \tfrac{7}{6}z^{-1}\right\},$$

$$Y(z) - \tfrac{1}{6}z^{-1}Y(z) - \tfrac{1}{6}z^{-2}Y(z) = -X(z) - \tfrac{7}{6}z^{-1}X(z),$$

$$y(n) - \tfrac{1}{6}y(n-1) - \tfrac{1}{6}y(n-2) = -x(n) - \tfrac{7}{6}x(n-1),$$

$$y(n) = -x(n) - \tfrac{7}{6}x(n-1) + \tfrac{1}{6}y(n-1) + \tfrac{1}{6}y(n-2).$$

When equating coefficients with the standard form of an Mth order causal difference equation

$$y(n) = \sum_{k=0}^{M} b_k x(n-k) - \sum_{k=1}^{M} a_k y(n-k),$$

we obtain:

$$b_0 = -1, \ b_1 = -\tfrac{7}{6}, \ b_2 = 0, \ -a_1 = \tfrac{1}{6}, \ -a_2 = \tfrac{1}{6}.$$

Plugging these into the network diagram in Fig. 7.17 on p. 7-28:

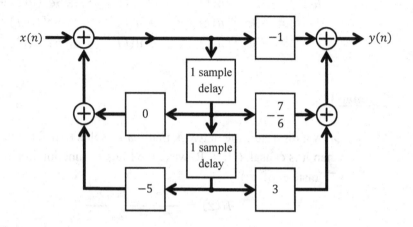

Chapter 12

A12.1 Design a simple RL analog lowpass filter having bandwidth of 2
rad/sec. Use a 1 Henry inductor.

a) Sketch your circuit. Label input/output signals and
component values.

Answer:
Since bandwidth $W = R/L$, we find that the value of $R = 2\Omega$:

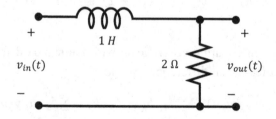

b) Write an expression for $H(s)$, your circuit's s-domain transfer function.

Answer:

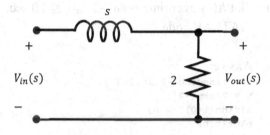

+

$V_{in}(s)$

s

2

+

$V_{out}(s)$

−

From the impedance-labelled circuit in the s-domain shown above, we see that $(s) = \frac{2}{s+2}$.

c) Draw a pole-zero diagram for $H(s)$.

Answer:

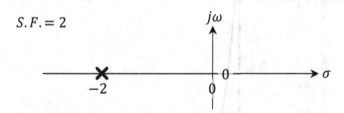

$S.F. = 2$

$j\omega$

−2

0

0

σ

d) Assuming a causal impulse response, what is the region of convergence of $H(s)$?

Answer:
The region of convergence for the Laplace transform of causal $h(t)$ is to the right of the right-most pole. Therefore, $ROC: \sigma = Re\{s\} > -2$.

e) Find $h(t) = \mathcal{L}^{-1}\{H(s)\}$ when it is causal.

Answer: $h(t) = 2e^{-2t}u(t)$.

f) Plot $h(t)$ over time span $-1 \leq t \leq 10$ sec. Show your
    MATLAB code.

Answer:

```
t = linspace(-1,10,1e3);
h = 2*exp(-2*t);
h(find(t<0)) = 0;
plot(t,h)
xlabel('time in sec')
ylabel('amplitude values of impulse response h(t)')
title('h(t) = 2e^-^2^tu(t)')
```

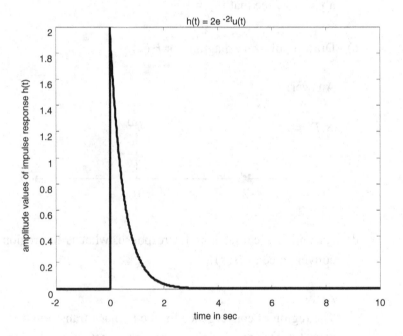

g) Based on your answer to (b), find this filter's frequency
    response function $H(\omega)$.

Answer:

Since all poles of $H(s)$ are in the left-half s plane and $h(t)$ is causal, the $j\omega$ axis lies in the region of convergence. Thus, $H(\omega)$ may be found as

$$H(\omega) = H(s)|_{s=j\omega} = \left.\frac{2}{s+2}\right|_{s=j\omega} = \frac{2}{j\omega+2}.$$

h) Plot $|H(\omega)|^2$ in dB vs. ω on a log scale, over the range $0.01 < \omega < 100$ rad/sec. Show your MATLAB code.

Answer:
```
w = logspace(log10(0.01),log10(100),1e3);
s = j*w;
H = 2./(s+2);
semilogx(w,20*log10(abs(H)))
xlabel('frequency in rad/sec')
ylabel('|H(\omega)|^2 in dB')
title('H(s) = 2/(s+2)')
grid on
```

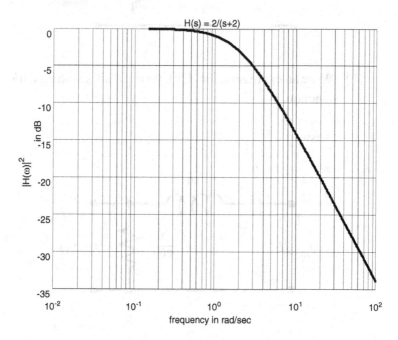

A12.2 Design a simple bandpass analog filter circuit using three
components: R, L, C. Choose capacitor value = 1F. The filter
should have 2 rad/sec bandwidth and 20 rad/sec passband center
frequency.

a) Sketch your circuit. Label input/output signals and
component values.

Answer:
Since $\omega_0^2 = (20)^2 = 400 = \frac{1}{LC} = \frac{1}{L}, L = 1/400.$
$W = 2 = R/L$, so $R = 2L = 1/200$:

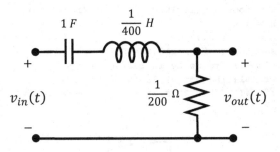

b) Write an expression for $H(s)$, your circuit's s-domain transfer
function.

Answer:

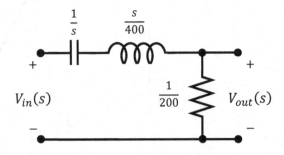

By voltage division we obtain $H(s)$:

$$H(s) = \frac{\frac{1}{200}}{\frac{1}{200}+\frac{s}{400}+\frac{1}{s}} = \frac{2s}{2s + s^2 + 400} = \frac{2s}{s^2 + 2s + 400}.$$

c) Draw a pole-zero diagram for $H(s)$.

Answer:

$$H(s) = \frac{2s}{s^2+2s+400} = 2 \frac{(s-0)}{\left(s-(-1+j\sqrt{399})\right)\left(s-(-1-j\sqrt{399})\right)}.$$

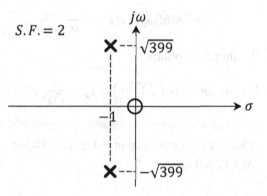

$S.F. = 2$

d) Assuming a causal impulse response, what is the region of convergence of $H(s)$?

Answer:
The region of convergence for the Laplace transform of causal $h(t)$ is to the right of the right-most pole. Therefore,
$ROC: \sigma = Re\{s\} > -1.$

e) Find $h(t) = \mathcal{L}^{-1}\{H(s)\}$ if it is causal.

Answer:

$$H(s) = \frac{2s}{s^2+2s+400} = \frac{2(s+1)-2}{(s+1)^2+399}$$

$$= 2\left\{\frac{s+1}{(s+1)^2+399}\right\} - \frac{2}{\sqrt{399}}\left\{\frac{\sqrt{399}}{(s+1)^2+399}\right\}.$$

From Table 10.1, we know the causal Laplace transform pairs:

$$e^{at}\cos(\omega_0 t)\,u(t) \Leftrightarrow \frac{s-a}{(s-a)^2+\omega_0^2},$$

$$e^{at}\sin(\omega_0 t)u(t) \Leftrightarrow \frac{\omega_0}{(s-a)^2+\omega_0^2}.$$

We therefore obtain:

$$h(t) = 2e^{-t}\cos\left(\sqrt{399}t\right)u(t) - \frac{2}{\sqrt{399}}e^{-t}\sin\left(\sqrt{399}t\right)u(t).$$

f) Plot $h(t)$ over time span $-1 \le t \le 10$ sec. Show your MATLAB code.

Answer:

```
t = linspace(-1,10,1e3);
h = 2*exp(-2*t).*(cos(sqrt(399)*t)-
sin(sqrt(399)*t)/sqrt(399));
h(find(t<0)) = 0;
plot(t,h)
xlabel('time in sec')
ylabel('amplitude values of impulse response h(t)')
title(['h(t) = 2e^-^2^t(cos(19.98t)-sin(19.98t)', ...
    '/19.98)u(t)'])
```

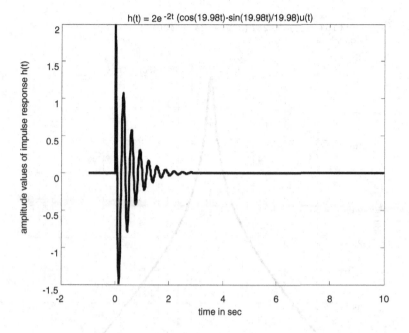

h(t) = 2e $^{-2t}$ (cos(19.98t)-sin(19.98t)/19.98)u(t)

g) Based on your answer to (b), find this filter's frequency response function $H(\omega)$.

Since all poles of $H(s)$ are in the left-half s plane and $h(t)$ is causal, the $j\omega$ axis lies in the region of convergence. As a result $H(\omega)$ may be found as

$$H(\omega) = H(s)|_{s=j\omega} = \left.\frac{2s}{s^2+2s+400}\right|_{s=j\omega} = \frac{2j\omega}{-\omega^2+2j\omega+400}.$$

h) Plot $|H(\omega)|^2$ in dB vs. ω on a log scale, over the range $2 \le \omega \le 200$ rad/sec. Show your MATLAB code.

Answer:
```
w = logspace(log10(2),log10(200),1e3);
s = j*w;
H = (2*s)./(s.^2 + 2*s + 400);
semilogx(w,20*log10(abs(H)))
xlabel('frequency in rad/sec')
```

```
ylabel('|H(\omega)|^2 in dB')
title('H(s) = 2s/(s^2 + 2s + 400)')
grid on
```

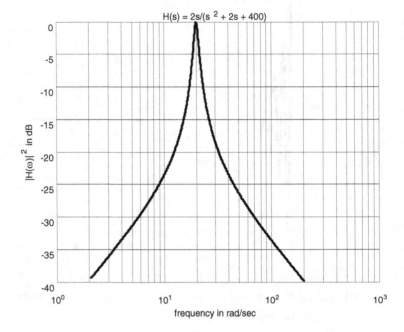

Bibliography

Abramowitz, Milton, and Irene A. Stegun. Handbook of mathematical functions: with formulas, graphs, and mathematical tables. Vol. 55. Courier Corporation, 1964.

Arsac, Jacques. Fourier transforms and the theory of distributions. Englewood Cliffs, NJ: Prentice-Hall, 1966.

Bateman, Harry. "Tables of integral transforms," 1954.

Bracewell, Ronald N., Fourier Transforms and Its Applications, Second Edition, McGraw Hill, 1978.

Brigham, E. Oran, The fast Fourier transform and its applications. Prentice Hall, 1988.

Chen, Wai-Kai. Passive and active filters: theory and implementations. Wiley, 1986.

Dwight, Herbert Bristol, Tables of Integrals and Other Mathematical Data, Fourth Edition, The MacMillan Company, 1961.

Jackson, Leland B. Digital Filters and Signal Processing: With MATLAB® Exercises. Springer Science & Business Media, 2013.

Lathi, B. P. Signal processing and linear systems. New York: Oxford University Press, 1998.

Lathi, B. P. Linear systems and signals. Vol. 2. New York: Oxford University Press, 2005.

Lathi, B. P. and Ding, Zhi, Modern Digital and Analog Communication Systems, Oxford University Press, 4th Edition, 2009.

Lighthill, M. J., Introduction to Fourier Analysis and Generalized Functions, Cambridge University Press, 1962.

McClellan, James H., Schafer, Ronald W., and Yoder, Mark A., Signal Processing First, Pearson Education, Inc., 2003.

Oppenheim, Alan V., and Schafer, Ronald W., Discrete-time Signal Processing, Prentice Hall signal processing series, 3rd Edition, 2010.

Oppenheim, Alan V., Willsky Alan, S., and Nawab, S. Hamid, Signals and Systems, Second Edition, Prentice Hall, 1997.

Papoulis, Athanasios, The Fourier Integral and its Applications, McGraw-Hill, 1962.

Papoulis, Athanasios, Signal Analysis, McGraw Hill, 1977.

Roberts, M. J., Signals and Systems, McGraw Hill, 2004.

Stuller, John. Introduction to signals and systems. Thomson-Engineering, 2007.

Tan, S. M., Lecture notes on Linear Systems, found at: https://sites.google.com/site/szemengtan/home/linear-systems.

Tervo, Richard, J., Practical Signal Theory and with MATLAB® Applications, Wiley, 2014.

Wiener, Norbert, The Fourier Integral and certain of its Applications, Cambridge University Press, 1988 (first published in 1933).

Index

Printed in the United States
By Bookmasters